T0155660

Wind Loading of Structures

Wind Loading of Structures

Fourth Edition

John D. Holmes and Seifu A. Bekele

CRC Press
Taylor & Francis Group
Boca Raton London New York

CRC Press is an imprint of the
Taylor & Francis Group, an **informa** business

Fourth edition published 2021
by CRC Press
6000 Broken Sound Parkway NW, Suite 300, Boca Raton, FL 33487-2742

and by CRC Press
2 Park Square, Milton Park, Abingdon, Oxon, OX14 4RN

© 2021 Taylor & Francis Group, LLC

First edition published by Spon Press 2001
Second edition published by Taylor and Francis 2007
Third edition published by CRC Press 2015

CRC Press is an imprint of Taylor & Francis Group, LLC

Library of Congress Cataloging-in-Publication Data
Names: Holmes, John D., 1942-, author. | Bekele, Seifu, author.
Title: Wind loading of structures / John D. Holmes and Seifu Bekele.
Description: Fourth edition. | Boca Raton : CRC Press, 2021. |
Includes bibliographical references and index.
Identifiers: LCCN 2020031521 (print) | LCCN 2020031522 (ebook) |
ISBN 9780367273262 (hbk) | ISBN 9780429296123 (ebk) | ISBN 9781000220834 (epub) |
ISBN 9781000220797 (mobi) | ISBN 9781000220759 (adobe pdf)
Subjects: LCSH: Wind-pressure. | Structural dynamics. |
Buildings—Aerodynamics. | BISAC: TECHNOLOGY & ENGINEERING / Civil /
General. | TECHNOLOGY & ENGINEERING / Civil / Earthquake. |
TECHNOLOGY & ENGINEERING / Structural.
Classification: LCC TA654.5 .H65 2021 (print) | LCC TA654.5 (ebook) |
DDC 624.1/75—dc23
LC record available at https://lccn.loc.gov/2020031521
LC ebook record available at https://lccn.loc.gov/2020031522

ISBN: 978-0-367-27326-2 (hbk)
ISBN: 978-0-429-29612-3 (ebk)

Typeset in Sabon
by codeMantra

Contents

15 Wind-loading codes and standards   453

16 Application of computational fluid dynamics to wind loading

Preface

Since the third edition of this book, severe wind storms have continued to wreak havoc on many parts of the earth. There is evidence that hurricanes and other tropical cyclones are increasing in strength, as indicated by three severe hurricanes in the Atlantic and Caribbean in the 2017 season, and the winter storms in western Europe in 2018. This further emphasizes the need for structural engineers to take wind loading seriously; the fourth edition of this book will hopefully assist in that task.

As in previous editions, Chapters 1–7 cover fundamental aspects such as meteorology, extreme value probability, bluff-body aerodynamics and structural dynamics that are required to understand the wind loading of structures. Chapters 8–14 discuss the particular characteristics of wind loading of structures of all types, not just buildings.

In terms of written content, the length of the book has been increased by about 24% since the third edition. Highlights of the expansion are as follows:

- Updates in Chapter 1 on the observed and predicted effects of global warming.
- An expansion of Chapter 3 to accommodate more information on the relationship between upper-level (geostrophic) and surface winds, wind profiles over water and the effects of topography.
- An increase of Chapter 6 on internal pressure prediction by about 40%. This important topic continues to be neglected elsewhere, and this chapter remains the best comprehensive source.
- Updating of Chapter 15 to reflect new editions of the major wind-loading codes and standards in the last 5 years.
- A new chapter (Chapter 16), contributed by Dr. Seifu Bekele, on the applicability and methods of computational fluid dynamics to wind loading of structures.
- A massive expansion in Appendix D (Extreme wind climates – a world survey). There is no other single source for this information that now covers most of the world.
- A new Appendix G that gives several methods of fitting of the Generalized Pareto distribution to extreme wind data, with examples.

There has also been expansion of nearly every other chapter from earlier editions.

The authors wish to thank the publishers (Taylor and Francis/CRC) for continuing to support this book into another edition. Hopefully, it will continue to be useful to practising engineers and students alike. Readers are welcome to contact the authors with suggestions and corrections (hopefully there are few of the latter...).

John D. Holmes (john.holmes@jdhconsult.com)
Seifu Bekele (seifu@gwts.com.au)
Melbourne, Australia
May 2020

Authors

Dr John D. Holmes has over 40 years of experience in wind engineering. As a consultant, he has also been involved in the determination of design wind loads for numerous structures, including major structures in several different countries. As a researcher and former academic, he has authored over 500 journal and conference papers, and consulting reports.

Dr Seifu A. Bekele has about 25 years of experience in wind engineering and related fields, and has an extensive consulting practice that specializes in computational techniques.

Chapter I

The nature of windstorms and wind-induced damage

I.I INTRODUCTION

Wind loading competes with seismic loading as the dominant environmental loading for structures. Wind loading and seismic loading have created almost equal damage over a long time period, although the frequency with which large and damaging earthquakes occur is much lower than severe windstorms. On almost every day of the year a severe windstorm occurs somewhere on the earth – although many storms are small in scale and their effects are localized. In the tropical oceans, the most severe of all wind events – tropical cyclones (including hurricanes and typhoons) – are generated. When these storms make landfall on populated coastlines, their effects can be devastating.

In this introductory chapter, the meteorology of severe windstorms – *gales* produced by large extra-tropical depressions, *tropical cyclones*, and *downbursts*, squall lines and *tornados* associated with thunderstorms – is explained, including the expected horizontal variation in wind speed that occurs during these events. The history of damaging wind events, particularly in the last fifty years, is discussed. The behaviour of flying debris, a major source of damage in severe windstorms, is also analysed. Insurance aspects are discussed, along with recent development of loss models that are based on historical data on the occurrences of large severe storms, the spatial characteristics for the wind speeds within them, and assumed relationships between building damage and wind speed.

The effects of global warming on natural hazards such as droughts, flooding and bushfires is well known. The evidence for the effects of severe windstorms is less clear; this is partly due to poor historical records maintained both by developed and developing countries. However, Section 1.8 summarizes some recent studies of climate-change trends affecting severe windstorms.

1.2 METEOROLOGICAL ASPECTS

Wind is air movement relative to the earth that is driven by several different forces, such as pressure differences in the atmosphere, which are themselves produced by differential solar heating of different parts of the earth's surface, and forces generated by the rotation of the earth. The differences in solar radiation between the poles and the equator, produce temperature and pressure differences. These, together with the effects of the earth's rotation, set up large-scale circulation systems in the atmosphere, with both horizontal and vertical orientations. The result of these circulations is that the prevailing wind directions in the tropics, and near the poles, tend to be easterly. Westerly winds dominate in the temperate latitudes.

Local severe winds may also originate from local convective effects (*thunderstorms*), or from the uplift of air masses produced by mountain ranges (*downslope winds*). Severe tropical cyclones, known in some parts of the world as *hurricanes*, and as *typhoons* in others, generate extremely strong winds over some parts of the tropical oceans and coastal regions, in latitudes from 10° to about 30°, both north and south of the equator.

For all types of severe storms, the wind is highly turbulent or gusty. The turbulence or gustiness is produced by eddies or vortices within the air flow which are generated by frictional interaction at ground level, or by shearing action between air moving in opposite directions at altitude. These processes are illustrated in Figure 1.1 for larger storms, such as gales or tropical cyclones, which are of the 'boundary-layer' type, and for downdrafts generated by thunderstorms.

1.2.1 Pressure gradient

The two most important forces acting on the upper-level air in the 'free atmosphere', that is, above the frictional effects of the earth's boundary layer are: the pressure gradient force and the Coriolis force.

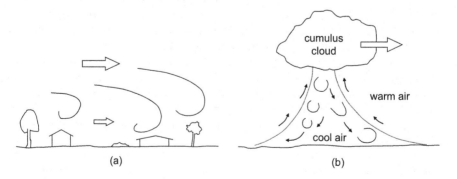

Figure 1.1 The generation of turbulence in (a) boundary-layer winds and (b) thunderstorm downdrafts.

It is shown in elementary texts on fluid mechanics that, at a point in a fluid in which there is a pressure gradient, $\partial p/\partial x$, in a given direction, x, in a Cartesian coordinate system, there is a resulting force per unit mass given by Equation (1.1):

$$\text{Pressure gradient force per unit mass} = -\left(\frac{1}{\rho_a}\right)\frac{\partial p}{\partial x} \qquad (1.1)$$

where ρ_a is the density of air.

This force acts from a high-pressure region to a low-pressure region.

1.2.2 Coriolis force

The *Coriolis* force is an apparent force due to the rotation of the earth. It acts to the right of the direction of motion in the northern hemisphere, and to the left of the velocity vector, in the case of the southern hemisphere. At the equator, the Coriolis force is zero. Figure 1.2 gives a simple explanation of the Coriolis force by observing the motion of particles of air northwards from the South Pole.

Consider a parcel of air moving horizontally away from the South Pole, P, with a velocity U, in the direction of point A (Figure 1.2a). Since the earth is rotating clockwise with angular velocity, Ω, the point originally at A, will have moved to B, and a point originally at A', will have moved to A, as the air parcel arrives. *Relative to the earth's surface*, the particle will appear to follow the path PA', i.e. a continuous deflection to the left. At the North Pole, the deflection is to the right. These deflections can be associated with an apparent acceleration acting at right angles to the velocity of the parcel, called the Coriolis acceleration.

Consider a short time interval, δt, (Figure 1.2b); AA' is then small compared with PA. In this case,

$$\text{AA}' = \Omega\, U\, (\delta t)^2 \qquad (1.2)$$

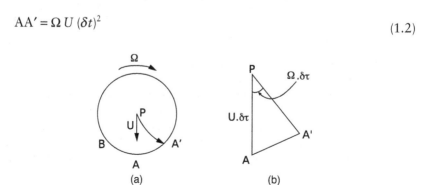

Figure 1.2 (a and b) The apparent (Coriolis) force due to the earth's rotation (Southern Hemisphere).

Let the Coriolis acceleration be denoted by a. Since AA' is the distance travelled under this acceleration, then it can also be expressed by:

$$AA' = (\tfrac{1}{2})a\,(\delta t)^2 \tag{1.3}$$

Equating the two expressions for AA', Equations (1.2) and (1.3),

$$a = 2\,U\,\Omega \tag{1.4}$$

This gives the Coriolis acceleration, or force per unit mass, at the poles.

At other points on the earth's surface, the angular velocity is reduced to Ω sin λ, where λ is the latitude. Then, the Coriolis acceleration becomes equal to $2U\,\Omega$ sin λ. The term $2\,\Omega$ sin λ is a constant for a given latitude, and is called the 'Coriolis parameter', often denoted by the symbol, f. The Coriolis acceleration then becomes equal to fU.

Thus, the Coriolis force is an apparent, or effective force acting to the right of the direction of air motion in the Northern Hemisphere, and to the left of the air motion in the Southern Hemisphere. At the Equator, the Coriolis force is zero, and in the equatorial region, within about 5° of the Equator is negligible in magnitude. The latter explains why tropical cyclones (Section 1.3.2), or other cyclonic systems, do not form in the equatorial regions.

1.2.3 Geostrophic wind

Steady flow under equal and opposite values of the *pressure gradient* and the *Coriolis* force is called 'balanced geostrophic flow'. Equating the pressure gradient force per unit mass from Equation (1.1), and the Coriolis force per unit mass, given by fU, we obtain:

$$U = -\left(\frac{1}{\rho_a f}\right)\frac{\partial p}{\partial x} \tag{1.5}$$

This is the equation for the *geostrophic wind speed*, which is proportional to the magnitude of the pressure gradient, $(\partial p/\partial x)$.

The directions of the pressure gradient and Coriolis forces, and of the flow velocity are shown in Figure 1.3, for both Northern and Southern hemispheres. It may be seen that the flow direction is parallel to the isobars (lines of constant pressure) in both hemispheres. In the Northern Hemisphere, the high pressure is to the right of an observer facing the flow direction; in the Southern Hemisphere, the high pressure is on the left. This results in anti-clockwise rotation of winds around a low-pressure centre in the Northern Hemisphere, and a clockwise rotation in the Southern Hemisphere. In both hemispheres, rotation about a low-pressure centre (which usually produces

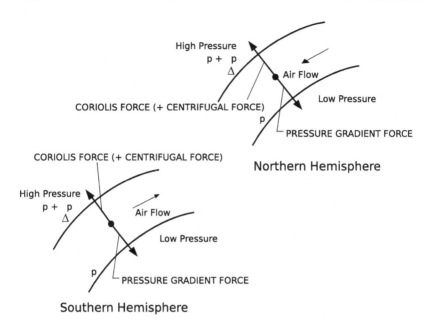

Figure 1.3 Balanced geostrophic flow in northern and southern hemispheres.

strong winds) is called a 'cyclone' by the meteorologists. Conversely, rotation about a high-pressure centre is called an 'anti-cyclone'.

1.2.4 Gradient wind

If the isobars have significant curvature (as for example near the centre of a tropical cyclone), then the *centrifugal* force acting on the air particles cannot be neglected. The value of the centrifugal force per unit mass is (U^2/r), where U is the resultant wind velocity and r is the radius of curvature of the isobars.

The direction of the force is away from the centre of curvature of the isobars. If the path of the air is around a high-pressure centre (anti-cyclone), the centrifugal force acts in the same direction as the pressure gradient force, and in the opposite direction to the Coriolis force. For flow around a low-pressure centre (cyclone), the centrifugal force acts in the same direction as the Coriolis force, and opposite to the pressure gradient force.

The equation of motion for a unit mass of air moving at a constant velocity, U, is then Equation (1.6) for an anti-cyclone, and (1.7) for a cyclone:

$$\frac{U^2}{r} - |f|U + \frac{1}{\rho_a}\left|\frac{\partial p}{\partial r}\right| = 0 \tag{1.6}$$

$$\frac{U^2}{r} + |f|U - \frac{1}{\rho_a}\left|\frac{\partial p}{\partial r}\right| = 0 \tag{1.7}$$

Equations (1.6) and (1.7) apply to both hemispheres. Note that the pressure gradient, $\frac{\partial p}{\partial r}$, is negative in an anti-cyclone and that f is negative in the Southern Hemisphere. These equations are quadratic equations for the *gradient wind speed*, U. In each case, there are two theoretical solutions, but if the pressure gradient is zero, then U must be zero, so that the solutions become:

$$U = \frac{|f|r}{2} - \sqrt{\frac{f^2 r^2}{4} - \frac{r}{\rho_a}\left|\frac{\partial p}{\partial r}\right|} \qquad (1.8)$$

for an anti-cyclone

$$U = -\frac{|f|r}{2} + \sqrt{\frac{f^2 r^2}{4} + \frac{r}{\rho_a}\left|\frac{\partial p}{\partial r}\right|} \qquad (1.9)$$

for a cyclone.

Examining Equation (1.8), it can be seen that the maximum value of U occurs when the term under the square root sign is zero. This value is $\frac{|f|r}{2}$, which occurs when $\left|\frac{\partial p}{\partial r}\right|$ is equal to $\frac{\rho_a f^2 r}{4}$. Thus, in an anti-cyclone, there is an upper limit to the gradient wind; *anti-cyclones are normally associated with low wind speeds*.

Now considering Equation (1.9), it is clear that the term under the square root sign is always positive. The wind speed in a cyclone is therefore only limited by the pressure gradient; *cyclones are therefore associated with strong winds*.

1.2.5 Frictional effects

As the earth's surface is approached, *frictional* forces, transmitted through shear between layers of air in the atmospheric boundary layer, gradually play a larger role. This force acts in a direction opposite to that of the flow direction, which, in order to achieve a vector balance, is now not parallel to the isobars, but directed towards the low-pressure region. Figure 1.4 shows the new balance of forces in the boundary layer.

Thus, as the ground surface is approached from above, the wind vector gradually turns towards the low-pressure centre with a reduction in height. This effect is known as the *Ekman Spiral*. The total angular change between gradient height and the surface is about 30°. However, the angular change over the height of even the tallest structures is quite small.

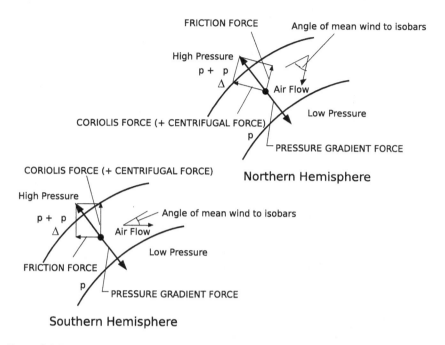

Figure 1.4 Force balance in the atmospheric boundary layer.

1.3 TYPES OF WIND STORMS

1.3.1 Gales from large depressions

In the mid-latitudes of 40°– 60°, the strongest winds are gales, generated by large and deep depressions or (extra-tropical) cyclones of synoptic
scale. They can also be significant contributors to winds in lower latitudes.
Navigators, particularly in sailing ships, are familiar with the strong westerlies of the 'roaring forties', of which the winds of the North Atlantic, and at
Cape Horn, are perhaps the most notorious. As shown in Section 1.4, severe
structural damage has been caused by winter gales in north-west Europe.

These systems are usually large in horizontal dimension – they can extend
for more than 1,000 km, so can influence large areas of land during their
passage – several countries in the case of Europe. They may take several
days to pass, although winds may not blow continuously at their maximum intensity during this period. The winds tend to be quite turbulent
near the ground, as the flow adjusts to the frictional effects of the earth's
surface over hundreds of kilometres. The direction of the winds remains
nearly constant over many hours. These features are illustrated in a typical
anemograph (wind speed and direction versus time) from this type of event
reproduced in Figure 1.5.

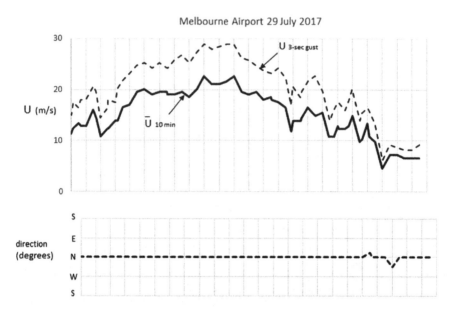

Figure 1.5 Anemograph for synoptic winds from large extra-tropical depression. Time unit: hours.

1.3.2 Tropical cyclones

Tropical cyclones are intense cyclonic storms that occur over the tropical oceans, mainly in late summer and autumn. They are driven by the latent heat of the oceans, and require a minimum sea temperature of about 26° to sustain them; they rapidly degenerate when they move over land, or into cooler waters. They will not form within about 5° of the equator, and do not reach full strength until they reach at least 10° latitude. They are usually at full strength when they are located between 20° and 30° latitude, but can travel to higher latitudes if there are warm ocean currents to sustain them.

The strongest tropical cyclones have occurred in the Caribbean, where they are known as *hurricanes*, in the South China Sea, where they are called *typhoons*, and off the North-West coast of Australia. Areas of medium tropical cyclone activity are, the eastern Pacific Ocean off the coast of Mexico, the southern Indian Ocean, the Bay of Bengal, the South Pacific, southern Japan, the Coral Sea (off eastern Australia) and the south-east Atlantic Ocean. Regions of lesser activity or weaker storms are: the Arabian Sea, the Gulf of Thailand, and the north coast of Australia (including the Gulf of Carpentaria).

A developed tropical cyclone has a three-dimensional vortex structure which is shown schematically in Figure 1.6. The horizontal dimensions of these storms are less than the extra-tropical cyclones, or depressions

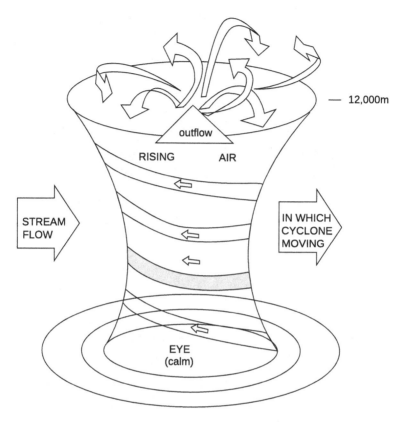

Figure 1.6 Three-dimensional structure in a developed tropical cyclone.

discussed earlier, but their effects can extend for several hundred kilometres. The circulation flows with a radial component towards the 'eye', outside of which is a region of intense thermal convection with air currents spiralling upwards. Inside the eye is a region of relative calm, with slow sinking air; the diameter of the eye can range between 8 and 80 km. Often clear skies have been observed in this region. The strongest winds occur just outside the eye wall.

Figure 1.7 gives an example of an anemograph measured at a height of 10 m above the ground for a tropical cyclone. This example shows a fortuitous situation when the eye of the storm passed directly over the recording station, resulting in a period of about two hours of very low winds. The wind direction changed from south-east to west during the passage of the vortex over the measuring station.

Outside of the eye of a tropical cyclone, the wind speed at upper levels decays with the radial distance from the storm centre. The gradient wind equation (Equation (1.9)) can be used to determine this wind speed:

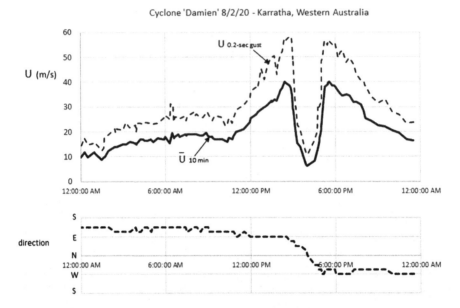

Figure 1.7 Wind speed (upper graph), and wind direction (lower graph), during the passage of a tropical cyclone (Australian Bureau of Meteorology).

$$U = -\frac{|f|r}{2} + \sqrt{\frac{f^2 r^2}{4} + \frac{r}{\rho_a}\left|\frac{\partial p}{\partial r}\right|}$$

(1.9)

where

f is the Coriolis parameter $(= 2\,\Omega \sin \lambda)$,

r is the radius from the storm centre,

ρ_a is the density of air,

p is the atmospheric pressure.

To apply Equation (1.9), it is necessary to establish a suitable function for the pressure gradient. A commonly assumed expression is Equation (1.10) (Holland, 1980):

$$\frac{p - p_o}{p_n - p_o} = \exp\left(\frac{-A}{r^B}\right)$$

(1.10)

where

p_o is the central pressure of the tropical cyclone,

p_n is the atmospheric pressure at the edge of the storm,

A and B are scaling parameters.

The pressure difference $(p_n - p_o)$ can be written as Δp, and is an indication of the strength of the storm.

Differentiating Equation (1.10) and substituting in (1.9), we have:

$$U = -\frac{|f|r}{2} + \sqrt{\frac{f^2 r^2}{4} + \frac{\Delta p}{\rho_a}\frac{AB}{r^B}\exp\left(-\frac{A}{r^B}\right)} \qquad (1.11)$$

This is an equation for the mean wind field at upper levels in a tropical cyclone as a function of radius from the storm centre, r, the characteristic parameters, A and B, the pressure drop across the cyclone, Δp and the Coriolis parameter, f.

Near the centre of a tropical cyclone, the Coriolis forces, i.e. the first two terms in Equations (1.9) and (1.11), are small, and it can be shown, by differentiating the remaining term, that the maximum value of U occurs when r equals $A^{1/B}$. Thus, $A^{1/B}$ is to a good approximation, the radius of maximum winds in the cyclone. The exponent B is found to be in the range of 1.0–2.5, and to reduce with increasing central pressure, p_o (Holland, 1980).

Figure 1.8 shows the profiles of pressure and gradient wind speed with radial distance from the centre of the storm calculated from Equations (1.10) and (1.11), for Cyclone 'Tracy' which severely damaged Darwin,

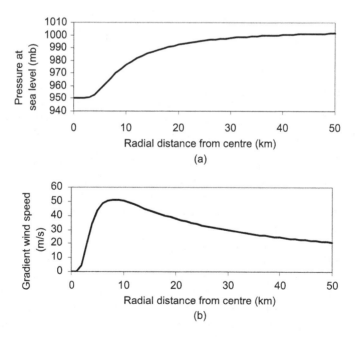

Figure 1.8 Pressure and gradient wind speeds for Cyclone 'Tracy', 1974: (a) sea level pressure and (b) gradient wind speed.

Table 1.1 Saffir-Simpson intensity scale for hurricanes

Category	Central pressure (millibars)	Wind-speed range (3-second gust, m/s)
I	>980	42–54
II	965–979	55–62
III	945–964	63–74
IV	920–944	75–88
V	<920	>88

Australia, in 1974. The parameters *A* and *B* were taken as 23 and 1.5, respectively (where *r* is measured in kilometres), following Holland (1980). The gradient wind speed in Figure 1.8b is approximately equal to the gust wind speed near ground level. The radius of maximum winds, in this case about 8 km, approximately coincides with the maximum pressure gradient.

The forward motion of the moving storm adds an additional vector component to the wind speed given by Equation (1.11), which gives the wind speed *relative* to the moving storm.

An intensity scale for North Atlantic and Caribbean hurricanes has been proposed by Saffir and Simpson. This is reproduced in Table 1.1.

This scale is widely used for forecasting and emergency management purposes. However, the wind speed ranges given in Table 1.1 should be used with caution, as the estimated wind speeds in hurricanes are usually obtained from upper-level aircraft readings. A similar, but not identical scale is used in the Australian region.

1.3.3 Thunderstorms

Thunderstorms, both isolated storms and those associated with advancing cold fronts, are small disturbances compared to extra-tropical depressions and tropical cyclones, but they are capable of generating severe winds, through tornadoes and downbursts. They contribute significantly to the strong gusts recorded in many countries, including the United States, Russia, Brazil, Australia and South Africa. They are also the main source of high winds in the equatorial latitudes (within about 10° of the equator), although their strength may not be very high in these regions.

Like tropical cyclones, thunderstorms derive their energy from heat. Warm moist air is convected upwards to mix with the drier upper air. With evaporation, rapid cooling occurs and the air mass loses its buoyancy, and starts to sink. Condensation then produces heavy rain or hail which falls, dragging cold air with it. A strong downdraft reaches the ground, and produces a strong wind for a short period of time – perhaps five to ten minutes. The strongest winds produced by this mechanism are known as *downbursts*, which are further sub-divided into *microbursts* and *macrobursts*, depending on their size. The strongest winds produced by these events have a large component of wind speed due to the forward motion of the convection cell.

The conditions for generation of severe thunderstorms are:

- water vapour in the atmosphere at low levels, i.e. high humidity
- instability in the atmosphere, i.e. a negative temperature gradient with height that is greater than the adiabatic rate of the neutral atmosphere
- a lifting mechanism that promotes the initial rapid convection – this may be provided by a mountain range, or a cold front, for example.

1.3.4 Tornadoes

The strongest convection cells, that often generate tornadoes, are known as *supercells*. They are larger and last longer than 'ordinary' convection cells. The tornado is a vertical, funnel-shaped vortex, created in thunderclouds. It is the most destructive of all windstorms. Fortunately, they are quite small in their horizontal extent – of the order of 100 m – but they can travel for quite long distances, up to 50 km before dissipating, leaving behind a long and narrow path of destruction. They occur mainly in large continental plains in countries such as the United States, Argentina, Russia and South Africa.

Periodically, atmospheric conditions in the central United States are such that severe outbreaks with many damaging tornadoes can occur in a short period. For example, this has occurred in April 1974, May 2003 and April 2011. In 1974, 335 fatalities, and destruction of about 7,500 dwellings resulted from the 'super-outbreak' of 148 tornadoes within a two-day period (April 3–4, 1974) that affected thirteen states of the United States. In 2003, a total of 393 tornadoes were reported in 19 states of the United States in a period of about a week. Of these, 15 resulted in 41 fatalities. In 2011, 325 deaths occurred in the outbreak of April 25–28 in the south-eastern states. 2011 was also notable for the Joplin, Missouri, tornado of May 22 which killed 158 persons. A detailed survey of tornadoes in South Africa, where they are known as 'inkanyamba', has been given by Goliger *et al.* (1997). They occur in that country at the rate of about four per year, with a concentration in Gauteng Province in the north of the country, with an occurrence rate of 1×10^{-4} per square kilometre per year. This compares with a rate of about 2×10^{-4} per square kilometre per year in the mid-West of the United States. More information on tornado occurrences in the United States is given in Section 1.4 and Table 1.4.

Tornadoes are sometimes confused with downbursts (described in Section 1.3.5); however, tornadoes can be identified by the appearance of the characteristic funnel vortex, a long narrow damage 'footprint', and evidence of varying wind directions.

The wind speed in a tornado can be related to the radial pressure gradient by neglecting the Coriolis term in the equation of motion. Hence, from either Equation (1.7) or (1.9):

$$U = \sqrt{\frac{r}{\rho_a}\left|\frac{\partial p}{\partial r}\right|} \tag{1.12}$$

This is known as the *cyclostrophic* wind speed. Assuming that the pressure is constant along the edge of a tornado funnel (actually a line of condensed water vapour), Equation (1.12) has been used to estimate wind speeds in tornados.

Measurement of wind speeds in tornadoes is very difficult. Because of their small size, they seldom pass over a weather recording station. If one does, the anemometer is quite likely to be destroyed. For many years, photogrammetric analyses of movie film shot by eyewitnesses was used to obtain reasonable estimates (Fujita *et al.*, 1976; Golden, 1976). The method involves the tracking of clouds, dust and solid debris from the film frames, and was first applied to the Dallas, Texas, tornado of April 2, 1957 by Hoecker (1960). The method is subject to a number of errors – for example, distortion produced by the camera or projector lenses or tracked large objects not moving with the local wind speed. Also, this method is not able to detect velocities that are normal to the image plane.

However, the photogrammetric method has enabled several significant features of tornados, such as 'suction vortices,' that are smaller vortex systems rotating around the main vortex core, and the high vertical velocities. In the latter case, analysis of a tornado at Kankakee, Illinois in 1963 (Golden, 1976) indicated vertical velocities of 55–60 m/s, at a height above the ground of less than 200 m.

Analyses of failures of engineered buildings in tornadoes have generally indicated lower maximum wind speeds in tornadoes than those obtained by photogrammetric or other methods (e.g. Mehta, 1976). After considering all the available evidence at that time, Golden (1976) estimated the maximum wind speeds in tornadoes to be no more than 110 m/s.

In recent years, portable Doppler radars have been successfully used in the United States for more accurate measurement of wind speeds in tornadoes.

An intensity scale for tornadoes was originally proposed in 1971 (Fujita, 1971). Several F-scale classifications are associated with wind speed ranges, although, in practice, classifications are applied based on observed damage to buildings and other structures. The original scaling was criticized by engineers for several reasons: e.g. for failing to account for variations in quality of construction, and that it has not been based on a proper correlation of damage descriptions and wind speeds. The wind speed ranges of the original Fujita Scale and the enhanced Fujita Scale (McDonald and Mehta, 2004) for F0 to F5 categories are given in Table 1.2. Ranges are given for

Table 1.2 Fujita intensity scale for tornadoes

Category	Original Fujita wind speed range (m/s)	Enhanced Fujita wind speed range (m/s)
F0	20–35	29–38
F1	35–52	39–49
F2	53–72	50–60
F3	72–93	61–74
F4	94–117	75–89
F5	117–142	over 90

3-second average gust speeds in metres per second. The Enhanced Fujita Scale is now used for operational purposes in the United States.

An engineering model of wind speed distributions in a tornado is discussed in Section 3.2.8 in Chapter 3.

1.3.5 Downbursts

Downbursts are severe convective downdrafts, usually associated with thunderstorm super-cells. Figure 1.9 shows an anemograph from a severe downburst, recorded at the Andrews Air Force Base, near Washington, D.C., U.S.A. in 1983, with a time scale in minutes. The short duration of the storm is quite apparent, and there is also a rapid change of wind direction during its passage across the measurement station. Such events typically produce a damage 'footprint' that are 2–6 km wide and 3–10 km long. The average 'footprint' aspect ratio (length/width) of downbursts in the United States is 1.5 (Fujita, 1978).

The horizontal wind speed in a thunderstorm downburst with respect to the moving storm is similar to that in a jet of fluid impinging on a plain surface. It varies approximately linearly from the centre of impact to a radius where the wind speed is maximum, and then decays with increasing radius. The forward velocity of the moving storm can be a significant component of the total wind speed produced at ground level, and must be added as a vector component to that produced by the jet.

1.3.6 Downslope winds

In certain regions such as those near the Rocky Mountains of the United States, Switzerland, Croatia, and the Southern Alps of New Zealand, extreme winds can be caused by thermal amplification of synoptic winds on the leeward slopes of mountains. The regions affected are usually quite small but are often identified as 'special regions' in wind loading codes and standards (see Chapter 15 and Appendix D).

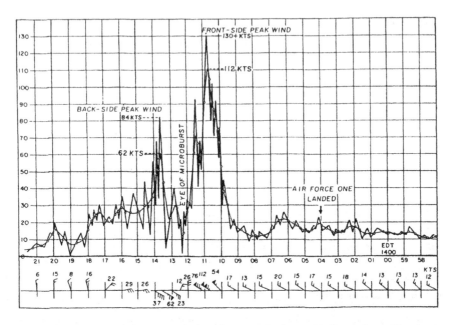

Figure 1.9 Anemograph for a severe downburst at Andrews Air Force Base, Maryland, U.S.A., 1983 (Fujita, 1985). Time units: minutes, wind speed: knots.

1.4 WIND DAMAGE

Damage to buildings and other structures by windstorms has been a fact of life for human beings from the time they moved out of cave dwellings to the present day. Trial and error has played an important part in the development of construction techniques and roof shapes for small residential buildings, which have usually suffered the most damage during severe winds. In past centuries, heavy masonry construction, as used for important community buildings such as churches and temples, was seen, by intuition, as the solution to resist wind forces (although somewhat less effective against seismic action). For other types of construction, windstorm damage was generally seen as an 'Act of God', as it is still viewed today by some insurance companies.

The nineteenth century was important as it saw the introduction of steel and reinforced concrete as construction materials, and began using stress-analysis methods for the design of structures. The methods were further developed in the twentieth century, especially in the second half, with the development of computer technology. During the last two centuries, major structural failures due to wind action have occurred periodically and provoked much interest in wind forces by engineers. Long-span bridges often produced the most spectacular of these failures, with the Brighton Chain Pier, England (1836) (Figure 1.10), the Tay Bridge, Scotland (1879)

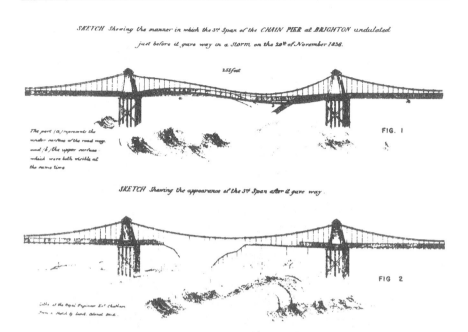

Figure 1.10 Failure of the Brighton Chain Pier, U.K., 1836.

and Tacoma Narrows Bridge, Washington State, U.S.A. (1940) being among the most notable, with the dynamic action of wind playing a major role.

Other large structures have experienced failures as well – for example, the collapse of the Ferrybridge Cooling towers in the U.K. in 1965 (Figure 1.11), and the permanent deformation of the columns of the Great Plains Life Building in Lubbock, Texas that occurred during a tornado (1970). These events were notable, not only as events in themselves, but also for the part they played, acting as a stimulus to the development of research into wind loading in the respective countries. Another type of structure which has proved to be dynamically sensitive to wind, is the guyed mast; it has also suffered a high failure rate. In a forty-year period up to 2007, there were 319 recorded failures of guyed masts worldwide, of which 44 were caused by wind, or by a combination of ice and wind (Smith, 2007).

Some major windstorms, which have caused large scale damage to residential buildings, as well as some engineered structures, are also important for the part they have played in promoting research and understanding of wind loads on structures. The Yorkshire (U.K.) storms of 1962 produced a huge amount of building damage, with as many as 101,500 houses damaged in the city of Sheffield alone (Page, 1968). The 1962 storms in northern England resulted in greatly increased research in wind loading in the United Kingdom, and eventually in a much more detailed code of practice for wind loads.

Figure 1.11 Ferrybridge Cooling Tower failures, U.K., 1965.

Cyclone 'Tracy' in Darwin, Australia in 1974, and Hurricane 'Andrew' in Florida, U.S.A., in 1992, can also be mentioned as seminal events of this type. However, these extreme events occur intermittently, and it is unfortunate that the collective human memory after them is only about ten years, and often, old lessons have to be relearned by a new generation. However, an encouraging sign is the current interest of major insurance and re-insurance groups in natural hazards, the changing climate (see Section 1.8), the estimation of the potential financial losses, and the realization that any structure can be made wind-resistant at a small additional cost, with appropriate knowledge of the forces involved, and suitable design approaches.

Figure 1.12 shows the annual insured losses in billions of US dollars from all major natural disasters, from 1970 to 2018. Windstorms account for about 70% of total insured losses. Bearing in mind that property insurance is much less common in the less-developed economies, Figure 1.12 does not show the total property damage from natural events, and in fact is biased towards losses in Europe and North America. However, the graph does show that the level of insured losses from natural disasters has increased dramatically after 1989. The major contributors to the increase were windstorms, especially tropical cyclones such as Hurricanes 'Hugo' (1989) 'Andrew' (1992), 'Charley' (2004), 'Ivan' (2004) 'Katrina' (2005), and 'Harvey, 'Irma'

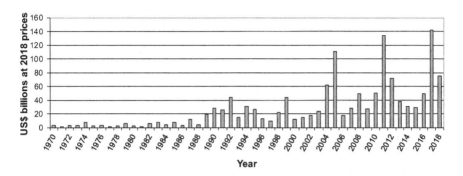

Figure 1.12 World insurance losses from major natural disasters (1970–2018) (Swiss Reinsurance Company).

and 'Maria' (2017) in the United States and Caribbean, and winter gales in Europe in 1990, 1999, 2007 and 2018. The European Environment Agency reported total losses from storms in Europe (EU and EEA member states) between 1980 and 2017 of €173 billion (about US$190 billion).

Some notable windstorms and the losses resulting from them are listed in Table 1.3. Events such as those listed in Table 1.3 have had a major influence on the insurance industry, and structural engineering profession.

Table 1.3 does not include tornadoes. However, the aggregate damage from multiple events can be substantial. For example, in the 'super-outbreak' of April 3–4, 1974, the total damage in the state of Ohio alone was estimated to be US$100 million. The estimated total damage from tornados in the United States, in the active year of 2011 was US$10 billion, with about US$3 billion arising from the Joplin, Missouri, tornado of May 22, 2011.

Tornados are also notable for the loss of life resulting from them in the United States. Table 1.4 summarizes the number of confirmed tornados in the U.S. and the loss of life for the period 2001–2019. The average number of tornados per annum is about 1,250, with an average number of deaths per year of about 100. However, the number of deaths caused by tornados in the United States fortunately appears to be reducing. The availability of increasing numbers of in-house shelters may be contributing to this.

1.5 WIND-GENERATED DEBRIS

As well as damage to buildings produced by direct wind forces – either overloads caused by overstressing under peak loads, or fatigue damage under fluctuating wind loads of a lower level, a major cause of damage in severe windstorms is impacts by flying debris. Penetration of the building envelope by flying missiles has a number of undesirable results: high internal pressures threatening the building structure, wind and rain penetration

Table 1.3 Some disastrous wind storms of the last fifty years

Year	Name	Country or Region	Approximate economic losses ($US mill)	Lives lost
1974	Cyclone 'Tracy'	Australia	1,000	52
1987	Gales	W. Europe	7,000	17
1989	Hurricane 'Hugo'	Caribbean, U.S.A.	18,000	61
1990	Gales	W. Europe	30,000	230
1992	Hurricane 'Andrew'	U.S.A.	48,000	44
1999	Gales ('Lothar', 'Martin')	France, Germany	15,000	140
2003	Typhoon 'Maemi'	Japan, Korea	6,000	131
2004	Hurricane 'Ivan'	Caribbean, U.S.A.	27,000	124
2005	Hurricane 'Katrina'	Southern U.S.A.	160,000	1,830
2007	Winter storm 'Kyrill'	Western Europe	7,000	47
2011	Cyclone 'Yasi'	Australia	3,500	1
2012	Hurricane 'Sandy'	U.S.A.	70,000	117
2017	Hurricane 'Harvey'	U.S.A. (Texas)	125,000	110
2017	Hurricane 'Irma'	U.S.A. (Florida), Leeward Is., Cuba	65,000	134
2017	Hurricane 'Maria'	Dominica, Puerto Rico	90,000	3,000
2018	Winter storm ('David'/'Friedrike')	Benelux countries, Germany	3,000	15
2019	Cyclone 'Idai'	Mozambique	1,000+	700+

Source: Cyclone Tracy and Hurricane Katrina: Munich Reinsurance and Swiss Reinsurance, AIR-Worldwide – losses adjusted to 2017 US$.

of the inside of the building, the generation of additional flying debris, and flying missiles inside the building endangering the occupants.

The area of a building that is most vulnerable to impact by missiles is the windward wall region, although impacts could also occur on the roof and side walls. As the wind approaches the windward wall, its horizontal velocity reduces rapidly. Heavier objects in the flow with higher inertia continue with high velocities until they impact on the wall. Lighter and smaller objects may lose velocity in this region, or even be swept around the building with the flow if they are not directed at the stagnation point (see Chapter 4).

1.5.1 Threshold of flight

Wills et al. (1998) carried out an analysis of debris flight conditions and its impact in terms of damage to buildings in severe winds. They considered 'compact' objects, sheet objects, and rods and poles (Figure 1.13), and established relationships between the body dimensions, and the wind speed, U_f, at which flight occurs and the objects become missiles. For each of the three categories, these relationships are:

Table 1.4 Annual count of tornados in the United States and deaths caused by them

Year	Number	Deaths
2001	1,219	40
2002	938	55
2003	1,394	54
2004	1,820	35
2005	1,262	38
2006	1,117	67
2007	1,102	81
2008	1,685	126
2009	1,305	21
2010	1,543	45
2011	1,894	553
2012	1,119	70
2013	903	55
2014	928	47
2015	1,178	36
2016	976	18
2017	1,418	35
2018	1,123	10
2019	1,390	42

Source: National Oceanic and Atmospheric Administration (NOAA and Wikipedia) (tornado counts in recent years are preliminary).

$$\ell = \frac{\frac{1}{2}\rho_a U_f^2 C_F}{I\rho_m g} \tag{1.13}$$

$$t = \frac{\frac{1}{2}\rho_a U_f^2 C_F}{I\rho_m g} \tag{1.14}$$

$$d = \frac{\frac{2}{\pi}\rho_a U_f^2 C_F}{I\rho_m g} \tag{1.15}$$

where
  ℓ is a characteristic dimension for 'compact' objects,
  t is the thickness of sheet objects,

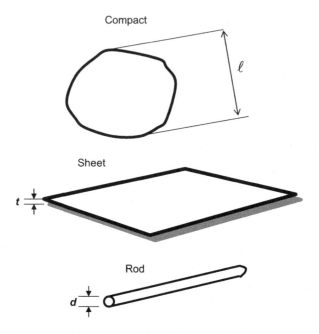

Figure 1.13 Three types of flying debris (after Wills et al., 1998).

d is the effective diameter of rod-type objects,
ρ_a is the density of air,
ρ_m is the density of the object material,
C_F is an aerodynamic force coefficient (see Section 4.2.2),
U_f is the wind speed at which flight occurs,
I is a fixing strength integrity parameter, i.e. the value of force required to dislodge the objects expressed as a multiple of their weight (for objects resting on the ground $I \cong 1$),
g is the gravitational constant.

Equations (1.13) – (1.15) illustrate the important point that the larger the value of the characteristic dimensions, ℓ, t or d, the higher the wind speed at which flight occurs. These equations also show that the higher the value of the density, ρ_m, the higher is the wind speed for lift off. Thus, as the wind speed in a cyclone builds up, the smaller and lighter, objects – for example, gravel, small loose objects in gardens and backyards, 'fly' first. At higher wind speeds, appurtenances on buildings are dislodged as the wind forces exceed their fixing resistance, and they also commence flight. At even higher wind speeds, substantial pieces of building structure, such as roof sheeting and purlins, may be removed, and become airborne.

As examples of the application of Equation (1.13), Wills *et al.* (1998) considered wooden compact objects ($\rho_m=500\,\text{kg/m}^3$) and stone objects ($\rho_m=2{,}700\,\text{kg/m}^3$). Assuming $C_F=1$, and $I=1$, Equation (1.13) gives ℓ equal to 110 mm for the wooden missile, but only 20 mm for the stone missile, for a lift-off speed of 30 m/s.

For sheet objects, Equation (1.14) shows that the wind speed for flight depends on the thickness of the sheet, but not on the length and width. Wills *et al.* expressed Equation (1.14) in a slightly different form:

$$\rho_m t = \frac{\frac{1}{2}\rho_a U_f^2 C_F}{Ig} \tag{1.16}$$

The left-hand side of Equation (1.16) is the mass per unit area of the sheet. This indicates the wind speed for flight for a loose object depends essentially on its mass per unit area. Thus, a galvanized iron sheet of 1 mm thickness with mass per unit area of 7.5 kg/m² will fly at about 20 m/s ($C_F =0.3$).

For 'rod'-like objects, which include timber members of rectangular cross-section, a similar formula to Equation (1.16) can be derived from Equation (1.15), with the 't' replaced by 'd', the equivalent rod diameter. Using this, Wills *et al.* calculated that a timber rod of 10 mm diameter will fly at about 11 m/s, and a 100 mm by 50 mm timber member, with an equivalent diameter of 80 mm, will fly at about 32 m/s, assuming C_F is equal to 1.0.

1.5.2 Trajectories of compact objects

A missile, once airborne, will continue to accelerate until its flight speed approaches the wind speed, or until its flight is terminated by impact with the ground or with an object such as a building. The trajectories of compact objects are produced by drag forces (Section 4.2.2), acting in the direction of the relative wind with respect to the body.

Consider first the aerodynamic force on a compact object (such as a sphere) in a horizontal wind of speed, U. Neglecting the vertical air resistance initially, the aerodynamic force can be expressed as:

$$\text{Accelerating force} = \frac{1}{2}\rho_a(U - v_m)^2 C_D A \tag{1.17}$$

where
  v_m is the horizontal velocity of the missile with respect to the ground,
  A is the reference area for the drag coefficient, C_D (Section 4.2.2).

Applying Newton's law, the instantaneous acceleration of the object (characteristic dimension, ℓ) is given by:

$$\text{Acceleration} = \frac{dv_m}{dt} = \frac{\frac{1}{2}\rho_a(U-v_m)^2 C_D A}{\rho_m \ell^3} = \frac{\frac{1}{2}\rho_a(U-v_m)^2 C_D}{\rho_m \ell} \tag{1.18}$$

taking A equal to ℓ^2.

Equation (1.18) shows that heavier and larger objects have lower accelerations, and hence their flight speeds are likely to be lower than smaller or lighter objects. The equation also shows that the initial acceleration from rest $(v_m=0)$ is high, but the acceleration rapidly reduces as the difference between the missile speed and the wind speed reduces, so that the missile speed approaches the wind speed very slowly. However, the missile speed cannot exceed the wind speed in steady winds.

Equation (1.18) can be integrated to obtain the time taken to accelerate to a given speed, v_m, and the distance travelled in this time. These equations are as follows:

Time taken to accelerate from 0 to v_m,

$$T = \frac{v_m}{kU(U-v_m)} \tag{1.19}$$

$$\text{Distance travelled} = U\left[T - \left(\frac{1}{kU}\right)\ln(1+kUT)\right] \tag{1.20}$$

where $k=(\rho_a C_D)/(2\rho_m \ell)$, with units of $(1/m)$.

Using Equation (1.20), the flight time and distance travelled by a steel ball of 8 mm diameter and 2 g mass, have been calculated, for a wind speed, U, of 32 m/s, and are given in Table 1.5.

The calculations show that it takes nearly a minute and 1.27 km for the steel ball to reach 30 m/s – i.e. 2 m/s less than the wind speed. In reality, such a long flight time and distance would not occur since the object would strike a building, or the ground, and lose its kinetic energy.

A more accurate analysis of the trajectories of compact objects requires the vertical air resistance to be included, as if we exclude it, then the calculated missile speed and distance travelled in a given time is underestimated (Holmes, 2004).

Table 1.5 Flight times and distances for a steel ball (neglecting vertical air resistance)

Object/speed	Time taken (seconds)	Horizontal distance travelled (m)
Steel ball to 20 m/s	5.4	71
Steel ball to 30 m/s	49	1,270

1.5.3 Trajectories of sheet and rod objects

Tachikawa (1983) carried out a fundamental study of the trajectories of missiles of the sheet type. Aerodynamic forces on auto-rotating plates were measured in a wind tunnel. These results were then used to calculate trajectories of the plates released into a wind stream. Free-flight tests of model plates with various aspect ratios were made in a small wind-tunnel and compared with the calculated trajectories. A distinct change in the mode of motion and the trajectory, with initial angle of attack of the plate was observed. The calculated trajectories predicted the upper and lower limits of the observed trajectories, with reasonable accuracy. A later study by Tachikawa (1990) extended the experiments to small prismatic models as well as flat plates and gave a method of estimating the position of a missile impact on a downstream building. The critical non-dimensional parameter for determination of trajectories is

$$K = \rho_a U^2 A / 2mg,$$

where:

ρ_a is the density of air,
U is the wind speed,
A is the plan area of a plate,
m is the mass of the missile,
g is the gravitational constant.

This parameter, known as the 'Tachikawa number' (Holmes *et al.*, 2006a), represents the ratio of aerodynamic forces to gravity forces, and can also be expressed as the product of three other non-dimensional parameters:

$$K = \frac{1}{2} \frac{\rho_a}{\rho_m} \frac{U^2}{gl} \frac{l}{t} \tag{1.21}$$

where

ρ_m is the missile density,
t is the plate thickness,
l is $\sqrt{A}$, i.e. a characteristic plan dimension.

In Equation (1.21), ρ_a/ρ_m is a density ratio and (U^2/gl) is a Froude number, both important non-dimensional quantities in aerodynamics (see also Section 7.4).

The equations of motion for horizontal, vertical and rotational motions of a flat plate moving in a vertical plane must be solved numerically. Good agreement has been obtained when comparing numerical solutions

with measurements of trajectories of many small plates in a wind tunnel
(Lin *et al.*, 2006; Holmes *et al.*, 2006b).

1.5.4 Standardized missile testing criteria

Standardized missile tests have been devised for those regions that are
prone to hurricanes and tropical cyclones (Section 1.3.2), and where the
occurrence of damage to buildings by wind-generated missiles has become
a major problem. These tests demonstrate the ability of wall claddings of
various types to resist penetration by flying debris or assist in the development of window protection screens.

When specifying appropriate test criteria for missile impact resistance,
the following principles should be followed:

- The missiles should be representative of actual objects available.
- The criteria should be physically realistic, i.e. if the flight threshold
  speed is greater than the expected wind speed in the storm, then the
  object should not be regarded as a potential missile.
- Realistic missile speeds should be specified for the expected separation distances between buildings.

Missile testing criteria were included in the Darwin Area Building Manual,
following Cyclone 'Tracy' in 1974, in Australia. These criteria specified
that windows and doors should withstand impact at any angle of a piece
of 100 mm by 50 mm timber weighing 4 kg, travelling at 20 m/s. A more
severe test was specified for cyclone refuge shelters. The refuge shelters
should be able to withstand an 'end-on' impact of a piece of 100 mm by
50 mm timber weighing 8 kg, travelling at 30 m/s. Later the test requirement for windows and doors of buildings was modified to a piece of
100 mm by 50 mm timber weighing 4 kg, travelling at 15 m/s. More recent
debris speeds given in Australian standards are linked to the regional gust
speed, used for wind loading design, with a horizontal debris speed of
40% of that wind speed.

Wind-borne debris impact test standards in the United States were discussed by Minor (1994). Following investigations of glass breakage (mainly
in high-rise buildings), during several U.S. hurricanes, Pantelides *et al.*
(1992) proposed a test protocol involving impacts from small spherical
missiles of 2 g. This was taken up in South Florida following Hurricane
'Andrew' in 1992. The Dade County and Broward County editions of the
South Florida Building Code required windows, doors and wall coverings
to withstand impacts from large and small debris. The large missile test,
which is similar to the Australian one, is only applicable to buildings below
9 m in height. The small missile test is only applicable to windows, doors

and wall coverings above 9 m, and differs between the two counties. The Dade County protocol uses ten 2 g pieces of roof gravel impacting simultaneously at 26 m/s, while the Broward County version uses ten 2 g steel balls impacting successively at 43 m/s.

1.6 WINDSTORM DAMAGE AND LOSS PREDICTION

The trend towards increased losses from windstorms has provoked concern in the insurance and re-insurance industries, and many of these groups require detailed assessments of the potential financial losses from the exposure of their portfolios of buildings to large-scale severe windstorms. Government bodies also require predictions of damage and economic losses to aid in planning for disaster and emergency management.

The prediction of average annual loss, or accumulated losses over an extended period, say fifty years, requires two major inputs: hazard models, and vulnerability curves. The hazard model focuses on the windstorm hazard itself, and makes use of historical meteorological data, and statistics to predict potential wind speeds at a site into the future. Vulnerability curves attempt to predict building (and sometimes contents) damage, given the occurrence of a particular wind speed.

1.6.1 Hazard models

The purpose of wind hazard models is to define the risk of occurrence of extreme wind speeds at the site of a single structure, on a system such as a transmission line, or on a complete city or region. The basis for these models is usually the historical record of wind speeds from anemometer stations, but often larger-scale storm parameters such as central pressures for tropical cyclones, and atmospheric stability indices for thunderstorm occurrences, are studied. The methods of statistics and probability are extensively used in the development of hazard models in wind engineering.

The application of statistical methods to the prediction of extreme wind speeds is discussed in Chapter 2 of this book.

An understanding of the structure of the wind within a storm enables predictions of 'footprints' such as that shown in Figure 1.14, (Holmes and Oliver, 2000). It shows simulated contours of maximum wind speeds, occurring at some time during the passage of a downburst (Section 1.3.5). A footprint of maximum gust speeds estimated to have occurred during the landfall of a tropical cyclone is shown in Figure 1.15 (Cyclone 'Yasi' which impacted the coast of Queensland, Australia in 2011).

This type of information, in combination with knowledge of the strength or 'vulnerability' of structures, enables predictions of potential damage to be made.

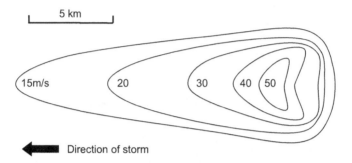

Figure 1.14 Wind speed threshold 'footprint' during the passage of a downburst (Holmes and Oliver, 1999).

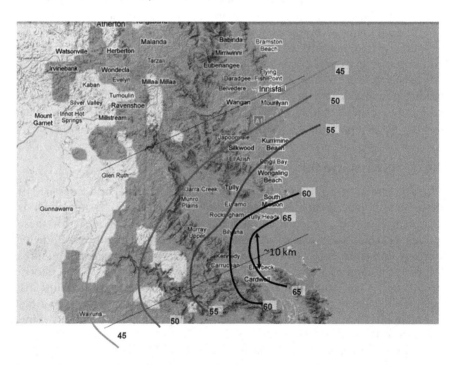

Figure 1.15 Maximum gust 'footprint' (m/s) at landfall of Tropical Cyclone 'Yasi' (Queensland, 2011).

1.6.2 Vulnerability curves

Insurance loss predictions are quite sensitive to the assumed variations of relative damage to building and contents, as a function of the local wind speed – usually a gust speed (see Section 3.3.3). Such graphs are known as 'vulnerability curves'. Vulnerability curves can be derived in a number of ways. Leicester (1981) proposed the simplified form, with straight-line

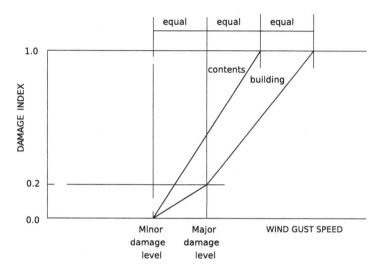

Figure 1.16 Form of vulnerability curve proposed by Leicester (1981).

segments, for Australian houses, shown in Figure 1.16. The ordinate is a 'damage index' defined as follows for the building:

Damage index(D) = (repair cost)/(initial cost of building)

For insurance purposes it may be more appropriate to replace the denominator with the insured value of the building. A similar definition can be applied to the building's contents, with 'replacement cost' in the numerator.

Separate lines are given for building and contents. Two parameters only need be specified – a threshold gust speed for the onset of minor damage, and a speed for the onset of major building damage (damage index>0.2).

Walker (1995) proposed the following relationships for housing in North Queensland, Australia. For pre-1980 buildings:

$$D = 0.2\left(\frac{U-30}{30}\right)^2 + 0.5\left(\frac{U-30}{30}\right)^6 \tag{1.22}$$

For post-1980 buildings:

$$D = 0.2\left(\frac{U-37.5}{37.5}\right)^2 + 0.5\left(\frac{U-37.5}{37.5}\right)^6 \tag{1.23}$$

Clearly in both cases *D* is limited to the range of 0–1.0.

The relationship of Equation (1.22) was also found to agree well with recorded damage and wind speed estimates in Hurricane 'Andrew' that impacted Florida in 1992 (see Table 1.3).

An alternative possible simple empirical form for the damage index is a Weibull function, corresponding to the equation for the cumulative probability distribution (Appendix C, Equation C.19):

$$D = 1 - \exp\left[-\left(\frac{U}{c}\right)^{w}\right] \tag{1.24}$$

This function correctly gives values between 0 and 1.0 and has an appropriate 'S' shape. w is a shape factor that governs the steepness of the main part of the curve, and c is a scale factor. A typical value for w in Equation (1.24) is 8. Once w is fixed, c can be determined selecting a value of U for a single damage state. For example, if w is taken as 8, and it is expected that a (gust) wind speed of 40 m/s would produce a damage index of 0.3 (i.e. the repair cost is 30% of the initial cost of the building), then c can be determined from:

$$c = \frac{40}{\left[-\log_{e}(1-0.3)\right]^{1/8}} = 45.5 \text{ m/s} \tag{1.25}$$

A simple form of a vulnerability curve for a fully-engineered structure consisting of a large number of members or components with strengths of known probability distribution can also be derived. The failure of each component is assumed to be independent of all the others, and they are all designed to resist the same wind load, or speed. Thus, the expected fractional damage to a complete structure, for a given wind speed, is the proportion of failed components expected at that wind speed. If all the components have the same probability distribution of strength, which would be true if they were all designed to the same codes, then the vulnerability curve can simply be derived from the cumulative distribution of strength of any element.

A curve derived in this way (Holmes, 1996) is shown in Figure 1.17, for a structure comprising of components with a lognormal distribution of strength, with a mean/nominal strength of 1.20, and a coefficient of variation of 0.13, values which are appropriate for steel components. The nominal design gust wind speed is taken as 65 m/s. This curve can be compared with that proposed by Walker, for post-1980 Queensland houses, in the tropical cyclone-affected coastal region (Equation 1.23). The theoretical curve, representing fully-engineered structures, is steeper than the Walker curve which has been derived empirically, and incorporates the greater variability in the components of housing structures.

Alternatively, vulnerability curves for individual building types can be derived through a probabilistic approach, based on assessing individual

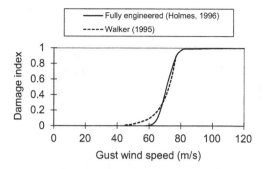

Figure 1.17 Theoretical and empirical vulnerability curves.

building component strengths, and on progressive failure simulation (e.g. Henderson and Ginger, 2007; Hamid *et al.*, 2010), following methods that have been adopted for predicting damage due to earthquakes. This approach often introduces intermediate 'fragility curves' which plot the probability of exceedance of a particular extent of damage against wind speed. Several of these are then combined, with a weighting based on relative cost of repair of each state, to give a vulnerability curve.

Walker (2011) and Pita *et al.* (2015) have reviewed recent developments in modelling of the vulnerability of buildings to wind loads, which are tending towards structural engineering models, rather than heuristic or empirical models, as has been the case in the past.

1.6.3 Damage produced by flying debris

Wills *et al.* (1998) carried out an analysis of the damage potential of flying missiles, based on the assumption that the damage of a given missile is proportional to its kinetic energy in flight. A number of interesting conclusions arose from this work:

- For compact objects, lower density objects have more damage potential,
- Sheet and rod objects have generally more damage potential than compact objects,
- Very little energy is required to break glass (e.g. 5-g steel ball travelling at 10 m/s is sufficient to break 6 mm annealed glass),
- Based on an assumed distribution of available missile dimensions, Wills et al. found that the total damage is proportional to U^n, where n is a power equal to about 5.

Probabilistic damage modelling has recently been extended to incorporate modelling of damage produced by flying debris, as well as that due to direct wind forces. The main features required for an engineering windborne debris damage model are summarized as follows (Twisdale *et al.*, 1996):

- A *windfield model* – similar to those discussed in Section 1.6.1.
- A *debris generation model* – this is required to establish the source, numbers and generic shapes of debris items.
- A *debris trajectory model* – once they become airborne, a debris trajectory model is required to predict consequent damage due to impact on downwind buildings.
- A *debris impact model* – this represents the magnitude of damage produced by an impacting windborne missile.

Twisdale *et al.* (1996) observed missiles after the 1995 Hurricanes 'Erin' and 'Opal' in the United States, and found the clearly dominant contributors were roofing tiles, shingles, sheathing, and structural members from roof trusses, with lesser contributions from wall cladding, miscellaneous house materials like guttering, vents, and yard items and accessories. This led to the development of a windborne debris damage model, with the generation part focused on the generation of windborne roofing elements.

Twisdale *et al.* (1996) then described a debris generation model which is essentially a wind load failure model for roofing elements of low-rise buildings. It is based on simulating wind loads on elements of a simple representative gable-roof building and checking whether these exceed the resistance of roof sheeting and roof truss elements. The resistance is based on the pull-out strength of nails, and an assessment of errors in construction – i.e. an assessment of the number of nails attaching plywood roof sheathing to the underlying roof trusses.

Damage produced by windborne debris is primarily generated by the horizontal energy and momentum at impact. The trajectories of missiles of the generic 'compact', 'rod' and 'sheet' types were discussed in Sections 1.5.2 and 1.5.3.

The horizontal velocities attained by all three types are mainly determined by the mass of the debris object, and product of the average drag coefficient and the exposed area, during the trajectories.

It was shown, numerically and experimentally (Baker, 2007; Lin *et al.*, 2007), that the *horizontal* velocity component of a windborne missile can be well represented by the following function:

$$\frac{u_m}{U_s} \cong 1 - \exp\left[-b\sqrt{x}\right] \tag{1.26}$$

where
  u_m is the horizontal missile velocity,
  U_s is the local (gust) wind speed,

x is the horizontal distance travelled (this can be related to average building spacing),

b is a dimensional parameter depending on the shape of the missile and its drag coefficient, and its mass (Equation (1.27)).

$$b = \sqrt{\frac{\rho_a C_{D,av} A}{m}} \qquad (1.27)$$

In Equation (1.27), $C_{D,\,av}$ is an *average* drag coefficient, averaged over the rotations of the body with respect to the relative wind. Note that the right-hand side of Equation (1.26) does not include the wind speed and is only a function of horizontal distance travelled (or building spacing in the case of an impact), and the missile properties in Equation (1.27).

Following from Equation (1.26), the momentum and kinetic energy at impact can be represented by Equations (1.28) and (1.29), respectively.

$$m \cdot u_m \cong m U_s \left\{ 1 - \exp\left[-b\sqrt{x} \right] \right\} \qquad (1.28)$$

$$E = \frac{1}{2} m \cdot u_m^2 \cong \frac{1}{2} m U_s^2 \left\{ 1 - \exp\left[-b\sqrt{x} \right] \right\}^2 \qquad (1.29)$$

where m is the mass of the missile.

The previous discussion has indicated that the horizontal velocities (and hence impact momenta and kinetic energy) of windborne debris are relatively simple functions of wind gust speeds and distance travelled and hence of building separation. The damage produced by an impacting missile on a building surface is dependent on the component of momentum normal to the surface and/or its kinetic energy at impact. The change of momentum at impact is directly related to the force applied to a surface – it is equal to the impulse applied – the integral of force with respect to time. A perfectly elastic surface (i.e. with a coefficient of restitution at impact of 1.0) would not absorb any of the kinetic energy of the missile; it would be retained as kinetic energy of the missile moving away from the surface. The total energy at impact must be conserved, and for many building materials suffering plastic deformation. Most of the kinetic energy of the debris item will be dissipated through deformation of the material.

1.7 HURRICANE DAMAGE MODELLING

The prediction of losses resulting from hurricane impact on buildings and facilities has become a major activity that several companies have embraced for the service of the insurance and re-insurance industries. While the details of most of these are commercial-in-confidence, some useful discussions of the background methodologies are in the public domain (e.g. Vickery *et al.*, 2000a).

A publicly-available model has been funded by the State of Florida (Hamid *et al.*, 2010), and will be used to illustrate the main components and features of these models in the following. The Florida Public Hurricane Loss Model (FPHLM) consists of three components:

- an 'atmospheric science component',
- an 'engineering' component, and
- an 'actuarial' component.

The atmospheric science component is essentially a form of 'hazard model' as introduced in Section 1.6.1. The first function of this component is to model the annual hurricane occurrences within the defined area chosen to encompass the historical origin points of hurricanes affecting Florida – a circle with a radius of 1,000 km, centred on a particular location just off the south-west coast of Florida. The Poisson Distribution (Section C3.5) was used to model the annual rate of hurricane occurrences within the defined area of interest. The points of origin of the simulated hurricanes were derived from the historical record of landfalling Atlantic tropical cyclones known as 'HURDAT' with small random perturbations. In this way, thousands of years of simulated hurricane tracks were generated. The intensities of each storm, represented by the difference in barometric pressure at sea level at the centre of the storm, and that at the periphery, was also varied at 24-hour intervals by sampling from an appropriate probability distribution.

The *wind-field model* is implemented when the simulated hurricane is close to a coastline. In the FPHLM, the slab boundary-layer model of Shapiro (1983) is used; a similar model was used by Vickery *et al.* (2000b). The model is initialized by a vortex in gradient balance (see Section 1.2.4), with a radially symmetric pressure profile given by the expression due to Holland (1980) (Equation 1.10).

As part of the engineering component, vulnerability curves (Section 1.6.2) were generated by Monte Carlo simulations for 168 cases for every combination of structural type (timber frame or masonry), geographical location in Florida, and roof type (gable, hip, tile, shingle, etc.). The strength of building components is determined as a function of gust wind speed through a detailed wind and structural engineering approach that includes an allowance for windborne debris damage. The latter is included empirically rather than through the detailed approach outlined in Section 1.6.3.

In the actuarial model, expected annual insurance losses for building structure, contents and additional living expenses using the vulnerability matrices are derived as discussed above. The probability distribution of gust wind speeds for each zip (postal) code is derived from the simulated set of hurricanes. These wind speeds are applied to the vulnerability curves and using the insured values the expected losses are estimated for each policy.

Another public-domain hurricane-damage model is HAZUS-MH described by Vickery *et al.* (2006a,b). The hurricane hazard model is based on the methods described by Vickery *et al.* (2000a,b). HAZUS-MH includes a detailed mechanics-based model of windborne debris impacts, and the hazard component includes estimates of rainfall rates to enable prediction of damage due to water ingress into buildings.

1.8 PREDICTED EFFECTS OF CLIMATE CHANGE

It has often been suggested that global warming is having significant effects on the numbers and strengths of windstorms – particularly on *tropical cyclones* (including hurricanes and typhoons). There have been a number of studies of the effects of global warming on tropical cyclones (including hurricanes) in recent years (e.g. Webster *et al.*, 2005; Emanuel, 2005; Klotzbach, 2006; Kossin *et al.*, 2007; Holland and Bruyere, 2014, Knutson *et al.*, 2020).

As discussed in Section 1.3.2, a sea surface temperature of 26° is required for tropical cyclone formation in the current climate; hence it might be expected that there would be an increase in the number of tropical cyclones worldwide with increasing average sea temperatures. In fact, Webster *et al.* (2005) found there was no significant trend in global cyclones of all strengths. However, they did show a statistically significant increasing trend in Category 4and 5 storms from the 1970s to the decade 1995 to 2004. These mainly seem to have occurred in the North Atlantic basin.

Klotzbach (2006) extended the analysis to all basins with tropical cyclone activity, and excluded data before 1986 on the basis that, before the mid-1980s, only *visible* satellite information was available and hence night-time observations were excluded; also, the quality and resolution of satellite imagery had improved greatly by the later period. Klotzbach's analysis, using the more recent (and more reliable) data, found only a small increase in Category 4–5 hurricanes in the North Atlantic and Northwest Pacific during the twenty-year study period.

However, the recent increases in severe hurricanes in the north Atlantic and Caribbean (see Table 1.3) may be an indication of a trend produced by climate change. Indeed, Holland and Bruyere (2014) studied hurricane occurrences in several basins in the Northern Hemisphere and found a significant increase in the proportion of Category 4 and 5 hurricanes, after accounting for observing system changes. This was balanced by a decrease in the proportion of Category 1 and 2 hurricanes.

An increase in the latitude of the maximum intensities of tropical cyclones was attributed to the poleward shift of the sub-tropical jet and the associated reduction in wind shear (Kossin *et al.*, 2014).

Fewer studies have been made of tropical cyclonic activity in the Southern Hemisphere, which has a much greater area of ocean, suggesting a different reaction to increasing water temperatures. Callaghan and Power (2010) found a *decrease* in the number of severe tropical cyclones making landfall on the eastern Australian coast since the late nineteenth century. However, their survey predated the decade 2011–2020, during which four, well-documented, Category 4 cyclones impacted the coast of the state of Queensland.

The following summary statement in the Special Report of the U.N. International Panel on Climate Change (IPCC) of October 2018 is relevant:

> Tropical cyclones are projected to increase in intensity (with associated increases in heavy precipitation) although not in frequency.

This statement was based mainly on simulation studies and was qualified by a statement: 'low confidence, limited evidence'.

However, a recent review for Australia by the insurance group IAG and the US National Center for Atmospheric Research (Bruyere *et al.*, 2019) suggests that a 5% increase in wind speeds from tropical cyclones, due to global temperature rises since the mid-nineteenth century, has already occurred, with 'high confidence'. The report also notes a poleward shift in the latitude of maximum intensity of 1.6°.

Figure 1.18 shows the numbers of severe tropical cyclones in the Coral Sea to the east of Australia for each of the five decades starting in 1969. 'Severe' in this case is defined as those with a central pressure (a good indicator of cyclone strength) of 950 hecto-Pascals, or less, at some stage in their lives. The figure shows an increasing trend in the numbers that seems to support the result of the simulations from climate modelling and from the IAG/NCAR document.

Gregow *et al.* (2017) studied damage to forests in Europe between 1951 and 2010 and identified a statistically-significant increase in storm intensity

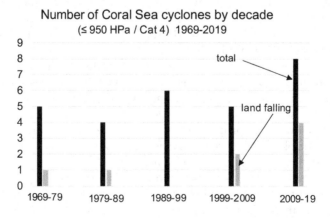

Figure 1.18 Numbers of severe tropical cyclones per decade observed off eastern Australia, 1969–2019.

since 1990. All but one of seven catastrophic *north Atlantic storms* have occurred since that year.

The effects of global warming on severe winds from smaller storms, such as *tornadoes* (Section 1.3.4) and *thunderstorm downbursts* (Section 1.3.5) are difficult to determine as their scales are too small to resolve with current computational climate models. However, since these events are driven by strong convection of moisture-laden air, followed by latent heat release, it would be expected that warming would enhance these processes, and hence generate stronger storms. For example, Allen and Karoly (2014) have identified possible increasing trends in numbers of severe thunderstorms in Eastern Australia, by studying changes in convective available potential energy (CAPE).

Although trends do seem to be emerging, clearly there is some uncertainty in the predictions of global warming effects on severe windstorms, and some regulators may be reluctant to impose additional economic costs of higher design wind loads on the community without more confidence in the trends. A rational approach to structural design in these circumstances would be an increase in wind-load factor, or a dedicated 'climate change factor' to cover these uncertainties.

1.9 SUMMARY

In this chapter, the physical mechanisms and meteorology of strong windstorms of all types have been described. The balance of forces in a large-scale synoptic system were established, and the gradient wind equation have been derived. Smaller scale storms, such as tornadoes and downbursts, were also discussed.

The history of several windstorms that have produced significant damage has been discussed. The mechanics of wind-generated flying debris was considered, and vulnerability curves relating fractional damage potential to wind speed, for insurance loss prediction, were derived. The modelling of wind damage from direct wind forces and from windborne debris, for disaster management and insurance purposes, has been outlined. Two hurricane loss models that are publicly available in the United States have been described.

The predicted effects of climate change on the frequencies and strengths of tropical cyclones and other storm types have also been briefly discussed.

1.10 THE FOLLOWING CHAPTERS AND APPENDICES

Following this introductory chapter, Chapters 2 – 7 are directed towards fundamental aspects of wind-loading that are common to all or most structures, for example, atmospheric wind structure and turbulence (Chapter 3), bluff-body aerodynamics (Chapter 4), resonant dynamic response of structures (Chapter 5), and wind-tunnel techniques (Chapter 7). Chapters 8 – 14 deal with aspects of wind loading for particular types of structures: buildings, bridges, towers, etc.

Chapter 15 discusses contemporary wind-loading codes and standards – the most common point of contact of practising structural engineers with wind loads. Chapter 16 (prepared by Dr. S.A. Bekele) gives an overview of the application of computational fluid dynamics to wind loading of structures.

Appendices A and B cover the terminology of wind engineering and the symbols used in this book, respectively. Appendix C describes probability distributions relevant to wind loading. Appendix D summarizes the extreme wind climates of over 140 countries and jurisdictions and gives numerous references for further information on basic design wind speeds for most of the world.

Appendix E gives some approximate formulae for calculating natural frequencies of structures. Appendix F gives a simple example of the calculation of effective static wind load distributions. In Appendix G, several methods of fitting the Generalized Pareto probability distribution to peak-over threshold wind data are outlined, with examples.

REFERENCES

Allen, J.T. and Karoly, D.J. (2014) A climatology of Australian severe thunderstorms environments 1979–2011: inter-annual variability and ENSO influence. *International Journal of Climate*, 34: 81–97.

Baker, C.J. (2007) The debris flight equations. *Journal of Wind Engineering and Industrial Aerodynamics*, 95: 329–353.

Bruyere, C.L., Holland, G.J., Prein, A., Done, J., Buckley, B, Chan, P., Leplastrier, M. and Dyer, A. (2019) Severe weather in a changing climate. Insurance Australia Group and National Center for Atmospheric Research (USA), November. (www.iag.com.au/sites/default/files/documents/Severe-weather-in-a-changing-climate-report-011119.pdf).

Callaghan, J. and Power, S.B. (2010) Variability and decline in the number of severe tropical cyclones making land-fall over eastern Australia since the late nineteenth century. *Climate Dynamics*, 37: 647–662, doi: 10.1007/s00382-013-010-0883-2.

Emanuel, K.A. (2005) Increasing destructiveness of tropical cyclones over the past thirty years. *Nature* 436: 686–688.

Fujita, T.T. (1971) Proposed characterization of tornadoes and hurricanes by area and intensity. SMRP Report. 91, Department of the Geophysical Sciences, University of Chicago.

Fujita, T.T. (1978) Manual of downburst identification for project Nimrod. SMRP Research paper 156, Department of the Geophysical Sciences, University of Chicago.

Fujita, T.T. (1985) *The Downburst*. Report on Projects NIMROD and JAWS, Published by the author at the University of Chicago.

Fujita, T.T., Pearson, A.D., Forbes, G.S., Umenhofer, T.A., Pearl, E.W. and Tecson, J.J. (1976) Photogrammetric analyses of tornados. *Symposium on Tornadoes: Assessment of Knowledge and Implications for Man*, Texas Tech University, Lubbock, Texas, USA, 22–24 June 1976, pp. 43–88.

Golden, J.H. (1976) An assessment of windspeeds in tornados. *Symposium on Tornadoes: Assessment of Knowledge and Implications for Man*, Texas Tech University, Lubbock, Texas, USA, 22–24 June 1976, pp. 5–42.

Goliger, A.M., Milford, R.V., Adam, B.F. and Edwards, M. (1997) *Inkanyamba: Tornadoes in South Africa*. CSIR Building Technology and S.A. Weather Bureau.

Gregow, H., Laaksonen, A. and Alper, M.E. (2017) Increasing large scale windstorm damage in Western, Central and Northern European forests, 1951–2010. *Scientific Reports*, 7: 46397. doi: 10.1038/srep46397

Hamid, S., Kibria, B.M.G., Gulati, S., Powell, M., Annane, B., Cocke, S., Pinelli, J-P., Gurley, K. and Chen, S-C. (2010) Predicting losses of residential structures in the state of Florida by the public hurricane loss evaluation model. *Statistical Methodology*, 7: 552–573.

Henderson, D.J. and Ginger, J.D. (2007) Vulnerability model of an Australian high-set house subjected to cyclonic wind loading. *Wind and Structures*, 10: 269–285.

Hoecker, W.H. (1960) Wind speed and airflow patterns in the Dallas tornado of April 2, 1957. *Monthly Weather Review*, 88: 167–80.

Holland, G.J. (1980) An analytic model of the wind and pressure profiles in a hurricane. *Monthly Weather Review*, 108: 1212–1218.

Holland, G.J. and Bruyere, C.L. (2014) Recent intense hurricane response top global climate change. *Climate Dynamics*, 42: 617–627. doi: 10.1007/s00382-013-1713-0.

Holmes, J.D. (1996) Vulnerability curves for buildings in tropical-cyclone regions for insurance loss assessment. *ASCE EMD/STD Seventh Specialty Conference on Probabilistic Mechanics and Structural Reliability*, Worcester, Massachusetts, USA, 7–9 August.

Holmes, J.D. (2004) Trajectories of spheres in strong winds with applications to wind-borne debris. *Journal of Wind Engineering and Industrial Aerodynamics*, 92: 9–22.

Holmes, J.D. and Oliver, S.E. (2000) An empirical model of a downburst. *Engineering Structures*, 22: 1167–1172.

Holmes, J.D., Baker, C.J. and Tamura, Y. (2006a). Tachikawa number: a proposal. *Journal of Wind Engineering and Industrial Aerodynamics*, 94: 41–47.

Holmes, J.D., Letchford, C.W. and Lin, Ning (2006b). Investigations of plate-type windborne debris. II. Computed trajectories. *Journal of Wind Engineering and Industrial Aerodynamics*, 94: 21–39.

Klotzbach, P.J. (2006) Trends in global tropical cyclone activity in the last twenty years (1986–2005). *Geophysical Research Letters*, 33: L10805.

Knutson, T., Camargo, S.J., Chan, J.C.L., Emanuel, K., Ho, C.H., Kossin, J., Mohapatra, M., Satoh, M., Sugi, M., Walsh, K. and Wu, L. (2020) Tropical cyclones and climate change assessment: Part II: Projected response to anthropogenic warming. Bulletin of the American Meteorological Society, 101:303–322.

Kossin, J.P., Knapp, K.R., Vimont, D.J., Murnane, R.J. and Harper B.A. (2007) A globally consistent reanalysis of hurricane variability and trends. *Geophysical Research Letters*, 34: L04815.

Kossin, J.P., Emanuel, K. and Vecchi, G.A. (2014) The poleward migration of the location of tropical cyclone maximum intensity. *Nature*, 509: 349–362

Leicester, R.H. (1981) A risk model for cyclone damage to dwellings. *Proceedings, 3rd International Conference on Structural Safety and Reliability*, Trondheim, Norway.

Lin, N., Letchford, C.W. and Holmes, J.D (2006) Investigations of plate-type windborne debris. I. Experiments in full scale and wind tunnel. *Journal of Wind Engineering and Industrial Aerodynamics*, 94: 51–76.

Lin, N., Holmes, J.D. and Letchford, C.W. (2007) Trajectories of windborne debris in horizontal winds and applications to impact testing. *ASCE Journal of Structural Engineering*, 133: 274–282.

McDonald, J.R. and Mehta, K.C. (2004). *A recommendation for an enhanced Fujita Scale*. Wind Science and Engineering Research Center, Texas Tech University, Lubbock, TX.

Mehta, K.C. (1976) Windspeed estimates: engineering analyses. *Symposium on Tornadoes: Assessment of Knowledge and Implications for Man*, Texas Tech University, Lubbock, Texas, USA, 22–24 June 1976, pp. 89–103.

Minor, J.E. (1994) Windborne debris and the building envelope. *Journal of Wind Engineering and Industrial Aerodynamics*, 53: 207–227.

Page, J.K. (1968) Field investigation of wind failures in the Sheffield gales of 1962. *Symposium on Wind effects on Buildings and Structures*, Loughborough University, England, UK, 2–4 April 1968, pp. 12.1–12.24.

Pantelides, C.P., Horst, A.D. and Minor, J.E. (1992) Post-breakage behaviour of architectural glazing in wind storms. *Journal of Wind Engineering and Industrial Aerodynamics*, 41–44: 2425–2435.

Pita, G.L., Pinelli, J-P., Gurley, K.R., and Mitrani-Reiser, J. (2015) State of the art of hurricane vulnerability estimation methods – a review. *Natural Hazards Review*, 16: 04014022. doi: 10.1061/(ASCE)NH.1527–6996.0000153.

Shapiro, L. (1983) The asymmetric boundary layer flow under a translating hurricane. *Journal of Atmospheric Sciences*, 40: 1984–1998.

Smith, B.W. (2007) Communication structures. Thomas Telford Publishing, London, UK.

Tachikawa, M. (1983) Trajectories of flat plates in uniform flow with application to wind-generated missiles. *Journal of Wind Engineering and Industrial Aerodynamics*, 14: 443–453.

Tachikawa, M. (1990) A method for estimating the distribution range of trajectories of windborne missiles. *Journal of Wind Engineering and Industrial Aerodynamics*, 29: 175–184.

Twisdale, L.A., Vickery, P.J. and Steckley, A.C. (1996) Analysis of hurricane windborne debris impact risk for residential structures. Report 5503, prepared for State Farm Mutual, Applied Research Associates, Rayleigh, North Carolina, USA.

Vickery, P.J., Skerjl, P.F. and Twisdale, L.A. (2000a) Simulation of hurricane risk in the United States using an empirical storm track modeling technique. *Journal of Structural Engineering*, 126: 1222–1237.

Vickery, P.J., Skerjl, P.F., Steckley, A.C. and Twisdale, L.A. (2000b) A hurricane wind field model for use in simulations. *Journal of Structural Engineering*, 126: 1203–1222.

Vickery, P.J., Lin, J., Skerjl, P.F., Twisdale, L.A. and Huang, K. (2006a) HAZUS-MH hurricane model methodology. I: hurricane hazard, terrain and wind load modeling. *Natural Hazards Review*, 7: 82–93.

Vickery, P.J., Skerjl, P.F., Lin, J. Twisdale, L.A., Young, M.A. and Lavelle, F.M. (2006b) HAZUS-MH hurricane model methodology. II: damage and loss estimation. *Natural Hazards Review*, 7: 94–103.

Walker, G.R. (1995) Wind vulnerability curves for Queensland houses. Alexander Howden Insurance Brokers (Australia) Ltd.

Walker, G.R. (2011) Modelling of the vulnerability of buildings to wind – a review. *Canadian Journal of Civil Engineering*, 38: 1031–1039.

Webster, P.J., Holland, G.J., Curry, J.A. and Chang, H.R. (2005) Changes in tropical cyclone number, duration and intensity in a warming environment. *Science*, 309: 1844–1846.

Wills, J., Wyatt, T. and Lee, B.E. (1998) Warnings of high winds in densely populated areas. United Kingdom National Coordination Committee for the International Decade for Natural Disaster Reduction.

Chapter 2

Prediction of design wind speeds and structural safety

2.1 INTRODUCTION AND HISTORICAL BACKGROUND

The establishment of appropriate design wind speeds is a critical first step towards the calculation of design wind loads for structures. It is also usually the most uncertain part of the design process for wind loads, and requires the statistical analysis of historical data on recorded wind speeds.

In the 1930s, the *Gaussian* distribution (Appendix C3.1), with a symmetrical bell-shaped probability density, was used to represent extreme wind speeds, and for the prediction of long-term design wind speeds. However, this failed to take note of the earlier theoretical work of Fisher and Tippett (1928), establishing the limiting forms of the distribution of the largest (or smallest) value in a fixed sample, that depends on the form of the tail of the parent distribution.

The use of extreme value analysis for design wind speeds lagged behind the application to flood analysis. E. J. Gumbel (1954) promoted the use of the simpler Type I extreme value distribution for such analysis. However, von Mises (1936) and Jenkinson (1955) showed that the three asymptotic distributions of Fisher and Tippett could be represented as a single *Generalized Extreme Value (GEV) Distribution*; this is discussed in one of the following sections. By the 1950s and the early 1960s, several countries had applied extreme value analyses to predict design wind speeds. Mainly, Type I (by now also known as the 'Gumbel Distribution'), was used for these analyses. The concept of *return period*, the reciprocal of the probability of exceedance of an epoch (usually annual) extreme was defined originally by Gumbel (1941) and became widely adopted in the following twenty years.

The use of probability and statistics as the basis for the modern approach to wind loads was, to a large extent, a result of the work of A.G. Davenport in the 1960s. This was recorded in many papers (e.g. Davenport 1961), although there were others, such as Shellard (1958, 1963) in the United Kingdom, and Whittingham (1964) in Australia, who also applied Gumbel's methods to make predictions of extreme wind speeds.

In the 1970s and 1980s, the enthusiasm was tempered for the then standard 'Gumbel analysis' due to events such as Cyclone 'Tracy' in Darwin, Australia (1974) and severe gales in Europe (1987). In those events, the previous design wind speeds that had been determined by a Gumbel fitting procedure were exceeded by quite an extent. This highlighted the importance of the following factors:

- sampling errors inherent in the recorded data base, usually less than 50 years, and
- the need for separation of data originating from different storm types.

The need to separate the recorded data by storm type was recognized in the 1970s by Gomes and Vickery (1977a).

The development of probabilistic methods in structural design was parallel with their use in wind engineering, followed by pioneering work by Freudenthal (1947, 1956) and Pugsley (1966). This field of research and development is known as 'structural reliability' theory. Limit-states design, based on probabilistic concepts, was steadily introduced into design practice from the 1970s onwards.

This chapter discusses modern approaches to the use of extreme value analysis for prediction of extreme wind speeds for the design of structures. Related aspects of structural design and safety are discussed in Section 2.9.

2.2 PRINCIPLES OF EXTREME VALUE ANALYSIS

The theory of extreme value analysis of wind speeds, or other geophysical variables, such as flood heights, or earthquake ground accelerations, is based on the application of one or more of the three asymptotic extreme value distributions identified by Fisher and Tippett (1928). These are discussed in the following section. They are asymptotic in the sense that they are the correct distributions for the largest of an *infinite* population of independent random variables of known probability distribution. In practice, of course, there will be a finite number in a population, but in order to make predictions, the asymptotic extreme value distributions are still used as empirical fits to the extreme data. Out of the three variables, which one is theoretically 'correct', depends on the form of the tail of the underlying parent distribution. However, unfortunately this form is not usually known with certainty due to lack of data. Physical reasoning has sometimes been used to justify the use of one or the other of asymptotic extreme value distributions.

Gumbel (1954, 1958) covered the theory of extremes and the state of the art in the 1950s; although dated, this work is still widely referenced for its methods of fitting the Type I extreme value distribution (see Section 2.3.1

following). A useful review of the various methodologies available for the prediction of extreme wind speeds, including those discussed in this chapter, was given by Palutikof *et al.* (1999). Coles (2001) discussed the principles of extreme value analysis from the point of view of a statistician. Torrielli *et al.* (2013) reviewed many known methods of extreme value analysis, including all of those discussed in the following sections, and tested them with a simulated data series representing more than 12,000 years of wind speeds.

2.2.1 The Generalized extreme value distribution

The *GEV Distribution* introduced by von Mises (1936), and later re-discovered by Jenkinson (1955), combines the three extreme value distributions into a single mathematical form:

$$F_U(U) = \exp\left\{-[1 - k(U - u)/a]^{1/k}\right\} \tag{2.1}$$

where $F_U(U)$ is the cumulative probability distribution function (see Appendix C) of the maximum wind speed in a defined period (e.g. one year).

In Equation (2.1), k is a shape factor, a is a scale factor, and u is a location parameter. When $k < 0$, the GEV is known as the *Type II Extreme Value* (or *Frechet*) Distribution; when $k > 0$, it becomes a *Type III Extreme Value Distribution* (a form of the *Weibull* Distribution). As k tends to 0, Equation (2.1) becomes Equation (2.2) in the limit. Equation (2.2) is the *Type I Extreme Value Distribution*, or *Gumbel* Distribution.

$$F_U(U) = \exp\left\{-\exp[-(U - u)/a]\right\} \tag{2.2}$$

The GEV with k equal to -0.2, 0 and 0.2 are plotted in Figure 2.1, in a form that the Type I Distribution appears as a straight line. As can be seen in the Figure, the Type III Distribution ($k = +0.2$) approaches a limiting value; it is therefore appropriate for variables that are 'bounded' on the high side. It should be noted that the Type I and Type II predict unlimited values; they are therefore suitable distributions for variables that are 'unbounded'. Since it would be expected that there is an upper limit to the values that can be produced by the atmosphere, the Type III Distribution may be more appropriate for wind speeds.

2.2.2 Return period and average recurrence interval

At this point, it is appropriate to introduce the term *return period, R_P*. It is the inverse of the complementary cumulative distribution of the extremes (Gumbel, 1941, 1958).

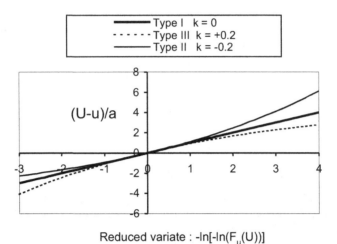

Figure 2.1 The generalized extreme value distribution ($k=-0.2$, 0. $+0.2$).

$$\text{i.e. Return period, } R_P = \frac{1}{\text{probability of exceedance}} = \frac{1}{1 - F_U(U)} \qquad (2.3)$$

Thus, if the annual maximum is being considered, then the return period is measured in years. A 50-year return period wind speed has a probability of exceedance of 0.02 (1/50) in any one year. The probability of wind speed, of given return period, being exceeded in the lifetime of a structure is discussed in Section 2.9.3.

The 'average (or mean) recurrence interval', R_I, is the average interval between exceedances of high thresholds, and the reciprocal of the average crossing rate. Average recurrence interval is related to return period through Equation (2.4).

$$\frac{1}{R_P} = 1 - \exp\left(-\frac{1}{R_I}\right) \qquad (2.4)$$

A proof of Equation (2.4) is given in Appendix C, and a graph is provided showing the relationship between R_P and R_I. The values converge at high levels and are virtually identical for values beyond 10 years (although interestingly there remains a difference of 0.5 years).

They diverge for values approaching unity, and R_P cannot be less than 1.0, as this would correspond to a probability of exceedance, exceeding 1.0 by Equation (2.3). On the other hand, R_I can take fractional values, with a lower limit of 0.

From Equation (2.3), Equation (2.1) for the GEV can be written as:

$$1 - \frac{1}{R_P} = \exp\left\{-\left[1 - k\left(\frac{U-u}{a}\right)\right]^{1/k}\right\}$$

leading to, $\quad \left[-\log_e\left(1 - \frac{1}{R_P}\right)\right]^k = \left[1 - k\left(\frac{U-u}{a}\right)\right]$

$$\left(\frac{U-u}{a}\right) = \frac{1}{k}\left\{1 - \left[-\log_e\left(1 - \frac{1}{R_P}\right)\right]^k\right\}$$

$$U = u + \frac{a}{k}\left\{1 - \left[-\log_e\left(1 - \frac{1}{R_P}\right)\right]^k\right\} \qquad (2.5)$$

Then replacing R_P in Equation (2.5) with R_I using Equation (2.4),

$$U = u + \frac{a}{k}\left\{1 - \left[\frac{1}{R_I}\right]^k\right\}$$

$$U = u + \frac{a}{k}\left\{1 - R_I^{-k}\right\}$$

$$U = C - DR_I^{-k} \qquad (2.6)$$

where

$$C = u + \frac{a}{k} \text{ and } D = \frac{a}{k}$$

Equation (2.6) is a simplified form of the GEV that will be adopted in subsequent sections and chapters. Note that C in Equation (2.6) is the maximum value of U for any R_I; i.e. it is an upper limit. $(C-D)$ is the value of U for a value of R_I of 1 year.

2.2.3 Separation by storm type

In Chapter 1, the various types of windstorm that are capable of generating winds strong enough to be important for structural design, were discussed. These different event types will have different probability distributions, and therefore should be statistically analyzed separately; however, this is quite a difficult task as weather bureaus, or meteorological offices do not always record the necessary information. If anemograph records such as those shown in Figures 1.5 and 1.7 are available for older data, these can be used for identification purposes. Modern automatic weather stations (AWS) can generate wind speed and direction data at short intervals of as low as one minute. These can be used to reconstruct time histories similar to those in Figures 1.5 and 1.7 and assist in identifying storm types. Identification criteria to separate storm types have been proposed (e.g. de Gaetano *et al.*, 2014).

The relationship between the combined return period, $R_{P,c}$ for a given extreme wind speed due to winds of either type, and for those calculated separately for storm types 1 and 2, ($R_{P,1}$ and $R_{P,2}$) is:

$$\left(1 - \frac{1}{R_{P,c}}\right) = \left(1 - \frac{1}{R_{P,1}}\right)\left(1 - \frac{1}{R_{P,2}}\right) \quad \text{AND} \quad \frac{1}{R_{I,C}} = \frac{1}{R_{I,1}} + \frac{1}{R_{I,2}} \quad (2.7)$$

Equation (2.7) relies on the assumption that exceedance of wind speeds from the two different storm types in a given year are independent events. Equation (2.7) also shows the relationship between average recurrence intervals, which follows from Equation (2.4).

2.2.4 Simulation methods for tropical-cyclone wind speeds

The winds produced by severe tropical cyclones, also known as 'hurricanes' and 'typhoons', are the most severe on earth (apart from those produced by tornados which affect very small areas). However, their infrequent occurrence at particular locations often makes the historical data of recorded wind speeds an unreliable predictor for design wind speeds. An alternative approach, which gained popularity from the 1970s and early 1980s, was the simulation or 'Monte-Carlo' approach, introduced originally for offshore engineering by Russell (1971). In this procedure, satellite and other information on storm size, intensity and tracks are made use of to enable a computer-based simulation of wind speed (and in some cases direction) at particular sites. Usually, established probability distributions are used for parameters such as: central pressure and radius of maximum winds. A recent use of these models is for damage prediction for insurance companies. The disadvantage of this approach is the subjective aspect resulting from the complexity of the problem. Significantly varying predictions could be obtained by adopting different assumptions. Clearly, available recorded wind-speed data should be used to calibrate these models.

2.2.5 Compositing data from several stations

No matter what type of probability distribution is used to fit historical extreme wind series, or what fitting method is used, extrapolations to high return periods or average recurrence intervals, for ultimate limit-states design (either explicitly or implicitly through the application of a wind load factor) are usually subject to significant sampling errors (see Section 2.6). This results from the limited record lengths available to the analyst. In attempts to reduce the sampling errors, a recent practice has been to combine records from several stations with perceived similar wind climates to increase the available records for extreme value analysis. Thus, 'superstations' with long records can be generated in this way.

For example, the extreme wind climates across stations in several European countries have been judged to have similar statistical behaviour, at least as far as the all-direction extreme wind speeds are concerned. As a result, a single design wind speed has been specified for each of these countries (Appendix D). A similar approach has been adopted in the United States (ASCE, 1998, 2016; Peterka and Shahid, 1998) and Australia (Standards Australia, 1989, 2011; Holmes, 2002).

2.2.6 Correction for gust duration

Wind data are recorded, and supplied by meteorological offices, after being averaged over certain time intervals. The most common averaging time is 10 minutes. Gust data obtained by modern AWS are typically averaged over 3 seconds. However, earlier data obtained from older anemometers and analogue recording systems are usually of a shorter duration (e.g. Miller *et al.*, 2013). A time series of gust data may need to be corrected to a common gust duration (see Section 3.3.3), before extreme value analysis is carried out. Correction factors can be derived using random process theory, with some assumptions on the form of the spectral density of the turbulence (Section 3.3.4), as a function of mean wind speed and turbulence intensity (e.g. Holmes and Ginger, 2012; Holmes *et al.*, 2014). More empirical approaches, based on gust factors (Section 3.3.3), have also been used (e.g. Durst 1960; Deacon, 1965; Wieringa, 1973).

2.2.7 Wind direction effects

Increased knowledge of the aerodynamics of buildings and other structures, through wind-tunnel and full-scale studies, has revealed the variation of structural response as a function of wind direction and speed. The approaches to probabilistic assessment of wind loads including direction, can be divided into those based on the parent distribution of wind speed, and those based on extreme wind speeds. In many countries, the extreme winds are produced by rare severe storms such as thunderstorms and tropical cyclones, and there is no direct relationship between the parent population of regular everyday winds, and the extreme winds. For such locations (which would include many countries at latitudes of less than 40 degrees), the latter approach may be more appropriate. Where a separate analysis of extreme wind speeds by direction sector has been carried out, the relationship between the return period, $R_{P,a}$, for exceedance of a specified wind speed from *all* direction sectors, and the return periods for the same wind speed from direction sectors θ_1, θ_2, etc., as given in Equation (2.8).

$$\left(1 - \frac{1}{R_{P,a}}\right) = \prod_{i=1}^{N}\left(1 - \frac{1}{R_{P,\theta_i}}\right) \quad \text{AND} \quad \frac{1}{R_{I,a}} = \sum_{i=1}^{N} \frac{1}{R_{I,\theta_i}} \tag{2.8}$$

Equation (2.8) follows from an assumption that the wind speeds from each direction sector are statistically independent of each other, and is a statement of the following:

> Probability that a wind speed U is *not* exceeded for all wind directions = (probability that U is not exceeded from direction 1)×(probability that U is not exceeded from direction 2)×(probability that U is not exceeded from direction 3)......etc.
> The equivalent relationship between average recurrence intervals is also shown in Equation (2.8).

Equation (2.8) is a similar relationship to Equation (2.7) for combining extreme wind speeds from different types of storms. A similar approach is adopted in the 'multi-sector' method for combining the extreme wind *pressure*, or a structural *response*, contributed from different directions (e.g. Section 9.11; Holmes, 1990; Holmes and Bekele, 2015; Holmes, 2020).

An alternative approach, based on an 'outcrossing', or 'level crossing' analysis (Davenport 1977; Lepage and Irwin, 1985), assumes that extreme winds are derived from a homogeneous parent population that can be represented as a stationary, random process (see also Appendix C, Section C5.2). This method requires calculation of the rate of change of wind speed, $\dot{U}(t)$, which is sensitive to the sampling interval between recorded wind speeds (e.g. Harris, 2017). Also, as noted previously, the assumption of a homogeneous parent population is doubtful for locations sited at lower latitudes. For these reasons, the 'outcrossing' approach may lead to underprediction of extreme wind speeds and structural responses in some cases (Holmes, 2020).

Use of Equation (2.8), or other full probabilistic methods of treating the directional effects of wind, adds a considerable degree of complexity when applied to structural design for wind. To avoid this, simplified methods of deriving direction multipliers, or directional factors have been developed (e.g. Cook, 1983; Melbourne, 1984; Cook and Miller, 1999; Holmes and Bekele, 2015). These factors, which have been incorporated into some codes and standards, allow climatic effects on wind direction to be incorporated into wind load calculations in an approximate way. Generally, the maximum (or minimum) value of a response variable, from any direction sector, is used for design purposes.

Three methods that have been suggested for deriving directional wind speeds, or direction multipliers, are:

a. The extreme wind speeds, given a direction sector, are fitted with an extreme value probability distribution, in the same way as the all-direction winds. Then direction multipliers are obtained by dividing the wind speed with a specified exceedance probability by the all-direction wind speed with the same exceedance probability. This is an *unconservative* approach when applied to structural response, as it ignores contributions to the combined probability of response from more than one direction sector.
b. The extreme wind speeds within a direction sector are fitted with an extreme value probability distribution in the same way as the all-direction winds. The target probability of exceedance of the all-direction wind is then divided by the number of direction sectors. The wind speed within each direction sector is calculated for the reduced exceedance probability. These wind speeds are then divided by the all direction wind speed, with the original target probability, to give wind direction multipliers. This conservative approach (proposed by Cook (1983)) clearly gives higher direction multipliers than Method (a) when applied to codes and standards for the design of buildings.
c. As an empirical approach that lies between the unconservative Approach (a) and the conservative Method (b), Melbourne (1984) suggested a modification to Method (a) to render it less unconservative, by specifying a probability of exceedance for the directional winds of the target probability of the all-direction wind speeds, divided by *one quarter* of the total number of direction sectors. In effect, Melbourne (1984) suggested that only two sectors out of eight in Equation (2.8), when $N=8$, are effective when contributing to the combined probability of a structural response, i.e. a 90° sector from the total of 360°. Wind direction multipliers are then obtained by dividing the all-direction wind speed with the target probability of exceedance, as for Methods (a) and (b).

Kasperski (2007) derived direction wind speeds and multipliers, M_d, for Dusseldorf, Germany, using Methods (a) and (b). These values are shown in Table 2.1.

For this location, the south-westerly and westerly sectors (i.e. 210° – 270°) are the dominant ones for strong winds, due to the effect of Atlantic gales. It will be noted that Method (a) gives values of M_d that are all less than 1.0. The conservative nature of Method (b) is shown by the values of 1.02 for wind directions from 210° to 270°.

Method (c) represents an empirical compromise between Methods (a) and (b), with maximum values of M_d close to 1.0. This approach has been

Table 2.1 Direction multipliers for Dusseldorf, Germany (re-calculated from data of Kasperski, 2007)

Wind direction (°)	M_d Method (a)	M_d Method (b)
0	0.70	0.86
30	0.75	0.91
60	0.75	0.91
90	0.73	0.89
120	0.73	0.89
150	0.83	0.97
180	0.90	1.00
210	0.97	1.02
240	0.97	1.02
270	0.95	1.02
300	0.84	0.98
330	0.83	0.96

used to derive wind direction multipliers for Australia and New Zealand in the Standard AS/NZS 1170.2 (Standards Australia, 2011).

Methods of treating the varying effects of wind direction, on the response of tall buildings to wind loads, are discussed in Section 9.11. Chapter 15 discusses the application of wind direction multipliers in some codes and standards for wind loading.

2.3 EXTREME-WIND ESTIMATION BY THE TYPE I EXTREME VALUE DISTRIBUTION

2.3.1 Gumbel's method

Gumbel (1954) gave a usable methodology for fitting recorded annual maxima to the Type I extreme value distribution. This distribution is a special case of the GEV distribution discussed in Section 2.2.1. The Type I distribution takes the form of Equation (2.2) for the cumulative distribution $F_U(U)$:

$$F_U(U) = \exp\{-\exp[-(U-u)/a]\}$$

where u is the mode of the distribution, and a is a scale factor

The return period, R_P, is directly related to the cumulative probability distribution, $F_U(U)$, of the annual maximum wind speed at a site by Equation (2.9).

$$R_P = \frac{1}{1 - F_U(U)} \tag{2.9}$$

Substituting for $F_U(U)$ from Equation (2.9) in (2.2), we obtain:

$$U_R = u + a \left\{ -\log_e \left[-\log_e \left(1 - \frac{1}{R_P} \right) \right] \right\} \tag{2.10}$$

Substituting for R_P from Equation (2.4) in Equation (2.10):

$$U_R = u + a \log_e R_I \tag{2.11}$$

where R_I is the average recurrence interval (Section 2.2.2).

According to Gumbel's original extreme value analysis method (applied to flood prediction as well as extreme wind speeds), the following procedure is adopted:

- the largest wind speed in each calendar year of the record is extracted
- the series is ranked in order of smallest to largest: 1, 2, ... m ... to N
- each value is assigned a probability of non-exceedance, p, according to the Gumbel plotting position formula:

$$p \approx m / (N + 1) \tag{2.12}$$

- a 'reduced variate', y, is formed from:

$$y = -\log_e \left(-\log_e p \right) \tag{2.13}$$

  y is an estimate of the term in {} brackets in Equation (2.10)
- the wind speed, U, is plotted against y, and a line of "best fit" is drawn, usually by means of linear regression.

From Equations (2.10) and (2.11), the Type I Extreme Value, or Gumbel Distribution, will predict unlimited values of U_R, as the return period, R_P, or the average recurrence interval, R_I, increases. That is, as R_P, or R_I, becomes larger, U_R as predicted by Equation (2.10) or (2.11) will also increase without limit. As discussed in Section 2.2.1, this is not consistent with the argument that there are physical upper limits to the wind speeds that can be generated in the atmosphere in different types of storms.

2.3.2 Gringorten method

The Gumbel procedure, as described in Section 2.3.1, has been used many times to analyse extreme wind speeds for many parts of the world.

Assuming that the Type I Extreme Value Distribution is in fact the correct one, the fitting method, due to Gumbel, is biased, that is, Equation (2.13) gives distorted values for the probability of non-exceedance, especially for high values of p near 1. Several alternative fitting methods have been devised which attempt to remove this bias. However, most of these are more difficult to apply, especially if N is large, and some of these methods require the use of computer programs to implement. A simple modification to the Gumbel procedure, which gives nearly unbiased estimates for this probability distribution, is due to *Gringorten* (1963). Equation (2.12) is replaced by the following modified formula:

$$p \approx (m-0.44)/(N+1-0.88) = (m-0.44)/(N+0.12) \qquad (2.14)$$

Fitting of a straight line to U versus the plotting parameter, p, then proceeds as for the Gumbel method.

2.3.3 Method of Moments

Another relatively simple method of fitting the Type I Extreme Value Distribution to a set of extreme data is known as the 'Method of Moments'. It is based on the following relationships between the mean and standard deviation of the distribution, and the mode and scale factor (or slope).

$$\text{mean} = u + 0.5772a \qquad (2.15)$$

$$\text{standard deviation} = \left(\frac{\pi}{\sqrt{6}}\right)a \qquad (2.16)$$

The method to estimate the parameters, u and a of the distribution simply entails the calculation of the sample mean, μ, and standard deviation, σ, from the data, then estimating u and a by use of the inverse of Equations (2.15) and (2.16), i.e.

$$a \cong \left(\frac{\sqrt{6}}{\pi}\right)\sigma \qquad (2.17)$$

$$u \cong \mu - 0.5772a \qquad (2.18)$$

Once the parameters u and a have been determined, predictions of the extreme wind speed for a specified return period, R_P, or annual recurrence interval, R_I, are made using Equation (2.10) or (2.11).

Another procedure is the 'best linear unbiased estimators' proposed by Lieblein (1974), in which the annual maxima are ordered, and the parameters of the distribution are obtained by weighted sums of the extreme values.

2.3.4 Example of fitting the Type I distribution to annual maxima

Wind gust data has been obtained from open terrain at a military airfield at East Sale, Victoria, Australia continuously since late 1951. The anemometer position has been constant throughout that period, and the height of the anemometer head has always been the standard meteorological value of 10 m. Thus, in this case no corrections for height and terrain are required. Also, the largest gusts have almost entirely been produced by gales from large synoptic depressions (Section 1.3.1). However, the few gusts that were produced by thunderstorm downbursts were eliminated from the list, in order to produce a statistically consistent population (see Section 2.2.3).

The annual maxima for the 47 calendar years from 1952 to 1998 are listed in Table 2.2. The values in Table 2.2 are sorted in order of increasing magnitude (Table 2.3) and assigned a probability, p, according to (i) the Gumbel formula (Equation (2.12)), and (ii) the Gringorten formula

Table 2.2 Annual maximum gust speeds from East Sale, Australia 1952–1998

Year	Maximum gust speed (m/s)
1952	31.4
1953	33.4
1954	29.8
1955	30.3
1956	27.8
1957	30.3
1958	29.3
1959	36.5
1960	29.3
1961	27.3
1962	31.9
1963	28.8
1964	25.2
1965	27.3
1966	23.7
1967	27.8
1968	32.4
1969	27.8
1970	26.2
1971	30.9
1972	31.9

(Continued)

Table 2.2 (Continued) Annual maximum gust speeds
from East Sale, Australia 1952–1998

Year	Maximum gust speed (m/s)
1973	27.3
1974	25.7
1975	32.9
1976	28.3
1977	27.3
1978	28.3
1979	28.3
1980	29.3
1981	27.8
1982	27.8
1983	30.9
1984	26.7
1985	30.3
1986	28.3
1987	30.3
1988	34.0
1989	28.8
1990	30.3
1991	27.3
1992	27.8
1993	28.8
1994	30.9
1995	26.2
1996	25.7
1997	24.7
1998	42.2
mean	29.27
Standard deviation	3.196

(Equation (2.14)). The reduced variate, $-\log_e(-\log_e p)$, according to Equation (2.13), is formed for both cases. These are tabulated in Table 2.3. The wind speed is plotted against the reduced variates, and straight lines are fitted by linear regression ('least squares' method). The results of this are shown in Figures 2.2 and 2.3, for the Gumbel and Gringorten methods respectively. The intercept and slope of these lines give the mode, u, and slope, a, of the fitted Type I Extreme Value Distribution, according to Equation (2.2).

u and a can also be estimated from the calculated mean and standard deviation (shown in Table 2.2) by the Method of Moments using Equations (2.17) and (2.18).

Table 2.3 Processing of East Sale data

Rank	Gust speed (m/s)	Reduced variate (Gumbel)	Reduced variate (Gringorten)
1	23.7	−1.354	−1.489
2	24.7	−1.156	−1.226
3	25.2	−1.020	−1.069
4	25.7	−0.910	−0.949
5	25.7	−0.816	−0.848
6	26.2	−0.732	−0.759
7	26.2	−0.655	−0.679
8	26.7	−0.583	−0.604
9	27.3	−0.515	−0.534
10	27.3	−0.450	−0.467
11	27.3	−0.388	−0.403
12	27.3	−0.327	−0.340
13	27.3	−0.267	−0.279
14	27.8	−0.209	−0.220
15	27.8	−0.151	−0.161
16	27.8	−0.094	−0.103
17	27.8	−0.037	−0.045
18	27.8	0.019	0.013
19	27.8	0.076	0.071
20	28.3	0.133	0.129
21	28.3	0.190	0.187
22	28.3	0.248	0.246
23	28.3	0.307	0.306
24	28.8	0.367	0.367
25	28.8	0.427	0.428
26	28.8	0.489	0.492
27	29.3	0.553	0.556
28	29.3	0.618	0.623
29	29.3	0.685	0.692
30	29.8	0.755	0.763
31	30.3	0.827	0.837
32	30.3	0.903	0.914
33	30.3	0.982	0.995
34	30.3	1.065	1.081
35	30.3	1.152	1.171
36	30.9	1.246	1.268
37	30.9	1.346	1.371
38	30.9	1.454	1.484
39	31.4	1.572	1.607

(Continued)

Table 2.3 (Continued) Processing of East Sale data

Rank	Gust speed (m/s)	Reduced variate (Gumbel)	Reduced variate (Gringorten)
40	31.9	1.702	1.744
41	31.9	1.848	1.898
42	32.4	2.013	2.075
43	32.9	2.207	2.285
44	33.4	2.442	2.544
45	34.0	2.740	2.885
46	36.5	3.157	3.391
47	42.2	3.861	4.427

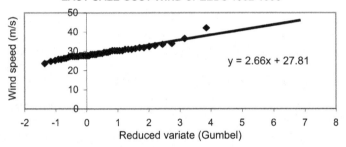

Figure 2.2 Analysis of annual maximum wind gusts from East Sale using the Gumbel method.

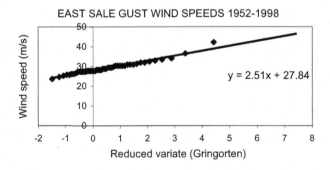

Figure 2.3 Analysis of annual maximum wind gusts from East Sale, using the Gringorten fitting method.

Predictions of extreme wind speeds for various return periods can then readily be obtained by application of Equation (2.10) or (2.11). Table 2.4 lists these predictions based on the Gumbel and Gringorten fitting methods, and by the Method of Moments. For return periods up to 500 years, the predicted values by the three methods are within 1 m/s of each other. However, these small differences are swamped by sampling errors, i.e. the

Table 2.4 Prediction of extreme wind speeds for East Sale (synoptic winds)

Return period (years)	Predicted gust speed (m/s) (Gumbel)	Predicted gust speed (m/s) (Gringorten)	Predicted gust speed (m/s) (method of moments)
10	33.8	33.5	33.4
20	35.7	35.3	35.2
50	38.2	37.6	37.6
100	40.0	39.4	39.3
200	41.9	41.1	41.0
500	44.3	43.5	43.3
1,000	46.2	45.2	45.0

errors inherent in trying to make predictions for return periods of 100 years or more from less than 50 years of data. The problem of high sampling errors can often be circumvented by compositing data, as discussed in Section 2.2.5.

A review by Torrielli et al. (2013), using more than 12,000 years of simulated wind data, representative of the climate of central Italy, found that use of the Type I Extreme Value Distribution will generally give conservative predictions of long-return-period wind extremes, especially when short record lengths (25 years or less) are processed. However, this conservatism may not necessarily be a bad thing in structural engineering applications.

2.3.5 General penultimate distribution

For extreme wind speeds that are derived from a Weibull parent distribution (see Section 2.8), Cook and Harris (2004) have proposed a 'General penultimate' Type I, or Gumbel Distribution. This takes the form of Equation (2.19).

$$F_U(U) = \exp\left\{-\exp[-(U^w - u^w)/a^w]\right\} \qquad (2.19)$$

where w is the Weibull exponent of the underlying parent distribution (see Equation (2.21).

Comparing Equation (2.19) with Equation (2.2), it can be seen that Equation (2.19) represents a Gumbel Distribution for a transformed variable, Z, equal to U^w.

If the parent wind speed data are available for a site, w can be obtained directly from fitting a Weibull Distribution to that. Alternatively, the penultimate distribution of Equation (2.19) can be treated as a three-parameter (u, a and w) distribution and fitted directly to the extreme wind data, without knowing the parent distribution directly.

The Weibull exponent, w, is typically in the range of 1.3–2.0; in that case, when Equation (2.19) is plotted in the Gumbel form (Figure 2.2), the resulting line curves downwards, and is similar in shape to the Type III Extreme Value Distribution. The main difference is that the latter has a finite upper limit, whereas for the penultimate distribution, U^w and U are unlimited. However, for practical design situations, the two distributions give very similar predictions (Holmes and Moriarty, 2001).

The review and simulations by Torrielli *et al.* (2013), found that the predictions from the penultimate distribution for return periods of 50, 200 and 1,000 years were 1–2 m/s below the known reference values for the climate of central Italy.

2.4 PEAKS OVER THRESHOLD APPROACHES

The approach of extracting a single maximum value of wind speed from each year of historical data, has obvious limitations in that there may be many storms during any year, but only one value from all these storms is being used. A shorter reference period than a year could also be used to increase the amount of data. However, it is important for extreme value analysis that the data are statistically independent; however, this may not be the case if the sampling period is as short as one day. An alternative approach which makes use only of the data of relevance to extreme wind prediction is the peaks, or *excesses, over threshold* approach. This approach makes use of all independent extremes above defined thresholds; for example, an analysis of gust wind speeds might make use of all the values of gust above 25 m/s, though, they may, or may not, be annual maxima.

When the GEV Distribution (Equation 2.1), with a shape factor, k, is found to be the appropriate distribution for a population of extremes from a defined epoch, such as annual maximum wind speeds, the Generalized Pareto Distribution (GPD) (Appendix C4.2), with the same shape factor is the appropriate distribution for the excesses over high thresholds of the underlying population of extremes.

Whereas most methods for fitting the GEV or GPD (e.g. Hosking *et al.*, 1985; Davison and Smith, 1990; Holmes and Moriarty, 1999) allow the shape factor to be determined by the data itself, in Appendix G, the shape factor, k, is presumed to be fixed at an appropriate value. This avoids the wide variability in shape factors that can occur when applying the methods with a variable shape factor to relatively short data series of wind speeds, often leading to widely varying predictions for high return periods/average recurrence intervals.

Typically, the specified shape factor for wind speeds would be a small positive value, such as 0.1. While such a value could be regarded as arbitrary, it is no more so than fixing it at 0.0, which is implicit when the Type I, Extreme Value Distribution is adopted (see Section 2.2.1).

Several methods of fitting the GPD with a fixed shape factor, or effectively its closely related variable, GEV are discussed in Appendix G with an example to illustrate each method.

Torrielli *et al.* (2013) found that peaks-over threshold approaches generally give accurate predictions of high return period wind speeds, with desirably narrow confidence limits (see Section 2.6). However, the results are sensitive to the method of fitting of the GPD, and to the chosen thresholds.

2.5 EXTREME WINDS BY DIRECTION SECTOR

The peaks-over-threshold approach can be applied to winds separated by direction sector. This is done in Figure 2.4 which shows a wind speed versus return period for a number of wind directions using combined gust data from Melbourne Airport, Australia, for synoptic wind events:

- Gust wind speeds separately analysed for 11 direction sectors of 22.5° width. For the remaining 5 sectors (ENE to SSE), there was insufficient data for a meaningful analysis to be carried out.
- Combined data from all direction sectors, analysed as a single data set.
- A combined distribution obtained by combining distributions for directional sectors, according to Equation (2.8).

A peaks-over-threshold approach was used, with the shape factor, k, fixed at 0.1 (see Appendix G). This resulted in distributions forming lines, curving slightly on the wind speed (linear) versus return period (logarithmic) graphs.

There is good agreement for the 'all-directions' line, and the 'combined directions' line, indicating the independence of the data from the various direction sectors, and the validity of Equation (2.8)

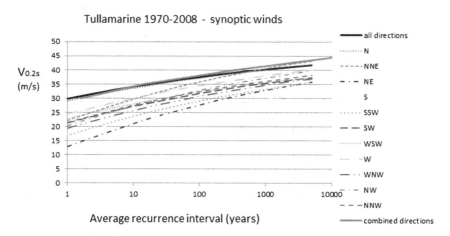

Figure 2.4 Probability distributions fitted to gust data for direction sectors for Melbourne Airport, 1970–2008.

2.6 BOOTSTRAPPING AND CONFIDENCE LIMITS

As far back as the 1950s, Gumbel (1958) emphasized the importance of calculating confidence limits when making predictions for long average recurrence intervals from limited data sets. A convenient way of doing this is by
use of a simulation or 'bootstrapping' technique, in which a large number
of samples of random data are generated using random numbers representative of the cumulative distribution function, with parameters determined
from the actual data (e.g. Naess and Clausen, 2001).

Figure 2.5 shows the 10% and 90% percentile limits obtained by simulating 50 samples of 50 years of annual maxima (2,500 in total), and fitting
each sample, using the Method of Moments (Section 2.3.3) to the Gumbel
Distribution of Equation (2.11), with u equal to 29.2 m/s and a equal to
3.0 m/s, It can be seen that the predictions of $V_{1,000}$ can be up to 4 m/s from
the correct value of 50 m/s. Even the V_{50} predictions can depart from the
correct value of 41 m/s by up to 2.5 m/s.

The standard deviations of extreme wind predictions can also be
obtained using the bootstrapping approach. This contributes to uncertainty
estimates for wind loading required for structural reliability studies (see
Table 2.6). For the example in Figure 2.5, the standard deviation of the
50-year ARI (V_{50}) estimates (based on 50-year samples) is 1.62 m/s. For
$V_{1,000}$, the standard deviation is 2.82 m/s. Note that these are uncertainties determined assuming that the underlying windspeed measurements are
accurate, and that the correct probability distribution has been chosen, and

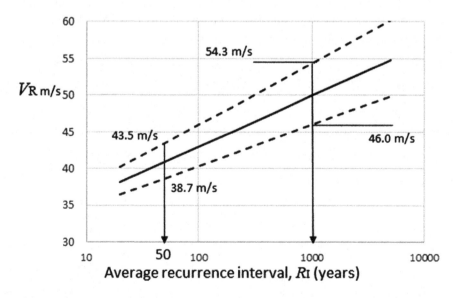

Figure 2.5 Confidence limits obtained by 'bootstrapping', for predictions using the
Gumbel Distribution.

other sources of uncertainty need to be included in reliability studies and load factor determination.

2.7 PREDICTION OF EXTREME WINDS FROM TORNADOS

The prediction of extreme wind speeds from small, rare extreme events like tornadoes is very difficult, as they seldom strike anemometers. However, damage surveys after such events can often be used to estimate the area of their 'footprints' with wind speeds exceeding defined values (e.g. Fujita, 1971).

The probability of a tornado strike on the site of a structure, with a wind speed exceeding a value U_t, in a time period, T, years, can be written as Equation (2.20) (Wen and Chu, 1973).

$$P_t(>U_t) = 1 - F_U(U_t) = \upsilon T a_U \tag{2.20}$$

where υ is the average rate of occurrence of tornados, per square kilometre, per year, in the region where the structure is located, and a_U is the average 'footprint' area of tornadoes with wind speed greater than U_t. υ can also be written as n/A_R, where n is the average number of tornados in a region (such as a state, county or shire, or a $1°$ square) of area A_R km^2.

Equation (2.20) applies to a 'point target'. A transmission line represents a 'line target' and the tornado-risk model for a complete transmission line is discussed in Section 13.3.1.

2.8 PARENT WIND DISTRIBUTIONS

For some design applications, it is necessary to have information on the distribution of the complete population of wind speeds at a site. An example is the estimation of fatigue damage for which account must be taken of damage accumulation over a range of windstorms (see Section 5.6). The population of wind speeds produced by synoptic windstorms at a site is usually fitted with a distribution of the *Weibull* type:

$$f_U(\bar{U}) = \frac{w\bar{U}^{w-1}}{c^w} \exp\left[-\left(\frac{\bar{U}}{c} \right)^w \right] \tag{2.21}$$

Equation (2.21) represents the probability density function for mean wind speeds produced by synoptic events. There are two parameters: a *scale factor*, c, which has units of wind speed, and a *shape factor*, w, which is dimensionless (see also Appendix C3.4). The probability of *exceedance* of any given wind speed is given by Equation (2.22):

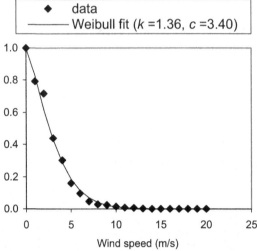

Loxton, 1984-2000 (all directions)

◆   data
——— Weibull fit (k =1.36, c =3.40)

Figure 2.6 Example of a Weibull distribution fit to parent population of synoptic winds.

$$1 - F(\bar{U}) = \exp\left[-\left(\frac{\bar{U}}{c}\right)^{w}\right] \tag{2.22}$$

Typical values of c are $3 - 10\,\text{m/s}$, and w usually falls in the range of $1.3 - 2.0$. An example of a Weibull fit to recorded meteorological data is shown in Figure 2.6.

Several attempts have been made to predict extreme winds from knowledge of the parent distribution of wind speeds, and thus make predictions from quite short records of wind speed at a site (e.g. Gomes and Vickery, 1977b). The 'asymptotic' extreme value distribution for a Weibull parent distribution is the Type I, or Gumbel, distribution. However, for extremes drawn from a finite sample (e.g. annual maxima), the 'penultimate' Type I, as discussed in Section 2.3.5, is the more appropriate extreme value distribution. However, it should be noted that the Weibull Distribution, the Type I Extreme Value Distribution, and the 'penultimate' distribution will all give unlimited wind speeds with reducing probability of exceedance.

2.9 WIND LOADS AND STRUCTURAL SAFETY

The development of structural reliability concepts – i.e. the application of probabilistic methods to the structural design process – has accelerated the adoption of such methods into wind engineering since the 1970s. The assessment of wind loads is only one part of the total structural design process, which also includes the determination of other loads and the resistance of structural materials. The structural engineer must proportion the structure so that collapse or overturning has a very low risk of occurring, and defined serviceability limits on deflection, acceleration, etc. are not exceeded very often.

2.9.1 Limit-states design

Limit-states design is a rational approach to the design of structures, which has now become accepted around the world. Explicitly defining the ultimate and serviceability limit states for design, the method takes a more rational approach to structural safety by defining 'partial' load factors ('gamma' factors) for each type of loading, and a separate resistance factor ('phi' factor) for the resistance. The application of the limit states design method is not, in itself, a probabilistic process, but probability is usually used to derive the load and resistance factors.

A typical ultimate limit-states design relationship involving wind loads, is as follows:

$$\varphi R \geq \gamma_D\, D + \gamma_W\, W \tag{2.23}$$

where

φ is a resistance factor,
R is the nominal structural resistance,
γ_D is the dead load factor,
D is the nominal dead load,
γ_W is the wind load factor,
W is the nominal wind load.

In this relationship, the partial factors, φ, γ_D, and γ_W are adjusted separately to take account of the variability and uncertainty in the resistance, dead load and wind load. The values used also depend on what particular nominal values have been selected. Often a final calibration of a proposed design formula is carried out by evaluating a 'safety', or 'reliability', index as discussed in the following section, for a range of design situations, e.g. various combinations of nominal dead and wind loads.

2.9.2 Probability of failure and the safety index

A quantitative measure of the safety of the structural design process, the *safety index*, or *reliability index*, is used as a method of calibration of existing and future design methods for structures. As will be explained

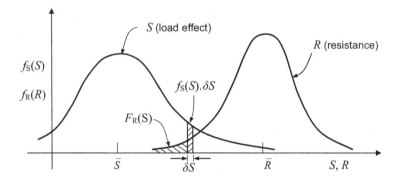

Figure 2.7 Probability densities for load effects and resistance.

in this section, there is a one-to-one relationship between this index and a probability of failure, based on the exceedance of a design resistance by an applied load (but not including failures by human errors and other accidental causes).

The structural design process is shown in its simplest form in Figure 2.7. The process consists of comparing a structural load effect, S, with the corresponding resistance, R. In the case of limit states associated with structural strength or collapse, the load effect could be an axial force in a member or a bending moment, or the corresponding stresses. In the case of serviceability limit states, S and R may be deflections, accelerations or crack widths, or their acceptable limits.

The probability density functions $f_s(S)$ and $f_R(R)$ for a load effect, S, and the corresponding structural resistance, R are shown in Figure 2.7. (Probability density is defined in Section C2.1 in Appendix C.) Clearly, S and R must have the same units. The dispersion or 'width' of the two distributions represents the uncertainty in S and R.

Failure (or unserviceability) occurs when the resistance of the structure is less than the load effect. The probability of failure will now be determined, assuming S and R are statistically independent.

The probability of failure occurring at a load effect between S and $S+\delta S =$ [probability of load effect lying between S and $S + \delta S$] × [probability of resistance, R, being less than S]

$$= [f_S(S) \cdot \delta S] \times F_R(S) \tag{2.24}$$

where $F_R(R)$ is the cumulative probability distribution of R, and,

$$F_R(S) = \int_{-\infty}^{S} f_R(R) \, dR \tag{2.25}$$

The terms in the product in Equation (2.24) are the areas shown in Figure 2.7.

The total probability of failure is obtained by summing, or integrating, Equation (2.24) over all possible values of S (between $-\infty$ and $+\infty$):

$$p_f = \int_{-\infty}^{+\infty} f_S(S) \cdot F_R(S)\, dS \qquad (2.26)$$

Substituting for $F_R(S)$ from Equation (2.25) into Equation (2.26),

$$p_f = \int_{-\infty}^{\infty} \int_{-\infty}^{S} f_S(S) \cdot f_R(R) \cdot dR \cdot dS = \int_{-\infty}^{\infty} \int_{-\infty}^{S} f(S, R) \cdot dR \cdot dS \qquad (2.27)$$

where $f(S, R)$ is the *joint* probability density of S, R.

The acceptable values of the probability of failure in practice, computed from Equation (2.27) are normally very small numbers, typically 1×10^{-2} to 1×10^{-5}.

The safety, or reliability index is defined according to Equation (2.28), and normally takes values in the range of 2 –5.

$$\beta = -\Phi^{-1}(p_f) \qquad (2.28)$$

where $\Phi^{-1}()$ is the inverse cumulative probability distribution of unitary normal (Gaussian) variate, i.e. a normal variate with a mean of zero and a standard deviation of one.

The relationship between the safety index, β, and the probability of failure, p_f, according to Equation (2.28) is shown plotted in Figure 2.8.

Equations (2.26) and (2.27) can be evaluated exactly when S and R are assumed to have Gaussian (normal) or lognormal (Appendix C3.2) probability distributions. However, in other cases, (which includes those involving wind loading), numerical methods must be used. Numerical methods must also be used when, as is usually the case, the load effect, S, and resistance, R, are treated as combinations (sums and products) of separate random variables, with separate probabilistic characteristics.

Details of structural reliability theory and practice can be found in a number of texts on the subject (e.g. Blockley (1980), Melchers (1987), Ang and Tang (1990)). Reliability concepts as applied to wind loading were addressed by Rojiani and Wen (1981), Davenport (1983), Pham et al. (1983), and Kasperski and Geurts (2005).

2.9.3 Nominal return periods for design wind speeds

The return periods (or annual recurrence intervals) for the nominal design wind speeds in various wind loading codes and standards are discussed in

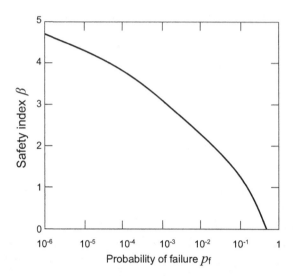

Figure 2.8 Relationship between safety index and probability of failure.

Chapter 15. The most common choice is 50 years. There should be no confusion between return period, R, and expected lifetime of a structure, L. The return period is just an alternative statement of annual risk of exceedance, e.g. a wind speed with a 50-year return period is one with an expected risk of exceedance of 0.02 (1/50) in any one year. An annual recurrence interval, R_I, can easily be converted to the equivalent return period, R_P, through application of Equation (2.4), although they are essentially equal for values greater than 10 years.

Assuming a stationary (unchanging) wind climate, the risk, r, of exceedance of a wind speed *over the lifetime,* can be determined by assuming that all years are statistically independent of each other. Then,

$$r = 1 - \left[1 - \left(\frac{1}{R_P}\right)\right]^L \tag{2.29}$$

Equation (2.29) is very similar to Equation (2.9) in which the combined probability of exceedance of a wind speed occurring over a range of wind directions was determined.

Setting both R_P and L as 50 years in Equation (2.28), we arrive at a value of r of 0.636. There is thus a nearly 64% chance that the 50-year return period wind speed will be exceeded at least once during a 50-year lifetime – i.e. a better than even chance that it *will occur.* Wind loads derived from wind speeds with this level of risk must be factored up when used for ultimate limit states design. Typical values of wind load factor, γ_W, are in the range of 1.4–1.6. Different values may be required for regions with different wind speed / return period relationships, as discussed in Section 2.9.

The use of 'lifetime exceedance probability' (LEP), rather than annual risk, for design, enables a changing climate (Section 1.8) to be incorporated into the design process (Xu *et al.*, 2020). Equation (2.30) is a modification of Equation (2.29) in which the structural lifetime, L, is divided into n discrete periods, in each of which the probability of exceedance (return period) of the specified wind speed is evaluated based on the expected state of the climate at that time.

$$r = 1 - \prod_{n=1}^{L}\left[1 - \left(\frac{1}{R_{P,n}}\right)\right] = 1 - exp\left[\sum_{n=1}^{L}\left(-\frac{1}{R_{I,n}}\right)\right] \qquad (2.30)$$

The use of a return period, or annual recurrence interval, substantially higher than the traditional 50 years, for the nominal design wind speed, avoids the need to have different wind load factors in different regions. This was a consideration in the revision of the Australian Standard for Wind Loads in 1989 (Standards Australia, 1989), which in previous editions, required the use of a special 'Cyclone Factor' in the regions of northern coastline affected by tropical cyclones. The reason for this factor was the greater rate of change of wind speed with return period in the cyclone regions. A similar 'hurricane importance factor' appeared in some editions of the American National Standard (ASCE, 1993), but was later incorporated into the specified basic wind speed (ASCE, 1998).

In AS1170.2-1989, the wind speeds for ultimate limits-states design had a nominal probability of exceedance of 5% in a lifetime of 50 years (a return period of 1,000 years, approximately). In later versions of this Standard, a range of annual recurrence intervals are provided, with values of 100–2,000 years specified in other documents, depending on the assessed importance level of a structure, and the perceived risk to human life. A similar approach has been adopted in recent editions of ASCE-7, with a range of mean recurrence intervals from 300 to 1,700 years.

However, a load factor of 1.0 is normally applied to the wind loads derived in this way – and this factor is the same in both cyclonic/hurricane and non-cyclonic/non-hurricane regions.

2.9.4 Uncertainties in wind load specifications

A reliability study of structural design involving wind loads requires an estimation of all the uncertainties involved in the specification of wind loads – wind speeds, multipliers for terrain, height, topography, pressure coefficients, local and area averaging effects, etc. Some examples of this type of study for buildings and communication towers were given by Pham *et al.* (1983, 1992).

Table 2.5 Variability of wind loading parameters

Parameter	Mean/nominal	Coefficient of variation	Assumed distribution
Wind speed (50-year maximum)	1.12	0.28	Gumbel
Directionality	0.9	0.05	Lognormal
Exposure	0.8	0.15	Lognormal
Pressure coefficient	0.8	0.15	Lognormal
Local & area reduction effects	0.85	0.10	Lognormal

Source: From Pham et al. (1983).

Table 2.5 shows estimates by Pham *et al.* (1983) of mean-to-nominal values of various parameters associated with wind loading calculations for regions affected by tropical cyclones from the Australian Standard of that time. It can be seen from Table 2.5 that the greatest assessed contributor to the variability and uncertainty in wind load estimation is the wind speed itself - particularly as it is raised to a power of two (or greater, when dynamic effects are important) when wind loads and effects are calculated. A secondary contributor is the uncertainty in the 'exposure' parameter in Table 2.5, which is also squared, and includes uncertainties in the vertical profile of mean and gust speeds as discussed in earlier sections of this chapter.

Kasperski and Geurts (2005) have also estimated uncertainties (mean and coefficients of variation) for various wind loading parameters, including the expected reductions in uncertainty in the estimation of aerodynamic coefficients by the use of wind-tunnel tests. Those values are comparable to those given in Table 2.5, and both sets are applicable to buildings, for which resonant dynamic response is not significant.

The uncertainties associated with wind loading of wind-sensitive structures, such as long-span bridges and communication towers, are somewhat different and should be treated separately.

2.10 WIND LOAD FACTORS

As discussed in the previous section, wind load factors, γ_w, in the range of 1.4–1.6 have traditionally been adopted for use with wind loads derived on the basis of 50-year return period wind speeds. Wind load factors in this range have usually been derived on the basis of two assumptions:

a. Wind loads vary as the square of wind speeds. This is valid for small structures of high frequency with high natural frequencies. However, this assumption is not valid for tall buildings, and other wind-sensitive structures, for which the effective wind loads vary to a higher power of wind speed, due to the effect of resonant dynamic response. This power can be up to 2.5 for along-wind response, and up to 3 for cross-wind response (see Chapter 9).

b. The wind load factors have been derived for non-tropical-cyclone wind loads, i.e. climates for which the rate of change of wind speed with return period R_p is relatively low, compared with those regions affected by tropical cyclones, typhoons or hurricanes.

For structures with significant dynamic response to wind such as tall buildings, long-span bridges and tall towers, wind loads and effects vary with wind speed, raised to a power greater than 2. Holmes and Pham (1993) considered the effect of the varying exponent on the safety index and showed that use of the traditional approach of 'working stress' wind loads based on a 50-year return period wind speed, with a fixed wind load factor of 1.5, resulted in a significant reduction in safety as the exponent increased. However, a near-uniform safety (insensitive to the effects of dynamic response) was achieved by the use of a high-return period, nominal design wind speed together with a wind load factor of 1.0.

The need for higher load factors in locations affected by tropical cyclones is illustrated by Table 2.6. This shows the equivalent return period of wind speeds when a wind load factor of 1.4 is applied to the 50-year return period value for several different locations. For two locations in the United Kingdom, the application of the 1.4 factor is equivalent to calculating wind loads from wind speeds with return periods considerably greater than 1,000 years. However, in Penang, Malaysia, where design wind speeds are governed by winds from relatively infrequent severe thunderstorms, the factor of 1.4 is equivalent to using a wind speed of about 500 years return period.

At Port Hedland (Western Australia), and Hong Kong, where tropical cyclones and typhoons are dominant, and extreme wind events are even rarer, the factored wind load based on V_{50} is equivalent to applying wind loads of only 150 and 220 years return period, respectively. In Australia, previous editions of AS1170.2 adjusted the effective return period by use of

Table 2.6 Return periods of factored wind loads with a wind-load factor of 1.4

Location	Wind type	V_{50} (m/s)	$\sqrt{(1.4 V_{50}{}^2)}$ (m/s)	Return period of factored wind speed (years)
Cardington (U.K.)	Atlantic gales	42.9	50.7	1,130
Jersey (U.K.)	Atlantic gales	45.6	53.9	1,470
Penang (Malaysia)	Thunderstorms	27.6	32.7	510
Port Hedland (W.A.)	Tropical cyclones	60.5	71.6	150
Hong Kong	Typhoons	61.5	72.8	220

Notes: i) Values of wind speed shown are gust speeds at 10 m height over flat open terrain, except for Hong Kong for which the gust speed is adjusted to 50 m height above the ocean.
ii) Probability distributions for Cardington and Jersey obtained from Cook (1985).

the 'Cyclone Factor' as discussed above. However, in AS/NZS1170:2010, a nominal design wind speed with 500 years return period (80 m/s basic gust wind speed at Port Hedland) with a wind load factor of 1.0, is used for design of most structures for ultimate limit states.

2.11 SUMMARY

In Chapter 2, the application of extreme value analysis to the prediction of design wind speeds has been discussed. In particular, the Gumbel, and 'peaks-over-threshold', approaches were described in detail. The need to separate wind speeds caused by windstorms of different types was emphasized, and wind direction effects were considered.

The main principles of the application of probability to structural design and safety under wind loads were also introduced.

EXERCISES

1. Using the Gumbel distribution and method, re-analyse the annual maximum gust wind speeds for East Sale for the years 1952–1997 (Table 2.2), i.e. ignore the high value recorded in 1998. Compare the resulting predictions of design wind speeds for (a) 50-years return period, and (b) 1,000-years return period, and comment.
2. Non-synoptic (thunderstorm) gusts greater than 25 m/s recorded at Kalgoorlie, Western Australia, during the years 1961–1994 are tabulated below. Using any two of the methods described in Appendix G, fit the data to the distribution given by: $V_R = C - D$ (Equation 2.6 with $k = 0.1$).

   For both methods, determine the parameters C and D, and predicted values of V_R for average recurrence intervals (R_I) of: (a) 20 years and (b) 500 years.

Year	Gust speed (m/s)
1961	28.8
1966	28.8
1967	33.0
1968	29.9
1970	33.0
1970	33.0
1971	27.8
1973	28.8
1974	32.4
1975	34.0

(Continued)

Year	Gust speed (m/s)
1978	26.8
1979	25.8
1979	38.6
1979	26.3
1982	27.8
1982	28.3
1982	34.5
1984	25.8
1986	31.4
1988	25.2
1990	25.8
1992	25.2
1994	39.1

REFERENCES

American Society of Civil Engineers. (1993) *Minimum design loads for buildings and other structures. ASCE Standard, ANSI/ASCE 7-93.* American Society of Civil Engineers, New York.

American Society of Civil Engineers. (1998) *Minimum design loads for buildings and other structures. ASCE Standard, ANSI/ASCE 7-98.* American Society of Civil Engineers, New York.

American Society of Civil Engineers. (2016) *Minimum design loads for buildings and other structures. ASCE/SEI 7-16.* American Society of Civil Engineers, New York.

Ang, A.H. and Tang, W. (1990) *Probability concepts in engineering planning and design.* Volume II. Decision, risk and reliability. Published by the authors.

Blockley, D. (1980) *The nature of structural design and safety.* Ellis Horwood, Chichester.

Coles, S.G. (2001) An introduction to statistical modeling of extreme values. Springer, Berlin and New York.

Cook, N.J. (1983) Note on directional and seasonal assessment of extreme wind speeds for design. *Journal of Wind Engineering and Industrial Aerodynamics,* 12: 365–372.

Cook, N.J. (1985) *The designer's guide to wind loading of building structures.* BRE-Butterworths, Watford, UK.

Cook, N.J. and Harris, R.I. (2004) Exact and general FT1 penultimate distributions of extreme winds drawn from tail-equivalent Weibull parents. *Structural Safety,* 26: 391–420.

Cook, N.J. and Miller, C.A.M. (1999) Further note on directional assessment of extreme wind speeds for design. *Journal of Wind Engineering and Industrial Aerodynamics,* 79: 201–208.

Davenport, A.G. (1961) The application of statistical concepts to the wind loading of structures. *Proceedings Institution of Civil Engineers,* 19: 449–471.

Davenport, A.G. (1977) The prediction of risk under wind loading. *2nd International Conference on Structural Safety and Reliability*, Munich, Germany, 19–21 September.

Davenport, A.G. (1983) The relationship of reliability to wind loading. *Journal of Wind Engineering and Industrial Aerodynamics*, 13: 3–27.

Davison, A.C. and Smith, R.L. (1990) Models for exceedances over high thresholds. *Journal of the Royal Statistical Society, Series B*, 52: 339–442.

Deacon, E.L. (1965) Wind gust speed: averaging time relationship. *Australian Meteorological Magazine*, No. 51, 11–14.

De Gaetano, P., Repetto, M.P., Repetto, T. and Solari, G. (2014) Separation and classification of extreme wind events from anemometric records. *Journal of Wind Engineering and Industrial Aerodynamics*, 126: 132–143.

Durst, C.S. (1960) Wind speeds over short periods of time. *Meteorological Magazine (UK)*, 89, 181–186.

Fisher, R.A. and Tippett, L.H.C. (1928) Limiting forms of the frequency distribution of the largest or smallest member of a sample. *Proceedings, Cambridge Philosophical Society Part 2*, 24: 180–190.

Freudenthal, A.M. (1947) The safety of structures. *ASCE Transactions*, 112: 125–159.

Freudenthal, A.M. (1956) Safety and the probability of structural failure. *ASCE Transactions*, 121: 1337–1397.

Fujita, T.T. (1971) Proposed characterization of tornadoes and hurricanes by area and intensity. SMRP Research Report 91, Department of the Geophysical Sciences, University of Chicago.

Gomes, L. and Vickery, B.J. (1977a) Extreme wind speeds in mixed wind climates. *Journal of Industrial Aerodynamics*, 2: 331–344.

Gomes, L. and Vickery, B.J. (1977b) On the prediction of extreme wind speeds from the parent distribution. *Journal of Industrial Aerodynamics*, 2: 21–36.

Gringorten, I.I. (1963) A plotting rule for extreme probability paper. *Journal of Geophysical Research*, 68: 813–814.

Gumbel, E.J. (1941) The return period of flood flows, *Annals of Mathematical Statistics*, 12:163-90.

Gumbel, E.J. (1954) *Statistical theory of extreme values and some practical applications*. Applied Math Series 33. National Bureau of Standards, Washington, DC.

Gumbel, E.J. (1958) *Statistics of Extremes*. Columbia University Press, New York.

Harris, R.I. (2017) The level crossing method applied to mean wind speeds from 'mixed' climates. *Structural Safety*, 67: 54–61.

Holmes, J.D. (1990) Directional effects on extreme wind loads. *Civil Engineering Transactions, Institution of Engineers, Australia*, 32: 45–50.

Holmes, J.D. (2002) A re-analysis of recorded extreme wind speeds in Region A. *Australian Journal of Structural Engineering*, 4: 29–40.

Holmes, J.D. (2020) Comparison of two probabilistic methods for the effect of wind direction on structural response. submitted to: *Structural Safety*.

Holmes, J.D. and Bekele, S. (2015). Directionality and wind-induced response - calculation by sector methods. *14th International Conference on Wind Engineering*, Porto Alegre, Brazil, 21–26 June.

Holmes, J.D. and Ginger, J.D. (2012) The gust wind speed duration in AS/NZS 1170.2. *Australian Journal of Structural Engineering*, 13: 207–217.

Holmes, J.D. and Moriarty, W.W. (1999) Application of the generalized Pareto distribution to extreme value analysis in wind engineering. *Journal of Wind Engineering and Industrial Aerodynamics*, 83: 1–10.

Holmes, J.D. and Moriarty, W.W. (2001) Response to discussion by N.J. Cook and R.I. Harris of: "Application of the generalized Pareto distribution to extreme value analysis in wind engineering". *Journal of Wind Engineering and Industrial Aerodynamics*, 89: 225–227.

Holmes, J.D. and Pham, L. (1993). Wind-induced dynamic response and the safety index. *Proceedings 6th International Conference on Structural Safety and Reliability, Innsbruck*, Austria, 9–13 August, pp. 1707–1709, A.A. Balkema, Publishers.

Holmes, J.D., Allsop, A.C. and Ginger, J.D. (2014) Gust durations, gust factors and gust response factors in wind codes and standards. *Wind and Structures*, 19: 339–352.

Hosking, J.R.M., Wallis, J.R. and Wood, E.F. (1985) Estimates of the generalized extreme value distribution by the method of probability-weighted moments. *Technometrics*, 27: 251–261.

Jenkinson, A.F. (1955) The frequency distribution of the annual maximum (or minimum) values of meteorological elements. *Quarterly Journal of the Royal Meteorological Society*, 81: 158–171.

Kasperski, M. (2007) Design wind speeds for a low-rise building taking into account directional effects. *Journal of Wind Engineering and Industrial Aerodynamics*, 95: 1125–1144.

Kasperski, M. and Geurts, C.W. (2005) Reliability and code level. *Wind and Structures*, 8: 295–307.

Lepage, M.F. and Irwin, P.A. (1985) A technique for combining historical wind data with wind tunnel tests to predict extreme wind loads. *5th U.S. National Conference on Wind Engineering*, Lubbock, Texas, 6–8 November.

Lieblein, J. (1974) Efficient methods of extreme-value methodology. Report NBSIR 74-602, National Bureau of Standards, Washington.

Melbourne, W.H. (1984) Designing for directionality. *Workshop on Wind Engineering and Industrial Aerodynamics*, Highett, Victoria, Australia, July.

Melchers, R. (1987) *Structural reliability – analysis and prediction*. Ellis Horwood, Chichester.

Miller, C.A., Holmes, J.D., Henderson, D.J., Ginger, J.D. and Morrison, M. (2013). The response of the Dines anemometer to gusts and comparisons with cup anemometers. *Journal of Atmospheric and Oceanic Technology*, 30: 1320–1336.

Naess, A. and Clausen, P.H. (2001). Combination of peaks-over-threshold and bootstrapping methods for extreme value prediction. *Structural Safety*, 23: 315–330.

Palutikof, J.P., Brabson, B.B., Lister, D.H. and Adcock, S.T. (1999) A review of methods to calculate extreme wind speeds. *Meteorological Applications*, 6: 119–132.

Peterka, J.A. and Shahid, S. (1998) Design gust wind speeds in the United States. *Journal of Structural Engineering (ASCE.)*, 124: 207–214.

Pham, L., Holmes, J.D. and Leicester, R.H. (1983) Safety indices for wind loading in Australia. *Journal of Wind Engineering and Industrial Aerodynamics*, 14: 3–14.

Pham, L., Holmes, J.D. and Yang, J. (1992) Reliability analysis of Australian communication lattice towers. *Journal of Constructional Steel Research*, 23: 255–272.

Pugsley, A.G. (1966) *The safety of structures*. Edward Arnold, London.

Rojiani, K. and Wen, Y-K. (1981) Reliability of steel buildings under winds. *Journal of the Structural Division, ASCE*, 107: 203–221.

Russell, L.R (1971) Probabilistic distributions for hurricane effects. *Journal of Waterways, Harbours and Coastal Engineering, ASCE*, 97: 139–154.

Shellard, H.C. (1958) Extreme wind speeds over Great Britain and Northern Ireland. *Meteorological Magazine (UK)*, 87: 257–265.

Shellard, H.C. (1963) The estimation of design wind speeds. *1st International Symposium on Wind effects on Buildings and Structures*, Teddington, England, UK.

Standards Australia. (1989) *SAA loading code. Part 2: wind loads*. Standards Australia, North Sydney. Australian Standard, AS1170.2-1989.

Standards Australia. (2011) *Structural design actions. Part 2: wind actions*. Standards Australia, Sydney. Australian/New Zealand Standard, AS/NZS1170.2: 2011.

Torrielli, A., Repetto, M.P and Solari, G. (2013) Extreme wind speeds from long-term synthetic records. *Journal of Wind Engineering and Industrial Aerodynamics*, 115: 22–38.

Von Mises, R. (1936) La distribution de la plus grande de *n* valeurs. Reprinted in Selected Papers of Richard von Mises, 2: 271–294, American Mathematical Society, Providence R.I., USA.

Wen, Y-K. and Chu, S-L. (1973) Tornado risks and design wind speeds. *Journal of the Structural Division, ASCE*, 99: 2409–2421.

Whittingham, H.E. (1964) Extreme wind gusts in Australia, Bulletin No. 46, Bureau of Meteorology, Melbourne, Victoria, Australia, 1964.

Wieringa, J. (1973) Gust factors over open water and built-up country. *Boundary-Layer Meteorology*, 3: 424–441.

Xu, H., Lin, N., Huang, M. and Lou, W. (2020) Design tropical cyclone wind speed when considering climate change. *Journal of Structural Engineering (ASCE)*, 146: 04020063.

Chapter 3

Strong wind characteristics and turbulence

3.1 INTRODUCTION

As the earth's surface is approached, the frictional forces play an important role in the balance of forces on the moving air. For larger storms such as extra-tropical depressions, this zone extends up to 500–1,000 m in height. For thunderstorms, the boundary layer is much smaller – probably around 100 m (see Section 3.2.7). The region of frictional influence is called the 'atmospheric boundary layer' and is similar in many respects to the turbulent boundary layer on a flat plate or airfoil at high wind speeds.

Figure 3.1 shows recordings of wind speeds recorded at three heights on a tall mast at Sale in southern Australia (as measured by sensitive cup anemometers, during a period of strong wind produced by a gale from a synoptic extra-tropical depression (Deacon, 1955). The records show the main characteristics of fully-developed "boundary-layer" flow in the atmosphere:

- the increase of the average wind speed as the height increases;
- the gusty or turbulent nature of the wind speed at all heights;
- the broad range of frequencies in the gusts in the air flow;
- there is some similarity in the patterns of gusts at all heights, especially for the more slowly changing gusts, or lower frequencies.

The term "boundary-layer" means the region of wind flow affected by friction at the earth's surface, which can extend up to 1 km. The Coriolis forces (Section 1.2.2) become gradually less in magnitude as the wind speed falls near the earth's surface. This causes the geostrophic balance, as discussed in Section 1.2.3 to be disturbed, and the mean wind vector turns from being parallel to the isobars to having a component towards the low pressure, as the height above the ground reduces. Thus, the mean wind speed may change in direction slightly with height, as well as magnitude. This effect is known as the *Ekman Spiral*. However, the direction change is small over the height range of normal structures, and is normally neglected in wind engineering.

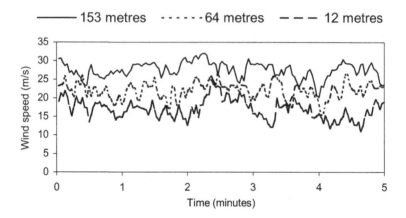

Figure 3.1 Wind speeds at three heights during a gale (Deacon, 1955).

The following sections will mainly be concerned with the characteristics of the mean wind and turbulence, near the ground, produced by severe gales, of synoptic scale, in the higher latitudes. These winds have been studied in detail for more than 50 years and are generally well understood, at least over flat homogeneous terrain. The wind and turbulence characteristics in tropical cyclones (Section 1.3.2) and thunderstorm downbursts (Section 1.3.5), which produce the extreme winds in the lower latitudes, are equally important, but are much less well understood. However, existing knowledge of their characteristics is presented in Sections 3.2.6 and 3.2.7. Tornadoes are rare events, but can produce significant damage in some parts of the world. A simple horizontal profile of wind components in a tornado vortex is discussed in Section 3.2.8.

3.2 MEAN WIND SPEED PROFILES

3.2.1 The logarithmic law

Here, we will consider the variation of the mean or time-averaged wind speed with height above the ground near the surface (say in first 100–200 m – the height range of most structures). In strong wind conditions, the most accurate mathematical expression is the 'logarithmic law'. The logarithmic law was originally derived for the turbulent boundary layer on a flat plate by Prandtl; however, it has been found to be valid in an unmodified form in strong wind conditions in the atmospheric boundary layer near the surface. It can be derived in a number of different ways. The following derivation is the simplest, and is a form of dimensional analysis.

We postulate that the wind shear, i.e. the rate of change of mean wind speed, $\bar{U}$, with height is a function of the following variables:

- the height above the ground, z
- the retarding force per unit area exerted by the ground surface on the flow – known as the surface shear stress, τ_o
- the density of air, ρ_a

Note that near the ground, the effect of the earth's rotation (Coriolis forces) is neglected. Also, because of the turbulent flow, the effect of molecular viscosity can be neglected.

Combining the wind shear with the above quantities, we can form a non-dimensional wind shear:

$$\frac{d\bar{U}}{dz} z \sqrt{\frac{\rho_a}{\tau_o}}$$

$\sqrt{(\tau_o/\rho_a)}$ has the dimensions of velocity, and is known as the *friction velocity*, u_* (note that this is not a physical velocity). Then, since there are no other non-dimensional quantities involved,

$$\frac{d\bar{U}}{dz} \frac{z}{u_*} = \text{a constant, say } \frac{1}{k} \tag{3.1}$$

Integrating,

$$\bar{U}(z) = \frac{u_*}{k}\left(\log_e z - \log_e z_o\right) = \frac{u_*}{k}\log_e\left(z / z_o\right) \tag{3.2}$$

where z_o is a constant of integration, with the dimensions of length, known as the *roughness length*.

Equation (3.2) is the usual form of the logarithmic law. k is known as *von Karman's constant*, and has been found experimentally to have a value of about 0.4, z_o, the roughness length, is a measure of the roughness of the ground surface.

Another measure of the terrain roughness is the *surface drag coefficient*, κ, which is a non-dimensional surface shear stress, defined as:

$$\kappa = \frac{\tau_0}{\rho_a \bar{U}_{10}^2} = \frac{u_*^2}{\bar{U}_{10}^2} \tag{3.3}$$

where $\bar{U}_{10}$ is the mean wind speed at 10 m height.

For urban areas and forests, where the terrain is very rough, the height, z, in Equation (3.2) is often replaced by an effective height, $(z-z_h)$, where z_h is a 'zero-plane displacement'. Thus, in this case,

$$\bar{U}(z) = \frac{u_*}{k}\log_e\left[\frac{z - z_h}{z_o}\right] \tag{3.4}$$

In urban terrain, the zero-plane displacement can be taken as about three-quarters of the general rooftop height.

Usually, the most useful way of applying Equation (3.4) is to use it to relate the mean wind speeds at two different heights as follows:

$$\frac{\bar{U}(z_1)}{\bar{U}(z_2)} = \frac{\log_e\left[(z_1 - z_h)/z_o\right]}{\log_e\left[(z_2 - z_h)/z_o\right]} \tag{3.5}$$

In the application of Equation (3.3), the 10 m reference height should be taken as 10 m above the zero-plane displacement, or $(10+z_h)$ metres above the actual ground level.

By applying Equations (3.3) and (3.4) for z equal to 10 m, a relationship between the surface drag coefficient, κ, and the roughness length, z_o, is given by the following equation:

$$\kappa = \left[\frac{k}{\log_e\left(\dfrac{10}{z_o}\right)}\right]^2 \tag{3.6}$$

Table 3.1 gives the appropriate values of roughness length and surface drag coefficient for various types of terrain types.

Although the logarithmic law has a sound theoretical basis, at least for fully developed wind flow over homogeneous terrain, these ideal conditions are rarely met in practice. Also, the logarithmic law has some mathematical characteristics which may cause problems. Firstly, since the logarithms of negative numbers do not exist, it cannot be evaluated for heights, z, below the zero-plane displacement z_h, and if $z-z_h$ is less than z_o, a negative wind speed is given. Secondly, it is less easy to integrate. To avoid some of these problems, wind engineers have often preferred to use the power law (Section 3.2.3).

3.2.2 The Deaves and Harris mean wind profile

The logarithmic law is applicable to strong winds with thermally neutral stability for heights in the surface layer up to 100–200 m, a height range which includes most structures. The Deaves and Harris (D & H) model

Table 3.1 Terrain types, roughness length and surface drag coefficient

Terrain type	Roughness length (m)	Surface drag coefficient
Very flat terrain (snow, desert)	0.001–0.005	0.002–0.003
Open terrain (grassland, few trees)	0.01–0.05	0.003–0.006
Suburban terrain (buildings 3–5 m)	0.1–0.5	0.0075–0.02
Dense urban (buildings 10–30 m)	1–5	0.03–0.3

attempts to provide a consistent mathematical model for the complete atmospheric boundary layer up to the height, h, at which gradient winds (Section 1.2.4) occur, (Deaves and Harris, 1978).

The mean velocity profile in the D & H model is based on asymptotic similarity which matches the 'defect' profile in the outer boundary layer with the logarithmic law in the inner, or surface, layer (Csanady, 1967; Blackadar and Tennekes, 1968; Deaves and Harris, 1982).

The D & H mean wind profile is given by Equation (3.7):

$$\bar{U}(z) = \frac{u_*}{k}\left[\ln\frac{z}{z_o} + 5.75\left(\frac{z}{h}\right) - 1.875\left(\frac{z}{h}\right)^2 - 1.33\left(\frac{z}{h}\right)^3 + 0.25\left(\frac{z}{h}\right)^4\right] \quad (3.7)$$

where h is the gradient height, where the wind speed reaches its maximum value and, according to the D & H model, is given by $(u_*/6f)$. The other terms have been defined earlier.

It can be seen that when z is small, i.e. close to the ground surface, the second to fifth terms on the right-hand side become small, and the equation is closely approximated by the logarithmic law of Equation (3.4) (assuming z_h equal to zero).

The D & H profile is based on several assumptions, and data to calibrate it were carefully selected to ensure that these conditions were satisfied. The most significant conditions were:

   i. strong statistically stationary wind (i.e. steady),
   ii. neutral thermal stability,
   iii. flat, level terrain,
   iv. long fetch distances of uniform roughness, overland.

Condition (i), in particular, eliminates many types of strong winds from a variety of sources that are transient in nature, or 'non-stationary', including those from thunderstorms (Sections 1.3.3 and 1.3.5), as well as other mechanisms not discussed in this book, such as monsoon winds, and 'shamal' winds in the Arabian Gulf region of the Middle East.

Nevertheless, the D & H profile and other parts of the model, such as turbulence intensities, have been adopted in the Australian/New Zealand Standard AS/NZS 1170.2 (Standards Australia, 2011) and elsewhere (see Chapter 15).

Also, the use of only overland wind measurements in its calibration means that the D & H model should be avoided for strong winds over water (see Section 3.2.4).

3.2.3 The power law

The power law is a simplified empirical form that has no theoretical basis, but is easily integrated over height – a convenient property when wishing to determine bending moments at the base of a tall structure, for example.

To relate the mean wind speed at any height, z, with that at 10 m (adjusted, if necessary, for rougher terrains, as described in the previous section), the power law can be written as follows:

$$\bar{U}(z) = \bar{U}_{10}\left(\frac{z}{10}\right)^{\alpha} \tag{3.8}$$

The exponent, α, in Equation (3.8) will change with the terrain roughness, and also with the height range, when matched to the logarithmic law. A relationship that can be used to relate the exponent to the roughness length, z_o, is as follows:

$$\alpha = \left(\frac{1}{\log_e(z_{ref} / z_o)}\right) \tag{3.9}$$

where z_{ref} is a reference height at which the two 'laws' are matched. z_{ref} may be taken as the average height in the range over which matching is required, or half the maximum height over which the matching is required.

Figure 3.2 shows a matching of the two laws for a height range of 100 m, using Equation (3.9), with z_{ref} taken as 50 m. It is clear that the two relationships are extremely close, and that the power law is quite adequate for engineering purposes.

3.2.4 Mean wind profiles over water

Over land, the surface drag coefficient, κ, is found to be nearly independent of mean wind speed. This is not the case over water, where higher winds

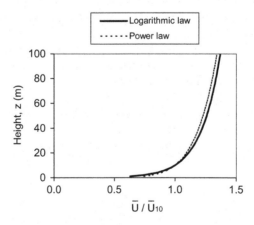

Figure 3.2 Comparison of the logarithmic (z_o=0.02 m) and power laws (α=0.128) for mean velocity profile.

create higher waves, and hence higher surface drag coefficients. The relationship between κ and $\bar{U}_{10}$ has been the subject of much study, and a large number of empirical relationships have been derived.

Amorocho and DeVries (1980) identified three wind-speed regions for wind over water, with mean speeds up to 30 m/s:

   i. a lower region, with wind speeds less than about 7 m/s, in which waves have not begun to break, and in which the surface drag coefficient, κ (Equation 3.3), is approximately constant,

  ii. a transition region, after the onset of breakers, in which κ varies non-linearly with $\bar{U}_{10}$,

 iii. a high wind-speed region, with $\bar{U}_{10}$ greater than about 20 m/s, corresponding to a region of 'breaker saturation', in which κ tends towards another (higher) constant value.

The two values of κ corresponding to (i) and (iii) are 0.00104 and 0.00254, respectively. By applying Equation (3.6), these values correspond to roughness lengths, z_o, of 0.04 mm and 3.6 mm respectively.

Charnock (1955), using dimensional arguments, proposed a mean wind profile over water, that implies that the roughness length, z_o, should be given by Equation (3.10).

$$z_o = \frac{a u_*^2}{g_0} = \frac{a \kappa \bar{U}_{10}^2}{g_0} \tag{3.10}$$

where g_0 is the gravitational constant, and a is an empirical constant.

Substituting for the surface drag coefficient, κ, from Equation (3.6) into Equation (3.10), Equation (3.11) is obtained:

$$z_o = \frac{a}{g_0} \left[\frac{k \bar{U}_{10}}{\log_e \left(\dfrac{10}{z_o} \right)} \right]^2 \tag{3.11}$$

(z_h is taken as zero over water)

Garratt (1977) examined a large amount of experimental data and suggested a constant value for a of 0.014. However, other studies (e.g. Amorocho and DeVries, 1980, Powell et al., 2003; Holmes, 2017) indicate a variety of values of a in the range of 0.01–0.04 for mean wind speeds greater than 5 m/s. In particular, Amorocho and DeVries (1980), using a large amount of experimental data, derived a complex non-linear relationship between a and $\bar{U}_{10}$.

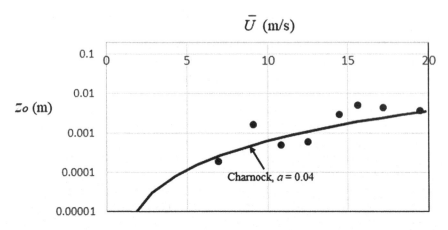

Figure 3.3 Roughness length, z_o, versus mean wind speed over water (Holmes, 2017).

Table 3.2 Roughness length over water (based on the model of Amorocho and DeVries, 1980, for $\overline{U}_{10} \leq 30\,\text{m/s}$)

$\overline{U}_{10}$ (m/s)	Roughness length, z_o (mm)
5	0.04
10	0.15
15	2.4
20	3.5
25	3.6
30	3.6
40[a]	1–4[a]
50[a]	0.5–1[a]

[a] Measurements in hurricanes from Powell *et al.* (2003).

Figure 3.3 shows roughness lengths derived from gust factor measurements, making use of Equation (3.22) for wind over a large bay, to which the Charnock relationship with a value of a of 0.04 has been fitted.

Taking g_0 equal to $9.8\,\text{m/s}^2$ and k equal to 0.4, the range of z_o versus $\overline{U}_{10}$ up to $30\,\text{m/s}$, given in Table 3.2, is obtained based on the model of Amorocho and DeVries (1980).

Mean wind speeds above $40\,\text{m/s}$ are characteristic of mature tropical cyclones. Analysis of dropwindsonde measurements (Powell *et al.*, 2003) suggests a significant reduction in roughness lengths at very high mean wind speeds. This was attributed to the lack of steep ocean waves and the development of a whitewater 'foam layer' (see also Section 3.2.6).

3.2.5 Relationship between upper-level and surface winds

For large-scale atmospheric boundary layers in synoptic winds, dimensional analysis suggests a functional relationship between a *geostrophic drag coefficient*, $C_g = u_*/U_g$, and the surface *Rossby Number*, $Ro = U_g/f z_o$. u_* is the friction velocity and U_g is the geostrophic wind (Section 1.2.3), f is the Coriolis parameter (Section 1.2.2), and z_o is the roughness length (Section 3.2.1). This relationship, originally suggested by Lettau (1959), enables surface conditions (roughness length, surface shear stress, mean wind speed near the ground) to be related to the upper-level winds, which are driven by the pressure gradient (Section 1.2.1) and Coriolis force (Section 1.2.2).

Davenport (1963), making use of analysis by Lettau (1950, 1959) of measurements near Leipzig, Germany and elsewhere, proposed a relationship of the form of Equation (3.12) with K equal to 0.16 and n equal to -0.09. Swinbank (1974) suggested a similar form, with values of K and n of 0.111 and -0.07, respectively.

$$C_g = K\,Ro^n \tag{3.12}$$

A version of Equation (3.12), with intermediate values of K and n, that appears to fit better to the available balloon measurements from the atmosphere, is Equation (3.13).

$$C_g = 0.12\,Ro^{-0.085} \tag{3.13}$$

An alternative expression for the geostrophic drag coefficient, C_g, based on the theory of 'asymptotic similarity' (Csanady, 1967; Blackadar and Tennekes, 1968; Deaves and Harris, 1978), is given by Equation (3.14):

$$C_g = \frac{u_*}{U_g} = \frac{0.4}{\left[\log_e\left(\dfrac{u_*}{fz_o}\right) - A\right]} = \frac{0.4}{\left[\log_e\left(\dfrac{u_*}{fz_o}\right) + 1.0\right]} \tag{3.14}$$

where the 'constant', A, has been set equal to -1.0 (Deaves and Harris, 1978).

Equation (3.14) requires an iterative solution. However, it is matched closely by the explicit Equation (3.15) (Jensen, 1978):

$$C_g = \frac{u_*}{U_g} \cong \frac{0.5}{\log_e(Ro)} \tag{3.15}$$

Figure 3.4 shows Equations (3.13) and (3.15) compared with data from the atmosphere reported by Csanady (1967) and Jensen (1978).

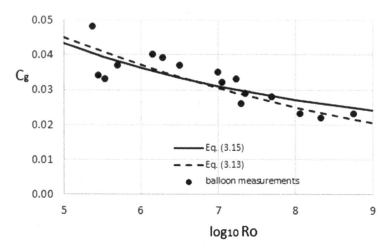

Figure 3.4 Geostrophic drag coefficient versus Rossby Number.

Example 3.1

Applying Equation (3.13) for a latitude of $40°$ ($f=0.935\times10^{-4}$ second^{-1}), a value of U_g equal to $40\,\text{m/s}$, and a roughness length of $20\,\text{mm}$ gives a friction velocity of $1.14\,\text{m/s}$ and from Equation (3.2), a value of $\bar{U}_{10}$ of $17.8\,\text{m/s}$. Thus, the wind speed near the surface is calculated to be equal to 0.44 times the geostrophic wind – the upper-level wind away from the frictional effects of the earth's surface.

Applying Equation (3.15) to the same case gives a friction velocity, u_*, of $1.18\,\text{m/s}$, a mean velocity at $10\,\text{m}$ height of $18.4\,\text{m/s}$, and a ratio $\bar{U}_{10}/U_g$, with a value of 0.46.

These estimates are within 4% of each other, and well within the error range of the available full-scale data.

3.2.6 Mean wind profiles in tropical cyclones

A number of low-level flights into Atlantic Ocean and Gulf of Mexico hurricanes have been made by the National Oceanic and Atmospheric Administration (NOAA) of the United States. However, the flight levels were not low enough to provide useful data on wind speed profiles below about 200 m. Measurements from fixed towers are also extremely limited. However, some measurements were made from a 390 m communications mast close to the coast near Exmouth, Western Australia, in the late 1970s (Wilson, 1979). SODAR (sonic radar) profiles have been obtained from typhoons on Okinawa, Japan (Amano *et al.*, 1999). These show similar characteristics near the regions of maximum winds: a logarithmic-type profile up to a certain height (60–200 m), followed by a layer of strong convection, with nearly constant mean wind speed. More recent probes, known as 'dropwindsondes', have been dropped from aircraft flying through hurricanes, and their positions have been tracked continuously by GPS satellites, enabling estimation of horizontal wind speeds to be made (Hock and Franklin, 1999).

Based on averages of the 'dropwindsonde' data, the following mean wind speed profile was proposed for the eye wall region (Franklin et al., 2003):

$$\bar{U}_z = \bar{U}_{10} \frac{\log_e(z / 0.0001)}{\log_e(10 / 0.0001)} \text{ for } z < 300 \text{ m}$$

$$U(z) = \bar{U}_{300} \text{ for } z \geq 300 \text{ m} \tag{3.16}$$

Equation (3.16) implies a very low roughness length, z_o, of 0.1 mm and does not account for any variation of z_o with mean wind speed over the ocean. For values of $\bar{U}_{10}$ up to 40 m/s, measurements derived from 'dropwindsonde' profiles over the ocean in Atlantic hurricanes (Powell et al., 2003) can be fitted by the Charnock expression (Equation 3.10) with a equal to about 0.01. This value is consistent with inferred values from gust factors obtained in an approaching tropical cyclone in the Coral Sea off Eastern Australia (Figure 3.5, from Holmes, 2017). Above 40 m/s, the Atlantic data indicated a *decrease* in roughness length with increasing wind speed, to values below 1 mm at a $\bar{U}_{10}$ value of 50 m/s. As indicated in Section 3.2.4, this suggests that the Charnock expression with a single value of the 'constant' a is not applicable in the very high wind speeds of mature tropical cyclones.

For regions over the ocean outside the eye wall, the relations given in Section 3.2.4 can be applied. As the tropical cyclone crosses the coast it weakens (see Section 1.3.2), and the mean wind profiles would be expected to adjust to the underlying ground roughness over the land.

3.2.7 Wind profiles in thunderstorm winds

The most common type of severe wind generated by a thunderstorm results from a severe downdraft, described by Fujita (1985) as a 'downburst', and discussed in Section 1.3.5. Downbursts produce severe winds for relatively short periods – typically a few minutes – and are transient in nature (see Figure 1.9). To define a 'mean' velocity requires a much shorter averaging time than the 10 minutes to 1 hour typically used for synoptic scale wind events. We can separate the slowly varying part (or 'running mean'), representing the downward airflow which becomes a horizontal 'outflow' near the ground, from any superimposed turbulence of higher frequency. A typical averaging time to produce quasi-stationary winds in a thunderstorm is in the range of 30–60 seconds; for example, Holmes et al. (2008) found that 40 seconds was the optimum averaging period for a well-documented event in Texas in 2002.

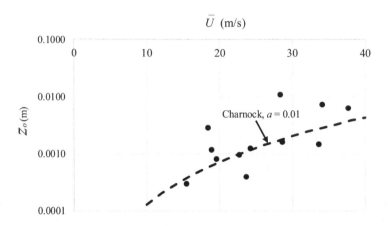

Figure 3.5 Roughness length, z_o, versus mean wind speed over the ocean in an approaching tropical cyclone (Holmes, 2017).

Thanks to Doppler radar measurements in the U.S.A., and some tower anemometer measurements in Australia and the U.S.A., there are some indications of the wind structure in the downburst type of thunderstorm wind, including the 'macroburst' and 'microburst' types identified by Fujita (1985). At the horizontal location where the maximum gust occurs, the wind speed increases from ground level up to a maximum value at a height of 50–100 m in a microburst. Above this height, the wind speed reduces relatively slowly.

A useful model of the velocity profiles in the vertical and horizontal directions in a downburst was provided by Oseguera and Bowles (1988). This model satisfies the requirements of fluid mass continuity, but does not include any effect of storm movement. The horizontal velocity component is expressed as Equation (3.17):

$$U = \left(\frac{\lambda R^2}{2r} \right) \left[1 - e^{-(r/R)^2} \right] \left(e^{-z/z^*} - e^{-z/\varepsilon} \right)$$

(3.17)

where,

r is the radial coordinate from the centre of the downburst
R is the characteristic radius of the downburst 'shaft'
z is the height above the ground
z^* is a characteristic height out of the boundary layer
ε is a characteristic height in the boundary layer
λ is a scaling factor, with dimensions of [time]$^{-1}$

According to this profile, maximum winds occur at a radius, r, of 1.121R. The vertical profile at this radius is shown in Figure 3.6. Radar and tower

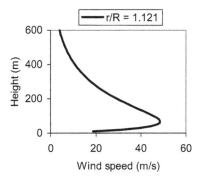

Figure 3.6 Profile of horizontal velocity near the ground during a stationary microburst (Oseguera and Bowles, 1988).

observations have shown that the height for maximum winds in microbursts is 50–100 m, but is greater in larger-scale downdraft events. As the outflow from a downdraft 'matures', it appears to adjust to the underlying terrain, and the profiles then appear more similar to those in synoptic winds, with maxima at a height of 500 m or greater (Gunter and Schroeder, 2013).

3.2.8 Wind profiles in tornados

There have been many studies of the wind structure in tornadoes based on full-scale studies using photogrammetry and portable Doppler radars (see also Section 1.3.4), along with laboratory studies of tornado-like vortices and theoretical analyses.

The simplest model of horizontal wind profile in a tornado is based on the Rankine, or combined, vortex (Figure 3.7). This consists of an inner 'core' with solid body rotation, in which the tangential wind velocity component, U_θ, is proportional to the radius from the centerline of the tornado, r. In the outer region $(r > R)$, the tangential velocity component is *inversely* proportional to the radius, r. This satisfies the equation of angular momentum (Lewellen, 1976), except for the discontinuity at r equal to R.

This model does not define the radial, U_r, or vertical, U_v, velocity components, but empirical values of these are shown on Figure 3.7.

Several alternative theoretical models for tornados are discussed by Lewellen (1976).

3.3 TURBULENCE AND GUST WIND SPEEDS

The general level of turbulence or 'gustiness' in the wind speed in storm winds, such as that shown in Figure 3.1, can be measured by its standard deviation or root-mean-square. First, we subtract out the steady or mean

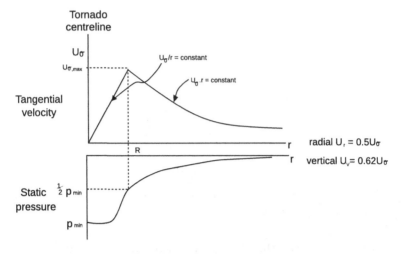

Figure 3.7 Velocity components in a tornado.

component (or the slowly varying component in the case of a transient storm, like a thunderstorm), and then quantify the resulting deviations. Since both positive and negative deviations can occur, we first square the deviations before averaging them, and finally the square root is taken to give a quantity with the units of wind speed. Equation (3.18) defines the standard deviation:

$$\sigma_u = \left\{ \frac{1}{T} \int_0^T \left[U(t) - \bar{U} \right]^2 dt \right\}^{1/2} \tag{3.18}$$

where $U(t)$ is the total velocity component in the direction of the mean wind, equal to $\bar{U} + u(t)$, and $u(t)$ is the 'longitudinal' turbulence component, i.e. the component of the fluctuating velocity in the mean wind direction.

Other components of turbulence in the lateral horizontal direction denoted by $v(t)$ and in the vertical direction denoted by $w(t)$ are quantified by their standard deviations, σ_v, and σ_w, respectively.

3.3.1 Turbulence intensities

The ratio of the standard deviation of each fluctuating component to the mean value is known as the *turbulence intensity* of that component:

$$\text{Longitudinal}: I_u = \sigma_u / \bar{U} \tag{3.19}$$

$$\text{Lateral}: I_v = \sigma_v / \bar{U} \tag{3.20}$$

$$\text{Vertical}: I_w = \sigma_w / \bar{U} \tag{3.21}$$

Near the ground in gales produced by large-scale depression systems, measurements have found that the standard deviation of longitudinal wind velocity, σ_u, is equal to 2.5 u_* to a good approximation, where u_* is the friction velocity (Section 3.2.1). Then, the turbulence intensity, I_u, is given by Equation (3.22):

$$I_u = \frac{2.5u_*}{(u_* / 0.4)\log_e (z / z_o)} = \frac{1}{\log_e (z / z_o)} \tag{3.22}$$

Thus, the turbulence intensity is simply related to the surface roughness, as measured by the roughness length, z_o. For a rural terrain, with a roughness length of 0.04 m, the longitudinal turbulence intensities for various heights above the ground are given in Table 3.3. It is clear from this table that the turbulence intensity decreases with height above the ground.

The lateral and vertical turbulence components are generally lower in magnitude than the corresponding longitudinal value. However, for well-developed boundary layer winds, simple relationships between standard deviation and the friction velocity u_* have been suggested. Thus, approximately the standard deviation of lateral (horizontal) velocity, σ_v, is equal to 2.20 u_*, and for the vertical component, σ_w is given approximately by 1.3–1.4 u_*. Then, equivalent expressions to Equation (3.22) for the variation with height of I_v and I_w can be derived:

$$I_v \cong 0.88 / \log_e (z / z_o) \tag{3.23}$$

$$I_w \cong 0.55 / \log_e (z / z_o) \tag{3.24}$$

The turbulence intensities in tropical cyclones (typhoons and hurricanes) are generally believed to be higher than those in gales in temperate latitudes. Choi (1978) found that the longitudinal turbulence intensity was about 50% higher in tropical-cyclone winds compared to synoptic winds. From

Table 3.3 Longitudinal turbulence intensities for rural terrain ($z_o = 0.04$ m)

Height, z (m)	I_u
2	0.26
5	0.21
10	0.18
20	0.16
50	0.14
100	0.13

measurements on a tall mast in north-western Australia during the passage of severe tropical cyclones, convective 'squall-like' turbulence was observed (Wilson, 1979). This was considerably more intense than the 'mechanical turbulence' seen closer to the ground, and was associated with the passage of bands of rain clouds.

Turbulence intensities in thunderstorm downburst winds are even less well defined than for tropical cyclones. However, the Andrews Air Force Base event of 1983 (Figure 1.9) indicates a turbulence 'intensity' of the order of 0.1 (10%) superimposed on the underlying transient flow (see also Section 3.3.7).

3.3.2 Probability density

As shown in Figure 3.1, the variations of wind speed in the atmospheric boundary layer are generally random in nature, and do not repeat in time. The variations are caused by eddies or vortices within the air flow, moving along at the mean wind speed. These eddies are never identical, and we must use statistical methods to describe the gustiness.

The probability density, $f_u(u_o)$, is defined so that the proportion of time that the wind velocity, $U(t)$, spends in the range $u_o + du$ is $f_u(u_o)\,du$ (see Section C2.1 in Appendix C). Measurements have shown that the wind velocity components in the atmospheric boundary layer follow closely the Normal or Gaussian probability density function, given by Equation (3.25):

$$f_u(u) = \frac{1}{\sigma_u \sqrt{2\pi}} \exp\left[-\frac{1}{2}\left(\frac{u - \bar{U}}{\sigma_u} \right)^2 \right] \qquad (3.25)$$

This function has the characteristic bell shape. It is defined only by the mean value, $\bar{U}$, and standard deviation, σ_u (see also Section C3.1 in Appendix C).

Thus, with the mean value and standard deviation, the probability of any wind velocity occurring can be estimated.

3.3.3 Gust wind speeds and gust factors

In many design codes and standards for wind loading (see Chapter 15), a peak gust wind speed is used for design purposes. The nature of wind as a random process means that the peak gust within a sample time, T, of, say, ten minutes is itself also a random variable. However, we can define an *expected*, or average, value within the ten minutes period.

Assuming that the longitudinal wind velocity has a Gaussian probability distribution, the expected peak gust, $\hat{U}$, is given by Equation (3.26):

$$\hat{U} = \bar{U} + g\sigma_u \qquad (3.26)$$

Table 3.4 Expected peak factors for various τ and T

Averaging time, τ (seconds)	Sample time, T (seconds)	g
3	3,600	3.0
3	600	2.5
1	3,600	3.4
1	600	2.9
0.2	3,600	3.8
0.2	600	3.4

In Equation (3.26), g is known as the *peak factor* and σ_u is the standard deviation of the fluctuating wind velocity

The expected peak factor, g, depends on the effective averaging time of the gust, τ, as well as the sample time T. Table 3.4 shows the expected values of g for various values of τ, and for T equal to 600 and 3,600 seconds, in synoptic scale wind events. These values can be obtained theoretically (ESDU 83045; Holmes, Allsop and Ginger, 2014) by 'filtering' the wind spectrum (Equation 3.30) by the theoretical moving average filter associated with the moving average time, τ.

Meteorological instruments used for long-term wind measurements do not have a perfect response, and the peak gust wind speed they measure is dependent on their response characteristics. The response can be measured by an equivalent averaging time, τ. The equivalent averaging time for the output of a rotating cup anemometer, the most common instrument used by meteorological agencies, depends on the distance constant of the instrument and on the mean wind speed, but typically lies in the range of 0.3–0.9 seconds at wind speeds of interest for structural design (Holmes *et al.*, 2014). However, it should be noted that after digitization, most meteorological agencies apply a digital 'moving average' filter, with an averaging time, τ, of 3 seconds, as recommended by the World Meteorological Organization (Beljaars, 1987). This is the source of the commonly used '3-second' gust in codes and standards. However, it should be noted that the effective frontal area associated with this gust is quite large (equivalent to the area of a typical tall building in a city centre), and loads on small structures, calculated from gusts with that duration, need to be increased by a factor greater than 1.0 (Holmes, Allsop and Ginger, 2014).

The Dines pressure tube anemometer used in many countries up to the 1990s had an effective averaging time of around 0.2 seconds, depending on the type and the mean wind speed (Miller *et al.*, 2013), and hence recorded generally higher wind gusts than modern automatic weather systems with cup anemometers and 3-second digital filtering (Holmes and Ginger, 2012).

For various terrains, a profile of peak gust with height can be obtained. Note, however, that gusts do not occur simultaneously at all heights, and such a profile would represent an envelope of the expected gust wind speed with height.

The *gust factor*, G, is the ratio of the expected maximum gust speed within a specified period to the mean wind speed (Equation (3.27).

The relationship between the expected gust factor, G, and the peak factor, g, is also shown in Equation (3.27):

$$G = \frac{\hat{U}}{\bar{U}} = \frac{\bar{U} + \sigma_u}{\bar{U}} = 1 + g\ I_u \tag{3.27}$$

where I_u is the longitudinal turbulence intensity.

For gales (synoptic winds in temperate climates), the magnitudes of gusts for various averaging times, τ, were studied by Durst (1960), Deacon (1965), Wieringa (1973) and Ashcroft (1994). Based on a 10-minute mean wind speed, Deacon gave gust factors of about 1.45 for 'open country with few trees' and 1.96 for suburban terrain at a height of 10 m.

Wieringa (1973) proposed Equation (3.28) for calculation of gust factors in synoptic winds over both land and water, as a function of height, z, roughness length of the terrain, z_o, mean velocity, $\bar{U}$, and the gust averaging time, τ. The derivation of this equation made use of Equations (3.22) and (3.27), with a semi-empirical expression for the peak factor, g, based on a Gaussian probability distribution for the turbulent fluctuations. Equation (3.28) is applicable to a sample time, T, of 10 minutes:

$$G = 1 + \frac{1.42 + 0.30 \log_e\left[\left(990/\bar{U}\tau\right) - 4\right]}{\log_e\left(z/z_o\right)} \tag{3.28}$$

Several authors, such as Ishizaki (1983), Krayer and Marshall (1992), Black (1992) and Vickery and Skerjl (2005), have provided estimates of gust factors over land for tropical cyclones (hurricanes and typhoons). The study of four U.S. hurricanes by Krayer and Marshall (1992) gave a value of 1.55 for the ratio of peak 2-second gust to 10-minute mean for 10-m height in open country terrain. This value is based on tropical cyclone winds with a wide range of wind speeds, to values as low as 10 m/s. The analysis by Black (1992), which appeared to be based on higher wind speeds in hurricanes, gave a higher value of 1.66 for the gust factor, $\dfrac{\hat{U}_{2\,sec,\,10\,m}}{\bar{U}_{10\,min,\,10\,m}}$.

Gust factors in tropical cyclones were also extensively reviewed by Harper *et al.* (2010), including the references mentioned previously. They found considerable scatter in the experimental data and relationships suggested. This is not surprising, as gust factors are dependent on the mean wind speed, the length scale of turbulence (Section 3.3.4), the height above ground, and the surface roughness length (Section 3.2.1), as well as the gust duration or averaging time, τ.

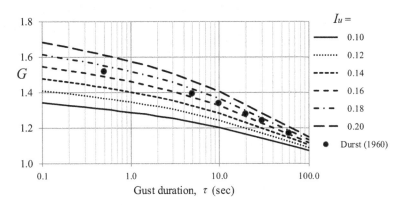

Figure 3.8 Gust factor $(\hat{U}_{\tau sec}/\bar{U}_{10min})$ as a function of turbulence intensity for $\ell_u/\bar{U} =$ 10 second.

Figure 3.8 shows gust factors, G, for synoptic winds (including tropical cyclones) for a sample time, T, of 600 seconds, plotted against the gust duration, τ, for a value of the ratio of turbulence length scale, ℓ_u, to mean wind speed $\bar{U}$ of 10 (units in seconds). The values of G were derived using an approach described by Holmes, Allsop and Ginger (2014). Figure 3.8 shows that gust factors are dependent on turbulence intensity (see Equation 3.27) which, in turn, depends on the height above ground and the roughness of the underlying terrain.

Also shown in Figure 3.8 are points derived from Durst (1960), which form the well-known 'Durst curve'. This curve is applicable to a turbulence intensity, I_u, of about 0.17, and should not be regarded as a 'universal' curve for all terrains and heights above the ground.

3.3.4 Wind spectra and turbulence length scales

The probability density function (Section 3.3.2) tells us something about the magnitude of the wind velocity, but nothing about how slowly or quickly it varies with time. In order to describe the distribution of turbulence with frequency, a function called the *spectral density*, often abbreviated to 'spectrum', is used. It is defined so that the contribution to the variance $(\sigma_u^2$, or square of the standard deviation), in the range of frequencies from n to $n + dn$, is given by $S_u(n)dn$, where $S_u(n)$ is the spectral density function for $u(t)$. Then integrating over all frequencies,

$$\sigma_u^2 = \int_0^\infty S_u(n)dn \tag{3.29}$$

There are many mathematical forms that have been used for $S_u(n)$ in meteorology and wind engineering. One of the most common and mathematically correct of these for the longitudinal velocity component (parallel to the mean wind direction) is the von Karman form (developed for laboratory turbulence by von Karman (1948) and adapted for wind engineering by Harris (1968)). This may be written in several forms; Equation (3.30) is a commonly used non-dimensional form:

$$\frac{n \cdot S_u(n)}{\sigma_u^2} = \frac{4\left(\dfrac{n\ell_u}{\bar{U}}\right)}{\left[1+70.8\left(\dfrac{n\ell_u}{\bar{U}}\right)^2\right]^{5/6}} \tag{3.30}$$

where ℓ_u is the turbulence *integral length scale*.

In this form, the curve of $n \cdot S_u(n)/\sigma_u^2$ versus $n/\bar{U}$ has a peak; the value of ℓ_u determines the value of $(n/\bar{U})$ at which the peak occurs – the higher the value of ℓ_u, the higher the value of $(\bar{U}/n)$ at the peak, or λ, known as the 'peak wavelength'. For the von Karman spectrum, λ is equal to $6.85\ell_u$. The length scale, ℓ_u, varies with both terrain roughness and height above the ground. The form of the von Karman spectrum is shown in Figure 3.9.

The other orthogonal components of atmospheric turbulence have spectral densities with somewhat different characteristics. The spectrum of vertical turbulence is the most important of these, especially for horizontal structures such as bridges. A common mathematical form for the spectrum of vertical turbulence (w') is the Busch and Panofsky (1968) form which can be written as Equation (3.31):

$$\frac{n \cdot S_w(n)}{\sigma_w^2} = \frac{2.15\left(\dfrac{nz}{\bar{U}}\right)}{\left[1+11.16\left(\dfrac{nz}{\bar{U}}\right)^{5/3}\right]} \tag{3.31}$$

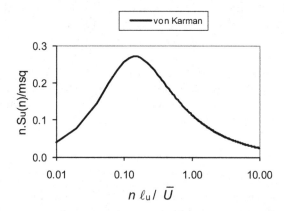

Figure 3.9 Normalized spectrum of longitudinal velocity component (von Karman).

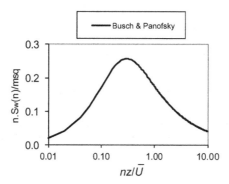

Figure 3.10 Normalized spectrum of vertical velocity component (Busch and Panofsky, 1968).

In this case, the length scale is directly proportional to the height above ground, z.

The Busch and Panofsky spectrum for vertical turbulence (w') is shown in Figure 3.10.

3.3.5 Correlation

Covariance and *correlation* are two important properties of wind turbulence in relation to wind loading. The latter is the same quantity that is calculated in linear regression analysis. In the present context, it relates the fluctuating wind velocities at two points in space, or wind pressures at two points on a building (such as a roof).

For example, consider the wind speed at two different heights on a tower (for example, Figure 3.1). The covariance between the fluctuating (longitudinal) velocities at two different heights, z_1 and z_2, is defined in Equation (3.32):

$$\overline{u'(z_1)u'(z_2)} = \frac{1}{T}\int_0^T \left[U(z_1,t)-\bar{U}(z_1)\right]\left[U(z_2,t)-\bar{U}(z_2)\right]dt \qquad (3.32)$$

Thus, the covariance is the product of the fluctuating velocities at the two heights, averaged over time. Note that the mean values, $\bar{U}(z_1)$ and $\bar{U}(z_2)$, are subtracted from each velocity in the right-hand side of Equation (3.31). Note that in the special case when z_1 is equal to z_2, the right-hand side is then equal to the variance (σ_u^2) of the fluctuating velocity at the single height.

The correlation coefficient, ρ, is defined by Equation (3.33):

$$\rho = \frac{\overline{u'(z_1)u'(z_2)}}{\sigma_u(z_1)\cdot\sigma_u(z_2)} \qquad (3.33)$$

where z_1 is equal to z_2 and the value of ρ is +1 (i.e. we have full correlation). It can be readily shown that ρ must lie between −1 and +1. A value of 0 indicates no correlation (i.e. no statistical relationship between the wind velocities) – this usually occurs when the heights z_1 and z_2 are widely separated.

The covariance and correlation are very useful in calculating the fluctuating wind loads on tall towers, and for estimating span reduction factors for transmission lines. In the latter case, the points would be separated horizontally, rather than vertically.

A mathematical function which is useful for describing the correlation, ρ, is the exponential decay function:

$$\rho \approx \exp\left[-C|z_1 - z_2|\right] \tag{3.34}$$

This function is equal to +1 when z_1 is equal to z_2 and tends to zero when $|z_1 - z_2|$ becomes very large (very large separations).

$(1/C)$ is the value of separation length for the cross-correlation to decay to a value of $1/e$, and is a measure of the lateral length scale of turbulent eddies.

Figure 3.11 shows Equation (3.34) with C equal to $(1/40)$ m^{-1}. It is compared with some measurements of longitudinal velocity fluctuations in the atmospheric boundary, at a height of 13.5 m, with horizontal separations, over urban terrain (Holmes, 1973).

The lateral length scale $(1/C)$ of about 40 m for urban terrain, indicated by Figure 3.11, compares with an average value of 24 m found by Teunissen (1980), for rural terrain in Canada.

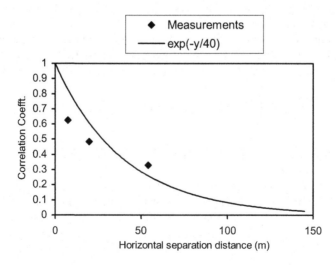

Figure 3.11 Cross-correlation of longitudinal velocity fluctuations in the atmospheric boundary layer at a height of 13.5 m (Holmes, 1973).

3.3.6 Co-spectrum and coherence

When considering the resonant response of structures to wind (Chapter 5), the correlation of wind velocity fluctuations from separated points *at different frequencies* is important. For example, the correlations of vertical velocity fluctuations with span-wise separation at the natural frequencies of vibration of a large-span bridge are important in determining its response to buffeting.

The frequency-dependent correlation can be described by functions known as the *cross-spectral density, co-spectral density* and *coherence*. Mathematical definitions of these functions are given by Bendat and Piersol (1999) and others. The cross-spectral density, as well as being a function of frequency, is a complex variable, with real and imaginary components. The co-spectral density is the real part, and may be regarded as a frequency-dependent covariance (Section 3.3.5). The coherence is a normalized magnitude of the cross-spectrum, approximately equivalent to a frequency-dependent correlation coefficient. The normalized co-spectrum is very similar to coherence, but does not include the imaginary components; this is the relevant quantity when considering the effect of wind forces from turbulence on structures.

The normalized co-spectrum and coherence are often represented by an exponential function of separation distance and frequency:

$$\rho(\Delta z,\, n) = \exp\left[-\left(\frac{k\, n \cdot \Delta z}{\bar{U}}\right)\right] \tag{3.35}$$

where k is an empirical constant, used to fit measured data; a typical range of values of k for atmospheric turbulence is 8–10 (e.g. Davenport, 1961; Brook, 1975). Δz is the vertical separation distance. A similar function is used to represent the co-spectrum when lateral (horizontal) separations, Δy, are considered.

Some measurements have suggested the alternative expression for $\rho(\Delta z,\, n)$ and $\rho(\Delta y,\, n)$ of Equation (3.36) (Bowen *et al.*, 1983):

$$\rho(\Delta r,\, n) = \exp\left[-\left(\frac{6\, n \cdot \Delta r}{\bar{U}}\right)\right]\exp\left[-\left(\frac{5 \cdot 5\, n \cdot (\Delta r)^2}{\bar{U} \cdot \bar{z}}\right)\right] \tag{3.36}$$

where Δr is either Δy or Δz, and $\bar{z}$ is the mean height of the measurements.

As for Equation (3.34), Equations (3.35) and (3.36) do not allow negative values – a theoretical problem – but of little practical significance. A more important disadvantage is that it implies full correlation at very low frequencies, no matter how large the separation distance, Δz, or Δy. Since the equations usually only need to be evaluated at the high frequencies corresponding to resonant frequencies, this is also not a great disadvantage.

More mathematically acceptable (but more complex) expressions for the normalized co-spectrum and coherence are available (e.g. Deaves and Harris, 1978).

3.3.7 Turbulence in a downdraft

The 'rear-flank' thunderstorm downdraft recorded by several towers near Lubbock, Texas on June 4, 2002, gave a unique opportunity to study the fluctuating wind characteristics, near the ground, in a severe event of this type (Orwig and Schroeder, 2007; Holmes *et al.*, 2008).

An individual time history from one tower during this event is shown in Figure 3.12. By applying a simple moving-average filter, a smoothed time history that shows the main features of the event can be extracted. This is shown in Figure 3.12, in which a 40-second moving average has been applied; this record can be called a 'running mean'. Subtracting the 'filtered' history from the original 'unfiltered' history results in a residual time history that is more or less random in nature, and can be described as 'turbulence' (Figure 3.12c). This is a non-stationary time history, and the conventional 'turbulence intensity' (Section 3.3.1), as defined for stationary synoptic winds, cannot be used here in the same way. However, Figure 3.12a shows that the level of random fluctuation varies with the running mean (Figure 3.12b), with an approximate 'intensity' of 10%. This is somewhat lower than the level obtained in stationary boundary-layer winds at this height in open country (for example, Table 3.3 gives a value of 18%), but is similar to that obtained in the Andrews AFB downburst (Figure 1.9).

Analysis of the turbulence from the event shown in Figure 3.12 suggests that the turbulence intensity with respect to the running mean was lower than that for similar terrain for synoptic boundary-layer winds. The high-frequency spectral density was similar to that in boundary-layer winds (i.e. Equation 3.30). However, correlations of the fluctuating components, with lateral separation, were considerably higher than those in boundary-layer winds, as a result of the large underlying 'running mean'. This leads to high values of 'span reduction factor' for transmission lines (see Section 13.2.3) and provides a convincing reason for the many failures of transmission line structures in thunderstorm downdraft events (Holmes *et al.*, 2008).

3.4 MODIFICATION OF WIND FLOW BY TOPOGRAPHY

Mean and gust wind speeds can be increased considerably by natural and man-made topography in the form of escarpments, embankments, ridges, cliffs and hills. These effects were the subject of much research from the

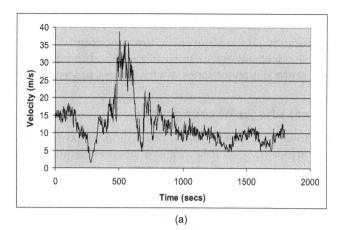

(a)

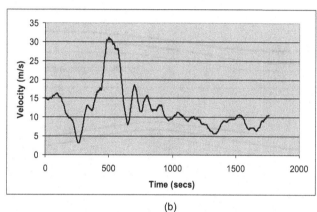

(b)

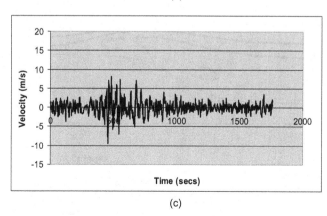

(c)

Figure 3.12 Time histories from a rear-flank downdraft, June 4, 2002, Lubbock, Texas: (a) velocities as recorded (unfiltered record), (b) time history filtered with 40-second moving average filter and (c) residual 'turbulence' obtained by subtraction.

1970s to the 1990s, with the incentive of the desire to exploit wind power, and to optimize the siting of wind turbines. This work greatly improved the prediction of mean wind speeds over shallow topography. Less well defined are the speed-up effects on turbulence and gust wind speeds, and the effects of steep topography – often of interest with respect to structural design.

3.4.1 General effects of topography

Figure 3.13 shows the general features of boundary-layer wind flow over a shallow escarpment, a shallow ridge, a steep escarpment, and a steep ridge.

As the wind approaches a shallow feature, its speed first reduces slightly as it encounters the beginning of the slope upwards. It then gradually increases in speed as it flows up the slope towards the crest. The maximum

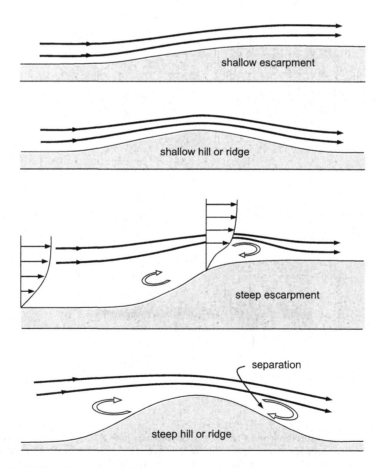

Figure 3.13 Wind flow over shallow and steep topography.

speed-up occurs at the crest, or slightly upwind of it. Beyond the crest, the flow speed gradually reduces to a value close to the value that occurs well upwind of the topographic feature; the adjustment is somewhat faster for a feature with a downwind slope, such as a ridge, than for an escarpment with a plateau downwind of the crest.

On steeper features, flow 'separation' (see also Section 4.1) may occur, as the flow is not able to overcome the increasing pressure gradients in the along-wind direction. Separations may occur at the start of the upwind slope, immediately downwind of the crest, and on the downwind slope for a ridge.

For steeper slopes (greater than about 0.3), the upwind separation 'bubble' presents an 'effective slope' of approximately constant value, independent of the actual slope underneath. This is often used in codes and standards to specify an upper limit to the speed-up effects of an escarpment or ridge.

The speed-up effects are greatest near the surface, and reduce with height above the ground. This can have the effect of producing mean velocity profiles near the crest of a topographic feature that are nearly constant, or have a peak (see Figure 3.13).

The above discussion relates to topographic features, which are two-dimensional in nature, that is they extend for an infinite distance normal to the wind direction. This may be a sufficient approximation for many long ridges and escarpments. Three-dimensional effects occur when air flow can occur around the ends of a hill, or through gaps or passes. These alternative air paths reduce the air speeds over the top of the feature, and generally reduce the speed-up effects. For structural design purposes, it is often convenient, and usually conservative, to ignore the three-dimensional effects, and to calculate wind loads only for the speed-up effects of the upwind and downwind slopes parallel to the wind direction of interest.

3.4.2 Topographic multipliers

The definition of topographic multiplier used in this book is as follows:

$$\frac{\text{Topographic}}{\text{Multiplier}} = \frac{\text{Wind speed at height, } z, \text{ above the feature}}{\text{Wind speed at height, } z, \text{ above the flat ground upwind}} \quad (3.37)$$

This definition applies to mean, peak gust and standard deviation wind speeds, and these will be denoted by $\bar{M}_t$, $\hat{M}_t$ and M'_t, respectively.

Topographic multipliers measured in full scale, or in wind tunnels, or calculated by computer programs, can be either greater or less than one. However, in the cases of most interest for structural design, we are concerned with speed-up effects for which the topographic multiplier for mean or gust wind speeds will exceed unity.

3.4.3 Shallow hills

An analysis by Jackson and Hunt (1975) of the mean boundary-layer wind flow over a shallow hill produced the following form for the mean topographic multiplier:

$$\bar{M}_t = 1 + k_t \cdot s\phi \tag{3.38}$$

where,

> ϕ is the upwind slope of the hill,
> k_t is a constant for a given shape of topography,
> s is a position factor.

The Jackson and Hunt linear theory, for hills of low slope, and subsequent refinements by others, was largely supported by full-scale measurements, particularly those from the Askervein Hill, a 116-m high hill on an island in the Outer Hebrides off the coast of Scotland, where the effects on boundary-layer winds, including turbulence, were monitored by multi-national groups in 1982 and 1983. During these tests, more than 50 towers were installed and instrumented for wind measurements (Walmsley and Taylor, 1996). These experiments were carried out primarily to support developments in wind energy, but the full-scale results from Askervein and subsequent wind-tunnel and computational studies also provide much useful data for wind-loading applications.

Equation (3.38) has been used in various forms for specifying topographic effects in several codes and standards. It indicates that the 'fractional speed-up', equal to $(\bar{M}_t - 1)$, is directly proportional to the upwind slope, ϕ. The latter is defined as $H/2L_u$, where H is the height of the crest above the level ground upwind, and L_u is the horizontal distance from the crest where the ground elevation drops to $H/2$.

Taylor and Lee (1984), based on the Jackson-Hunt theory and full-scale data from hills such as Askervein, proposed the following values of the constant, k_t, for various types of topography:

- 4.0 for two-dimensional ridges
- 1.6 for two-dimensional escarpments
- 3.2 for three-dimensional (axisymmetric) hills

The position factor, s, is 1.0 close to the crest of the feature, and falls upwind and downwind of the crest. There is a reduction of s with height, z, above the local ground level, which can be approximated by an exponential decay function (Equation 3.39):

$$s = \exp\left(\frac{-A \cdot z}{L_u}\right) \tag{3.39}$$

where A is 3 for two-dimensional ridges, 2.5 for two-dimensional escarpments, and 4 for three-dimensional hills (Taylor and Lee, 1984).

To a first approximation, the longitudinal turbulence component, σ_u, does not change over a shallow hill or escarpment near the ground surface (Walmsley and Taylor, 1996). This results in Equation (3.40) for the *gust* topographic multiplier, $\hat{M}_t$:

$$\hat{M}_t = 1 + k_t'\, s\phi \tag{3.40}$$

where k_t' is a constant for the gust multiplier, related to k_t by Equation (3.41):

$$k_t' = \frac{k_t}{1 + g\left(\dfrac{\sigma_u}{\bar{U}}\right)} \tag{3.41}$$

$\left(\dfrac{\sigma_u}{\bar{U}}\right)$ is the longitudinal turbulence intensity (over flat level ground) defined in Section 3.3.1, and g is the peak factor (Section 3.3.3).

Equations (3.38)–(3.41) show that the gust topographic multiplier is lower than the mean topographic multiplier for the same type of topography and height above the ground.

There is a slight dependence of topographic multipliers on the ratio (H/z_o), where z_o is the surface roughness length (Section 3.2.1). (H/z_o) is a form of the Jensen Number (Section 4.4.5).

3.4.4 Steep hills, cliffs and escarpments

Once the effective upwind slope of a hill or escarpment reaches a value of about 0.3 (about 17°), separations occur on the upwind face (Figure 3.13) and the simple formulae given in Section 3.4.3 become less accurate.

For slopes between 0.3 and 1 (17° and 45°), the separation bubble on the upwind slope presents an effective slope to the wind which is relatively constant, as discussed in Section 3.4.1. The topographic multipliers, at or near the crest, are therefore also fairly constant with upwind slope in this range. Thus, for this range of slopes, Equations (3.38) and (3.40) can be applied with ϕ replaced by an effective slope ϕ', equal to about 0.3 (Figure 3.14).

For slopes greater than about 1, for example steep cliffs, the flow stream lines near ground level at the crest originate from the upwind flow at levels near cliff height above the upwind ground level, rather than near ground level upwind (Figure 3.15). The concept of the topographic multiplier as defined by Equation (3.37) is less appropriate in such cases. Some of the apparent speed-up is caused by the upstream boundary layer profile rather than a perturbation produced by the hill or cliff.

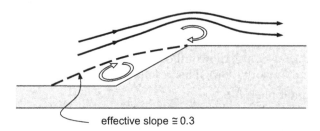

effective slope ≅ 0.3

Figure 3.14 Effective upwind slope for steep escarpments.

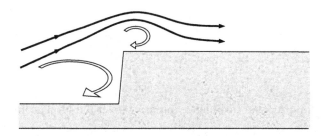

Figure 3.15 Wind flow over a steep cliff.

An additional complication for steep features is that separations can occur at or downwind of the crest (see Figure 3.13). Separated flow was found within the first 50 m height above the crest of a 480 m high feature, with an upwind slope of 0.48 (average angle of 26°), in both full scale and 1/1,000 scale wind-tunnel measurements (Glanville and Kwok, 1997). This has the effect of decreasing the mean velocity and increasing the turbulence intensity, as shown in Figure 3.16.

3.4.5 Valleys and gorges

For wind normal to a valley or gorge, there can be some shielding by the valley sides and a reduction in mean wind speed at the base of the valley. For many valleys, there would be a separated flow within the valley, and hence an increase in turbulence. Hence, there may be less reduction in peak gust wind speeds.

Taylor and Lee (1984) suggested treating a valley as a ridge with negative height, H, so that the value of k_t of 4.0 given in Section 3.4.3 would be applicable. However, there is little available experimental evidence to justify this, and wind codes and standards generally do not allow a reduction in wind speeds due to topography.

Taylor and Lee (1984) also gave estimates of mean wind speeds in rolling topography consisting of a sequence of hills and valleys. These are shown in Equations (3.42) and (3.43).

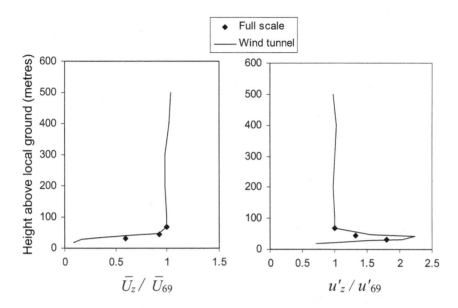

Figure 3.16 Mean velocity profile and root-mean-square longitudinal turbulence velocity near the crest of a steep escarpment (H=480 m, upwind slope=0.48).

For wind normal to two-dimensional hills and valleys:

$$\frac{U_H - U_V}{\frac{1}{2}(U_H + U_H)} = 3.1 \frac{H}{L} \tag{3.42}$$

For wind normal to three-dimensional 'rolling' hills:

$$\frac{U_H - U_V}{\frac{1}{2}(U_H + U_H)} = 2.2 \frac{H}{L} \tag{3.43}$$

In Equations (3.42) and (3.43), H is the vertical height of the hilltops above the valley floors, and L is one-quarter of the distance between ridge crests, or hilltops. U_H and U_V are the mean wind speeds on the hill or ridge tops and on the valley bottoms, respectively. If the average wind speed, $\frac{1}{2}(U_H + U_V)$, can be estimated, then these equations can be used to estimate both U_H and U_V separately. However, caution should be used for steep-sided valleys in which a separated flow may occur. Also, the applicability of these equations to peak gust wind speeds is doubtful.

For wind directions nearly parallel to valleys, channelling or funnelling with re-direction of the wind direction along the valley can occur, leading to increases in wind speeds of up to 20%.

3.4.6 Case studies

From the 1970s to the 1990s, several well-documented case studies of speed-up effects on wind flow over hills and escarpments of varying steepness were made in full-scale boundary-layer winds. The use of full-scale data avoids the issue of Reynolds Number scaling (Section 4.2.4), that often arises when wind-tunnel tests of topography, at small scale, are used. Table 3.5 shows the effective slopes and maximum measured topographic multipliers, as defined by Equation (3.37), for several of these cases, including Askervein (Section 3.4.3), and the escarpment of Figure 3.16. The values of maximum mean speed-up, i.e. $\bar{M}_t - 1$, in Table 3.5, for the shallower slopes, show general agreement with the simple estimates of Taylor and Lee (1984) discussed in Section 3.4.3, and it is clear that even shallow slopes can produce significant and measurable increases in mean wind speeds. A doubling of mean wind speeds for the steeper cases corresponds to a *quadrupling* of mean wind pressures and forces, emphasizing the importance of accounting for the effects of topography in the design of structures.

3.4.7 Effects of topography on tropical cyclones and thunderstorm winds

The effects of topographic features on wind near the ground in tropical cyclones and thunderstorm downbursts are much less clearly understood

Table 3.5 Some topographic multipliers measured in full scale

Location	Type	Author(s)	Effective upwind slope	Maximum $\bar{M}_t$
Riso (Denmark)	escarpment	Jensen and Peterson (1978)	0.02	1.07
Bungendore (Australia)	2-d ridge	Bradley (1983)	0.05	1.20 (neutral stability)
Pouzauges (France)	3-d hill	Sacré (1979)	0.06	1.15
Kettles Hill (Canada) 245°	3-d hill	Teunissen (1983)	0.12	1.40
Kettles Hill 220°	3-d hill	Teunissen (1983)	0.16	1.64
Blashaval (Scotland, U.K.)	3-d hill	Mason and King (1985)	0.22	1.70
Askervein (Scotland, U.K.)	3-d hill	Mickle et al. (1988); Salmon et al. (1988)	0.26	1.87
Black Mountain (Australia)	3-d hill	Bradley (1980)	0.31	2.11
Mount Dandenong (Australia)	escarpment	Holmes et al. (1997)	0.48	1.97
Rakaia River (New Zealand)	Escarpment (cliff)	Bowen and Lindley (1974)	→ ∞	2.10

than those in the well-developed boundary layers of large-scale synoptic systems.

Tropical cyclones are large storms with similar boundary layers to extra-tropical depressions on their outer edges. Near the region of the strongest winds, they appear to have much lower boundary-layer heights – of the order of 300 m (see Section 3.2.6). Topographic features greater than this height would therefore be expected to interact with the structure of the storm itself.

Thunderstorm downdrafts have 'boundary-layers' with peaks in the velocity profiles at 50–100 m. They also do not have fully developed boundary-layer velocity profiles. There have been some basic studies using wind-tunnel jets impinging on a flat board (Letchford and Illidge, 1999; Wood *et al.*, 1999) to indicate considerably lower topographic multipliers compared with developed thick boundary-layer flows. However, the effect of forward motion of the storm is uncertain.

3.5 CHANGE OF TERRAIN

When strong winds in a fully developed boundary layer encounter a change of surface roughness, for example winds from open country flowing over the suburbs of a town or city, a process of adjustment in the turbulent boundary-layer flow properties develops. The adjustment starts at ground level and gradually moves upwards. The result is the development of an internal boundary layer over the new terrain as shown in Figure 3.17.

Deaves (1981), from numerical studies, developed the following relationships for the horizontal position of the inner boundary layer as a function of its height, z:

For flow from smooth terrain (roughness length z_{o1}) to rougher terrain (z_{o2}) with $z_{o2} > z_{o1}$:

$$x_i(z) = z_{o2}\left(\frac{z}{0.36z_{o2}}\right)^{4/3} \tag{3.44}$$

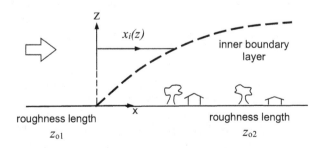

Figure 3.17 Internal boundary-layer development at a change of terrain roughness.

For flow from rough terrain (roughness length z_{o1}) to smoother terrain (z_{o2}) with $z_{o1} > z_{o2}$:

$$x_i(z) = 14z\left(\frac{z_{o1}}{z_{o2}}\right)^{1/2} \tag{3.45}$$

Setting z_{o2} equal to 0.2 m, approximately the value for suburban terrain with low rise buildings 3–5 m high (see Table 3.1), and z equal to 10 m, Equation (3.44) gives a value for x_i (10) of 144 m. Beyond this distance, the shape of the mean velocity profile below 10 m has the characteristics of the new terrain. However, the *magnitude* of the mean velocity continues to reduce for many kilometres, until the complete atmospheric boundary layer has fully adjusted to the rougher terrain.

Melbourne (1992) found the *gust* wind speed at a height of 10 m, adjusts to a new terrain approximately exponentially with a distance constant of about 2,000 m. Thus, the peak gust at a distance x (in metres) into the new terrain (2) can be represented by Equation (3.46):

$$\hat{U}_{2,x} = \hat{U}_1 + \left(\hat{U}_2 - \hat{U}_1\right)\left[1 - \exp\left(\frac{-x}{2,000}\right)\right] \tag{3.46}$$

where $\hat{U}_1$ and $\hat{U}_2$ are the asymptotic gust velocities over fully developed terrain of types 1 (upstream) and 2 (downstream).

Equation (3.46) was found to fit data from a wind tunnel for flow from rough to smooth, as well as smooth to rough, and when there were several changes of roughness.

3.6 WEAKENING OF A TROPICAL CYCLONE AFTER A COAST CROSSING

As the high convection regions of a tropical cyclone, including a hurricane or typhoon (see Section 1.3.2), cross a coastline from a warm ocean, the storm loses its source of energy and starts to weaken. There is also a reduction due to surface friction for winds blowing over land. These effects result in maximum wind speeds falling progressively with distance from the coastline.

Kaplan and DeMaria (1995, 2001) developed a simple empirical model for predicting the decay of tropical-cyclone winds after landfall. The model is based on a least-squares fit to the maximum sustained (1-minute) surface wind estimates made by the National Hurricane Center of the United States to land-falling tropical storm and hurricanes. The equation for maximum sustained wind as a function of time takes the form of Equation (3.47):

$$U(t) = U_b + (RU_o - U_b)e^{-\alpha t} \tag{3.47}$$

where,

U_b is a 'background' wind speed,
U_o is the maximum wind speed just before landfall,
R is a reduction factor to account for the immediate effect of surface roughness immediately on landfall,
α is a decay parameter.

For storms making landfall closer to the Equator than 37° latitude, Kaplan and DeMaria established values for U_b, α and R of 26.7 knots, 0.095 hour^{-1} and 0.9, respectively. For storms land-falling on the Atlantic coast north of 37°N, the values obtained were: U_b=29.6 knots, α=0.187 hour^{-1} and R=0.9.

Equation (3.47) can be modified in terms of decay with *distance* instead of time:

$$U(x) = U_b + (RU_o - U_b)e^{-\beta x}$$

$$U(w) = U_b + (RU_o - U_b)e^{-(\beta/\cos\theta)w} \tag{3.48}$$

where,

β is a decay constant based on the distance inland x
w is the shortest distance from the smoothed coastline equal to $x \cos\theta$
θ is the angle of the cyclone track from the normal to the smoothed coastline
β can be estimated from the decay constant of Kaplan and DeMaria as follows:

$$\beta = \frac{\alpha}{c_{av}} \tag{3.49}$$

where c_{av} is an average translation speed of the storm.

An average value of the angle to the normal can be calculated as follows:

$$(\cos\theta)_{av} = \left(\frac{1}{\pi}\right)\int_{-\pi/2}^{+\pi/2} \cos\theta \, d\theta = \frac{2}{\pi} \tag{3.50}$$

Hence, substituting in Equation (3.48), we get:

$$U(w) = U_b + (RU_o - U_b)e^{-(\pi\alpha w/2c_{av})} \tag{3.51}$$

3.7 OTHER SOURCES

A well-documented and detailed description of the atmospheric boundary layers in temperate synoptic systems, for wind-loading purposes, is given in a series of data items published by the *Engineering Sciences Data Unit* (ESDU, 1974–1999). These include the effects of topographic and terrain changes. Teunissen (1970) and Counihan (1975) also provided a comprehensive review of adiabatic (i.e. neutrally stable) atmospheric boundary layer data. Cook (1985) described, for the designer, a structure of the atmospheric boundary layer, which is consistent with the ESDU models. Assessment of roughness parameters for terrains of various types (Wieringa, 1993), and the effects of terrain changes and topography on boundary-layer winds have been studied and reviewed extensively in journals such as *Boundary-Layer Meteorology* (Springer) and the *Quarterly Journal of the Royal Meteorological Society* (UK).

The references in the previous paragraph are recommended for descriptions of strong-wind structure for *large-scale, synoptic, neutrally stable, boundary-layer* winds in temperate zones. However, as discussed in this chapter, the structure of extreme wind in tropical and semi-tropical locations, such as those produced by *thunderstorms* and *tropical cyclones*, is different, and such models should be used with caution in these regions. Mason (2017) reviewed the available information on the structure of localized strong wind events, such as thunderstorm downdrafts and tornados, with a view to incorporating this into design codes and standards. However, full-scale data on these events remain sparse, and presenting it in a form suitable for designers is still a 'work in progress'.

3.8 SUMMARY

In this chapter, the structure of strong winds near the earth's surface, relevant to wind loads on structures, has been described. The main focus has been the atmospheric boundary layer in large-scale synoptic winds over land. The mean wind speed profile and some aspects of the turbulence structure have been described. Aspects of wind over the oceans, and in tropical cyclones, thunderstorm downbursts and tornadoes, have also been discussed.

The modifying effects of topographic features, including hills, ridges, escarpments and valleys, and of changes in terrain, and the weakening of tropical cyclones after landfall, have also been covered.

REFERENCES

Amano, T., Fukushima, H., Ohkuma, T., Kawaguchi, A. and Goto, S. (1999) The observation of typhoon winds in Okinawa by Doppler sodar. *Journal of Wind Engineering and Industrial Aerodynamics*, 83: 11–20.

Amorocho, J. and DeVries, J.J. (1980) A new evaluation of the wind stress coefficient over water surfaces. *Journal of Geophysical Research*, 85: 433–422.

Ashcroft, J. (1994) The relationship between the gust ratio, terrain roughness, gust duration and the hourly mean wind speed. *Journal of Wind Engineering and Industrial Aerodynamics*, 53: 331–355.

Beljaars, A.C.M. (1987) The measurement of gustiness at routine wind stations – a review. Instruments and Observing Methods Report WMO No. 31, World Meteorological Organization, Geneva, Switzerland.

Bendat, J.S. and Piersol, A.G. (1999) *Random data: analysis and measurement procedures*, 3rd Edition. John Wiley & Sons, New York.

Black, P.G. (1992) Evolution of maximum wind estimates in typhoons. *ICSU/WMO Symposium on Tropical Cyclone Disasters*, Beijing, China, 12–18 October.

Blackadar, A.K. and Tennekes, H. (1968) Asymptotic similarity in neutral barotropic boundary layers. *Journal of the Atmospheric Sciences*, 25: 1015–1020.

Bowen, A.J. and Lindley, D. (1974) Measurement of the mean flow over various escarpment shapes. *5th Australasian Conference on Hydraulics and Fluid Mechanics*, Christchurch, New Zealand, 9–13 December.

Bowen, A.J., Flay, R.G.J. and Panofsky, H.A. (1983) Vertical coherence and phase delay between wind components in strong winds below 20m. *Boundary-Layer Meteorology*, 26: 313–324.

Bradley, E.F. (1980) An experimental study of the profiles of wind speed, shearing stress and turbulence at the crest of a large hill. *Quarterly Journal of the Royal Meteorological Society*, 106: 101–123.

Bradley, E.F. (1983) The influence of thermal stability and angle of incidence on the acceleration of wind up a slope. *Journal of Wind Engineering and Industrial Aerodynamics*, 15: 231–241.

Brook, R.R. (1975) A note on vertical coherence of wind measured in an urban boundary layer. *Boundary-Layer Meteorology*, 9: 11–20.

Busch, N. and Panofsky, H. (1968) Recent spectra of atmospheric turbulence. *Quarterly Journal of the Royal Meteorological Society*, 94: 132–148.

Charnock, H. (1955) Wind stress on a water surface. *Quarterly Journal of the Royal Meteorological Society*, 81: 639–640.

Choi, E.C.C. (1978) Characteristics of typhoons over the South China Sea. *Journal of Industrial Aerodynamics*, 3: 353–365.

Cook, N.J. (1985) *The designer's guide to wind loading of building structures. Part 1 Background, damage survey, wind data and structural classification.* Building Research Establishment and Butterworths, London.

Counihan, J. (1975) Adiabatic atmospheric boundary layers: a review and analysis of data from the period 1880–1972. *Atmospheric Environment*, 9: 871–905.

Csanady, G.T. (1967) On the resistance law of a turbulent Ekman layer. *Journal of the Atmospheric Sciences*, 24: 467–471.

Davenport, A.G. (1961) The spectrum of horizontal gustiness in high winds. *Quarterly Journal of the Royal Meteorological Society*, 87: 194–211.

Davenport, A.G. (1963) The relationship of wind structure to wind loading. *Proceedings, International Conference on Wind Effects on Buildings and Structures*, Teddington UK, 26–28 June, pp. 53–102.

Deacon, E.L. (1955) Gust variation with height up to 150 metres. *Quarterly Journal of the Royal Meteorological Society*, 81: 562–573.

Deacon, E.L. (1965) Wind gust speed: averaging time relationship. *Australian Meteorological Magazine*, No. 51, 11–14.

Deaves, D.M. (1981) Computations of wind flow over changes in surface roughness. *Journal of Wind Engineering and Industrial Aerodynamics*, 7: 65–94.

Deaves, D.M. and Harris, R.I. (1978) A mathematical model of the structure of strong winds. Construction Industry Research and Information Association (UK), Report 76.

Deaves, D.M. and Harris, R.I. (1982) A note on the use of asymptotic similarity theory in neutral atmospheric boundary layers. *Atmospheric Environment*, 16: 1889–1893.

Durst, C.S. (1960) Wind speeds over short periods of time. *Meteorological Magazine (UK)*, 89, 181–186.

ESDU. (1974–1999) Wind speeds and turbulence. Engineering Sciences Data Unit (ESDU International), Wind Engineering series Volumes 1a and 1b. ESDU Data Items 74030, 82026, 83045, 84011, 84031, 85020, 86010, 86035, 91043, 92032.

Franklin, J.L., Black, M.L., Valde, K. (2003) GPS dropwindsonde wind profiles in hurricanes and their operational implications. *Weather and Forecasting*, 18: 32–44.

Fujita, T.T. (1985) The downburst. Report on Projects NIMROD and JAWS, Published by the author at the University of Chicago.

Garratt, J.R. (1977) Review of drag coefficients over oceans and continents. *Monthly Weather Review*, 105: 915–929.

Gunter, W.S. and Schroeder, J.L. (2013). High-resolution full-scale measurements of thunderstorm outflow winds. *12th Americas Conference on Wind Engineering*, Seattle, Washington, DC, USA, 16–20 June.

Glanville, M.J. and Kwok, K.C.S. (1997) Measurements of topographic multipliers and flow separation from a steep escarpment. Part II. Model-scale measurements. *Journal of Wind Engineering and Industrial Aerodynamics*, 69–71: 893–902.

Harper, B.A., Kepert, J.D. and Ginger, J.D. (2010). Guidelines for converting between various wind averaging periods in tropical cyclone conditions. World Meteorological Organization Report, WMO/TD No.1555, August 2010.

Harris, R.I. (1968) On the spectrum and auto-correlation function of gustiness in high winds. Electrical Research Association. Report 5273.

Hock, T.F. and Franklin, J.L. (1999) The NCAR GPS dropwindsonde. *Bulletin, American Meteorological Society*, 80: 407–420.

Holmes, J.D. (1973) Wind pressure fluctuations on a large building. *Ph.D. thesis*, Monash University, Australia.

Holmes, J.D. (2017) Roughness lengths and turbulence intensities for wind over water. *9th Asia-Pacific Conference on Wind Engineering*, Auckland, New Zealand, 3–7 December.

Holmes, J.D. and Ginger, J.D. (2012) The gust wind speed duration in AS/NZS 1170.2. *Australian Journal of Structural Engineering*, 13: 207–217.

Holmes, J.D., Banks, R.W. and Paevere, P. (1997) Measurements of topographic multipliers and flow separation from a steep escarpment. Part I. Full-scale measurements. *Journal of Wind Engineering and Industrial Aerodynamics*, 69–71: 885–892.

Holmes, J.D., Hangan, H.M., Schroeder, J.L., Letchford, C.W. and Orwig, K.D. (2008) A forensic study of the Lubbock-Reese downdraft of 2002. *Wind and Structures*, 11: 137–152.

Holmes, J.D., Allsop, A. and Ginger, J.D. (2014) Gust durations, gust factors and gust response factors in wind codes and standards. *Wind and Structures*, 19: 339–352.

Ishizaki, H. (1983) Wind profiles, turbulence intensities and gust factors for design in typhoon-prone regions. *Journal of Wind Engineering and Industrial Aerodynamics*, 13: 55–66.

Jackson, P.S. and Hunt, J.C.R. (1975) Turbulent flow over a low hill. *Quarterly Journal of the Royal Meteorological Society*, 101: 929–955.

Jensen, N.O. (1978) Change of surface roughness and the planetary boundary layer. *Quarterly Journal of the Royal Meteorological Society*, 104: 351–356.

Jensen, N.O. and Peterson, E.W. (1978) On the escarpment wind profile. *Quarterly Journal of the Royal Meteorological Society*, 104: 719–728.

Kaplan, J. and DeMaria, M. (1995) A simple empirical model for predicting the decay of tropical cyclone winds after landfall. *Journal of Applied Meteorology*, 34: 2499–2512.

Kaplan, J. and DeMaria, M. (2001) On the decay of tropical cyclone winds after landfall in the New England area. *Journal of Applied Meteorology*, 40: 280–286.

Krayer, W.R. and Marshall, R.D. (1992) Gust factors applied to hurricane winds. *Bulletin, American Meteorological Society*, 73: 613–17.

Letchford, C.W. and Illidge, G. (1999) Turbulence and topographic effects in simulated thunderstorm downdrafts by wind tunnel jet. *Proceedings, 10th International Conference on Wind Engineering*, Copenhagen, Denmark, 21–24 June 1999, Balkema, Rotterdam.

Lettau, H.H. (1950) A re-examination of the "Leipzig Wind Profile" considering some relations between wind and turbulence in the frictional layer. *Tellus*, 2: 125–129.

Lettau, H.H. (1959) Wind profile, surface stress and geostrophic drag coefficients in the atmospheric boundary layer. *Advances in Geophysics*, 6: 241–257.

Lewellen, W.S. (1976) Theoretical models of the tornado vortex. *Symposium on Tornadoes: Assessment of Knowledge and Implications for Man*, Texas Tech University, Lubbock, Texas, USA, 22–24 June 1976, pp. 107–143.

Mason, M.S. (2017) Towards codification of localized windstorms: progress and challenges. *9th Asia-Pacific Conference on Wind Engineering*, Auckland, New Zealand, 3–7 December.

Mason, P.J. and King, J.C. (1985) Measurements and predictions of flow and turbulence over an isolated hill of moderate slope. *Quarterly Journal of the Royal Meteorological Society*, 111: 617–640.

Melbourne, W.H. (1992) Unpublished course notes, Monash University, Australia.

Mickle, R.E., Cook, N.J., Hoff, A.M., Jensen, N.O., Salmon, J.R., Taylor, P.A., Tetzlaff, G. and Teunissen, H.W. (1988) The Askervein Hill Project: Vertical profiles of wind and turbulence. *Boundary-Layer Meteorology*, 43: 143–169.

Miller, C.A., Holmes, J.D., Henderson, D.J., Ginger, J.D. and Morrison, M. (2013). The response of the Dines anemometer to gusts and comparisons with cup anemometers. *Journal of Atmospheric and Oceanic Technology*, 30: 1320–1336.

Orwig, K.D. and Schroeder, J.L. (2007) Near-surface wind characteristics of extreme thunderstorm outflows. *Journal of Wind Engineering and Industrial Aerodynamics*, 95: 565–584.

Oseguera, R.M. and Bowles, R.L. (1988) A simple analytic 3-dimensional downburst model based on boundary layer stagnation flow. *NASA Technical Memorandum 100632*, National Aeronautics and Space Administration, Washington, DC, USA.

Powell, M.D., Vickery, P.J. and Reinhold, T.A. (2003) Reduced drag coefficients for high wind speeds in tropical cyclones. *Nature*, 422: 279–283.

Sacré, C. (1979) An experimental study of the airflow over a hill in the atmospheric boundary layer. *Boundary-Layer Meteorology*, 17: 381–401.

Salmon, J.R., Bowen, A.J., Hoff, A.M., Johnson, R., Mickle, R.E., Taylor, P.A., Tetzlaff, G. and Walmsley, J.L. (1988) The Askervein hill project: mean wind variations at fixed heights above the ground. *Boundary-Layer Meteorology*, 43: 247–271.

Standards Australia. (2011) *Structural design actions. Part 2: Wind actions*. Standards Australia, Sydney. Australian/New Zealand Standard AS/NZS1170.2:2011 (amended 2012 to 2016).

Swinbank, W.C. (1974) The geostrophic drag coefficient. *Boundary-Layer Meteorology*, 7: 125–127.

Taylor, P.A. and Lee, R.J. (1984) Simple guidelines for estimating windspeed variation due to small scale topographic features. *Climatological Bulletin (Canada)*, 18: 3–32.

Teunissen, H.W. (1970) Characteristics of the mean wind and turbulence in the planetary boundary layer. Institute of Aerospace Sciences, Review No 32, University of Toronto, Canada.

Teunissen, H.W. (1980) Structure of mean winds and turbulence in the planetary boundary layer over rural terrain. *Boundary-Layer Meteorology*, 19: 187–221.

Teunissen, H.W. (1983) Wind-tunnel and full-scale comparisons of mean flow over an isolated low hill. *Journal of Wind Engineering and Industrial Aerodynamics*, 15: 271–286.

Vickery, P.J. and Skerlj, P.F. (2005) Hurricane gust factors revisited. *Journal of Structural Engineering*, 131: 825–832.

von Karman, T. (1948) Progress in the statistical theory of turbulence. *Proceedings of the National Academy of Sciences (US)*, 34: 530–539.

Walmsley, J.L. and Taylor, P.A. (1996) Boundary-layer flow over topography: impacts of the Askervein study. *Boundary-Layer Meteorology*, 78: 291–320.

Wieringa, J. (1973) Gust factors over open water and built-up country. *Boundary-Layer Meteorology*, 3: 424–441.

Wieringa, J. (1993) Representative roughness parameters for homogeneous terrain. *Boundary-Layer Meteorology*, 63: 323–363.

Wilson, K.J. (1979) Characteristics of the subcloud layer wind structure in tropical cyclones. *International Conference on Tropical Cyclones*, Perth, Western Australia, November 1979.

Wood, G.S., Kwok, K.C.S., Motteram, N., and Fletcher, D.F. (1999) Physical and numerical modelling of thunderstorm downbursts. *Proceedings, 10th International Conference on Wind Engineering*, Copenhagen, Denmark, 21–24 June 1999, Balkema, Rotterdam.

Chapter 4

Basic bluff-body aerodynamics

4.1 FLOW AROUND BLUFF BODIES

Structures of interest in this book can generally be classified as *bluff* bodies with respect to the air flow around them, in contrast to *streamlined* bodies, such as aircraft wings and yacht sails (when the boat is sailing across the wind). Figure 4.1 shows the flow patterns around an airfoil (at low angle of attack), and around a two-dimensional body of rectangular cross section. The flow patterns are shown for steady free-stream flow; turbulence in the approaching flow, which occurs in the atmospheric boundary layer, as discussed in Chapter 3, can modify the flow around a bluff body, as will be discussed later.

It can be seen in Figure 4.1 that the flow streamlines around the airfoil follow closely the contours of the body. The free-stream flow is separated from the surface of the airfoil only by a thin boundary layer, in which the tangential flow is brought to rest at the surface. The flow around the rectangular section (a typical bluff body) in Figure 4.1 is characterized by a 'separation' of the flow at the leading edge corners. The separated flow region is divided from the outer flow by a thin region of high shear and vorticity, a region known as a free shear layer, which is similar to the boundary layer on the airfoil, but not attached to any surface. These layers are unstable in a sheet form and will roll up towards the wake, to form concentrated vortices, which are subsequently shed downwind.

In the case of the bluff body with a long 'after-body' in Figure 4.1, the separated shear layer 're-attaches' on to the surface. However, the shear layer is not fully stabilized and vortices may be formed on the surface, and subsequently roll along the surface.

4.2 PRESSURE AND FORCE COEFFICIENTS

4.2.1 Bernoulli's Equation

The region outside the boundary layers in the case of the airfoil and the outer region of the bluff-body flow are *inviscid* (zero viscosity) and *irrotational*

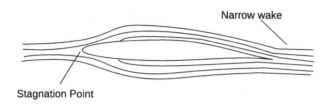

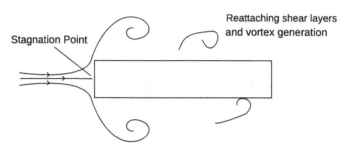

Figure 4.1 Flow around streamlined and bluff bodies.

(zero vorticity) flows, and the pressure, p, and velocity, U, in the fluid are related by *Bernoulli's Equation*:

$$p + \frac{1}{2}\rho_a U^2 = a \text{ constant} \tag{4.1}$$

Denoting the pressure and velocity in the region outside the influence of the body by p_o and U_0, we have:

$$p + \frac{1}{2}\rho_a U^2 = p_o + \frac{1}{2}\rho_a U_0^2$$

$$\text{Hence} \quad p - p_o = \frac{1}{2}\rho_a\left(U_0^2 - U^2\right)$$

The surface pressure on the body is usually expressed in the form a non-dimensional *pressure coefficient*:

$$C_p = \frac{p - p_o}{\frac{1}{2}\rho_a U_0^2} \tag{4.2}$$

In the region in which Bernoulli's Equation holds:

$$C_p = \frac{\frac{1}{2}\rho_a\left(U_0^2 - U^2\right)}{\frac{1}{2}\rho_a U_0^2} = 1 - \left(\frac{U}{U_0}\right)^2 \tag{4.3}$$

At the stagnation point, where U is zero, Equation (4.3) gives a pressure coefficient of one. This is the value measured by a *total* pressure or *pitot* tube pointing into a flow. The pressure $(1/2)\rho_a U_0^2$ is known as the *dynamic* pressure. Values of pressure coefficient near 1.0 also occur at the stagnation point on a circular cylinder, but the largest (mean) pressure coefficients on the windward faces of buildings are usually less than this theoretical value.

In the regions where the flow velocity is greater than U_0, the pressure coefficients are negative. Strictly, Bernoulli's Equation is not valid in the separated flow and wake regions, but reasonably good predictions of surface pressure coefficients can be obtained from Equation (4.3), by taking the velocity, U, as that just outside the shear layers and wake region.

4.2.2 Force coefficients

Force coefficients are defined in a similar non-dimensional way to pressure coefficients:

$$C_F = \frac{F}{\frac{1}{2}\rho_a U_0^2 A} \tag{4.4}$$

where F is the total aerodynamic force and A is a reference area (not necessarily the area over which the force acts). Often A is a projected frontal area.

In the case of long, or two-dimensional, bodies, a force coefficient per unit length is usually used:

$$C_f = \frac{f}{\frac{1}{2}\rho_a U_0^2 b} \tag{4.5}$$

where f is the aerodynamic force per unit length, and b is a reference length, usually the breadth of the structure normal to the wind.

Aerodynamic forces are conventionally resolved into two orthogonal directions. These may be parallel and perpendicular to the wind direction (or mean wind direction in the case of turbulent flow), in which case the axes are referred to as *wind axes*, or parallel and perpendicular to a direction related to the geometry of the body (*body axes*). These axes are shown in Figure 4.2.

Following the terminology of aeronautics, the terms 'lift' and 'drag' are commonly used in wind engineering for cross-wind and along-wind force components, respectively. Substituting 'L' and 'D' for 'F' in Equation (4.4) gives the definition of *lift* and *drag coefficients*.

The relationship between the forces and force coefficients resolved with respect to the two axes can be derived using trigonometry, in terms of the angle, α, between the sets of axes, as shown in Figure 4.3. α is called the *angle of attack* (or sometimes angle of incidence).

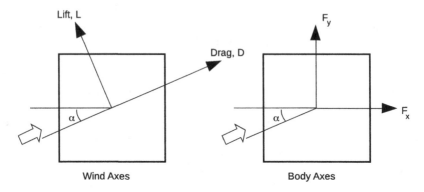

Figure 4.2 (a) Wind axes and (b) body axes.

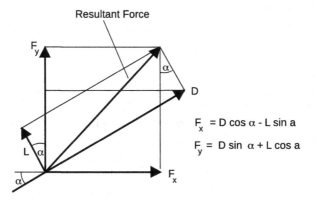

$$F_x = D \cos \alpha - L \sin a$$
$$F_y = D \sin \alpha + L \cos a$$

Figure 4.3 Relationship between resolved forces.

4.2.3 Functional dependence of pressure and force coefficients

Pressure and force coefficients are non-dimensional quantities, which are dependent on a number of variables related to the geometry of the body and to the upwind flow characteristics. These variables can be grouped together into non-dimensional groups, using processes of dimensional analysis, or by inspection.

Assuming that we have several bluff bodies of geometrically similar shape, which can be characterized by a single length dimension (for example, buildings with the same ratio of height, width and length, and with the same roof pitch, characterized by their height, h). Then the pressure coefficients for pressures at corresponding points on the surface of the body may be a function of a number of other non-dimensional groups: π_1, π_2, π_3, etc...

Thus, $C_p = f(\pi_1, \pi_2, \pi_3 \text{ etc}...)$ $\qquad$ (4.6)

Examples of relevant non-dimensional groups are as follows:

- h/z_o (Jensen Number)
  (where z_o is the roughness length of the ground surface, as discussed in Section 3.2.1)
- I_u, I_v, I_w the turbulence intensities in the approaching flow
- (ℓ_u/h), (ℓ_v/h), (ℓ_w/h) representing ratios of turbulence length scales in the approaching flow, to the characteristic body dimension
- (Ub/ν), Reynolds Number, where ν is the kinematic viscosity of air

Equation (4.6) is relevant to the practice of wind-tunnel model testing, in which geometrically scaled models are used to obtain pressure (or force) coefficients for application to full-scale prototype structures (see Section 7.4). The aim should be to ensure that all relevant non-dimensional numbers (π_1, π_2, π_3, etc.) should be equal in both model and full scale. This is difficult to achieve for all the relevant numbers, and methods have been devised for minimizing the errors resulting from this. Wind-tunnel testing techniques are discussed in Chapter 7.

4.2.4 Reynolds Number

Reynolds Number is the ratio of fluid inertia forces in the flow to viscous forces, and is an important parameter in all branches of fluid mechanics. In bluff-body flows, viscous forces are only important in the surface boundary layers and free shear layers (Section 4.1). The dependence of pressure coefficients on Reynolds Number is often overlooked for sharp-edged bluff bodies, such as most buildings and industrial structures. For these bodies, separation of flow occurs at sharp edges and corners, such as wall-roof junctions, over a very wide range of Reynolds Numbers. However, for bodies with curved surfaces such as circular cylinders or arched roofs, the separation points *are* dependent on Reynolds Number, and this parameter should be considered. Surface roughness has significant effects on flow around circular cylinders (see Section 4.5.1) and other bodies with curved surfaces. This may sometimes be used to advantage to modify the flow around these shapes on wind-tunnel models to approximate the flow around the full-scale body (see Section 7.4.4). The addition of turbulence to the flow also reduces the Reynolds Number dependence for bodies with curved surfaces.

In most references to Reynolds Number in this book, the breadth of the body, b (i.e. the diameter in the case of a circular cylinder), is used to form the Reynolds Number, denoted by Re_b. However, the average height of roughness on the body, k (not to be confused with ground roughness length, z_0), is also used as a length scale in Section 4.5.1 – forming the 'roughness Reynolds Number' Re_k, equal to Uk/ν.

4.3 FLAT PLATES AND WALLS

4.3.1 Flat plates and walls normal to the flow

The flat plate, with its plane normal to the air stream, is representative of a common situation for wind loads on structures. Examples are elevated hoardings and signboards, which are mounted so that their plane is vertical. Solar panels are another example, but, in this case, the plane is generally inclined to the vertical to maximize the collection of solar radiation. Free-standing walls are another example, but the fact that they are attached to the ground has a considerable effect on the flow and the resulting wind loading. In this section, some fundamental aspects of flow and drag forces on flat plates and walls are discussed.

For a flat plate or wall with its plane normal to the flow, the only aerodynamic force will be one parallel to the flow, i.e. a drag force. Then if p_W and p_L are the average pressures on the front (windward) and rear (leeward) faces, respectively, the drag force, D, will be given by:

$$D = (p_W - p_L)A$$

where A is the frontal area of the plate or wall.

Then dividing both sides by $(1/2)\rho_a U^2 A$, we have:

$$C_D = C_{p,W} - C_{p,L}$$
$$= C_{p,W} + (-C_{p,L}) \tag{4.7}$$

In practice, the windward wall pressure, p_W, and pressure coefficient, $C_{p,W}$, vary considerably with position on the front face. The leeward (or 'base') pressure, however, is nearly uniform over the whole rear face, as this region is totally exposed to the wake region, with relatively slow-moving air.

The mean drag coefficients for various plate and wall configurations are shown in Figure 4.4. The drag coefficient for a square plate in a smooth, uniform approach flow is about 1.1, slightly greater than the total pressure in the approach flow, averaged over the face of the plate. Approximately 60% of the drag is contributed by positive pressures (above static pressure) on the front face, and 40% by negative pressures (below static pressure) on the rear face (ESDU, 1970).

The effect of free-stream turbulence is to increase the drag on the normal plate slightly. The increase in drag is caused by a decrease in leeward, or base, pressure, rather than an increase in front face pressure. A hypothesis for this is that the free-stream turbulence causes an increase in the rate of entrainment of air into the separated shear layers. This leads to a reduced radius of curvature of the shear layers, and to a reduced base pressure (Bearman, 1971).

Figure 4.4 also shows the drag coefficient on a long flat plate with a theoretically infinite width into the paper – the 'two-dimensional' flat plate. The drag coefficient of 1.9 is higher than that for the square plate. The reason for the increase on the wide plates can be explained as follows. For a square plate, the flow is deflected around the plate equally around the four sides. The extended width provides a high-resistance flow path, thus forcing the flow to travel faster over the top edge, and under the bottom edge. This faster flow results in more entrainment from the wake into the shear layers, thus generating lower base, or leeward face, pressure and higher drag.

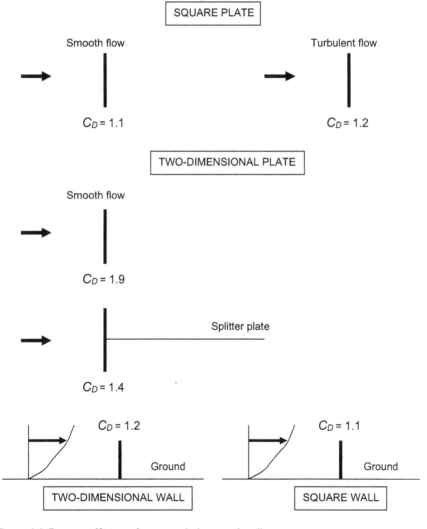

Figure 4.4 Drag coefficients for normal plates and walls.

Rectangular plates with intermediate values of width to height have intermediate values of drag coefficient. A formula given by ESDU (1970) for the drag coefficient on plates of height/breadth ratio in the range, $1/30 < h/b < 30$, in smooth uniform flow normal to the plate, is Equation (4.8):

$$C_D = 1.10 + 0.02[(h/b) + (b/h)] \qquad (4.8)$$

In the case of two-dimensional plate, strong vortices are shed into the wake alternately from top and bottom, in a similar way to the bluff-body flow shown in Figure 4.1. These contribute greatly to the increased entrainment into the wake of the two-dimensional plate. Suppression of these vortices by a splitter plate has the effect of reducing the drag coefficient to a lower value, as shown in Figure 4.4.

This suppression of vortex shedding is nearly complete when a flat plate is attached to a ground plane, and becomes a wall, as shown in the lower sketch in Figure 4.4. In this case, the approach flow will be of a boundary-layer form with a wind speed increasing with height as shown. The value of drag coefficient, with U taken as the mean wind speed at the top of the wall, $\bar{U}_h$, is very similar for the two-dimensional wall, and finite wall of square planform, i.e. a drag coefficient of about 1.2 for an infinitely long wall. The effect of finite length of wall is shown in Figure 4.5. Little change in the mean drag coefficient occurs, although a slightly lower value occurs for an aspect ratio (length/height) of about 5 (Letchford and Holmes, 1994).

The case of two thin normal plates in series, normal to the flow, as shown in Figure 4.6, is an interesting one. At zero spacing, the two plates act like a single plate with a combined drag coefficient (based on the frontal area of one plate) of about 1.1, for a square plate. For spacings in the range of 0–2h, the combined drag coefficient is actually *lower* than that for a single plate, reaching a value of 0.8 at a spacing of 1.5h, for two square plates. As the spacing increases, the combined drag coefficient then increases, so that, for very high spacings, the plates act like individual plates with no interference

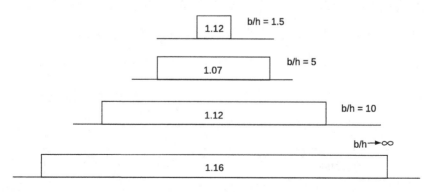

Figure 4.5 Mean drag coefficients on walls in boundary-layer flow.

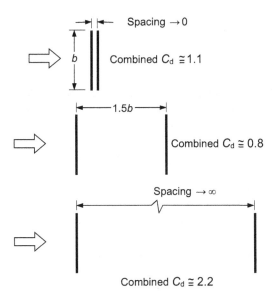

Figure 4.6 Drag coefficients for two square plates in series.

with each other, and a combined drag coefficient of 2.2. The mechanism that produces the reduced drag at the critical spacing of 1.5*b* has not been studied in detail, but clearly there is a large interference in the wake and in the vortex shedding, generated by the downstream plate.

The drag forces on two flat plates separated by small distances normal to the flow is also a relevant situation in wind engineering, with applications for clusters of lights or antennas together on a frame, for example. Experiments by Marchman and Werme (1982) found increases in drag of up to 15% when square, rectangular or circular plates were within half a width (or diameter) from each other.

If uniform porosity is introduced, the drag on a normal flat plate or wall reduces as some air is allowed to flow through the plate, and reduces the pressure difference between front and rear faces. The reduction in drag coefficient can be represented by the introduction of a porosity factor, K_p, which is dependent on the solidity of the plate, δ, being the ratio of the 'solid' area of the plate, to the total elevation area, as indicated in Equation (4.9):

$$C_{D,\delta} = C_D \cdot K_p \tag{4.9}$$

For two-dimensional plates (height to breadth approaching infinity), normal to the flow, with circular perforations, K_p is approximately equal to δ for values of solidity between 0.4 and 0.8 (Castro, 1971).

However, for plates and walls of finite aspect ratio, K_p is not linearly related to the solidity. An approximate expression for K_p, which fits the data quite well for plates and walls with ratios of height to breadth between 0.2 and 5, is given by Equation (4.10):

$$K_p \cong 1 - (1 - \delta)^2 \qquad (4.10)$$

Equation (4.10) has the required properties of equalling one for a value of δ equal to 1, i.e. an impermeable plate or wall, and tending to zero as the solidity tends to zero. For very small values of δ (for example an open-truss plate made up of individual members), K_p tends to a value of 2δ, since, from Equation (4.10):

$$K_p = 1 - (1 - 2\delta + \delta^2) \cong 2\delta,$$

noting that δ^2 is negligible in comparison with 2δ for very small δ.

Considering the application of this to the drag coefficient for an open-truss plate of square planform, we have the following equation from Equations (4.9) and (4.10),

$$C_{D,At} \cong 1.1 \, (2\delta) = 2.2 \, \delta$$

where $C_{D,At}$ denotes that the drag coefficient, defined as in Equation (4.4), is with respect to the total (enclosed) elevation area of A_t. With respect to the elevation area of the actual members in the truss A_m, the drag coefficient is larger, being given by:

$$C_{D,Am} = C_{D,At} \left(A_t / A_m \right) 4 = C_{D,At} \cdot (1/\delta) \cong 2.2$$

In this case of a very open plate, the members will act like isolated bluff bodies with individual values of drag coefficient of 2.2.

Cook (1990) has discussed in detail the effect of porosity on aerodynamic forces on bluff bodies.

4.3.2 Flat plates and walls inclined to the flow

Figure 4.7 shows the case with the wind at an oblique angle of attack, α, to a two-dimensional flat plate. In this case, the resultant force remains primarily at right angles to the plate surface, i.e. it is no longer a drag force in the direction of the wind. There is also a tangential component, or 'skin friction' force. However, this is not significant in comparison with the normal force, for angles of attack greater than about 10°.

For small angles of attack, α, (less than 10°), the normal force coefficient, C_N, with respect to the total plan area of the plate, viewed normal to its surface, is given approximately by:

$$C_N \cong 2\pi \cdot \alpha \qquad (4.11)$$

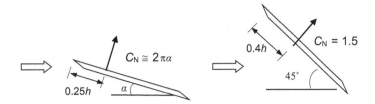

Figure 4.7 Normal force coefficients for an inclined two-dimensional plate.

where α is measured in radians, not in degrees.

Equation (4.11) comes from a theory used in aeronautics. The 'centre of pressure', denoting the position of the line of action of the resultant normal force, is at, or near, one quarter of the height h, from the leading edge, again a result from aeronautical theory.

As the angle of attack, α, increases, the normal force coefficient, C_N, progressively increases towards the normal plate case ($\alpha = 90°$), discussed in Section 4.3.1, with the centre of pressure at a height of $0.5h$. For example, the normal force coefficient for an angle of attack of 45° is about 1.5, with the centre of pressure at a distance of about $0.4h$ from the leading edge, as shown in Figure 4.7. The corresponding values for α equal to 30° are 1.2 and $0.38h$ (ESDU 1970).

Now, consider finite length walls and hoardings, at or near ground level, and hence in a highly sheared and turbulent boundary-layer flow. The mean net pressure coefficients at the windward end of the wall, for an oblique wind blowing at 45° to the normal, are quite high due to the presence of a strong vortex system behind the wall. Some values of area-averaged mean pressure coefficients are shown in Figure 4.8; these high values are usually the critical cases for the design of free-standing walls and hoardings for wind loads.

4.4 RECTANGULAR PRISMATIC SHAPES

4.4.1 Drag on two-dimensional rectangular prismatic shapes

Understanding of the wind forces on rectangular prismatic shapes is clearly of importance for many structures, especially buildings of all heights and bridge decks. We will consider first the drag coefficients for two-dimensional rectangular prisms.

Figure 4.9 shows how the drag coefficient varies for two-dimensional rectangular prisms with sharp corners, as a function of the ratio, d/b, where d is the along-wind or afterbody length, and b is the cross-wind dimension. The flow is normal to a face of width b and is 'smooth', i.e. the turbulence level is low. As previously shown in Figure 4.4, the value of the drag

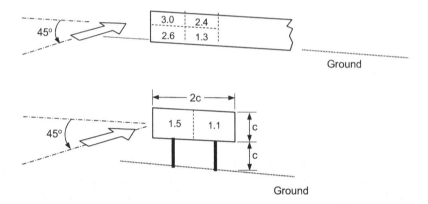

Figure 4.8 Area-averaged mean pressure coefficients on walls and hoardings for oblique wind directions.

coefficient is 1.9 for (d/b) close to zero, i.e. for a flat plate normal to a flow stream. As (d/b) increases to 0.65–0.70, the drag coefficient increases to about 2.9 (e.g. Bearman and Trueman, 1972). The drag coefficient then decreases with increasing (d/b), reaching 2.0 for a square cross section. The drag coefficient continues to decrease with further increases in (d/b), reaching about 1.0 for values of (d/b) of 5 or greater.

These variations can be explained by the behaviour of the free shear layers separating from the upstream corners. As depicted in Figure 4.1, these shear layers are unstable and eventually form discrete vortices. During the formation of these vortices, air is entrained from the wake region behind the prism; it is this continual entrainment process which sustains a base pressure lower than the static pressure. As (d/b) increases to the range of 0.65–0.70, the size of the wake decreases simply because of the increased volume of the prism occupying part of the wake volume. Thus, the same entrainment process acts on a smaller volume of wake air, causing the base pressure to decrease further, and the drag to increase. However, as (d/b) increases beyond 0.7, the rear, or downstream, corners interfere with the shear layers, and if the length d is long enough, the shear layers will stabilize, or 're-attach', on to the sides of the prisms. Although, the attached shear layers will eventually separate again from the *rear* corners of the prism, the wake is smaller for prisms with long afterbodies (high d/b), and the entrainment is weaker. The result is a lower drag coefficient, as shown in Figure 4.9.

4.4.2 Effect of aspect ratio

The effect of a finite aspect ratio (height/ breadth) is to introduce an additional flow path around the end of the body, and a means of increasing the pressure

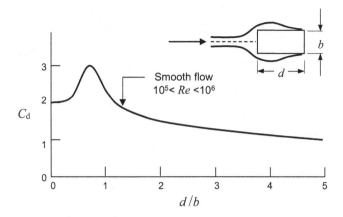

Figure 4.9 Drag coefficients for two-dimensional rectangular prisms in smooth flow.

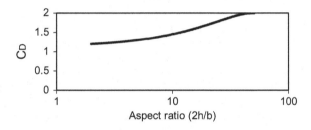

Figure 4.10 Effect of aspect ratio on drag coefficient for a square cross section.

in the wake cavity. The reduced airflow, normal to the axis results in a lower drag coefficient for finite length bodies in comparison to two-dimensional bodies of infinite aspect ratio. Figure 4.10 shows the drag coefficient for a square cross section with a free end exposed to the flow (Scruton and Rogers, 1972). The aspect ratio in this case is calculated as $2h/b$, where h is the height, since it is assumed that the flow is equivalent to that around a body with a 'mirror image' added to give an overall height of $2h$, with two free ends.

4.4.3 Effect of turbulence

Free-stream turbulence containing scales of similar size to the prism dimensions, or smaller, can have significant effects on the mean drag coefficients of rectangular prisms, and also produce fluctuating forces. As shown in Figure 4.4, the effect of free-stream turbulence on a flat plate, normal to an air stream, is to increase the drag coefficient slightly (Bearman, 1971). This results from increased mixing and entrainment into the free shear layers induced by the turbulence. Observations have also shown a reduction

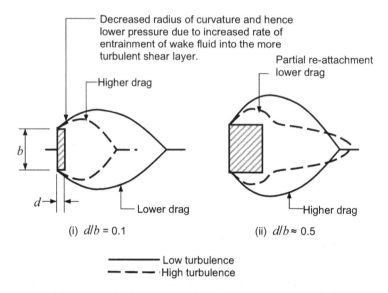

Figure 4.11 Effect of turbulence on shear layers from rectangular prisms (Laneville et al., 1975).

in the radius of curvature of the mean shear layer position (Figure 4.11). As the after-body length increases, the drag first increases and then decreases, as occurs in smooth flow. However, because of the decrease in the mean radius of curvature of the shear layers caused by the free-stream turbulence, the (d/b) ratio for maximum drag will decrease with increasing turbulence intensity, as shown in Figure 4.12 (Gartshore, 1973; Laneville *et al.*, 1975).

The drag coefficients for two-dimensional rectangular prisms on the ground in turbulent boundary-layer flow are shown in Figure 4.13. In comparison with rectangular prisms in smooth uniform flow (Figure 4.9), the drag coefficients, based on the mean wind speed at the height of the top of the prism, are much lower; because of the high turbulence in the boundary-layer flow, they do not show any maximum value.

Melbourne (1995) has discussed the important effects of turbulence on flow around bluff bodies in more detail.

4.4.4 Drag and pressures on a cube and finite-height prisms

The mean pressure distributions on a cube in a wind-tunnel flow with a mean velocity profile, similar to that in the atmospheric boundary layer, are shown in Figure 4.14 (Baines, 1963). However, no attempt was made to model the turbulence properties of the natural wind in these early tests. The pressure coefficients are based on the mean wind speed at the height of the

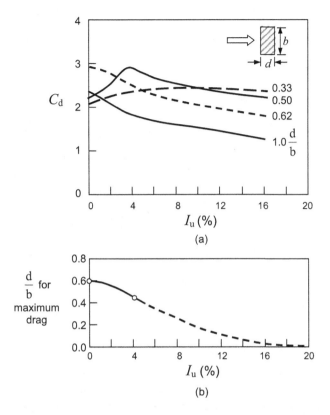

Figure 4.12 (a and b) Effect of turbulence on drag coefficients for rectangular prisms (Laneville et al., 1975).

top of the cube. The drag coefficient of 0.8 is lower than that of the two-dimensional square section prism (d/b equal to 1.0 in Figure 4.13). This is due to the three-dimensional flows that occur around the side walls of the block which increase the base pressure (decrease the negative pressure).

The mean pressure distribution on a tall prism of square cross section in a flow with a varying mean velocity profile, similar to that in the atmospheric boundary layer, is shown in Figure 4.15, (Baines, 1963). The mean pressure coefficients are again based on the dynamic pressure calculated from the mean wind speed at the top of the prism. The effect of the vertical velocity profile on the windward wall pressure is clearly seen. The maximum pressure occurs at about 85% of the height. On the windward face of unshielded tall buildings, the strong pressure gradient can cause a strong downwards flow, often causing high wind speeds which may cause problems for pedestrians at ground level.

Figure 4.16 shows effective sectional drag coefficients for finite-height prisms with rectangular cross sections, representative of tall buildings, in a turbulent atmospheric boundary layer simulation, representing wind flow

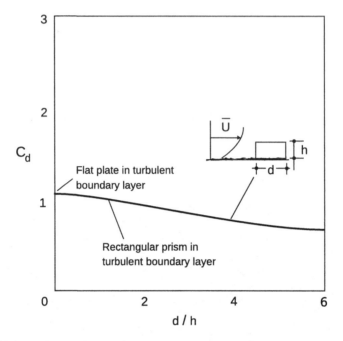

Figure 4.13 Mean drag coefficients for rectangular prisms in turbulent boundary-layer flow.

over suburban terrain. The effective drag coefficient is defined as the average sectional drag coefficient referenced to the upstream mean wind speed varying with height over the prism, which gives the measured, or predicted, base shear. It is of interest to compare these data with those in Figures 4.9 and 4.12, which show drag coefficients for two-dimensional rectangular cross sections in smooth and turbulent flow, respectively. Clearly the effect of the finite aspect ratio (defined as $h/\sqrt{(bd)}$ in Figure 4.16) is to greatly reduce the effective drag coefficient, as shown in Figure 4.10. In Figure 4.16, the maximum drag coefficients generally occur for the lowest d/b value of 0.2, although for the higher aspect ratios, the peak occurs at d/b equal to 0.67, a similar value as that for the peak drag on two-dimensional bodies in smooth flow (Figure 4.9).

4.4.5 Jensen Number

For bluff bodies such as buildings immersed in a turbulent boundary-layer flow, the ratio of characteristic body dimension, usually the height, h, in the case of a building, to the characteristic boundary-layer length, represented by the roughness length, z_o, is known as the Jensen Number. In a classic series of experiments, Jensen (1958) established the need for equality of (h/z_o) in order for wind-tunnel mean pressure measurements on a model of a small building to match those in full scale. The effect is greatest on the roof

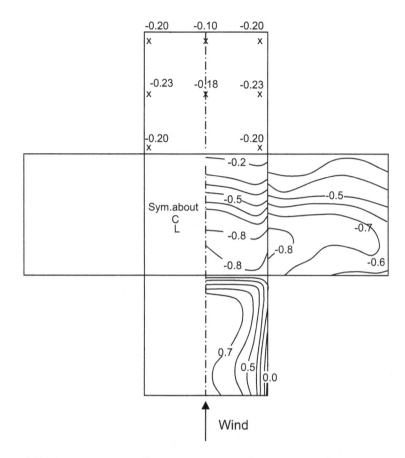

Figure 4.14 Mean pressure coefficients on a cube in a boundary-layer flow (Baines, 1963).

and side walls, where the increased turbulence in the flow over the rougher ground surfaces promotes shorter flow reattachment lengths.

For a given height, h, greater values of roughness length, z_o, and lower values of Jensen Number imply a rougher ground surface and hence greater turbulence intensities at the height of the body. Thus, fluctuating pressure coefficients also depend on Jensen Number – a decrease in Jensen Number generally leads to an increase in root-mean-square pressure coefficients.

4.5 CIRCULAR CYLINDERS

4.5.1 Effects of Reynolds Number and surface roughness

For bluff bodies with curved surfaces such as the circular cylinder, the positions of the separation of the local surface boundary layers are much

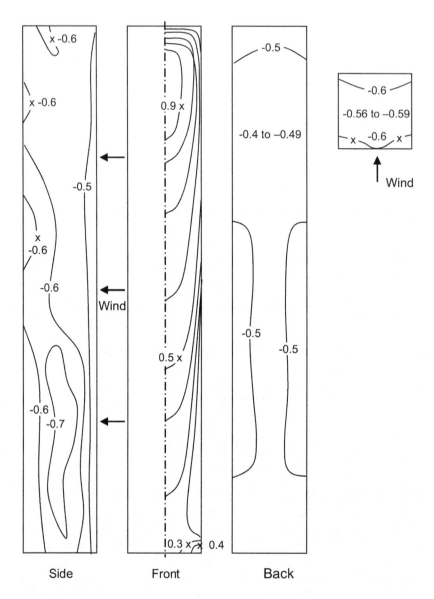

Figure 4.15 Mean pressure coefficients on a tall prism in a boundary-layer flow (Baines, 1963).

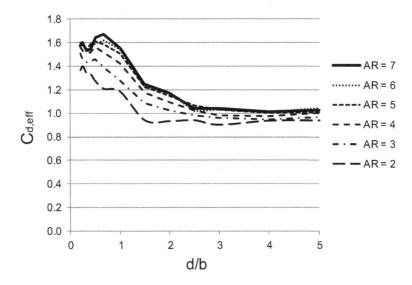

Figure 4.16 Effective drag coefficients for rectangular prisms in a boundary-layer flow. Aspect ratio, $AR=h/\sqrt{(bd)}$. (Data courtesy of Professor C-M. Cheng of Tamkang University, Taiwan.)

more dependent on viscous forces than is the case with sharp-edged bodies. This results in a variation of drag forces with Reynolds Number, which is the ratio of inertial forces to viscous forces in the flow (see Section 4.2.4). Figure 4.17 shows the variation of drag coefficient with Reynolds Number for square section bodies with various corner radii (Scruton, 1981). The appearance of a 'critical' Reynolds Number, at which there is a sharp fall in drag coefficient, occurs at a relatively low corner radius.

The various flow regimes for a circular cylinder with a smooth surface finish in smooth (low turbulence) flow are shown in Figure 4.18. The sharp fall in drag coefficient at a Reynolds Number of 2×10^5 is caused by a transition to turbulence in the surface boundary layers ahead of the separation points. This causes separation to be delayed to an angular position of $140°$ from the front stagnation point, instead of $90°$, which is the case for sub-critical Reynolds Numbers. This delay in the separation results in a narrowing in the wake, and an increased (less negative) base pressure, and hence a lower drag coefficient. The pressure distributions at sub-critical and super-critical Reynolds Numbers are shown in Figure 4.19.

Although information on flow and force coefficients for two-dimensional circular cylinders at sub-critical Reynolds Numbers (Re_b below about 1×10^5 for cylinders with nominally 'smooth' surfaces) is widely available, there is less data in the critical range up to the minimum drag coefficient, and very few studies in the super-critical range up to Re_b equal to 10^6 and beyond are available, as only a few wind tunnels have the required dimensions such

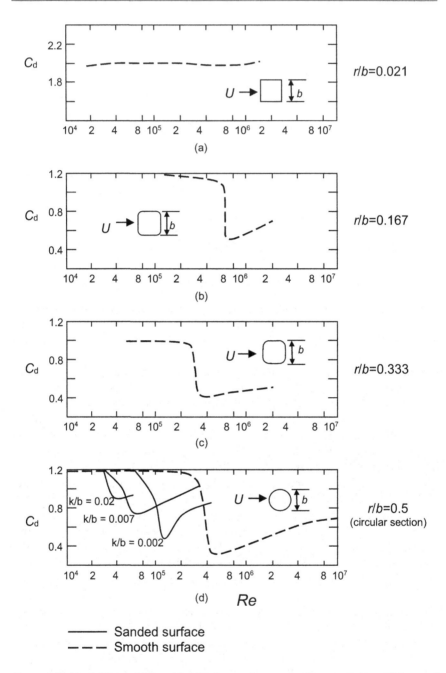

Figure 4.17 (a–d) Effect of Reynolds Number and corner radius on drag coefficients of square sections (Scruton, 1981).

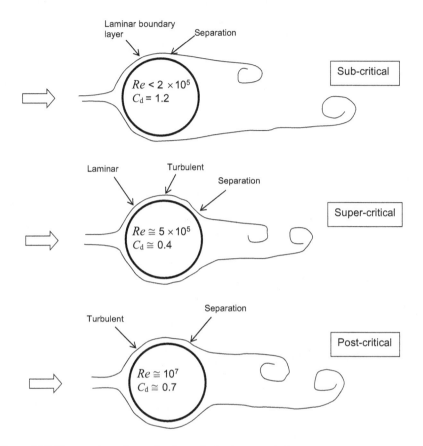

Figure 4.18 Flow regimes for a circular cylinder in smooth flow.

as a large working section, together with high wind speeds. Unfortunately, the flow around most structures with circular cross sections such as lighting poles, chimneys and observation towers at wind speeds for structural design at ultimate limit states, falls into the critical or super-critical ranges, where the information is most sparse.

As shown in Figure 4.17, the presence of a rough surface on a circular cylinder causes the critical Reynolds Number range to be lower than that for a smooth cylinder, as noted in the pioneering experiments of Fage and Warsap (1930). The minimum drag coefficient is also higher for the rougher surfaces.

The Reynolds Number based on diameter (Re_b) defining the lower limit of the super-critical flow range, can be taken to be that at which the drag coefficient reaches a minimum value, Re_{minCd}. The latter value, and the minimum drag coefficient C_{dmin}, are both functions of the ratio of average surface roughness height, k, to the diameter, b. As the ratio k/b increases, Re_{minCd} decreases and C_{dmin} increases.

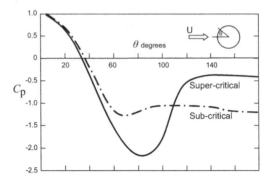

Figure 4.19 Pressure distributions around a two-dimensional circular cylinder at sub-critical and trans-critical Reynolds Numbers.

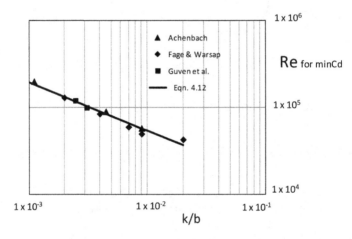

Figure 4.20 Reynolds Number (based on diameter) for minimum drag coefficient of roughened circular cylinders, as a function of the ratio of average roughness height to diameter.

Figure 4.20 shows Re_{minCd} plotted against k/b for the results of Achenbach (1971), Fage and Warsap (1930), and Guven *et al.* (1980). The values are well fitted with a straight line on log-log axes, suggesting the relationship in Equation (4.12) (from Holmes and Burton, 2016):

$$Re_{minCd} \cong 4,210(k/b)^{-0.555} \tag{4.12}$$

A slightly different relationship (Equation 4.13) was proposed by Achenbach and Heinecke (1981):

$$Re_{minCd} \cong 6,000(k/b)^{-0.5} \tag{4.13}$$

For (k/b) less than 10^{-3}, the value of Re_{minCd} is nearly constant – i.e. it is independent of (k/b) for relatively smooth cylinders (Adachi, 1995).

Szechenyi (1975) replotted the results of Fage and Warsap, together with some new data, and showed that the drag coefficients, in the super-critical range (i.e. for $Re_b > Re_{minCd}$), effectively collapsed on to a single line, when they were plotted against a *roughness Reynolds Number, Re_k*, defined as (Uk/ν), where U is the wind speed, k is the average roughness height, and ν is the kinematic viscosity. Data from a nominally smooth cylinder were included by Szechenyi by assuming a relative roughness (k/b) of 3.5×10^{-5}.

Figure 4.21a shows some more recent data (Holmes and Burton, 2016) on drag coefficients of three roughened circular cylinders of high aspect ratio plotted against Re_b. Figure 4.21b shows the data from the super-critical range from the same tests replotted against the roughness Reynolds Number, Re_k. This method of plotting does indeed 'collapse' the data, indicating the effectiveness of the Szechenyi (1975) approach.

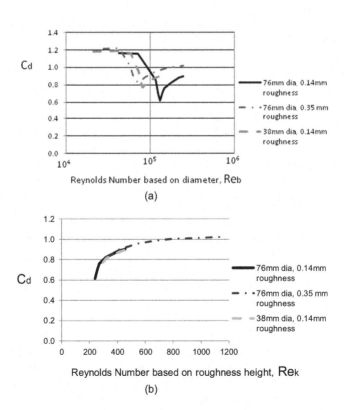

Figure 4.21 Drag coefficients of two-dimensional circular cylinders: (a) plotted against Reynolds Number Re_b and (b) plotted against Re_k.

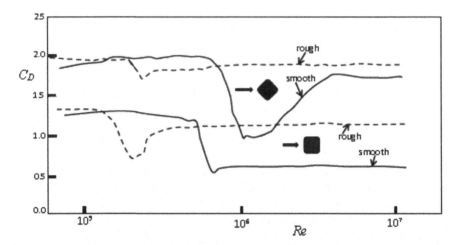

Figure 4.22 Effect of Reynolds Number and surface roughness on drag coefficients of a square section with rounded corners (r/b=0.16).

The effectiveness of surface roughness in the simulation of super-critical flow in wind-tunnel tests on structures with circular cross section is further explored in Chapter 7.

The effect of roughness (k/b=1×10^{-3}) on the drag coefficients of square sections, with a radius of curvature on the corners (r/b=0.16), is shown in Figure 4.22, in comparison with the values for a smooth surface (data from van Hinsberg, 2018). The effect of the roughness is to reduce the fall in drag at the critical Reynolds Number. For both wind directions shown, the drag of the rough cross section is nearly constant over a wide range of super-critical Reynolds Numbers.

4.5.2 Effect of aspect ratio

The reduction in drag coefficient for a smooth circular cylinder of finite aspect ratio (single free end) in smooth flow (sub critical) is shown in Figure 4.23 (Scruton and Rogers, 1972). This figure is analogous to Figure 4.10 for a square cross section. As for the square section, the reduction in drag for a circular cylinder results from the additional flow path provided by the free end on the body.

The mean pressure distribution around a circular cylinder with a height to diameter (aspect) ratio of 1, with its axis vertical in a turbulent boundary-layer flow is shown in Figure 4.24 (Macdonald *et al.*, 1988). The minimum mean pressure coefficient on the side occurs at angular position of about 80°, and is about –1.2, lower in magnitude than the value of about –2.0 for a two-dimensional cylinder in super-critical flow (see Figure 4.18). The minimum $\bar{C}_p$ increases in magnitude with increasing aspect ratio, reaching the two-dimensional value at an aspect ratio of about 2.0.

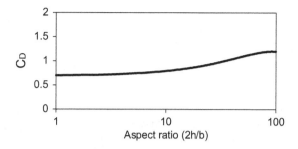

Figure 4.23 Effect of aspect ratio on drag coefficient of a circular cylinder (sub-critical Reynolds Number).

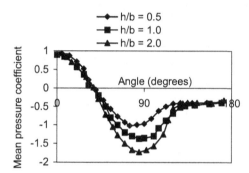

Figure 4.24 Effect of aspect ratio (height/diameter) on pressure distributions around circular cylinders.

4.6 FLUCTUATING FORCES AND PRESSURES

4.6.1 Introduction

The turbulent and fluctuating nature of wind flow in the atmospheric boundary layer has been described in Chapter 3. This and the unstable nature of flow around bluff bodies, which results in flow separations and sometimes re-attachments, produce pressures and forces on bodies in the natural wind which are also highly fluctuating.

The main sources of the fluctuating pressures and forces are as follows:

- Natural turbulence or gustiness in the free-stream flow. This is often called 'buffeting'. If the body dimensions are small, relative to the length scales of the turbulence, the pressure and force variations will tend to follow the variations in velocity (see Section 4.6.2 on the quasi-steady assumption).
- Unsteady flow generated by the body itself, by phenomena such as separations, re-attachments and vortex shedding.

- Fluctuating forces due to movement of the body itself (e.g. aerodynamic damping)

The third source arises only for relatively flexible vibration-prone "aeroelastic structures". In the following sections, only the first two sources will be considered.

4.6.2 The quasi-steady assumption

The 'quasi-steady' assumption is the basis of many wind loading codes and standards. The fluctuating pressure on a structure is assumed to follow the variations in longitudinal wind velocity upstream. Thus,

$$p(t) = C_{po}\left(\frac{1}{2}\right)\rho_a[U(t)]^2 \tag{4.14}$$

where C_{po} is a quasi-steady pressure coefficient

Expanding $U(t)$ into its mean and fluctuating components,

$$p(t) = C_{po}\left(\frac{1}{2}\right)\rho_a[\bar{U}+u'(t)]^2$$

$$= C_{po}\left(\frac{1}{2}\right)\rho_a\left[\bar{U}^2 + 2\bar{U}u'(t) + u'(t)^2\right] \tag{4.15}$$

Taking mean values,

$$\bar{p} = C_{po}\left(\frac{1}{2}\right)\rho_a\left[\bar{U}^2 + \sigma_u^2\right]$$

For small turbulence intensities, σ_u^2 is small in comparison with $\bar{U}^2$. Then the quasi-steady pressure coefficient, C_{po}, is approximately equal to the mean pressure coefficient, $\bar{C}_p$.

$$\text{Then, } \bar{p} \cong C_{po}\left(\frac{1}{2}\right)\rho_a\bar{U}^2 \cong \bar{C}_p\left(\frac{1}{2}\right)\rho_a\bar{U}^2 \tag{4.16}$$

Subtracting the mean values from both sides of (4.15), we have

$$p'(t) = C_{po}\left(\frac{1}{2}\right)\rho_a\left[2\bar{U}u'(t)+u'(t)^2\right]$$

Neglecting the second term in the square brackets (valid for low turbulence intensities), squaring and taking mean values, we get,

$$\overline{p'^2} \cong \bar{C}_p\left(\frac{1}{4}\right)\rho_a^2\left[4\bar{U}^2\overline{u'^2}\right] = \bar{C}_p^2\,\rho_a^2\,\bar{U}^2\,\overline{u'^2} \tag{4.17}$$

Equation (4.17) is a quasi-steady relationship between mean square pressure fluctuations and mean square longitudinal velocity fluctuations.

To predict peak pressures by the quasi-steady assumption,

$$\hat{p}, \check{p} = C_{po}\left(\frac{1}{2}\right)\rho_a\left[\hat{U}^2\right] \cong \bar{C}_p(1/2)\rho_a\left[\hat{U}^2\right] \tag{4.18}$$

Thus, according to the quasi-steady assumption, we can predict peak pressures (maxima and minima) by using mean pressure coefficients with a peak gust wind speed. This is the basis of many codes and standards that use a peak gust as a basic wind speed (see Chapter 15). Its main disadvantage is that building induced pressure fluctuations (the second source described in Section 4.6.1) are ignored. Also, when applied to wind pressures over large areas, it is conservative, because full correlation of the pressure peaks is implied.

4.6.3 Body-induced pressure fluctuations and vortex shedding forces

The phenomena of separating shear layers and vortex shedding have already been introduced in Sections 4.1, 4.3.1, 4.4.1 and 4.5 in descriptions of the flow around some basic bluff-body shapes. These phenomena occur whether the flow upstream is turbulent or not, and the resulting surface pressure fluctuations on a bluff body, can be distinguished from those generated by the flow fluctuations in the approaching flow.

The regular vortex shedding into the wake of a long bluff body results from the rolling-up of the separating shear layers alternately one side, then the other, and occurs on bluff bodies of all cross-sections. A regular pattern of decaying vortices, known as the 'von Karman vortex street', appears in the wake. Turbulence in the approaching flow tends to make the shedding less regular, but the strengths of the vortices are maintained, or even enhanced. Vibration of the body may also enhance the vortex strength, and the vortex-shedding frequency may change to the frequency of vibration, in a phenomenon known as *lock-in*.

As each vortex is shed from a bluff body, a strong cross-wind force is induced towards the side of the shed vortex. In this way, the alternate shedding of vortices induces a nearly harmonic (sinusoidal) cross-wind force variation on the structure.

For a given cross-sectional shape, the frequency of vortex shedding, n_s, is proportional to the approaching flow speed, and inversely proportional to the width of the body. It may be expressed in a non-dimensional form, known as the *Strouhal Number, St*.

$$St = \frac{n_s b}{\bar{\bar{U}}} \tag{4.19}$$

Where

b is the cross-wind body width,
$\bar{U}$ is the mean flow speed

The Strouhal Number varies with the shape of the cross section, and for circular and other cross sections with curved surfaces varies with Reynolds Number. Some representative values of Strouhal Number for a variety of cross sections are shown in Figure 4.25.

An inclined two-dimensional flat plate with an angle of attack, α (Figure 4.7), has a Strouhal number of about 0.15 based on the breadth, b, normal to the flow, or $(0.15/\sin \alpha)$ based on the chord, c, where $b=c\cdot\sin \alpha$ (Chen and Fang, 1996).

The variation with Reynolds Number for a smooth circular cylinder is shown in Figure 4.26 (Scruton, 1963; Schewe, 1983). In the sub-critical range, up to a Reynolds Number of 2×10^5, the Strouhal Number is quite constant at a value of 0.20. In the critical Reynolds Number range, coinciding with the sharp fall in drag coefficient (see Figure 4.16), the Strouhal Number jumps to 0.3 and then 0.48, although in this range, the vortex shedding is random, and not clearly defined. A slightly decreasing Strouhal Number to 0.4, in the super-critical range, is followed by a fall to 0.2 again, at a Reynolds Number of 2×10^6. Helical strakes (Figure 4.27) are often used to inhibit vortex shedding and the resulting cross-wind forces on structures with circular sections such as chimney stacks (Scruton and Walshe, 1957).

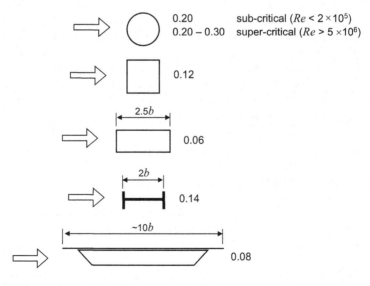

Figure 4.25 Strouhal numbers for vortex shedding for various cross sections.

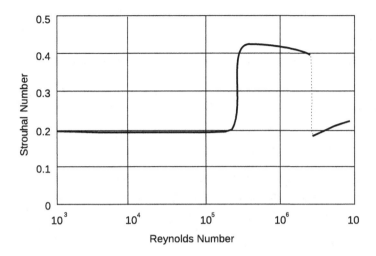

Figure 4.26 Strouhal Number versus Reynolds Number for circular cylinders.

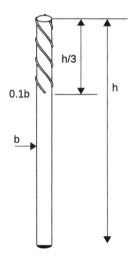

Figure 4.27 Helical strakes for inhibiting vortex shedding.

4.6.4 Universal wake Strouhal Number

In an attempt to unify the observed wake characteristics of two-dimensional bluff bodies of various cross sections, Roshko (1955) proposed a universal wake Strouhal Number, St^*, defined as follows:

$$St^* = \frac{n_s \cdot b'}{\bar{U}_s} \tag{4.20}$$

In Equation (4.20), b' is the breadth of the wake behind the body (usually greater than b), and $\bar{U}_s$ is the velocity of the flow just outside the separating shear layer.

The value of St^* depends weakly on Reynolds Number, but for the range of interest in wind engineering can be taken as 0.16 (Roshko, 1955, Figure 7).

The relationship between St^* and the normal Strouhal Number, St, defined in Equation (4.19), is:

$$St = St^* \frac{b}{b'} \frac{\bar{U}_s}{\bar{U}_0} \tag{4.21}$$

If the ratio between $\bar{U}_s$ and the free-stream velocity $\bar{U}_0$ is denoted by k, then by Bernoulli's Equation (Equation 4.3), the base pressure, i.e. the coefficient of the average pressure on the rear, or downwind, face of the body, C_{ps}, is given by:

$$C_{ps} = 1 - \left(\frac{\bar{U}_s}{\bar{U}_0}\right)^2 = 1 - k^2$$

Hence,

$$k = \sqrt{1 - C_{ps}} \tag{4.22}$$

k can then be determined from experimental measurements of base pressure and hence C_{ps}. A typical value of k for sections with no wake interference, and at high Reynolds Numbers, is 1.5.

The dimension b' can also be determined experimentally, or alternatively by the theoretical 'free-streamline' approach of Roshko (1954) and others, and Roshko (1954, Figures 3 to 5) gave a graph of the ratio (b'/b) as a function of k for three different bluff shapes.

Roshko (1955, Figure 2) also gave a graph of sectional drag coefficient, C_d, as a function of k for three cross sections. Roshko's approach thus provides a useful link between drag coefficients and vortex-shedding frequencies, which may enable one, or both, to be estimated from limited experimental data for new cross sections.

4.6.5 Fluctuating pressure and force coefficients

The root-mean-square fluctuating (standard deviation) pressure coefficient at a point on a bluff body is defined by:

$$C_p' = \frac{\sqrt{\overline{p'^2}}}{\frac{1}{2}\rho_a \bar{U}^2} \tag{4.23}$$

$\sqrt{\overline{p'^2}}$ is the root-mean-square fluctuating, or standard deviation, pressure (also denoted by σ_p).

The root-mean-square fluctuating sectional force coefficient per unit length of a two-dimensional cylindrical or prismatic body is defined by:

$$C_f' = \frac{\sqrt{\overline{f'^2}}}{\frac{1}{2}\rho_a\bar{U}^2 b} \tag{4.24}$$

$\sqrt{\overline{f'^2}}$ is the root-mean-square fluctuating force per unit length. b is a reference dimension, usually the cross-wind breadth.

For a whole body,

$$C_F' = \frac{\sqrt{\overline{F'^2}}}{\frac{1}{2}\rho_a\bar{U}^2 A} \tag{4.25}$$

$\sqrt{\overline{F'^2}}$ is the root-mean-square fluctuating force acting on the complete body. A is a reference area – usually the frontal area.

The total fluctuating force acting on a cylindrical body of finite length can be calculated from the fluctuating sectional force, knowing the correlation function, or correlation length.

With the quasi-steady assumption (Section 4.6.2), the root-mean-square fluctuating pressure coefficient can be estimated from Equations (4.17) and (4.23):

$$C_p' = \frac{\sqrt{\overline{p'^2}}}{\frac{1}{2}\rho_a\bar{U}^2} \approx \frac{\bar{C}_p\rho_a\bar{U}\sqrt{\overline{u'^2}}}{\frac{1}{2}\rho_a\bar{U}^2} = 2\bar{C}_p I_u \tag{4.26}$$

where I_u is the longitudinal turbulence intensity ($= \frac{\sqrt{\overline{u'^2}}}{\bar{U}}$ or $\frac{\sigma_u}{\bar{U}}$), as defined in Section 3.3.1.

Similarly the r.m.s. fluctuating drag coefficient can be estimated using the quasi-steady assumption:

$$C_D' \approx 2\bar{C}_D I_u \tag{4.27}$$

Fluctuating forces in the cross-wind direction are usually determined by experiment, however. Measurements have shown that square cross sections experience stronger cross-wind fluctuating forces due to vortex shedding, than do circular cross-sections. Figure 4.28 shows the variation of r.m.s. fluctuating cross-wind force per unit length, for a circular cylinder, as a function of Reynolds Number (Wootton and Scruton, 1970). The value is 0.5 at subcritical Reynolds Numbers, falling to much lower values in the critical and supercritical ranges, coinciding with a reduction in drag coefficient (Section 4.5.1).

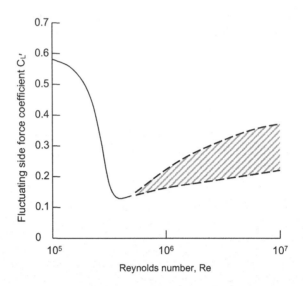

Figure 4.28 Variation of fluctuating cross-wind force coefficient per unit length with Reynolds Number for a circular cylinder (Wootton and Scruton, 1970). (Reproduced by permission of CIRIA, London.)

The fluctuating cross-wind force coefficient for a two-dimensional square cross-section, with sharp corners, is three to four times higher than that for a circular section, due to the greater strength of the shed vortices. In smooth flow, the root-mean-square fluctuating cross-wind force coefficient, for the square section, is about 1.3; this drops to 0.7 in turbulent flow of 10% intensity (Vickery, 1966).

4.6.6 Correlation length

The spatial correlation coefficient for fluctuating forces at two points along a cross-section is defined by:

$$\rho = \frac{\overline{f_1'(t)f_2'(t)}}{\overline{f'^2}} = \frac{\overline{f_1'(t)f_2'(t)}}{\sigma_f^2} \tag{4.28}$$

where $f_1'(t)$, $f_2'(t)$ are the fluctuating forces per unit length at two sections along a cylindrical or prismatic body. (Correlation was previously discussed in relation to atmospheric turbulence in Section 3.3.5.)

If we assume that the mean square fluctuating force per unit length is constant along the body, then:

$$\overline{f_1'^2} = \overline{f_2'^2} = \overline{f'^2}$$

As the separation distance, y, between the two sections 1 and 2 approaches zero, the correlation function, $\rho(y)$, approaches 1. As the separation distance becomes very large, $\rho(y)$ tends to zero; this means there is no statistical relationship between the fluctuating forces.

The *correlation length*, ℓ, is then defined as:

$$\ell = \int_0^\infty \rho(y)dy \qquad (4.29)$$

The correlation length is thus the area under the graph of $\rho(y)$ plotted against y.

Measurements of correlation length for a smooth circular cylinder in smooth flow are shown in Figure 4.29 (Wootton and Scruton, 1970). The correlation length falls from about five diameters to one diameter over the critical Reynolds Number range.

The correlation length for fluctuating cross-wind forces on a two-dimensional square section has been found to be $5.6d$ for smooth flow, and $3.3d$ for turbulent flow of 10% intensity, where d is the side length (Vickery, 1966). However, a finite aspect ratio reduces the spanwise correlations of fluctuating cross-wind forces considerably; for example, the effective correlation length for a square section with an aspect ratio of 14 is only about $1d$ (Whitbread and Scruton, 1965; Vickery, 1966).

4.6.7 Total fluctuating forces on a slender body

Consider a long cylindrical, or prismatic, body of length, L, subjected to fluctuating wind forces along its length. Divide the body into a large number, N, of sections of width, $\delta y_1, \delta y_2, \ldots, \delta y_N$, as shown in Figure 4.30. Assume that the mean square fluctuating force is the same at all sections.

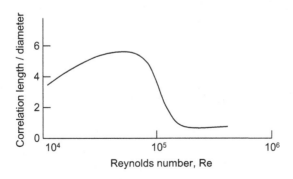

Figure 4.29 Variation of correlation length with Reynolds Number for a stationary circular cylinder (Wootton and Scruton, 1970). (Reproduced by permission of CIRIA, London.)

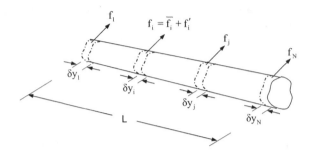

Figure 4.30 Sectional force fluctuations on a long slender body.

At any section, i, the total force per unit length can be separated into a mean, or time-averaged, component, and a fluctuating component with a zero mean:

$$f_i(t) = \bar{f}_i + f_i'(t) \tag{4.30}$$

The *total* mean force acting on the whole body is given by:

$$\bar{F} = \sum \bar{f}_i \delta y_i$$

where the summation is taken from i equal to 1 to N.

As we let the number of sections tend to infinity, δy_i tends to zero, and the right-hand side becomes an integral:

$$\bar{F} = \int_0^L \bar{f}_i dy_i \tag{4.31}$$

The instantaneous fluctuating force on the body as a whole is given by:

$$F'(t) = \sum f_i'(t) \cdot \delta y_i$$

$$= f_1'(t) \cdot \delta y_1 + f_2'(t) \cdot \delta y_2 + \cdots f_N'(t) \cdot \delta y_N$$

Squaring both sides,

$$[F'(t)]^2 = \left[f_1'(t) \cdot \delta y_1 + f_2'(t) \cdot \delta y_2 + \cdots f_N'(t) \cdot \delta y_N \right]^2$$

$$= \left[f_1'(t) \cdot \delta y_1 \right]^2 + \left[f_2'(t) \cdot \delta y_2 \right]^2 + \left[f_N'(t) \cdot \delta y_N \right]^2 +$$

$$f_1'(t) f_2'(t) \cdot \delta y_1 \delta y_2 + f_1'(t) f_3'(t) \cdot \delta y_1 \delta y_3 + \cdots$$

$$= \sum_i^N \sum_j^N f_i'(t) f_j'(t) \delta y_i \delta y_j$$

Now, taking means (time averages) of both sides,

$$\overline{F'^2} = \sum_i^N \sum_j^N \overline{f_i'(t)f_j'(t)} \, \delta y_i \delta y_j \tag{4.32}$$

As δy_i, δy_j tend to zero,

$$\overline{F'^2} = \int_0^L \int_0^L \overline{f_i'(t)f_j'(t)} dy_i \, dy_j \tag{4.33}$$

Equations (4.32) and (4.33) are important equations, illustrating how to obtain a total fluctuating force from the fluctuating force on small elements. The integrand in Equation (4.33) is the *covariance* of the sectional force fluctuations.

Now assume that the integrand can be written in the following form:

$$\overline{f_i'(t)f_j'(t)} = \overline{f'^2}\rho(y_i - y_j)$$

where $\rho(y_i-y_j)$ is the correlation coefficient for the fluctuating sectional forces, which is assumed to be a function of the separation distance, (y_i-y_j), but not of the individual positions y_i and y_j, i.e. we assume that the wind forces are horizontally, or vertically (for vertical structures or members), *homogeneous*.

$$\text{Then, } \overline{F'^2} = \overline{f'^2(t)}\int_0^L \int_0^L \rho(y_i - y_j) \, dy_i \, dy_j \tag{4.34}$$

This is the fundamental equation for the total mean square fluctuating force on the body, in terms of the mean square fluctuating force per unit length.

By introducing a new variable equal to (y_i-y_j), Equation (4.34) can be written:

$$\overline{F'^2} = \overline{f'^2}\int_0^L dy_j \int_{-y_j}^{L-y_j} \rho(y_i - y_j) d(y_i - y_j) \tag{4.35}$$

Equations (4.34) or (4.35) can be evaluated in two special cases:

i. *Full correlation*

This assumption implies that $\rho(y_i-y_j)$ equals 1 for all separations, (y_i-y_j). Then Equation (4.34) becomes:

$$\overline{F'^2} = \overline{f'^2(t)} \, L^2$$

In this case the fluctuating forces are treated like static forces.

ii. *Rapidly decreasing correlation length*

In this case, ℓ is much less than L, and the second part of Equation (4.35) can be approximated by:

$$\int_{-y_j}^{L-y_j} \rho\left(y_i - y_j\right) d\left(y_i - y_j\right) = \int_{-\infty}^{\infty} \rho\left(y_i - y_j\right) d\left(y_i - y_j\right) = 2\ell,$$

from Equation (4.29).

Then, from Equation (4.35),

$$\overline{F'^2} = \overline{f'^2(t)}\, L \cdot 2\ell \tag{4.36}$$

Thus, the mean square total fluctuating force is directly proportional to the correlation length, ℓ. This is an important result that is applicable to structures such as slender towers.

4.7 SUMMARY

This chapter has summarized the relevant aspects of bluff-body aerodynamics, itself a large subject with applications in many fields, to wind loads on structures. The basic fluid mechanics of stagnation, separation and wakes has been described, and pressure and force coefficients are defined. The characteristics of pressures and forces on the basic shapes of flat plates and walls, cubes and rectangular prisms, and circular cylinders have been described. The effect of turbulence and the ground surface on drag force coefficients are covered.

Fluctuating pressures and forces, particularly those generated by upwind turbulence, and the regular shedding of vortices by a bluff body are discussed. The concept of correlation length and the averaging process by which fluctuating total forces on a body can be calculated are described. Numerical values of fluctuating force coefficients and correlation lengths for basic circular and square cross sections are also given.

Bluff-body aerodynamics is an ongoing research topic, and more detailed and up-to-date information can be found in research papers in several international journals, such as the *Journal of Wind Engineering and Industrial Aerodynamics*, the *Journal of Fluid Mechanics* and the *Journal of Fluids and Structures*.

REFERENCES

Achenbach, E. (1971) Influence of surface roughness on the cross-flow around a circular cylinder. *Journal of Fluid Mechanics*, 46: 321–335.

Achenbach, E. and Heinecke, E. (1981) On vortex shedding from smooth and rough cylinders in the range of Reynolds Numbers 6 × 103 to 5 × 106. Journal of Fluid Mechanics, 109: 239–251.

Adachi, T. (1995) The effect of surface roughness of a body in high Reynolds Number flow. *Journal of Rotating Machinery*, 1: 187–197.

Baines, W.D. (1963) Effects of velocity distributions on wind loads and flow patterns on buildings. *Proceedings, International Conference on Wind Effects on Buildings and Structures*, Teddington, UK, 26–28 June, pp. 197–225.

Bearman, P.W. (1971) An investigation of the forces on flat plates normal to a turbulent flow. *Journal of Fluid Mechanics*, 46: 177–198.

Bearman, P.W. and Trueman, D.H. (1972) An investigation of the flow around rectangular cylinders. *Aeronautical Quarterly*, 23: 229–237.

Castro, I.P. (1971) Wake characteristics of two-dimensional perforated plates normal to an air stream. *Journal of Fluid Mechanics*, 46: 599–609.

Chen, J.M. and Fang, Y-C. (1996) Strouhal Numbers of inclined flat plates. *Journal of Wind Engineering and Industrial Aerodynamics*, 61: 99–102

Cook, N.J. (1990) *The designer's guide to wind loading of building structures. Part 2 Static structures*. Building Research Establishment and Butterworths, London.

ESDU (1970) Fluid forces and moments on flat plates. Engineering Sciences Data Unit (ESDU International), ESDU Data Item 70015.

Fage, A. and Warsap, J.H. (1930) The effects of turbulence and surface roughness on the drag of a circular cylinder. Reports and Memoranda No. 1283, Aeronautical Research Council, UK.

Gartshore, I.S. (1973) The effects of freestream turbulence on the drag of rectangular two-dimensional prisms. University of Western Ontario, Boundary Layer Wind Tunnel Report, BLWT-4-73.

Guven, O., Farell, C. and Patel, V.C. (1980) Surface-roughness effects on the mean flow past circular cylinders. *Journal of Fluid Mechanics*, 98: 673–701.

Holmes, J.D. and Burton, D. (2016). Drag coefficients for rough circular cylinders re-visited. *8th International Colloquium on Bluff-Body Aerodynamics and Applications*, Boston, Massachusetts, USA, 7–11 June.

Jensen, M. (1958) The model law for phenomena in the natural wind. *Ingenioren*, 2: 121–128.

Laneville, A., Gartshore, I.S., and Parkinson, G.V. (1975). An explanation of some effects of turbulence on bluff bodies. *4th International Conference on Wind Effects on Buildings and structures*, London, UK, September.

Letchford, C.W. and Holmes, J.D. (1994) Wind loads on free-standing walls in turbulent boundary layers. *Journal of Wind Engineering and Industrial Aerodynamics*, 51: 1–27.

Macdonald, P.A., Kwok, K.C.S., and Holmes J.D. (1988) Wind loads on circular storage bins, silos and tanks: i point pressure measurements on isolated structures. *Journal of Wind Engineering and Industrial Aerodynamics*, 31: 165–188.

Marchman, J.F. and Werme, T.D. (1982) Mutual interference drag on signs and luminaires. *A.S.C.E. Journal of the Structural Division*, 108: 2235–2244.

Melbourne, W.H. (1995) Bluff body aerodynamics for wind engineering. In *A state of the art in wind engineering*. ed. P. Krishna. Wiley Eastern Limited, New Delhi.

Roshko, A. (1954) A new hodograph for-free-streamline theory. National Advisory Committee for Aeronautics (NACA), Washington, DC, USA, Technical Note 3168, July.

Roshko, A. (1955) On the wake and drag of bluff bodies. *Journal of Aeronautical Sciences*, 22: 124–132.

Schewe, G. (1983) On the force fluctuations acting on a circular cylinder in cross-flow from subcritical up to transcritical Reynolds Numbers. *Journal of Fluid Mechanics*, 133: 265–285.

Scruton, C. (1963) On the wind-excited oscillations of stacks, towers and masts. *Proceedings, International Conference on Wind Effects on Buildings and Structures*, Teddington, UK, 26–28 June, pp. 798–832.

Scruton, C. (1981) *An introduction to wind effects on structures*. Oxford University Press, Oxford, England.

Scruton, C. and Rogers, E.W.E. (1972) Steady and unsteady wind loading of buildings and structures. *Philosophical Transactions, Royal Society, A*, 269: 353–383.

Scruton, C. and Walshe, D.E.J. (1957) A means for avoiding wind-excited oscillations of structures of circular or near-circular cross section. National Physical Laboratory (UK), NPL Aero Report 335 (unpublished).

Szechenyi, E. (1975) Supercritical Reynolds number simulation for two-dimensional flow over circular cylinders. *Journal of Fluid Mechanics*, 70: 529–542.

Van Hinsberg, N.P. (2018) On the effects of corner radius and surface roughness on square-section cylinders at very high Reynolds Numbers. *12th UK Conference on Wind Engineering*, Leeds, UK, 3–4 September.

Vickery, B.J. (1966) Fluctuating lift and drag on a long cylinder of square cross-section in a smooth and turbulent flow. *Journal of Fluid Mechanics*, 25: 481–494.

Whitbread, R.E. and Scruton, C. (1965) An investigation of the aerodynamic stability of a model of the proposed tower blocks for the World Trade Center. National Physical Laboratory (UK), NPL Aero Report 1165, July, (unpublished).

Wootton, L.R. and Scruton, C.P. (1970) Aerodynamic stability. *Proceedings, CIRIA Seminar on the Modern Design of Wind-Sensitive Structures*, C.I.R.I.A., 6 Storey's Gate, London, UK, 18 June, pp. 65–81.

Chapter 5

Resonant dynamic response and effective static load distributions

5.1 INTRODUCTION

Due to the turbulent nature of the wind velocities in storms of all types, the wind loads acting on structures are also highly fluctuating. There is a potential to excite resonant dynamic response for structures, or parts of structures, with natural frequencies less than 1 Hz. The resonant response of a structure introduces the complication of a time-history effect, in which the response at any time depends not just on the instantaneous wind gust velocities acting along the structure, but also on the previous time history of wind gusts.

This chapter introduces the principles and analysis of dynamic response to wind. Some discussion of aeroelastic and fatigue effects is included. Also in this chapter, the method of equivalent or effective static wind loading distributions is introduced.

Treatment of dynamic response is continued in Chapters 9–12 on tall buildings, large roofs and sports stadiums, slender towers and masts, and bridges, with emphasis on the particular characteristics of these structures. In Chapter 15, approaches in design codes and standards to dynamic response are considered.

5.2 PRINCIPLES OF DYNAMIC RESPONSE

The fluctuating nature of wind velocities, pressures and forces, as discussed in Chapters 3 and 4, may cause the excitation of significant resonant vibratory response in structures or parts of structures, provided their natural frequencies and damping are low enough. This resonant dynamic response should be distinguished from the background fluctuating response to which all structures are subjected to. Figure 5.1 shows the response spectral density of a dynamic structure under wind loading; the area under the entire curve represents the total mean-square fluctuating response (note that the mean response is not included in this plot). The resonant responses in the first two modes of vibration are shown hatched in this diagram. The background

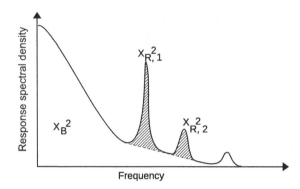

Figure 5.1 Response spectral density for a structure with significant resonant contributions.

response, made up largely of low-frequency contributions below the lowest natural frequency of vibration, is the largest contributor in Figure 5.1, and, in fact, is usually the dominant contribution in the case of along-wind loading. Resonant contributions become more and more significant, and will eventually dominate, as structures becomes taller or longer in relation to their width, and their natural frequencies become lower and lower.

Figure 5.2a shows the characteristics of the time histories of an along-wind (drag) force; the structural response for a structure with a *high* fundamental natural frequency is shown in Figure 5.2b, and the response with a *low* natural frequency in Figure 5.2c. In the former case, the resonant, or vibratory component clearly plays a minor role in the response, which generally follows closely the time variation of the exciting forces. However, in the latter case, the resonant response, in the fundamental mode of vibration, is important, although response in higher modes than the first can usually be neglected.

In fact, the majority of structures fall into the category of Figure 5.2b, and will *not* experience significant resonant dynamic response. A well known rule of thumb is that the lowest natural frequency should be below 1 Hz for the resonant response to be significant. However the amount of resonant response also depends on the damping, aerodynamic or structural, present. For example, high voltage transmission lines usually have fundamental sway frequencies which are well below 1 Hz; however, the *aerodynamic* damping is very high – typically around 25% of critical – so that the resonant response is largely damped out. Lattice towers, because of their low mass, also have high aerodynamic damping ratios. Slip-jointed steel lighting poles have high *structural* damping due to friction at the joints – this energy absorbing mechanism will limit the resonant response to wind.

Resonant response, when it does occur, may occasionally produce complex interactions, in which the movement of the structure itself results in additional aeroelastic forces being produced (Section 5.5). In some extreme cases, for example the Tacoma Narrows Bridge failure of 1940 (see Chapter 1),

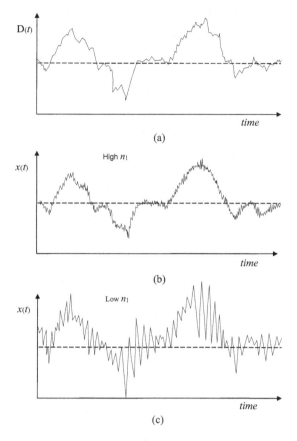

Figure 5.2 Time histories of (a) wind force, (b) response of a structure with a high natural frequency and (c) response of a structure with a low natural frequency.

catastrophic failure has resulted. These are exceptional cases, which of course must be avoided, but in the majority of structures with significant resonant dynamic response, the dynamic component is superimposed on a significant, or dominant mean and background fluctuating response.

The two major sources of fluctuating wind loads are discussed in Section 4.6. The first and obvious source, exciting resonant dynamic response, is the natural unsteady or turbulent flow in the wind, produced by shearing actions as the air flows over the rough surface of the earth, as discussed in Chapter 3. The other main source of fluctuating loads is the alternate vortex shedding which occurs behind bluff cross-sectional shapes, such as circular cylinders or square cross-sections. A further source is buffeting forces from the wakes of other structures upwind of the structure of interest.

When a structure experiences resonant dynamic response, counter-acting structural forces come into play to balance the wind forces:

- inertial forces proportional to the mass of the structure
- damping or energy-absorbing forces – in their simplest form, these are proportional to the velocity, but this is not always the case
- elastic or stiffness forces proportional to the deflections or displacements

When a structure does respond dynamically, i.e. the resonant response is significant, an important principle to remember is that the condition of the structure, i.e. stresses and deflections, at any given time depends not only on the wind forces acting at the time, but also on the *past history* of wind forces. In the case of quasi-static loading, the structure responds directly to the forces acting instantaneously at any given time.

The effective load distribution due to the resonant part of the loading (Section 5.4.4) is given to a good approximation by the distribution of inertial forces along the structure. This is based on the assumption that the fluctuating wind forces at the resonant frequency, approximately balance the damping forces, once a stable amplitude of vibration is established.

At this point, it is worth noting the essential differences between dynamic response of structures to wind and earthquake. The main differences between the excitation forces due to these two natural phenomena are:

- Earthquakes are of much shorter duration than windstorms (with the possible exception of the passage of a tornado), and are thus treated as transient loadings.
- The predominant frequencies of the earthquake ground motions are typically 10–50 times those of the frequencies in fully developed windstorms. This means that structures will be affected in different ways, e.g. buildings in a certain height range may not experience significant dynamic response to wind loadings, but may be prone to earthquake excitation.
- The earthquake ground motions will appear as *fully* correlated equivalent forces acting over the height of a tall structure. However, the eddy structure in windstorms results in *partially*-correlated wind forces acting over the height of the structure. Vortex-shedding forces on a slender structure also are not fully correlated over the height.

Figure 5.3 shows the various frequency ranges for excitation of structures by wind and earthquake actions.

5.3 THE RANDOM VIBRATION OR SPECTRAL APPROACH

In some important papers in the 1960s, Davenport outlined an approach to the wind-induced vibration of structures based on random vibration theory (Davenport, 1961, 1963, 1964). Other significant early contributions to

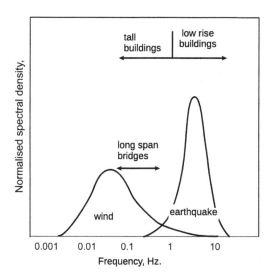

Figure 5.3 Dynamic excitation frequencies of structures by wind and earthquake.

the development of this approach were made by Harris (1963) and Vickery (1965, 1966).

The approach uses the concept of the stationary random process to describe wind velocities, pressures and forces. This assumes that the complexities of nature are such that we can never describe, or predict, perfectly (or 'deterministically') the forces generated by wind storms. However, we are able to use averaged quantities like standard deviations, correlations and spectral densities (or 'spectra') to describe the main features of both the exciting forces and the structural response. The *spectral density*, which has already been introduced in Section 3.3.4 and Figure 5.1, is the most important quantity to be considered in this approach, which primarily uses the *frequency domain* to perform calculations, and is alternatively known as the *spectral approach*.

Wind speeds, pressures and resulting structural response have generally been treated as stationary random processes in which the time-averaged or mean component is separated from the fluctuating component. Thus:

$$x(t) = \bar{x} + x'(t) \tag{5.1}$$

where $x(t)$ denotes either a wind velocity component, a pressure (measured with respect to a defined reference static pressure), or a structural response such as bending moment, stress resultant, and deflection; $\bar{x}$ is the mean or time-averaged component; and $x'(t)$ is the fluctuating component such that $\overline{x'(t)} = 0$. If x is a response variable, $x'(t)$ should include any resonant dynamic response resulting from excitation of any natural modes of vibration of the structure.

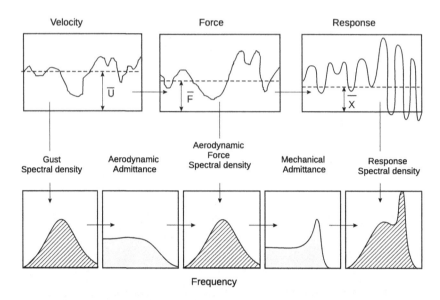

Figure 5.4 The random vibration (frequency domain) approach to resonant dynamic response (Davenport, 1963).

Figure 5.4 (after Davenport, 1963) illustrates graphically the elements of the spectral approach. The main calculations are done in the bottom row, in which the total mean square fluctuating response is computed from the spectral density, or 'spectrum', of the response. The latter is calculated from the spectrum of the aerodynamic forces, which are, in turn, calculated from the wind turbulence, or gust spectrum. The frequency-dependent *aerodynamic* and *mechanical admittance* functions form links between these spectra. The amplification at the resonant frequency, for structures with a low fundamental frequency, will result in a higher mean square fluctuating and peak response, than is the case for structures with a higher natural frequency, as previously illustrated in Figure 5.2.

The use of stationary random processes and Equation (5.1) is appropriate for large-scale windstorms such as gales in temperate latitudes and tropical cyclones. It may not be appropriate for some short-duration, transient storms, such as downbursts or tornadoes associated with thunderstorms. A method to compute the dynamic impact of these types of storms is discussed in Section 5.3.8.

5.3.1 Along-wind response of a single-degree-of-freedom structure

We will consider first the along-wind dynamic response of a small body, whose dynamic characteristics are represented by a simple mass-spring-damper

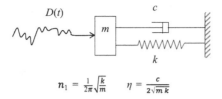

$$n_1 = \frac{1}{2\pi}\sqrt{\frac{k}{m}} \qquad \eta = \frac{c}{2\sqrt{mk}}$$

Figure 5.5 Simplified dynamic model of a structure.

(Figure 5.5), and which does not disturb the approaching turbulent flow significantly. This is a single-degree-of-freedom system and is reasonably representative of a structure consisting of a large mass supported by a column of low mass, such as a lighting tower or mast with a large array of lamps on top.

The equation of motion of this system under an aerodynamic drag force, $D(t)$, is given by Equation (5.2):

$$m\ddot{x} + c\dot{x} + kx = D(t) \tag{5.2}$$

The quasi-steady assumption (Section 4.6.2) for small structures allows the following relationship between mean square fluctuating drag force, and fluctuating longitudinal wind velocity, to be written as follows:

$$\overline{D'^2} = C_{Do}^2 \rho_a^2 \overline{U}^2 \overline{u'^2} A^2 \cong \overline{C_D}^2 \rho_a^2 \overline{U}^2 \overline{u'^2} A^2 = \frac{4\overline{D}^2}{\overline{U}^2}\overline{u'^2} \tag{5.3}$$

Equation (5.3) is analogous to Equation (4.17) for pressures.
Writing Equation (5.3) in terms of spectral density,

$$\int_0^\infty S_D(n)\cdot dn = \frac{4\overline{D}^2}{\overline{U}^2}\int_0^\infty S_u(n)\cdot dn$$

Hence $S_D(n) = \frac{4\overline{D}^2}{\overline{U}^2}S_u(n)$ \tag{5.4}

To derive the relationship between the fluctuating force and the response of the structure, represented by the simple dynamic system of Figure 5.5, the deflection is first separated into mean and fluctuating components, as in Equation (5.1):

$$x(t) = \overline{x} + x'(t) \tag{5.1}$$

The relationship between mean drag force, $\overline{D}$, and mean deflection, $\overline{X}$, is as follows:

$$\bar{D} = k \cdot \bar{x} \tag{5.5}$$

where k is the spring stiffness in Figure 5.5.

The spectral density of the deflection is related to the spectral density of the applied force as follows:

$$S_x(n) = \frac{1}{k^2} |H(n)|^2 S_D(n) \tag{5.6}$$

where $|H(n)|^2$ is known as the *mechanical admittance* for the single-degree-of-freedom dynamic system under consideration, given by Equation (5.7):

$$|H(n)|^2 = \frac{1}{\left[1 - \left(\dfrac{n}{n_1}\right)^2\right]^2 + 4\eta^2\left(\dfrac{n}{n_1}\right)^2} \tag{5.7}$$

$|H(n)|$, i.e. the square root of the mechanical admittance, may be recognized as the *dynamic amplification factor*, or *dynamic magnification factor*, which arises when the response of a single-degree-of-freedom system to a harmonic, or sinusoidal, excitation force is considered. n_1 is the undamped natural frequency, and η is the ratio of the damping coefficient, c, to critical damping, as shown in Figure (5.5).

By combining Equations (5.4) and (5.6), the spectral density of the deflection response can be related to the spectral density of the wind velocity fluctuations:

$$S_x(n) = \frac{1}{k^2} |H(n)|^2 \frac{4\bar{D}^2}{\bar{U}^2} S_u(n) \tag{5.8}$$

Equation (5.8) applies to those structures which have small frontal areas in relation to the length scales of atmospheric turbulence.

For larger structures, the velocity fluctuations do not occur simultaneously over the windward face and their correlation over the whole area, A, must be considered. To allow for this effect, an *aerodynamic admittance*, $X^2(n)$, is introduced:

$$S_x(n) = \frac{1}{k^2} |H(n)|^2 \frac{4\bar{D}^2}{\bar{U}^2} \cdot X^2(n) \cdot S_u(n)$$

Substituting for $\bar{D}$ from Equation (5.5),

$$S_x(n) = \frac{4\bar{x}^2}{\bar{U}^2} |H(n)|^2 \cdot X^2(n) \cdot S_u(n) \tag{5.9}$$

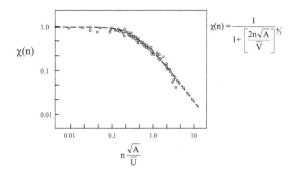

Figure 5.6 Aerodynamic admittance – experimental data and fitted function (Vickery, 1965, 1968).

For open structures, such as lattice frame towers, which do not disturb the flow greatly, $X^2(n)$ can be determined from the correlation properties of the upwind velocity fluctuations (see Section 3.3.6). This assumption is also made for solid structures, but $X^2(n)$ has also been obtained experimentally.

Figure 5.6 shows some experimental data with an empirical function fitted. Note that $X(n)$ tends towards 1.0 at low frequencies and for small bodies. The low-frequency gusts are nearly fully correlated, and fully envelope the face of a structure. For high frequencies, or very large bodies, the gusts are ineffective in producing total forces on the structure, due to their lack of correlation, and the aerodynamic admittance tends towards zero.

To obtain the mean square fluctuating deflection, the spectral density of deflection, given by Equation (5.9), is integrated over all frequencies:

$$\sigma_x^2 = \int_0^\infty S_x(n) \cdot dn = \int_0^\infty \frac{4\overline{x}^2}{\overline{U}^2} \left| H(n) \right|^2 \cdot X^2(n) \cdot S_u(n) \cdot dn \qquad (5.10)$$

The area underneath the integrand in Equation (5.10) can be approximated by two components, B and R, representing the 'background' and resonant parts, respectively (Figure 5.7).

Thus,

$$\sigma_x^2 = \frac{4\overline{x}^2 \sigma_u^2}{\overline{U}^2} \int_0^\infty \left| H(n) \right|^2 \cdot X^2(n) \cdot \frac{S_u(n)}{\sigma_u^2} \cdot dn \cong \frac{4\overline{x}^2 \sigma_u^2}{\overline{U}^2} \left[B + R \right] \qquad (5.11)$$

where

$$B = \int_0^\infty X^2(n) \cdot \frac{S_u(n)}{\sigma_u^2} \cdot dn \qquad (5.12)$$

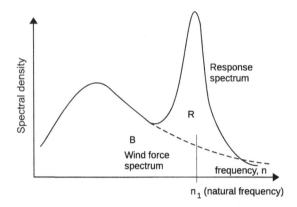

Figure 5.7 Background and resonant components of response.

$$R = X^2\left(n_1\right) \cdot \frac{S_u\left(n_1\right)}{\sigma_u^2} \int\limits_0^\infty \left|H\left(n\right)\right|^2 \cdot dn \tag{5.13}$$

The approximation of Equation (5.11) is based on the assumption that over the width of the resonant peak in Figure 5.7, the functions $X^2(n)$, $S_u(n)$ are constant at the values $X^2(n_1)$, $S_u(n_1)$. This is a good approximation for the relatively uniform spectral densities characteristic of wind loading, and when the resonant peak is narrow, as occurs when the damping is low (Ashraf Ali and Gould, 1985). The integral $\int |H(n)|^2.dn$ integrated for n from 0 to ∞, can be evaluated by the method of poles (Crandall and Mark, 1963) and shown to be equal to $(\pi n_1/4\eta)$.

The approximation of Equation (5.11) is used widely in code methods of evaluating along-wind response, and is discussed further in Chapter 15.

The background factor, B, represents the quasi-static response caused by gusts below the natural frequency of the structure. Importantly, it is independent of frequency, as shown by Equation (5.12), in which the frequency appears only in the integrand, and thus 'integrated out'. For many structures under wind loading, B is considerably greater than R, i.e. the background response is dominant in comparison with the resonant response. An example of such a structure is that whose response is shown in Figure 5.2(b).

5.3.2 Gust response factor

A commonly used term in wind engineering is *gust response factor*. The term *gust loading factor* was used by Davenport (1967), and *gust factor* by Vickery (1966). These essentially have the same meaning, although sometimes the factor is applied to the effective applied loading, and sometimes to the response of the structure. The term 'gust factor' is better applied to the wind speed itself (Section 3.3.3).

The *gust response factor*, *G*, may be defined as the ratio of the expected maximum response (e.g. deflection or stress) of the structure in a defined time period (e.g. 10 minutes or 1 hour), to the mean, or time-averaged response, in the same time period. It really only has meaning in stationary or near-stationary winds such as those generated by large scale synoptic wind events such as gales from depressions in temperate latitudes, or tropical cyclones (see Chapter 2).

The expected maximum response of the simple system described in Section 5.3.1 can be written:

$$\hat{x} = \bar{x} + g\sigma_x$$

where *g* is a *peak factor*, which depends on the time interval for which the maximum value is calculated, and the frequency range of the response.

From Equation (5.11),

$$G = \frac{\hat{x}}{\bar{x}} = 1 + g\frac{\sigma_x}{\bar{x}} = 1 + 2g\frac{\sigma_u}{\bar{U}}\sqrt{B+R} \qquad (5.14)$$

Equation (5.14), or variations of it, are used in several codes and standards for wind loading, for simple estimations of the along-wind dynamic loading of structures. The usual approach is to calculate *G* for the modal coordinate in the first mode of vibration, a_1, and then to apply it to a mean load distribution on the structure, from which all responses, such as bending moments, are calculated. This is an approximate approach which works reasonably well for some structures and load effects, such as the base bending moment of tall buildings. However, in other cases, it gives significant errors and should be used with caution (e.g. Holmes, 1994; Vickery, 1995 – see also Chapter 11).

5.3.3 Peak factor

The along-wind response of structures to wind has a probability distribution which is closely Gaussian. For this case, Davenport (1964) derived the following expression for the expected peak factor, *g*:

$$g = \sqrt{2 \log_e (vT)} + \frac{0.577}{\sqrt{2 \log_e (vT)}} \qquad (5.15)$$

where v is the 'cycling rate' or effective frequency for the response; this is often conservatively taken as the natural frequency, n_1. *T* is the time interval over which the maximum value is required.

5.3.4 Dynamic response factor

In transient or non-stationary winds such as downbursts from thunderstorms, for example, the use of a gust factor, or gust response factor, is rather meaningless. The gust response factor is also meaningless in cases when the mean response is very small or zero (such as cross-wind response). In these cases, use of a 'dynamic response factor' is more appropriate. This approach has been adopted recently in some codes and standards for wind loading. The dynamic response factor may be defined in the following way:

Dynamic response factor = (maximum response including resonant and correlation effects)/(maximum response calculated ignoring both resonant and correlation effects)

The denominator is in fact the response calculated using 'static' methods in codes and standards. As defined above, the dynamic response factor will usually have a value close to 1. A value greater than 1 can only be caused by a significant resonant response.

The use of the gust response factor and dynamic response factor in wind loading codes and standards, will be discussed further in Chapter 15.

5.3.5 Influence coefficient

When considering the action of a time-dependent and spatially varying load such as wind loading on a continuous structure, the *influence coefficient* or *influence line* is an important parameter. To appreciate the need for this, we must understand the concept, familiar to structural designers, of 'load effect'. A load effect is not the load itself but a parameter resulting from the loading which is required for comparison with design criteria. Examples are internal forces or moments such as bending moments or shear forces, stresses or deflections. The influence line represents the value of a single load effect as a unit (static) load is moved around the structure.

Two examples of influence lines are given in Figure 5.8. Figure 5.8a shows the influence lines for the bending moment and shear force at a level, s, halfway up a lattice tower. These are relatively simple functions; in the case of the shear force loads (or wind pressures) above the level s, have uniform effect on the shear force at that level. The influence line for the bending moment varies linearly from unity at the top to zero at the level s; thus wind pressures at the top of the structure have a much larger effect than those lower down, on the bending moment, which, in turn, is closely related to the axial forces in the leg members of the tower. It should be noted that loads or wind pressures below the level s have *no* effect on the shear force or bending moment at that level.

Figure 5.8b shows the influence line for the bending moment at a point in an arch roof. In this case, the sign of the influence line changes along the

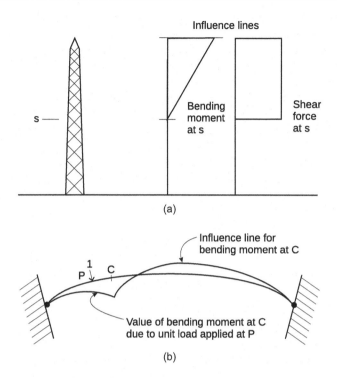

Figure 5.8 (a and b) Examples of influence lines for an arch roof and a tower.

arch. Thus wind pressures applied in the same direction at different parts of the roof may have opposite effects on the bending moment at C, M_c.

It is important to take into account these non-uniform influences when considering the structural effects of wind loads, even for apparently simple structures, especially for the fluctuating part of the loading.

5.3.6 Along-wind response of a structure with distributed mass – modal analysis

The usual approach to the calculation of the dynamic response of multi-degree-of freedom structures to dynamic forces, including resonance effects, is to expand the complete displacement response as a summation of components associated with each of the natural modes of vibration:

$$x(z,t) = \sum a_j(t)\phi_j(z) \tag{5.16}$$

where j denotes the natural modes; z is a spatial coordinate on the structure; $a_j(t)$ is a time-varying modal (or generalized) coordinate; and $\phi_j(z)$ is a mode shape for the jth mode.

Modal analysis is discussed in most texts on structural dynamics (e.g. Clough and Penzien (1975), Warburton (1976)).

The approach will be described here in the context of a two-dimensional or 'line-like' structure, with a single spatial coordinate, z, but it can easily be extended to more complex geometries.

Equation (5.16) can be used to determine the complete response of a structure to random forcing, i.e. including the mean component, $\bar{x}$, and the sub-resonant (background) fluctuating component, as well as the resonant responses.

The result of this approach is that separate equations of motion can be written for the modal coordinate $a_j(t)$, for each mode of the structure:

$$G_j \ddot{a}_j + C_j \dot{a}_j + K_j a_j = Q_j(t) \tag{5.17}$$

where

G_j is the generalized mass equal to $\int_0^L m(z)\phi_j^2(z)\,dz$

$m(z)$ is the mass per unit length along the structure

L is the length of the structure

C_j is the modal damping $(=2\eta_j G_j \omega_j)$

K_j is the modal stiffness

η_j is the damping as a fraction of critical for the jth mode

ω_j is the natural undamped circular frequency for the jth mode $(= 2\pi n_j = \sqrt{\dfrac{K_j}{G_j}})$

$Q_j(t)$ is the generalized force, equal to $\int_0^L f(z,t)\phi_j(z)\,dz$

$f(z, t)$ is the force per unit length along the structure

$f(z, t)$ can be taken as along-wind or cross-wind forces. For along-wind forces, applying a 'strip' assumption, which relates the forces on a section of the structure with the flow conditions upstream of the section, it can be written as:

$$f(z,t) = C_d(z) \cdot b(z) \frac{1}{2}\rho_a U^2(z,t) \tag{5.18}$$

where

$C_d(z)$ is a local drag coefficient

$b(z)$ is the local breadth

$U(z, t)$ is the longitudinal velocity upstream of the section. If the structure is moving, this should be a relative velocity, which then generates an aerodynamic damping force (Section 5.5.1 and Holmes (1996a)). However, at this point we will assume the structure is stationary, in which case $U(z, t)$ can be written as:

$$U(z,t) = \bar{U}(z) + u'(z,t)$$

where $u'(z, t)$ is the fluctuating component of longitudinal velocity (zero mean).

Then from Equation (5.18), we get:

$$f(z, t) = C_d(z) \cdot b(z) \rho_a \left[\frac{1}{2} \bar{U}^2(z) + \bar{U}(z) u'(z,t) + \frac{1}{2} u'^2(z,t) \right]$$

Neglecting the third term within the square brackets, the fluctuating sectional along-wind force is given by:

$$f'(z, t) = C_d(z) \cdot b(z) \rho_a \bar{U}(z) u'(z,t)$$

and the instantaneous fluctuating generalized force is therefore:

$$Q'_j(t) = \int_0^L f'(z,t) \phi_j(z) \, dz = \int_0^L C_d(z) \cdot b(z) \rho_a \bar{U}(z) u'(z,t) \phi_j(z) dz$$

Applying tbe same procedure used in Section 4.6.6, the mean square generalized force is:

$$\overline{Q_j'^2} = \int_0^L \int_0^L \iint \overline{f'(z_1) f'(z_2)} \phi_j(z_1) \phi_j(z_2) dz_1 dz_2$$

$$= \rho_a^2 \int_0^L \int_0^L \overline{u'(z_1) u'(z_2)} C_d(z_1) \cdot C_d(z_2) b(z_1) b(z_2) \bar{U}(z_1) \bar{U}(z_2) \phi_j(z_1) \phi_j(z_2) dz_1 dz_2$$

This can be simplified for a uniform cross-section, with $C_d(z)$ and $b(z)$ constant with z:

$$\overline{Q_j'^2} = (\rho_a C_d b)^2 \int_0^L \int_0^L \overline{u'(z_1) u'(z_2)} \; \bar{U}(z_1) \bar{U}(z_2) \phi_j(z_1) \phi_j(z_2) dz_1 dz_2 \qquad (5.19)$$

where $\overline{u'(z_1) u'(z_2)}$ is the covariance for the fluctuating velocities at heights z_1 and z_2. If the standard deviation of velocity fluctuations is constant with z, then the covariance can be written as:

$$\overline{u'(z_1) u'(z_2)} = \sigma_u^2 \rho_{uu}(z_1, z_2)$$

where $\rho_{uu}(z_1, z_2)$ is the correlation coefficient for fluctuating velocities at heights z_1 and z_2, defined in Section 3.3.5.

The spectral density of $Q'_j(t)$ can be obtained in analogous way to the mean square value:

$$S_{Qj}(n) = (\rho C_d b)^2 \int_0^L \int_0^L Co(z_1, z_2, n) \bar{U}(z_1) \bar{U}(z_2) \phi_j(z_1) \phi_j(z_2) dz_1 dz_2 \quad (5.20)$$

where $Co(z_1, z_2, n)$ is the co-spectral density of the longitudinal velocity fluctuations (Section 3.3.6), (defined in random process theory – e.g. Bendat and Piersol (2010)).

Analogously with Equation (5.6), the spectral density of the modal coordinate $a_j(t)$ is given by:

$$S_{aj}(n) = \frac{1}{K_j^2} |H_j(n)|^2 S_{Qj}(n) \quad (5.21)$$

where the mechanical admittance for the jth mode is:

$$|H_j(n)|^2 = \frac{1}{\left[1 - \left(\frac{n}{n_j}\right)^2\right]^2 + 4\eta_j^2 \left(\frac{n}{n_j}\right)^2} \quad (5.22)$$

The mean square value of $a_j(t)$ can then be obtained by integration of Equation (5.21) with respect to frequency:

$$\overline{a_j'^2} = \int_0^\infty S_{aj}(n) \cdot dn$$

Applying Equation (5.16), the mean square displacement is obtained from:

$$\overline{x'^2} = \sum_{j=1}^N \sum_{k=1}^N \overline{a_j' a_k'} \phi_j(z) \phi_k(z)$$

If cross-coupling between modes can be neglected (however, see Section 5.3.7), the above equation becomes:

$$\overline{x'^2} = \sum_{j=1}^N \overline{a_j'^2} \phi_j^2(z) \quad (5.23)$$

The mean square value of any other response, r (e.g. bending moment, stress), can similarly be obtained if the response, R_j for a unit value of the modal coordinate, a_j, is known. That is:

$$\overline{r'^2} = \sum_{j=1}^{N} \overline{a_j'^2} R_j^2$$

(5.24)

5.3.7 Along-wind response of a structure with distributed mass – separation of background and resonant components

In the case of wind loading, the method described in the previous section is not an efficient one. For the vast majority of structures, the natural frequencies are at the high end of the range of forcing frequencies from wind loading. Thus, the resonant components as j increases in Equation (5.16) become very small. However, the contributions to the mean and background fluctuating components for j greater than 1 in Equation (5.16) *may not be small*. Thus, it is necessary to include higher modes ($j>1$) in Equation (5.16) not for their resonant contributions, but to accurately determine the mean and background contributions. For example, Vickery (1995) found that over twenty modes were required to determine the mean value of a response, and over ten values were needed to compute the variance. Also, for the background response, cross coupling of modes cannot be neglected, i.e. Equation (5.23) is not valid.

A much more efficient approach is to separately compute the mean and background components, as for a quasi-static structure. Thus, the total peak response, $\hat{r}$, can be taken to be:

$$\hat{r} = r + \sqrt{\hat{r}_B^2 + \sum_j \left(\hat{r}_{R,j}^2 \right)}$$

(5.25)

where $\hat{r}_B$ is the peak background response equal to $g_B \sigma_B$; and $\hat{r}_{R,j}$ is the peak resonant response computed for the jth mode, equal to $g_j \sigma_{R,j}$. This approach is illustrated in Figure 5.1.

g_B and g_j are peak factors which can be determined from Equation (5.15); in the case of the resonant response, the cycling rate, ν, in Equation (5.15), can be taken as the natural frequency, n_j.

The mean square value of the quasi-static fluctuating (background) value of any reponse, r, is:

$$\overline{r_B^2} = \sigma_B^2 \int_0^L \int_0^L \overline{f'(z_1)f'(z_2)} \ i_r(z_1) i_r(z_2) dz_1 dz_2$$

$$= \rho_a^2 \int_0^L \int_0^L \overline{u'(z_1)u'(z_2)} \ C_d(z_1)$$

$$\cdot C_d(z_2) b(z_1) b(z_2) \bar{U}(z_1) \bar{U}(z_2) \ i_r(z_1) i_r(z_2) dz_1 dz_2$$

(5.26)

where $i_r(z)$ is the influence line for r, i.e. the value of r when a unit load is applied at z.

The resonant component of the response in mode j, can be written, to a good approximation as:

$$\overline{r_{R,j}'^2} = \frac{S_{Qj}(n_j)R_j^2}{K_j^2} \int_0^\infty |H_j(n)|^2 \cdot dn = \frac{\pi n_1 \cdot S_{Qj}(n_j)R_j^2}{4\eta_j K_j^2} \tag{5.27}$$

because the integral $\int_0^\infty |H_j(n)|^2 \cdot dn$, evaluated by the method of poles (Crandall and Mark, 1963), is equal to $(\pi n_j/4\eta_j)$.

5.3.8 Along-wind response to non-stationary (transient) winds

It is clear that downburst winds as generated by severe thunderstorms produce time histories which are non-stationary, as shown in Figure 3.9. Calculation of dynamic response to such winds requires a different approach to those described earlier in this chapter, for turbulent winds generated within the boundary layers of synoptic winds – which can be considered statistically stationary. One such approach has been 'borrowed' from earthquake engineering.

The use of Duhamel's Integral is a standard technique for calculation of the dynamic response of structures, to transient loadings such as blast loadings, or earthquakes. Since it represents the response to an arbitrary loading as the superposition of the response to many discrete impulses, this technique is limited to linear structures. However, structures with non-linear characteristics (e.g. stiffness and damping) can usually be linearized with sufficient accuracy to make use of this very convenient technique.

The displacement response of any linear system to an arbitrary force input $D(t)$, can be written as:

$$x(t) = \int_0^t h(t-\tau) \cdot D(\tau) d\tau \tag{5.28}$$

where $h(t)$ is the unit impulse response function. Equation (5.28) is a 'convolution integral'.

The unit impulse response function for a simple mass-spring-damper system (Figure 5.5), with an equation of motion given by Equation (5.2), depends on the value of the damping ratio, η, and the natural circular frequency, ω_1, where $\eta = c/2\sqrt{(mk)}$, and $\omega_1 = \sqrt{(k/m)}$.

For $\eta < 1$,

$$h(t) = \left(\frac{1}{m \cdot \omega_1\sqrt{1-\eta^2}}\right) \exp[-\eta\omega_1 t] \cdot \sin\left[\omega_1\sqrt{1-\eta^2} \cdot (t)\right] \tag{5.29}$$

Hence, from Equation (5.28),

$$x(t) = \left(\frac{1}{m \cdot \omega_1 \sqrt{1-\eta^2}}\right)$$

$$\times \int_0^t \exp\left[-\eta\omega_1(t-\tau)\right] \cdot \sin\left[\omega_1\sqrt{1-\eta^2}.(t-\tau)\right] \cdot D(\tau)d\tau \qquad (5.30)$$

The right-hand side of Equation (5.30) is known as *Duhamel's Integral*, e.g. Clough and Penzien (1975).

The effective quasi-static along-wind force, $D_{eff}(t)$, is then given by the product of the displacement response, $x(t)$, and the stiffness, k:

$$D_{eff}(t) = k \cdot x(t) = \left(\frac{\omega_1}{\sqrt{1-\eta^2}}\right)$$

$$\times \int_0^t \exp\left[-\eta\omega_1(t-\tau)\right] \cdot \sin\left[\omega_1\sqrt{1-\eta^2} \cdot (t-\tau)\right] \cdot D(\tau)d\tau \quad (5.31)$$

The dynamic response factor (Section 5.3.4) can then be obtained as the ratio of the maximum value of the effective static force, $D_{eff}(t)$, in the time history, to the maximum value of the applied force, $D(t)$, in the same time history. Note that these maxima will generally not occur at the same time, t.

Figure 5.9 shows calculated dynamic response factors to the downburst measured in 2002 at Lubbock, Texas (Figure 3.9), for structures with periods of 6–100 seconds (circular frequencies of 0.06 to 1 rad/sec), and damping ratios from 0.1 to 3 (Holmes *et al.*, 2005). For structures with shorter periods, more closely spaced time intervals are required in the recorded wind time histories. The information in Figure 5.9 resembles closely that provided in 'response spectra' for earthquake design.

5.4 EFFECTIVE STATIC LOADING DISTRIBUTIONS

5.4.1 Introduction

Effective static wind load distributions are those loadings that produce the correct expected values of peak load effects, such as bending moments, axial forces in members, or deflections, generated by the fluctuating wind loading. The effective peak loading distributions associated with the mean wind loading, the fluctuating quasi-static or background response, and the resonant response are identified, and combined to give a total effective peak wind loading distribution.

Following the procedure described in the previous sections, effective static peak loading distributions can be separately derived for the following three components:

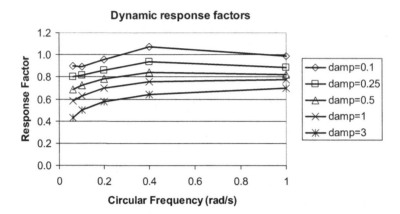

Figure 5.9 Dynamic response factors for the response of structures of various frequencies and damping to the Lubbock downburst of 2002 (Holmes et al., 2005).

a. mean component;
b. background or sub-resonant component; and
c. resonant components.

The background component is derived making use of a formula derived by Kasperski and Niemann (1992), and depends on the load effect in question. The resonant component comprises an inertial loading, similar to that used in earthquake engineering.

The approach will be illustrated by examples of buildings with long-span roofs and freestanding lattice towers and chimneys. Simplifications will be suggested to make the method more palatable to structural engineers used to analyzing and designing with static loadings.

The main advantage of the effective static load distribution approach is that the distributions can be applied to static structural analysis computer programs for use in detail structural design. This approach can be applied to any type of structure (Holmes and Kasperski, 1996).

5.4.2 Mean load distributions

The mean wind loading on a structure which does not distort the airflow significantly can be obtained simply by relating the mean local pressure or force per unit length to the mean wind speed. Thus, for the mean along-wind force per unit height acting on a tower:

$$\bar{f}(z) = \left[0.5\rho_a \bar{U}(z)^2\right] C_d b(z)$$

$$(5.32)$$

where ρ_a is the density of air, $\bar{U}(z)$ is the mean wind speed at height z, C_d is a drag coefficient and $b(z)$ is the reference breadth at the height z.

The mean value of any load effect (e.g. shear force, bending moment, deflection) can be obtained by integrating the local load with the influence line over the appropriate height. However, if the purpose is to derive an equivalent static loading, then Equation (5.32) is already in this form.

In the case of 'solid' structures (such as cooling towers and most buildings) with at least two dimensions comparable to the size of turbulent eddies in the atmosphere, Equation (5.32) cannot be used, but wind-tunnel tests can be employed to determine mean pressure coefficients, $\bar{C}_p$, which can then be used with a reference wind speed, $\bar{U}_h$, to determine local mean pressures on the structure:

$$\bar{p}(z) = \left[0.5\rho_a \bar{U}_h^2 \right] \bar{C}_p \tag{5.33}$$

5.4.3 Background-loading distributions

As discussed previously, the background wind loading is the quasi-static loading produced by fluctuations due to turbulence, but with frequencies too low to excite any resonant response. Over the duration of a windstorm, because of the incomplete correlations of pressures at various points on a structure, loadings varying in both space and time will be experienced. It is necessary to identify those instantaneous loadings which produce the critical load effects in a structure. The formula which enables this to be done is the 'Load-Response Correlation' formula, derived by Kasperski and Niemann (1992).

This formula gives the expected 'instantaneous' pressure distribution, associated with the maximum or minimum load effect. Thus, for the maximum value, $\hat{r}$, of a load effect, r:

$$(p_i)_{\hat{r}} = \bar{p}_i + g_B \cdot \rho_{r,pi} \cdot \sigma_{pi} \tag{5.34}$$

where $\bar{p}_i$ and σ_{pi} are the mean and root-mean-square (r.m.s.) pressures at point or panel, i.

$\rho_{r,pi}$ is the correlation coefficient between the fluctuating load effect, and the fluctuating pressure at point i (this can be determined from the correlation coefficients for the fluctuating pressures at all points on the tributary area, and from the influence coefficients); and g_B is the peak factor for the background response which normally lies in the range of 2.5–5.

A simple example of the application of this formula is given in Appendix F.

The second term on the right-hand side of Equation (5.34) represents the background fluctuating load distribution. This term can also be written in the form of a continuous distribution:

$$f_B(z) = g_B \rho(z) \sigma_p(z) \tag{5.35}$$

where $\rho(z)$ denotes the correlation coefficient between the fluctuating load at position z on the structure, and the load effect of interest; and $\sigma_p(z)$ is the r.m.s. fluctuating load at position z.

In Equation (5.34), the correlation coefficient, $\rho_{r,pi}$, can be shown to be given by:

$$\rho_{r,pi} = \sum_k \left[\overline{p_i(t)p_k(t)}\, i_k \right] \Big/ \left(\sigma_{pi}\sigma_r \right) \tag{5.36}$$

where i_k is the influence coefficient for a pressure applied at position, k.

The standard deviation of the structural load effect, σ_r, is given by (Holmes and Best, 1981):

$$\sigma_r^2 = \sum_i \sum_k \overline{p_i(t)p_k(t)}\, i_i i_k \tag{5.37}$$

When the continuous form is used, Equations (5.36) and (5.37) are replaced by an integral form (Holmes, 1996b):

$$\rho(z) = \frac{\int_s^h \overline{f'(z)f'(z_1)}\, i_r(z_1)b(z_1)dz_1}{\left\{ \int_s^h \int_s^h \overline{f'(z_1)f'(z_2)}\, i_r(z_1)i_r(z_2)b(z_1)b(z_2)dz_1\,dz_2 \right\}^{1/2} \sqrt{\overline{f'^2(z)}}} \tag{5.38}$$

where $i_r(z)$ now denotes the influence function for the load effect, r, as a function of position z, and $b(z)$ is the breadth of the structure at position z. For a vertical structure, the integrations in Equation (5.38) are carried out for the height range from s, the height at which the load effect (e, g. bending moment, shearing force, member force) is being evaluated, and the top of the structure, h.

Clearly, since the correlation coefficient, $\rho_{r,\,pi}$, calculated by Equation (5.36), or $\rho(z)$ calculated by Equation (5.38), are dependent on the particular load effect, then the background load distribution will also depend on the nature of the load effect.

Figures 5.10 and 5.11 give examples of background loading distributions calculated using these methods. Figure 5.10 shows examples of peak load (mean+background) distributions for a support reaction (dashed) and a bending moment (dotted) in an arch roof. These distributions fall within an envelope formed by the maximum and minimum pressure distributions along the arch. It should also be noted that the distribution for the bending moment at C includes a region of positive pressure.

Figure 5.11 shows the background pressure distribution for the base shear force and base bending moment on a lattice tower 160 m high, determined by calculation using Equation (5.35), (Holmes, 1996b). The maxima

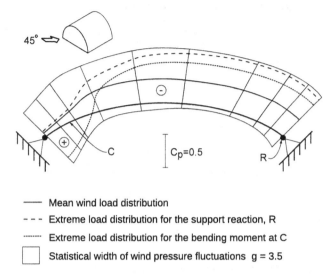

Figure 5.10 Mean and effective background loading distributions for an arch roof (after Kasperski and Niemann, 1992).

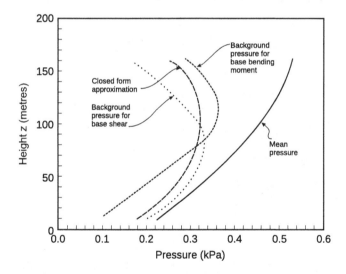

Figure 5.11 Mean and effective background load distributions for a 160 m tower (Holmes, 1996b).

for these distributions occur at around 70 m height for the base shear and about 120 m for the base bending moment. An approximation (Holmes, 1996b) to these distributions, which is independent of the load effect but dependent on the height at which the load effect is evaluated, is also shown in Figure 5.11.

5.4.4 Load distributions for resonant response

The equivalent load distribution for the resonant response in the first mode can be represented as a distribution of inertial forces over the length of the structure. Thus, an equivalent load distribution for the resonant response, $f_R(z)$, is given by:

$$f_R(z) = g_R m(z)(2\pi n_1)^2 \sqrt{\overline{a'^2}}\, \phi_1(z) \qquad (5.39)$$

where g_R is the peak factor for resonant response; $m(z)$ is a mass per unit length; n_1 is the first mode natural frequency; $\sqrt{\overline{a'^2}}$ is the r.m.s. modal coordinate (resonant contribution only), and $\phi_1(z)$ is the mode shape for the first mode of vibration.

Determination of the r.m.s. modal coordinate requires knowledge of the spectral density of the exciting forces, the correlation of those forces at the natural frequency (or aerodynamic admittance), and the modal damping and stiffness, as discussed in Sections 5.3.1 and 5.3.6.

5.4.5 Combined load distribution

The combined effective static load distribution for mean, background and resonant components (one mode) is obtained as follows:

$$f_c(z) = \overline{f}(z) + W_{\text{back}} f_B(z) + W_{\text{res}}(z) f_R(z) \qquad (5.40)$$

where the absolute values of the weighting factors W_{back} and W_{res} are given by:

$$|W_{\text{back}}| = \frac{g_B \sigma_{r,B}}{\left(g_B^2 \sigma_{r,B}^2 + g_R^2 \sigma_{r,R}^2\right)^{1/2}}$$

$$\qquad (5.41)$$

$$|W_{\text{res}}| = \frac{g_R \sigma_{r,R}}{\left(g_B^2 \sigma_{r,B}^2 + g_R^2 \sigma_{r,R}^2\right)^{1/2}}$$

The above assumes that the fluctuating background and resonant components are uncorrelated with each other, so that Equation (5.25) applies. W_{back} and W_{res} will be positive if the influence line of the load effect, r, and the mode shape are both positive, but either could be negative in many cases.

By multiplying by the influence coefficient and summing over the whole structure, Equations (5.40) will give Equation (5.25) for the total peak load effect.

An alternative to Equation (5.40) is to combine the background and resonant distributions in the same way that the load effects themselves were combined (Equation (5.25)), i.e.:

$$f_C(z) = \overline{f}(z) + \sqrt{\left[f_B(z)\right]^2 + \left[f_R(z)\right]^2} \tag{5.42}$$

The second term on the right-hand side is an approximation, to the correct combination formula (Equation 5.40), and is independent of the load effect or its influence line. Equation (5.42) with positive and negative signs taken in front of the square root is, in fact, an 'envelope' of the combined distributions for all load effects. However, it is a good approximation for cases where the influence line $I_r(z)$, and the mode shape have the same sign for all z (Holmes, 1996b).

Examples of the combined distribution, calculated using Equation (5.42), are given in Figure 5.12 for a 160 m lattice tower (Holmes, 1996b). When the resonant component is included, the combined loading can exceed the 'peak gust pressure envelope', i.e. the expected limit of non-simultaneous peak pressures, as is the case in Figure 5.12 for the bending moment at 120 m.

Equations (5.40) and (5.41) can be extended to cover more than one resonant mode by introducing an additional term for each participating mode of vibration. An example of combined equivalent static load distributions, when more than one resonant mode contributes significantly, is discussed in Section 12.3.6.

5.5 AEROELASTIC FORCES

For very flexible, dynamically wind-sensitive structures, the motion of the structure may itself generate aerodynamic forces. In extreme cases, the forces may be of such a magnitude and act in a direction to sustain or

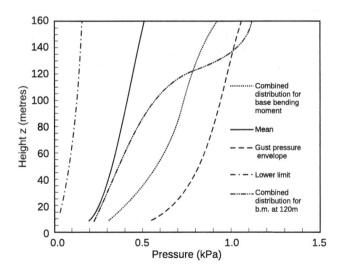

Figure 5.12 Mean and combined (including resonant contributions) load distributions for a 160 m tower (Holmes, 1996b).

increase the motion; in these cases, an unstable situation may arise such that a small disturbance may initiate a growing amplitude of vibration. This is known as 'aerodynamic instability' – examples of which are the 'galloping' of iced-up transmission lines and the flutter of long suspension bridges (such as the Tacoma Narrows Bridge failure of 1940).

On the other hand 'aerodynamic damping' forces may act to reduce the amplitude of vibration, induced by wind. This is the case with the along-wind vibration of tall structures such as lattice towers of relatively low mass.

The subject of aeroelasticity and aerodynamic stability is a complex one, and one which most engineers will not need to be involved with. However, some discussion of the principles will be given in this section. A number of general reviews are available of this aspect of wind loads (e.g. Scanlan, 1982).

5.5.1 Aerodynamic damping

Consider the along-wind motion of a structure with a square cross section, as shown in Figure 5.13. Ignoring initially the effects of turbulence, we will consider only the mean wind speed, $\bar{U}$. If the body itself is moving in the along-wind direction with a velocity, $\dot{x}$, the relative velocity of the air with respect to the moving body is $\bar{U} - \dot{x}$. We then have a drag force per unit length of the structure equal to:

$$D = C_D \frac{1}{2} \rho_a b \left(\bar{U} - \dot{x} \right)^2 \cong C_D \frac{1}{2} \rho_a b \bar{U}^2 \left(1 - \frac{2\dot{x}}{\bar{U}} \right)$$

$$= C_D \frac{1}{2} \rho_a b \bar{U}^2 - C_D \rho_a b \bar{U} \dot{x}$$

for small values of $\dot{x} / \bar{U}$. The second term on the right-hand side is a quantity proportional to the structure velocity, $\dot{x}$, and this represents a form of damping. When transferred to the left-hand side of the equation of motion

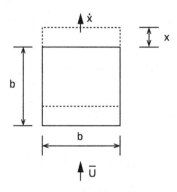

Figure 5.13 Along-wind relative motion and aerodynamic damping.

(Equation 5.2), it combines with the structural damping term, $c\dot{x}$, to reduce the aerodynamic response.

For a continuous structure, the along-wind aerodynamic damping coefficient in mode j can be shown to be (Holmes, 1996a):

$$C_{\mathrm{aero},j} = \rho_a \int_0^L C_d(z)b(z)\bar{U}(z)\phi_j^2(z)\,dz$$

giving a critical aerodynamic damping ratio, $\eta_{\mathrm{aero},j}$, equal to:

$$\eta_{\mathrm{aero},j} = \frac{\rho_a \displaystyle\int_0^L C_d(z)b(z)\bar{U}(z)\phi_j^2(z)\,dz}{4\pi n_j G_j} \tag{5.43}$$

where G_j is the generalised mass (see Equation (5.17)).

5.5.2 Galloping

Galloping is a form of single-degree-of-freedom aerodynamic instability, which can occur for long bodies with certain cross sections. It is a pure translational, cross-wind vibration. Consider a section of a body with a square cross section as shown in Figure 5.14.

The aerodynamic force per unit length in the z-direction, is obtained from the lift and drag by a change of axes (Figure 4.3):

$$F_z = D\sin\alpha + L\cos\alpha = \frac{1}{2}\rho_a\bar{U}^2 b\left(C_D\sin\alpha + C_L\cos\alpha\right)$$

Hence, $\quad \dfrac{dF_z}{d\alpha} = \dfrac{1}{2}\rho_a\bar{U}^2 b\left(C_D\cos\alpha + \dfrac{dC_D}{d\alpha}\sin\alpha - C_L\sin\alpha + \dfrac{dC_L}{d\alpha}\cos\alpha\right)$

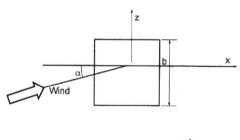

Figure 5.14 Cross-wind relative motion and galloping.

Setting α equal to zero (for flow in the x-direction),

$$\frac{dF_z}{d\alpha} = \frac{1}{2}\rho_a \bar{U}^2 b\left(C_D + \frac{dC_L}{d\alpha}\right) \qquad (5.44)$$

If the body is moving in the z direction with velocity, $\dot{z}$, there will be a reduction in the apparent angle of attack of the flow by $\dot{z}/\bar{U}$, or an increase in angle of attack by $-\dot{z}/\bar{U}$.

Equation (5.44) shows:

$$\Delta F_z \cong \frac{1}{2}\rho_a \bar{U}^2 b\left(C_D + \frac{dC_L}{d\alpha}\right)\Delta\alpha$$

Substituting, $\Delta\alpha = -\dot{z}/\bar{U}$,

$$\Delta F_z \cong \frac{1}{2}\rho_a \bar{U}^2 b\left(C_D + \frac{dC_L}{d\alpha}\right)\left(-\frac{\dot{z}}{\bar{U}}\right)$$

$$= -\frac{1}{2}\rho_a \bar{U} b\left(C_D + \frac{dC_L}{d\alpha}\right)\dot{z} \qquad (5.45)$$

If $\left(C_D + \dfrac{dC_L}{d\alpha}\right) < 0$, there will be an aerodynamic force in the z-direction, proportional to the velocity of motion, $\dot{z}$, or a *negative* aerodynamic damping term when it is transposed to the left-hand-side of the equation of motion. This is known as 'den Hartog's criterion'.

This situation can arise for a square section, with a negative slope $\dfrac{dC_L}{d\alpha}$, and with a magnitude greater than C_D, for α equal to zero (Figure 5.14).

Galloping instability will be initiated when the negative aerodynamic force overcomes the positive damping force due to structural damping. Hence, a critical wind speed for galloping instability, $\bar{U}_{crit}$, can be determined by equating the negative damping force from Equation (5.45), to the positive velocity-dependent structural damping:

$$\frac{1}{2}\rho_a \bar{U}_{crit} b\left(C_D + \frac{dC_L}{d\alpha}\right)\dot{z} + c\dot{z} = 0$$

Hence,

$$\bar{U}_{crit} = -2c\Big/\rho_a b\left(C_D + \frac{dC_L}{d\alpha}\right) = -2\eta_{struct}c_c\Big/\rho_a b\left(C_D + \frac{dC_L}{d\alpha}\right)$$

$$\bar{U}_{crit} = -8\pi\eta_{struct}n_1 m\Big/\rho_a b\left(C_D + \frac{dC_L}{d\alpha}\right) \qquad (5.46)$$

where η_{struct} is the ratio of the structural damping to critical damping $(c_c = 4\pi m n_1)$, m is the mass per unit length of the body, and n_1 is the natural frequency in the cross-wind (z) direction.

Equation (5.46) can be re-written as:

$$\frac{\bar{U}_{crit}}{n_1 b} = Sc.\frac{2}{-\left(C_d + \dfrac{dC_l}{d\alpha}\right)} \tag{5.47}$$

where Sc is known as the 'Scruton Number', or 'mass-damping' parameter:

$$Sc = \frac{4_{struct}m}{\rho_a b^2} \tag{5.48}$$

Low values of Scruton Number are a general indication of a propensity to wind-induced vibrations (see also Sections 11.5.1 and 11.5.2 which are concerned with the vortex-induced vibrations of towers and chimneys).

Equation (5.46) or (5.47) can be used to estimate the wind speed for the initiation of galloping instability, knowing the various aerodynamic and structural parameters on the right-hand side.

5.5.3 Flutter

Consider now a two-dimensional bluff body able to move, with elastic restraint, in both vertical translation and rotation (i.e. bending and torsion deflections).

The body shown in Figure (5.15) is being twisted, and the section shown is rotating with an angular velocity, $\dot{\theta}$, radians per second. This gives the relative wind, with respect to the rotating body, a vertical component of velocity at the leading edge of $\dot{\theta}d/2$, and hence a relative angle of attack between the apparent wind direction and the rotating body of $\dot{\theta}d/2$. This

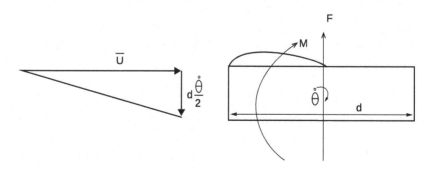

Figure 5.15 Aeroelastic forces generated by rotation of a cross section.

effective angle of attack can generate both a vertical force, and a moment if the centre of pressure is not collinear with the centre of rotation of the body. These aeroelastic forces can generate instabilities, if they are not completely opposed by the structural damping in the rotational mode. Aerodynamic instabilities involving rotation are known as 'flutter', using aeronautical parlance, and are a potential problem with the suspended decks of long-span bridges.

The equations of motion (per unit mass or moment of inertia) for two degrees-of-freedom of a bluff body can be written as follows (Scanlan and Tomko, 1971; Scanlan and Gade, 1977; Matsumoto, 1996):

$$\ddot{z} + 2\eta_z \omega_z \dot{z} + \omega_z^2 z = \frac{F_z(t)}{m} + H_1 \dot{z} + H_2 \dot{\theta} + H_3 \theta \tag{5.49}$$

$$\ddot{\theta} + 2\eta_\theta \omega_\theta \dot{\theta} + \omega_\theta^2 \theta = \frac{M(t)}{I} + A_1 \dot{z} + A_2 \dot{\theta} + A_3 \theta \tag{5.50}$$

The terms A_i, and H_i are linear aeroelastic coefficients, or *flutter derivatives*, which are usually determined experimentally for particular cross sections. They are functions of non-dimensional or *reduced* frequency. $F_z(t)$ and $M(t)$ are forces and moments due to other mechanisms which act on a static body (e.g. turbulent buffeting or vortex shedding). ω_z ($=2\pi n_z$) and ω_θ ($=2\pi n_\theta$) are the undamped circular frequencies in still air for vertical motion and rotation respectively.

Note that Equations (5.49) and (5.50) have been 'linearized', i.e. they only contain terms in $\dot{z}, \theta, \dot{\theta}$, etc. There could be smaller terms in $\dot{z}^2$, $\theta^2, \dot{\theta}^3$ etc. The two equations are 'coupled' second-order linear differential equations. The coupling arises from the ocurrence of terms in z and θ, or their derivatives in both equations. This can result in coupled aeroelastic instabilities, which are a combination of vertical (bending) and rotational (torsion) motions, depending on the signs and magnitudes of the A_i and H_i derivatives. All bridge decks will reach this state at a high enough wind speed.

Several particular types of instability for bluff bodies have been defined. Three of these are summarised in Table 5.1.

Coupled aeroelastic instabilities in relation to long-span bridge decks, and flutter derivatives, are further discussed in Chapter 12 – Bridges.

Table 5.1 Types of aerodynamic instabilities

Name	Conditions	Type of motion	Type of section
Galloping	$H_1 > 0$	Translational	Square section
'Stall' flutter	$A_2 > 0$	Rotational	Rectangle, H-section
'Classical' flutter	$H_2 > 0, A_1 > 0$	Coupled	Flat plate, airfoil

5.5.4 Lock-in

Motion-induced forces can occur during vibration produced by vortex shedding (Section 4.6.3). Through a feedback mechanism, the frequency of the shedding of vortices can 'lock-in' to the frequency of motion of the body. The strength of the vortices shed, and the resulting fluctuating forces are also enhanced. *Lock-in* has been observed many times during the vibration of lightly damped cylindrical structures such as steel chimneys, and occasionally during the vortex-induced vibration of long-span bridges.

5.6 FATIGUE UNDER WIND LOADING

5.6.1 Metallic fatigue

The 'fatigue' of metallic materials under cyclic loading has been well researched, although the treatment of fatigue damage under the random dynamic loading characteristic of wind loading is less well developed.

In the usual engineering failure model for the fatigue of metals, it is assumed that each cycle of a sinusoidal stress response inflicts an increment of damage which depends on the amplitude of the stress. Each successive cycle then generates additional damage which accumulates in proportion to the number of cycles until failure occurs. The results of constant amplitude fatigue tests are usually expressed in the form of an S-N curve, where S is the stress range (equal to twice the amplitude, s), and N is the number of cycles until failure. For many materials, the S-N curve is well approximated by a straight line when $\log S$ is plotted against $\log N$ (Figure 5.16). This implies an equation of the form:

$$NS^m = N(2s)^m = K \qquad (5.51)$$

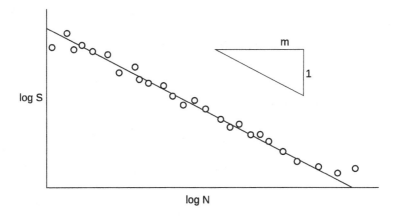

Figure 5.16 Form of a typical S-N curve relating stress range to cycles to failure.

where K is a constant which depends on the material, and the exponent m varies between 5 and 20.

A criterion for failure under repeated loading, for various stress ranges is given by Miner's Rule:

$$\sum \left(\frac{n_i}{N_i} \right) = 1 \qquad (5.52)$$

where n_i is the number of stress cycles with a range for which N_i cycles are required to cause failure. Thus, failure is expected when the sum of the fractional damage for all stress levels is unity.

Note that there is no restriction on the *order* in which the various stress ranges, or amplitudes, are applied in Miner's Rule. Thus we may apply it to a random loading process which can be considered as a series of cycles with randomly varying amplitudes.

5.6.2 Narrow-band fatigue loading

Some wind loading situations produce resonant 'narrow-band' vibrations. For example, the along-wind response of structures with low natural frequencies (Section 5.3.1), and cross-wind vortex-induced response of circular cylindrical structures with low damping. In these cases, the resulting stress variations can be regarded as quasi-sinusoidal with randomly varying amplitudes, as shown in Figure 5.17.

For a narrow-band random stress $s(t)$, the proportion of cycles with amplitudes in the range from s to $s+\delta s$, is $f_p(s)$. δs, where $f_p(s)$ is the probability density of the peaks. The total number of cycles in a time period, T, is $\nu_o^+ T$, where ν_o^+ is the rate of crossing of the mean stress. For narrow-band resonant vibration, ν_o^+ may be taken to be equal to the natural frequency of vibration.

Then the total number of cycles with amplitudes from s to δs is given by,

$$n(s) = \nu_o^+ T f_p(s) \cdot \delta s \qquad (5.53)$$

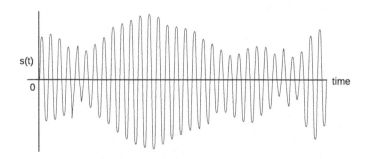

Figure 5.17 Stress-time history under narrow-band random vibrations

If $N(s)$ is the number of cycles at amplitude, s, to cause failure, then the fractional damage at this stress level is given by:

$$= \frac{n(s)}{N(s)} = \frac{v_0^+ T f_p(s)(2s)^m \, \delta \, s}{K}$$

where Equation (5.53) has been used for $n(s)$, and Equation (5.51) for $N(s)$.

The total expected fractional damage over all stress amplitudes is then, by Miner's Rule:

$$D = \sum_0^\infty \frac{n(s)}{N(s)} = \frac{v_0^+ T \int_0^\infty f_p(s)(2s)^m \, ds}{K} \tag{5.54}$$

Wind-induced narrow-band vibrations can be taken to have a normal or Gaussian probability distribution (Appendix C, Section C3.1). If this is the case, then the peaks or amplitudes, s, have a Rayleigh distribution (e.g. Crandall and Mark, 1963):

$$f_p(s) = \frac{s}{\sigma^2} \exp\left(-\frac{s^2}{2\sigma^2}\right) \tag{5.55}$$

where σ is the standard deviation of the entire stress history. Derivation of Equation (5.55) is based on the level-crossing formula of Rice (1944–1945) (see Appendix C, Section C5.2).

Substituting into Equation (5.54), we get:

$$D = \frac{v_0^+ T}{K^2} \int_0^\infty 2^m s^{m+1} \exp\left(-\frac{s^2}{2\sigma^2}\right) ds = \frac{v_0^+ T}{K} \left(2\sqrt{2}\sigma\right)^m \Gamma\left(\frac{m}{2}+1\right) \tag{5.56}$$

Here, the following mathematical result has been used (Dwight, 1961):

$$\int_0^\infty x^q \exp\left[-(rx)^p\right] dx = \frac{1}{pr^{q+1}}\left(\frac{q+1}{p}\right) \tag{5.57}$$

where $\Gamma(x)$ is the Gamma Function.

Equation (5.56) is a very useful 'closed-form' result, but it is restricted by two important assumptions:

- 'high-cycle' fatigue behaviour in which steel is in the elastic range, and for which an S-N curve of the form of Equation (5.51) is valid, has been assumed,

- narrow-band vibration in a single resonant mode of the form shown in Figure 5.17 has been assumed. In wind loading, this is a good model of the behaviour for vortex-shedding induced vibrations in low turbulence conditions. For along-wind loading, the background (sub-resonant) components are always important and result in a random wide-band response of the structure.

5.6.3 Wide-band fatigue loading

Wide-band random vibration consists of contributions over a broad range of frequencies, with a large resonant peak; this type of response is typical for wind loading (Figure 5.7). A number of cycle counting methods for wide-band stress variations have been proposed (Dowling, 1972). One of the most realistic of these is the 'rainflow' counting method, proposed by Matsuishi and Endo (1968). In this method, which uses the analogy of rain flowing over the undulations of a roof, cycles associated with complete hysteresis cycles of the metal are identified. Use of this method, rather than a simple level-crossing approach which is the basis of the narrow-band approach described in Section 5.6.2, invariably results in fewer cycle counts.

A useful empirical approach has been proposed by Wirsching and Light (1980). They proposed that the fractional fatigue damage under a wide-band random stress variation can be written as:

$$D = \lambda \cdot D_{nb} \tag{5.58}$$

where, D_{nb} is the damage calculated for narrow-band vibration with the same standard deviation, σ, (Equation 5.56). λ is a parameter determined empirically. The approach used to determine λ was to use simulations of wide-band processes with spectral densities of various shapes and band-widths, and rainflow counting for fatigue cycles.

The formula proposed by Wirsching and Light to estimate λ was:

$$\lambda = a + (1-a)(1-\varepsilon)^b \tag{5.59}$$

where a and b are functions of the exponent m (Equation 5.51), obtained by least-squares fitting, as follows:

$$a \cong 0.926 - 0.033m \tag{5.60}$$

$$b \cong 1.587m - 2.323 \tag{5.61}$$

ε is a spectral bandwidth parameter equal to:

$$\varepsilon = 1 - \frac{\mu_2^2}{\mu_0 \mu_4} \tag{5.62}$$

where, μ_k is the kth moment of the spectral density defined by:

$$\mu_k = \int_0^\infty n^k S(n)\, dn \tag{5.63}$$

For narrow-band vibration, ε tends to zero, and, from Equation (5.59), λ approaches 1. As ε tends to its maximum possible value of 1, λ approaches a, as given by Equation (5.60). These values enable upper and lower limits on the damage to be determined. Alternatively, the spectral bandwidth parameter, ε, can be calculated from the spectral density of the response, and Equation (5.59) is used to calculate λ.

5.6.4 Effect of varying wind speed

Equation (5.56) applies to a particular standard deviation of stress, σ, which, in turn, is a function of mean wind speed, $\bar{U}$. This relationship can be written in the following form:

$$\sigma = A\bar{U}^n \tag{5.64}$$

The mean wind speed, $\bar{U}$, itself, is a random variable. Its probability distribution can be represented by a Weibull distribution (see Sections 2.5 and C.3.4):

$$f_U(\bar{U}) = \frac{w\,\bar{U}^{w-1}}{c^w}\exp\left[-\left(\frac{\bar{U}}{c}\right)^w\right] \tag{5.65}$$

The total damage from narrow-band vibration for all possible mean wind speeds is obtained from Equations (5.56), (5.64) and (5.65) and integrating.

The fraction of the time T during which the mean wind speed falls between U and $U+\delta U$ is $f_U(U)\cdot\delta U$.

Hence, the amount of damage generated while this range of wind speed occurs is, from Equations (5.56) and (5.64):

$$D_U = \frac{v_o^+ T f_U(U)\delta U}{K}(2\sqrt{2}AU^n)^m\,\Gamma\left(\frac{m}{2}+1\right)$$

For wide-band loading, the average cycling rate of the stress cycles can be assumed to vary with mean wind speed according to Equation (5.66):

$$v_o^+ = v_1\bar{U}^p = v_c\left(\frac{\bar{U}}{c}\right)^p \tag{5.66}$$

where ν_c is the cycling rate when the mean wind speed is equal to c, and ν_1 is notionally the cycling rate when the mean wind speed is equal to $1\,\text{m/s}$. p is an exponent with a value of 0.1–0.5.

Then the expected amount of damage in a time, T, at mean wind speed, $\bar{U}$, is:

$$D_U = \frac{\nu_c T f_{\bar{U}}(\bar{U}) \bar{U}}{K c^p} \left(2 \sqrt{2}\ A\right)^m \bar{U}^{mn+p} \Gamma\left(\frac{m}{2}+1\right)$$

5.6.5 Accumulated fatigue damage, and fatigue life estimation

The total damage in time T during all mean wind speeds between 0 and ∞ can be calculated as:

$$D = \frac{\nu_c T \left(2 \sqrt{2}\ A\right)^m}{K c^p} \Gamma\left(\frac{m}{2}+1\right) \int_0^\infty \bar{U}^{mn+p} f_{\bar{U}}(\bar{U})\, d\bar{U} \tag{5.67}$$

$$= \frac{\nu_c T \left(2 \sqrt{2}\ A\right)^m}{K c^p} \Gamma\left(\frac{m}{2}+1\right) \int_0^\infty \bar{U}^{mn+p+w-1} \frac{w}{c^w} \exp\left[-\left(\frac{\bar{U}}{c}\right)^w\right] d\bar{U}$$

Hence,

$$D = \frac{w \nu_c T \left(2 \sqrt{2}\ A\right)^m}{K c^{p+w}} \Gamma\left(\frac{m}{2}+1\right) \int_0^\infty \bar{U}^{mn+p+w-1} \exp\left[-\left(\frac{\bar{U}}{c}\right)^w\right] d\bar{U}$$

This is now of the form of Equation (5.57), so that:

$$D = \frac{w \nu_c T \left(2 \sqrt{2}\ A\right)^m}{K c^{p+w}} \Gamma\left(\frac{m}{2}+1\right) \frac{c^{mn+p+w}}{w} \Gamma\left(\frac{mn+p+w}{w}\right)$$

$$= \frac{\nu_c T \left(2 \sqrt{2}\ A\right)^m c^{mn}}{K} \Gamma\left(\frac{m}{2}+1\right) \Gamma\left(\frac{mn+p+w}{k}\right) \tag{5.68}$$

This is a useful closed form expression for the fatigue damage over a lifetime of wind speeds, assuming narrow-band vibration.

For wide-band vibration, Equation (5.68) can be modified, following Equation (5.58), to:

$$D = \frac{\lambda \nu_c T \left(2 \sqrt{2}\ A\right)^m c^{mn}}{K} \Gamma\left(\frac{m}{2}+1\right) \Gamma\left(\frac{mn+p+w}{k}\right) \tag{5.69}$$

By setting D equal to 1 in Equations (5.68) and (5.69), we can obtain lower and upper limits to the fatigue life as follows:

$$T_{\text{lower}} = \frac{K}{v_c \left(2\sqrt{2}\ A\right)^m c^{mn} \Gamma\left(\frac{m}{2}+1\right) \Gamma\left(\frac{mn+p+w}{k}\right)} \tag{5.70}$$

$$T_{\text{upper}} = \frac{K}{\lambda v_c \left(2\sqrt{2}\ A\right)^m c^{mn} \Gamma\left(\frac{m}{2}+1\right) \Gamma\left(\frac{mn+p+w}{k}\right)} \tag{5.71}$$

Example 5.1

To enable the calculation of fatigue life of a welded connection at the base of a steel pole, using Equations (5.70) and (5.71), the following values are assumed:

$m=5;\ n=2;\ p=0.2;\ v_c=0.42\,\text{Hz}$

$K=6.4\times10^{16}\ [\text{MPa}]^5;\ c=8\,\text{m/s};\ w=2;\ A=0.1\dfrac{\text{MPa}}{(\text{m/s})^2}$

$\Gamma\left(\dfrac{m}{2}+1\right)=\Gamma(3.5)=e^{1.201}=3.323$

$\Gamma\left(\dfrac{mn+p+w}{k}\right)=\Gamma(6.1)=e^{4.959}=142.5$

Then from Equation (5.70),

$$T_{\text{lower}} = \frac{6.4\times10^{16}}{0.42\left(2\sqrt{2}\ \times0.1\right)^5 \times8^{10}\times3.323\times142.5}$$

$$= 1.656\times10^8\ \text{seconds} = 5.25\ \text{years}$$

From Equation (5.60), $a=0.926-0.033m=0.761$.
From Equation (5.59), this is a lower limit for λ.

$$T_{\text{upper}} = \frac{T_{\text{lower}}}{\lambda} = 6.90\ \text{years}$$

This example illustrates the sensitivity of the estimates of fatigue life to the values of both A and c. For example, increasing A to $0.15\dfrac{\text{MPa}}{(\text{m/s})^2}$ would decrease the fatigue life by 7.6 times, i.e. (1.5^5). Decreasing c from 8 to 7 m/s will increase the fatigue life by 3.8 times, i.e. $(8/7)^{10}$.

5.6.6 Number of cycles above a defined stress level

The number of cycles of stress with amplitude, s, or range, S, is often required when the fatigue damage relationship is undefined, or cannot be described by Equation (5.52). For a narrow-band process, application of Equations (5.53), (5.55) and (5.64) gives the expected number of cycles exceeding a stress range, S, at a mean wind speed, $\bar{U}$, in a time period, T:

$$N(S) = v_o^+ T \exp\left(-\frac{s^2}{2\sigma^2}\right) = v_o^+ T \exp\left(-\frac{S^2}{8A^2\bar{U}^{2n}}\right) \tag{5.72}$$

Then, for a wind speed represented by the Weibull distribution (Equation 5.63), the total number of cycles with a range exceeding, S, in a time period, T, for all wind speeds, is obtained by integration:

$$N(S) = \frac{wv_cT}{c^{p+w}} \int_0^\infty \bar{U}^{p+w-1} \exp\left[-\left(\frac{\bar{U}}{c}\right)^w - \left(\frac{S^2}{8A^2U^{2n}}\right)\right] d\bar{U} \tag{5.73}$$

Changing the variable of integration from $\bar{U}$ to $(\bar{U}/c)$ gives:

$$\frac{N(S)}{wv_cT} = \int_0^\infty \left(\frac{\bar{U}}{c}\right)^{p+w-1} \exp\left[-\left(\frac{\bar{U}}{c}\right)^w - \left(\frac{S^2/(8A^2c^{2n})}{\left(\frac{\bar{\bar{U}}}{c}\right)^{2n}}\right)\right] d\left(\frac{\bar{U}}{c}\right) \tag{5.74}$$

From Equation (5.74), it can be shown that the number of stress cycles in the time period, T, is a function of the following non-dimensional parameters:

$$N(s) = F\{b, w, n, \bar{N}, (S/S_{max})\} \tag{5.75}$$

where $\bar{N}$ is equal to $(v_c T)$ $[= (v_1 c^p T)]$ and is a characteristic number of cycles (or mean crossings) in the time period T.

S_{max} is the expected largest value of stress in the time T and can be closely approximated by setting $N(S)$ equal to 1 in Equation (5.74), and solving it numerically.

Thus, for a fixed set of four parameters, p, w, n and $\bar{N}$, a single relationship between $N(S)$ and (S/S_{max}) can be obtained. Note that Equation (5.75) is independent of the parameter A in Equation (5.64).

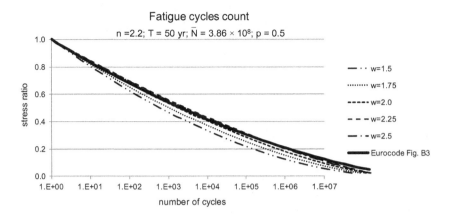

Figure 5.18 Wind-induced fatigue – normalized stress-range ratio versus expected number of cycles, variation with Weibull shape factor, *w* (Holmes, 2012).

It can also be shown that (S/S_{max}) is relatively insensitive to T, c, n and p (Holmes, 2012). There is more sensitivity to the Weibull shape factor, w (Kemper and Feldmann, 2013).

However, assuming a conservative value for w of about 2.5, provides a basis for a 'universal' relationship between normalized stress range (S/S_{max}), and the expected number of cycles under wind loading (Holmes, 2012). Such a relationship has been provided in several design codes and standards, such as Eurocode 1 (British Standards Institution, 2005). This is shown in Figure 5.18, which shows the variation of the stress range cycle count with w with some typical values of the other parameters, based on numerical calculations of Equations (5.74) and (5.75), and compared with that given in Eurocode 1.

5.7 SUMMARY

This chapter has covered a wide range of topics relating to the dynamic response of structures to wind forces. For wind loading, the sub-resonant or background response should be distinguished from the contributions at the resonant frequencies, and calculated separately.

The along-wind response of structures that can be represented as single- and multi-degree of freedom systems has been considered. The effective static load approach in which distributions of the mean, background and resonant contributions to the loading are considered separately, and assembled as a combined effective static wind load, has been presented.

Aeroelastic effects such as aerodynamic damping, and the instabilities of galloping and flutter have been introduced. Finally wind-induced fatigue has been treated resulting in usable formulae for the calculation of fatigue life of a structure under turbulent wind loading.

Cross-wind dynamic response from vortex shedding has not been treated in this chapter, but is discussed separately in Chapters 9 and 11.

REFERENCES

Ashraf Ali, M. and Gould, P.L. (1985) On the resonant component of the response of single degree-of-freedom systems under wind loading. *Engineering Structures*, 7: 280–282.

Bendat, J.S. and Piersol, A.G. (2010) *Random data: analysis and measurement procedures*, 4th Edition. Wiley, New York.

British Standards Institution. (2005) *Eurocode 1: Actions on structures - Part 1–4: General actions - Wind actions*. BS EN 1991-1-4.6, B.S.I., London.

Clough, R.W. and Penzien, J. (1975) *Dynamics of structures*. McGraw-Hill, New York.

Crandall, S.H. and Mark W.D. (1963) *Random vibration in mechanical systems*. Academic Press, New York.

Davenport, A.G. (1961) The application of statistical concepts to the wind loading of structures. *Proceedings, Institution of Civil Engineers*, 19: 449–471.

Davenport, A.G. (1963) The buffetting of structures by gusts. *Proceedings, International Conference on Wind Effects on Buildings and Structures*, Teddington, UK, 26–28 June, pp. 358–391.

Davenport, A.G. (1964) Note on the distribution of the largest value of a random function with application to gust loading. *Proceedings, Institution of Civil Engineers*, 28: 187–196.

Davenport, A.G. (1967) Gust loading factors. *ASCE. Journal of the Structural Division*, 93: 11–34.

Dowling, N.E. (1972) Fatigue failure predictions for complicated stress-strain histories. *Journal of Materials*, 7: 71–87.

Dwight, H.B. (1961) *Tables of integrals and other mathematical data*. Macmillan, Toronto.

Harris, R.I. (1963) The response of structures to gusts. *Proceedings, International Conference on Wind Effects on Buildings and Structures*, Teddington, UK, 26–28 June, pp. 394–421.

Holmes, J.D. (1994) Along-wind response of lattice towers: part I – derivation of expressions for gust response factors. *Engineering Structures*, 16: 287–292.

Holmes, J.D. (1996a) Along-wind response of lattice towers: part II – aerodynamic damping and deflections. *Engineering Structures*, 18: 483–488.

Holmes, J.D. (1996b) Along-wind response of lattice towers: part III – effective load distributions. *Engineering Structures*, 18: 489–494.

Holmes, J.D. (2012) Wind-induced fatigue cycle counts – sensitivity to wind climate and dynamic response. *7th International Congress of the Croatian Society of Mechanics (7th ICCSM)*, Zadar, Croatia, 22–25 May.

Holmes, J.D. and Best, R.J. (1981) An approach to the determination of wind load effects for low-rise buildings. *Journal of Wind Engineering & Industrial Aerodynamics*, 7: 273–287.

Holmes, J.D. and Kasperski, M. (1996) Effective distributions of fluctuating and dynamic wind loads. *Civil Engineering Transactions, Institution of Engineers, Australia*, CE38: 83–88.

Holmes, J.D., Forristall, G. and McConochie, J. (2005) Dynamic response of structures to thunderstorm winds. *10th Americas Conference on Wind Engineering*, Baton Rouge, Louisiana, USA, 1–4 June, 2005.

Kasperski, M. and Niemann, H.-J. (1992) The L.R.C. (Load-Response-Correlation) method: A general method of estimating unfavourable wind load distributions for linear and non-linear structural behavior. *Journal of Wind Engineering & Industrial Aerodynamics*, 43: 1753–1763.

Kemper, F.H. and Feldmann, M. (2013) Stress range spectra and damage equivalence factors for along-wind responses of arbitrary structures. *12th Americas Conference on Wind Engineering*, Seattle, Washington, DC, USA, 16–20 June, 2013.

Matsuishi, M. and Endo, T. (1968) Fatigue of metals subjected to varying stress. *Japan Society of Mechanical Engineers Meeting*, Fukuoka, March.

Matsumoto, M. (1996) Aerodynamic damping of prisms. *Journal of Wind Engineering and Industrial Aerodynamics*, 59: 159–175.

Rice, S.O. (1944–1945) Mathematical analysis of random noise. *Bell System Technical Journal*, 23: 282–332 (1944) and 24: 46–156. Reprinted in N. Wax, "Selected papers on noise and stochastic processes", Dover, New York, 1954.

Scanlan, R.H. (1982) Developments in low-speed aeroelasticity in the civil engineering field. *A.I.A.A. Journal*, 20: 839–844.

Scanlan, R.H. and Gade, R.H. (1977) Motion of suspended bridge spans under gusty winds. *ASCE. Journal of the Structural Division*, 103: 1867–1883.

Scanlan, R.H. and Tomko, J.J. (1971) Airfoil and bridge deck flutter derivatives. *ASCE Journal of the Engineering Mechanics Division*, 97: 1717–1737.

Vickery, B.J. (1965) On the flow behind a coarse grid and its use as a model of atmospheric turbulence in studies related to wind loads on buildings. National Physical Laboratory (UK) Aero Report 1143.

Vickery, B.J., (1966) On the assessment of wind effects on elastic structures. *Australian Civil Engineering Transactions*, CE8: 183–192.

Vickery, B.J., (1968) Load fluctuations in turbulent flow. *ASCE Journal of the Engineering Mechanics Division*, 94: 31–46.

Vickery, B.J. (1995) The response of chimneys and tower-like structures to wind loading. In *A state of the art in wind engineering*. ed. P. Krishna, Wiley Eastern Limited, New Delhi.

Warburton, G.B. (1976) *The dynamical behaviour of structures*, 2nd Edition. Pergamon Press Ltd, Oxford.

Wirsching, P.H. and Light, M.C. (1980) Fatigue under wide band random stresses. *Journal of the Structural Division, ASCE*, 106: 1593–1607.

Chapter 6

Internal pressures

6.1 INTRODUCTION

Pressures within the interior of a building, produced by external wind action, can form a high proportion of the total design wind load in some circumstances, for example, low-rise buildings when there are dominant openings in the walls. On high-rise buildings, a critical design case for a window at a corner may be an opening in the adjacent wall at the same corner - perhaps caused by glass failure due to flying debris.

In this chapter, the fundamentals of the prediction of wind-induced internal pressures within enclosed buildings are discussed. A number of cases are considered: a single dominant opening in one wall, multiple wall openings, and the effects of roof flexibility and background wall porosity. The possibility of Helmholtz resonance occurring is discussed. A 'risk-consistent' internal pressure, allowing for a number of possible opening scenarios in a building envelope, is introduced in Section 6.6.

6.2 SINGLE LARGE OPENING

We will first consider the case of a dominant windward wall opening, that is, a situation which often arises in severe windstorms, such as after failure of a glass window due to flying debris. In a steady flow situation, the internal pressure will quickly build up to equal the external pressure on the windward wall in the vicinity of the opening, and there may be some oscillations in internal pressure (Section 6.2.4), but these will die out after a short time. However, when a building is immersed in a turbulent boundary-layer wind, the external pressure will be highly fluctuating, and the internal pressure will respond in some way to these fluctuations. Since there is only a single opening, flow into the building resulting from an increase in external pressure will cause an increase in the density of the air within the internal volume. This, in turn, will produce an increase in internal pressure. The pressure changes produced by wind are only about 1% of atmospheric pressure (about 1,000 Pascals compared to atmospheric pressure of about

100,000 Pascals), and the relative density changes are of the same order. These small density changes can be maintained by small mass flows in and out of the building envelope, and consequently the internal pressure can be expected to respond quickly to changes in external pressure, except for very small opening areas.

6.2.1 Dimensional analysis

It is useful to first carry out a dimensional analysis for the fluctuating internal pressures, resulting from a single windward opening to establish the non-dimensional groups involved.

The fluctuating internal pressure coefficient, $C_{pi}(t)$, can be written as:

$$C_{pi} = \frac{p_i - p_o}{\frac{1}{2}\rho_a \bar{U}^2} = F\left(\Phi_1, \Phi_2, \Phi_3, \Phi_4, \Phi_5\right) \tag{6.1}$$

p_o is the reference atmospheric, or static, pressure, ρ_a is the density of air, and $\bar{U}$ is the mean wind velocity in the approach flow.

$\Phi_1 = A^{3/2}/V_o$ – where A is the area of the opening, and V_o is the internal volume.

$\Phi_2 = a_s/\bar{U} = \sqrt{\gamma p_o/\rho_a}/\bar{U}$ – where a_s is the speed of sound, and γ is the ratio of specific heats of air.

$\Phi_3 = \rho_a \bar{U} A^{\frac{1}{2}}/\mu$ – where μ is the dynamic viscosity of air (Reynolds Number).

$\Phi_4 = \dfrac{\sigma_u}{\bar{U}}$ – where σ_u is the standard deviation of the longitudinal turbulence velocity upstream – i.e. Φ_4 is the turbulence intensity (Section 3.3.1).

$\Phi_5 = \ell_u/\sqrt{A}$ – where ℓ_u is the length scale of turbulence (Section 3.3.4).

Φ_1 is a non-dimensional parameter related to the geometry of the opening and the internal volume, Φ_3 is a Reynolds Number (Section 4.2.4) based on a characteristic length of the opening, Φ_5 is a ratio between characteristic length scales in the turbulence in the approaching flow and of the opening. Φ_2, is the Mach Number.

Amongst these parameters, Φ_1, Φ_2 and Φ_5 are important (see Section 6.2.8).

6.2.2 Response time

If the inertial (i.e. mass times acceleration) effects are initially neglected, an expression can be derived for the time taken for the internal pressure to become equal to a sudden increase in pressure outside the opening, such as that caused by a sudden window failure. (Euteneur, 1970).

For conservation of mass, the rate of mass flow into the building through the opening must equal the rate of mass increase inside the volume:

$$\rho_i Q = \left(\frac{d\rho_i}{dt} \right) V_o \qquad (6.2)$$

where ρ_i denotes the air density within the internal volume, Q is the volume flow rate, and V_o is the internal volume.

For turbulent flow through an orifice, the following relationship between flow rate, Q, and the pressure difference across the orifice, $p_e - p_i$, applies:

$$Q = kA \sqrt{\frac{2(p_e - p_i)}{\rho_a}} \qquad (6.3)$$

where k is an orifice constant, or discharge coefficient, equal to about 0.6 for steady uni-directional flow (see also Section 6.2.7).

Assuming an adiabatic law relating the internal pressure and density,

$$\frac{p_i}{\rho_i^\gamma} = \text{constant} \qquad (6.4)$$

where γ is the ratio of specific heats of air.

Substituting from Equations (6.2) and (6.4) in Equation (6.3), and integrating the differential equation, Equation (6.5) can be obtained for the response, or equilibrium, time, τ, when the internal pressure becomes equal to the external pressure:

$$\tau = \frac{\rho_a V_o \bar{U}}{\gamma k A p_o} \sqrt{C_{pe} - C_{pio}} \qquad (6.5)$$

where the pressures have been written in terms of pressure coefficients:

$$C_{pe} = \frac{p_e - p_o}{\frac{1}{2}\rho_a \bar{U}^2} \text{ and } C_{pi} = \frac{p_i - p_o}{\frac{1}{2}\rho_a \bar{U}^2} \text{ and } C_{pio} \text{ is the initial value of } C_{pi}, \text{ (i.e. at } t=0)$$

Example 6.1

It is instructive to apply Equation (6.5) to a practical example. The following numerical values will be substituted:

$\rho_a = 1.20\,\text{kg/m}^3$; $V_o = 1,000\,\text{m}^3$; $\bar{U} = 40$ m/s
$\gamma = 1.4$; $k = 0.6$; $A = 1.0\,\text{m}^2$; $p_o = 10^5$ Pascals
$C_{pe} = +0.7$; $C_{pio} = -0.2$

Then the response time, $\tau = \dfrac{1.2 \times 1,000 \times 40}{1.4 \times 0.6 \times 1.0 \times 10^5} \sqrt{0.7 - (-0.2)}$

$= 0.54$ seconds

Thus, for an internal volume of $1,000\,\text{m}^3$, Equation (6.5) predicts a response time of half a second for the internal pressure to adjust to the external pressure, following the creation of an opening on the windward face of $1\,\text{m}^2$.

6.2.3 Helmholtz resonator model

In the previous section, inertial effects on the development of internal pressure following a sudden opening were neglected. These will now be included in a general model of internal pressure, which can be used for the prediction of the response to turbulent external pressures (Holmes, 1979).

The Helmholtz resonator is a well-established concept in acoustics (Rayleigh, 1896; Malecki, 1969), which describes the response of small volumes to the fluctuating external pressures. Although originally applied to the situation where the external pressures are caused by acoustic sources, it can be applied to the case of external wind pressures, 'driving' the internal pressures within a building. It is similar to the cavity resonance felt by occupants of a travelling motor vehicle, with an open window. Acoustic resonators made from brass or earthenware, based on this principle, were used to improve the acoustic quality in the amphitheatres of ancient Greece and Rome (Malecki, 1969).

Figure 6.1 illustrates the concept as applied to internal pressures in a building. It is assumed that a defined "slug" of air moves in and out of the opening in response to the changes in external pressure. Thus, mixing of the moving air, either with the internal air or the external air, is disregarded in this model of the situation.

A differential equation for the motion of the slug of air can be written as follows:

$$\rho_a A \ell_e \ddot{x} + \frac{\rho_a A}{2k^2} \dot{x}|\dot{x}| + \frac{\gamma p_o A^2}{V_o} x = A\, \Delta p_e(t) \tag{6.6}$$

The dependent variable, x, in this differential equation is the displacement of the air "slug" from its initial or equilibrium position. The first term on the left-hand side of Equation (6.6) is an inertial term proportional to the acceleration, $\ddot{x}$, of the air "slug", whose mass is $\rho_a A \ell_e$, in which ℓ_e is an effective length for the "slug". The second term is a loss term associated with energy losses for flow through the orifice, and the third term is a 'stiffness' associated with the resistance of the air pressure already in the internal volume to the movement of the "slug".

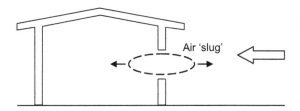

Figure 6.1 The Helmholtz resonator model of fluctuating internal pressures with a single windward opening.

A movement x in the air "slug", can be related to the change in density $\Delta\rho_i$, and hence pressure, Δp_i, within the internal volume:

$$\rho_a A x = V_o \Delta\rho_i = \frac{\rho_i V_o}{\gamma p_o} \Delta p_i \tag{6.7}$$

Making use of Equation (6.4) and converting the internal and external pressures to pressure coefficients, Equation (6.6) can be re-written in the form of a differential equation for the fluctuating internal pressure coefficient, $C_{pi}(t)$:

$$\frac{\rho_a \ell_e V_o}{\gamma p_o A} \ddot{C}_{pi} + \left(\frac{\rho_a V_o \bar{U}}{2k\gamma\; A p_o} \right)^2 \dot{C}_{pi} \left| \dot{C}_{pi} \right| + C_{pi} = C_{pe} \tag{6.8}$$

Equation (6.8) can also be derived (Vickery, 1986) by writing the discharge equation for unsteady flow through the orifice in the form:

$$p_e - p_i = \left(\frac{1}{k^2} \right) \frac{1}{2} \rho_a u_o^2 + \rho_a \ell_e \frac{du_o}{dt} \tag{6.9}$$

where ρ_a is taken as the air density within the volume (ρ_i), and u_o as the (unsteady) spatially averaged velocity through the opening.

Equations (6.6) and (6.8) give the following equation for the (undamped) natural frequency for the resonance of the movement of the air "slug", and of the internal pressure fluctuations. This frequency is known as the Helmholtz frequency, n_H.

$$n_H = \frac{1}{2\pi} \sqrt{\frac{\gamma A p_o}{\rho_a \ell_e V_o}} \tag{6.10}$$

Internal pressure resonances at, or near, the Helmholtz frequency have been measured both in wind-tunnel studies (Holmes, 1979; Liu and Rhee, 1986) and in full scale.

The effective length, ℓ_e, varies with the shape and depth of the opening, and is theoretically equal to $\sqrt{(\pi A/4)}$ for a thin circular orifice. For practical purposes (openings in thin walls), it is sufficiently accurate to take ℓ_e as equal to $1.0 \sqrt{A}$ (Vickery, 1986).

6.2.4 Sudden windward opening with inertial effects

Equation (6.8) can be solved numerically for the case of a step change in external pressure coefficient, C_{pe} (representative of the situation after a sudden window failure). Figure 6.2a and b shows the response of a 600 m³ volume (rigid walls and roof) with opening areas of 1 and 9 m², respectively (Holmes, 1979). For these simulations, the effective length, ℓ_e, was equivalent to $0.96 \sqrt{A}$, and the discharge coefficient, k, was taken as 0.6.

It is apparent from Figure 6.2b that the inertial effects are significant for the larger opening when the damping term in Equation (6.8) is much smaller (note that the area, A, is in the denominator in this term). Many oscillatory cycles in internal pressure occur before equilibrium conditions are reached in this case. However, the flexibility of the walls and roof of real buildings, discussed in the following section, also increases the damping term (Vickery, 1986), and hence cause more rapid attenuation of the oscillations.

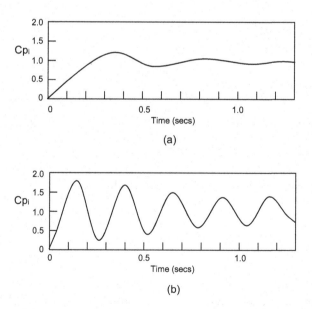

Figure 6.2 Response to a step change in external pressure, V_o=600 m³, $\bar{U}$ =30 m/s. (a) A= 1 m²; (b) A=9 m².

6.2.5 Effect of roof flexibility

Equations (6.8) and (6.10) assume that the building or enclosure has rigid walls and roof. Real buildings may have considerable flexibility.

Sharma (2008) studied the effect of roof flexibility and derived a modified equation for internal pressure fluctuations:

$$\frac{\rho_a \ell_e V_o}{\gamma p_o A}\left(\ddot{C}_{pi} + \dot{v}\dot{C}_{pi} + \frac{\gamma p_o}{q}\ddot{v}\right) + \frac{\rho_a q V_o^2}{2(k\gamma A p_o)^2}\left(\dot{C}_{pi} + \frac{\gamma p_o}{q}\dot{v}\right)\left|\dot{C}_{pi} + \frac{\gamma p_o}{q}\dot{v}\right|$$

$$+ C_{pi} = C_{pe} \tag{6.11}$$

where $q = \dfrac{1}{2}\rho_a \bar{U}^2$.

The fractional change in volume, $v = \dfrac{\Delta V}{V_o} = \dfrac{q}{K_B}(C_{pi} - C_{pr})$,

where C_{pr} is an area-averaged fluctuating external roof pressure coefficient.

K_B is the bulk modulus for the building, i.e. the internal pressure for a unit change in relative internal volume. Theoretically, it is the internal pressure required to double the internal volume.

Simplifying and neglecting second-order terms,

$$\frac{\rho_a \ell_e V_o}{\gamma p_o A}\left(\ddot{C}_{pi} - \frac{b}{(1+b)}\dot{C}_{pr}\right) + \frac{\rho_a q V_o^2 (1+b)^2}{2(k\gamma A p_o)^2}\left(\dot{C}_{pi} - \frac{b}{(1+b)}\dot{C}_{pr}\right)$$

$$\left|\dot{C}_{pi} - \frac{b}{(1+b)}\dot{C}_{pr}\right| + C_{pi} = C_{pe} \tag{6.12}$$

where b ($= K_A/K_B$) is the ratio of the bulk modulus of air to that of the building.

K_A, the bulk modulus of air $(\rho_a \Delta p)/\Delta \rho$, is equal to γp_o

The typical range of b for low-rise buildings is 0.2–5.

Equation (6.12) in general requires numerical solutions.

From Equation (6.12), the equation for the Helmholtz frequency becomes (Vickery, 1986):

$$n_H = \frac{1}{2\pi}\sqrt{\frac{\gamma A p_o}{\rho_a \ell_e V_o [1+b]}} \tag{6.13}$$

It can be seen from Equation (6.13) that the effect of building flexibility is equivalent to an increase in the effective internal volume.

6.2.6 Effect of background leakage

The effect of background leakage on the leeward side of the building on the internal pressures, for a building with a large windward wall opening, can be incorporated by assuming that:

- The area of the windward opening is large compared with that of the background leakage,
- internal pressure fluctuations are fully correlated throughout the internal volume,
- a single mean external pressure coefficient, $\overline{C}_{pL}$, applies to the leeward walls, and
- inertial effects on flow through the leakage are neglected.

With those assumptions, the governing equation for fluctuations in internal pressure coefficient becomes Equation (6.14) (Yu et al., 2008).

$$
\frac{\rho_a l_e V_o}{\gamma p_o A_W} \ddot{C}_{pi} + \frac{\rho_a l_e A_L \overline{U}_h}{2k' A_W q \sqrt{\left(C_{pi} - \overline{C}_{pL}\right)}} \dot{C}_{pi}
$$

$$
+ \frac{\rho_a q V_o^2}{2\left(k\gamma A_W p_o\right)^2}\left(\dot{C}_{pi} + \frac{A_L \overline{U}_h k' \gamma p_o}{q V_o} \sqrt{\left(C_{pi} - \overline{C}_{pL}\right)}\right)
$$

$$
\cdot \left|\dot{C}_{pi} + \frac{A_L \overline{U}_h k' \gamma p_o}{q V_o} \sqrt{\left(C_{pi} - \overline{C}_{pL}\right)}\right| + C_{pi} = C_{pW}
$$

(6.14)

where k' is the discharge coefficient for flow through the leakage, and A_W and A_L are the open areas of the windward opening and background leakage, respectively q is defined in Section 6.2.5.

It can be seen that Equation (6.14) reduces to Equation (6.8) when the area of the background leakage, A_L, is taken as zero, and $A_W = A$ and $C_{pW} = C_{pe}$.

In general Equation (6.14) must be solved numerically (Yu et al., 2008). The numerical solutions by Yu et al. (2008), supported by wind-tunnel experiments showed that background leakage of more than 10% of the windward opening area can significantly reduce the root-mean-square of fluctuating internal pressures. An area of background leakage of about 50% of that of the opening on the windward wall resulted in a reduction in the fluctuating internal pressure coefficient of about 25%.

6.2.7 Helmholtz resonance frequencies

Section 6.2.3 discussed the phenomenon of Helmholtz resonance in the interior of buildings, when there is a single opening, and Equations (6.10) and (6.13) gave formulas to calculate the Helmholtz frequency, given the opening area, internal volume and flexibility of the roof.

Applying Equation (6.13) for the Helmholtz resonance frequency, and setting $p_o = 10^5$ Pascals (atmospheric pressure), $\rho_a = 1.2$ kg/m³(air density), $\gamma = 1.4$ (ratio of specific heats), and ℓ_e equal to $1.0 \sqrt{A}$, we have the following approximate Equation (6.15) for n_H:

$$n_H \approx 55 \frac{A^{1/4}}{V_o^{1/2}[1+(K_A / K_B)]^{1/2}} \qquad (6.15)$$

Equation (6.15) can be used to calculate n_H for typical low-rise buildings in Table 6.1 (Vickery, 1986).

Table 6.1 indicates that for the two smallest buildings, the Helmholtz frequencies are greater than 1 Hz, and hence, significant resonant excitation of internal pressure fluctuations by natural wind turbulence is unlikely. However, for the large arena, this would certainly be possible. In this case the structural frequency of the roof is likely to be considerably greater than the Helmholtz resonance frequency of the internal pressures, and the latter will therefore not excite any structural vibration of the roof (Liu and Saathoff 1982). It is clear, however, that there could be an intermediate combination of area and volume (such as the 'concert hall' in Table 6.1), for which the Helmholtz frequency is similar to the natural structural frequency of the roof, and in a range which could be excited by the natural turbulence in the wind. However, such a situation has not yet been recorded.

6.2.8 Non-dimensional formulation

A non-dimensional formulation of the governing equations is useful to create groups to make a comparison of full-scale and wind-tunnel data on internal pressures produced by large openings (see Section 7.4.2), and to assist in development of reduction factors for large internal volumes for design purposes (Section 6.2.8).

Holmes (1979) and Holmes and Ginger (2012) showed that Equation (6.8) can be re-written as Equation 6.16, all the terms are non-dimensional:

$$C_I \frac{1}{\Phi_1 \Phi_2^2 \Phi_5^2} \frac{d^2 C_{pi}}{dt^{*2}} + \left(\frac{1}{4k^2}\right)\left[\frac{1}{\Phi_1 \Phi_2^2 \Phi_5}\right]^2 \frac{dC_{pi}}{dt^*}\left|\frac{dC_{pi}}{dt^*}\right| + C_{pi} = C_{pe} \qquad (6.16)$$

Table 6.1 Helmholtz resonance frequencies for some typical buildings

Type	Internal volume (m³)	Opening area (m²)	Stiffness ratio K_A/K_B	Helmholtz frequency (Hz)
house	600	4	0.2	2.9
warehouse	5,000	10	0.2	1.3
concert hall	15,000	15	0.2	0.8
arena (flexible roof)	50,000	20	4	0.23

where, $C_I = \dfrac{\ell_e}{\sqrt{A}}$ i.e. an 'inertial coefficient' representing the effective length of the "'slug" discussed in Section 6.2.3, non-dimensionalized by the area of the opening. t^* is a non-dimensional time given by (tU/ℓ_u)

Φ_1, Φ_2 and Φ_5 are defined in Section 6.2.1, and k is the discharge coefficient defined in Section 6.2.2. The product $\Phi_1\Phi_2{}^2$ can be replaced by a single non-dimensional variable and defined as the non-dimensional opening size to volume parameter, $S^* = \left(a_s/\bar{U}\right)^2 \left(A^{3/2}/V_o\right)$. S^* is related to the parameter, S adopted by Yu et al. (2006), by $S^*=S^{3/2}$. Liu (1982) defined an alternative parameter, N, that is equal to $(\gamma/2S^*)$. Vickery and Bloxham (1992) used a parameter, also denoted by S, that is equal to the square root of S^*.

S^* is the single most important non-dimensional parameter affecting internal pressure fluctuations. For example, when carrying out wind-tunnel studies of fluctuating internal pressures, it is desirable to have equality of this parameter between model and full scale to ensure correct scaling, (see Sections 6.5 and 7.4.2).

The discharge coefficient, k, is alternatively expressed as a 'loss coefficient', C_L, where

$$C_L = 1/k^2 \tag{6.17}$$

For highly fluctuating and reversing flow of air through a dominant opening in a building, k is much less than the theoretical value for steady potential flow of 0.61. The variable k reduces with increasing values of S^*, or equivalently, C_L increases with S^* (Ginger et al., 2010). Furthermore, C_L has been found to increase with increasing turbulence intensity (Xu et al., 2017).

6.2.9 Reduction factors for large volumes and small opening areas

After fitting the available experimental data, Holmes and Ginger (2012) proposed the following expressions for the ratio of the standard deviation of the internal pressure to that of the external pressure at the opening:

$$\frac{\sigma_{pi}}{\sigma_{pe}} = 1.1 \quad \text{for } S^* \geq 1.0 \tag{6.18}$$

$$\frac{\sigma_{pi}}{\sigma_{pe}} = 1.1 + 0.2\log_{10}(S^*) \quad \text{for } S^* < 1.0 \tag{6.19}$$

where the parameter S^* was defined in Section 6.2.8.

The amplification for values of S^* greater than 1.0 (i.e. for small volumes and relatively large opening areas) occurs because of Helmholtz resonance within the internal volume (see Section 6.2.7).

The ratios of the expected peak internal pressure to the peak external pressure at the opening on a windward wall can be written as (Holmes and Ginger, 2012):

$$\frac{\hat{p}_i}{\hat{p}_e} = \frac{1 + 2gI_u\left(\sigma_{pi}/\sigma_{pe}\right)}{1 + 2gI_u} \tag{6.20}$$

where I_u is the longitudinal turbulence intensity (Section 3.3.1), and g is a peak factor for the pressures (from 3.5 to 4).

Combining Equation (6.19) with (6.20), it can be shown that for $(A^{3/2}/V_o)$ less than about 0.006:

$$\frac{\hat{p}_i}{\hat{p}_e} = a_1 + a_2 \, \log_{10}\left(\frac{A^{3/2}}{V_o}\right) \tag{6.21}$$

where $a_1 = \dfrac{1 + 2gI_u\left[1.1 + 0.4 \, \log_{10}\left(\dfrac{a_s}{\bar{U}}\right)\right]}{1 + 2gI_u}$ and $a_2 = \dfrac{0.4gI_u}{1 + 2gI_u}$,

It can be shown that the values of a_1 and a_2 are relatively insensitive to a range of typical design values of $\bar{U}$ (20–35 m/s) and I_u (0.15–0.30), and to the peak factor, g (3.5–4.0). Then, taking average values of a_1 and a_2 of 1.33 and 0.12, respectively:

$$\frac{\hat{p}_i}{\hat{p}_e} \ 1.33 + 0.12 \, \log_{10}\left(\frac{A^{3/2}}{V_o}\right) \quad \text{for } (A^{3/2}/V_o) < 0.006 \tag{6.22a}$$

$$\frac{\hat{p}_i}{\hat{p}_e} \approx 1.06 \quad \text{for } (A^{3/2}/V_o) \geq 0.006 \tag{6.22b}$$

As illustrated by the following example, although a number of approximations have been made to derive it, Equation (6.22) can be used to indicate the approximate reduction in peak internal pressure expected within a building of large internal volume.

It should be noted that Equation (6.22) has been derived for full-scale design wind speeds, and hence generally is not applicable to wind-tunnel data obtained at lower wind speeds, for which Equation (6.21) should be used directly.

Example 6.2

Consider a rigid building of internal volume of 120,000 m², with a single dominant opening of 10 m² on the windward wall.
$A^{3/2}/V_o = 10^{1.5}/120,000 = 2.6 \times 10^{-4}$, i.e. less than 0.006

By Equation (6.22a), $\dfrac{\hat{p}_i}{\hat{p}_e} \approx 1.33 + 0.12 \log_{10}(2.6 \times 10^{-4})$

$= 1.33 - 0.43 = 0.90$

Thus, a reduction factor of about 0.9 should be applicable to obtain the peak internal pressure from the external pressure in the vicinity of the opening.

6.2.10 Side-wall cavity resonance

Helmholtz resonance is a phenomenon that can occur as a result of openings in *windward* walls. Wind blowing parallel to cavities in the *side* walls of buildings can induce cavity resonances through the shedding of vortices at the leading edge of the opening. This is the same phenomenon that produces annoying low-frequency vibrations in motor cars travelling with open windows, and also generates severe vibrations in bomber aircraft with open bomb bays. Unlike Helmholtz frequencies (Equation 6.10), the cavity frequencies are proportional to wind speed, and several modes can be generated (Rossiter, 1964). There have been cases of extreme vibration in the doors of balconies that form a partially open cavity on high-rise buildings. These cases have been attributed to cavity resonance, but it has not been investigated to any depth.

6.3 MULTIPLE WINDWARD AND LEEWARD OPENINGS

6.3.1 Mean internal pressures

The mean internal pressure coefficient inside a building with total areas (or effective areas if permeability is included) of openings on the windward and leeward walls of A_W and A_L, respectively, can be derived by using Equation (6.3) and applying mass conservation. The latter relation can be written for a total of N openings in the envelope:

$$\sum_1^N \rho_a Q_j = 0 \tag{6.23}$$

If quasi-steady and incompressible flow is given, then we can assume the density, ρ_a, to be constant. Applying Equation (6.3) for the flow through each of the N openings, Equation (6.23) becomes:

$$\sum_{j=1}^N A_j \sqrt{|p_{e,\,j} - p_i|} \cdot \text{sgn}(p_{e,\,j} - p_i) = 0 \tag{6.24}$$

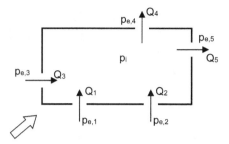

Figure 6.3 Inflows and outflows for a multiple openings.

where the modulus $|p_{e,j} - p_i|$ allows for the fact that for some openings, the flow is from the interior to the exterior.

Figure 6.3 shows a building (or a floor of a high-rise building) with five openings in the envelope. Applying Equation (6.24) to this case:

$$A_1\sqrt{|p_{e,1} - p_i|} + A_2\sqrt{|p_{e,2} - p_i|} + A_3\sqrt{|p_{e,3} - p_i|}$$
$$= A_4\sqrt{|p_{e,4} - p_i|} + A_5\sqrt{|p_{e,5} - p_i|} \tag{6.25}$$

In Equation (6.25), the inflows through the windward openings on the left-hand side balance the outflow through openings on the leeward and side walls, on the right-hand side. Equation (6.25), or similar equations for a large number of openings, can be solved by iterative numerical methods.

For the simpler case of a single windward opening with a single leeward opening, Equation (6.23) can be applied, with a conversion to pressure coefficients, to get:

$$A_W\sqrt{C_{pW} - C_{pi}} = A_L\sqrt{C_{pi} - C_{pL}}$$

This can be re-arranged to give Equation (6.26) for the coefficient of internal pressure:

$$C_{pi} = \frac{C_{pW}}{1 + \left(\dfrac{A_L}{A_W}\right)^2} + \frac{C_{pL}}{1 + \left(\dfrac{A_W}{A_L}\right)^2} \tag{6.26}$$

Equation (6.26) can be applied with A_W taken as the combined open area for several openings on a windward wall, and C_{pW} taken as an average mean pressure coefficient, with similar treatment for the leeward/side walls. It has been applied to give specified values of internal pressures in design codes and standards (see Chapter 15), where the coefficients are used with

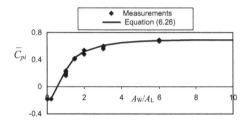

Figure 6.4 Mean internal pressure coefficient as a function of windward/leeward open area.

mean pressure coefficients to predict peak internal pressures, making use of the quasi-steady assumption (see Section 4.6.2).

Measurements of mean internal pressure coefficients for a building model with various ratios of windward/leeward opening area are shown in Figure 6.4. The solid line in this figure is Equation (6.26) with C_{pW} taken as +0.7, and C_{pL} taken as −0.2. These values are the values of mean external pressure coefficients on the walls at or near the windward and leeward openings, respectively. It may be seen that the agreement between the measurements and Equation (6.26) is good.

6.3.2 Fluctuating internal pressures

The analysis of fluctuating internal pressures is more difficult when there are large openings on more than one wall of a building. In general, for accurate solutions, numerical methods are required (Saathoff and Liu, 1983; Yu *et al.*, 2008). The case of a single large windward opening, and leakage through cracks and small openings in the other walls is discussed in Section 6.2.6.

However, for codes and standards, a quasi-steady approximation (Section 4.6.2) is sufficiently accurate – i.e. the *peak* internal pressure coefficients may be assumed to be given by Equation (6.26).

6.4 NOMINALLY CLOSED BUILDINGS

Internal pressures in buildings that are nominally closed, i.e. without large openings, but having some leakage distributed over all surfaces can be treated by neglecting the inertial terms, and lumping together windward and leeward leakage areas (Vickery, 1986, 1994; Harris, 1990).

Harris (1990) showed that when there is a combined open area on a windward wall of A_W and external pressure coefficient C_{pW}, with a total open area A_L and external pressure coefficient C_{pL} on a leeward wall, then there is a characteristic response time given by:

$$\tau = \frac{\rho_a V_o \bar{U} A_W A_L}{\gamma k p_o (A_W^2 + A_L^2)^{3/2}} \sqrt{C_{pW} - C_{pL}} \tag{6.27}$$

The effect of building wall and roof flexibility is to increase the response time according to Equation (6.28):

$$\tau = \frac{\rho_a V_o \bar{U} A_W A_L \left[1 + (K_A / K_B)\right]}{\gamma k p_o (A_W^2 + A_L^2)^{3/2}} \sqrt{C_{pW} - C_{pL}} \tag{6.28}$$

For 'normal' low-rise building construction, K_A/K_B is about 0.2 (Vickery, 1986, and Section 6.2.8), and the response time therefore increases by about 20%.

In a similar approach, Vickery (1986, 1994) obtained a characteristic frequency, n_c. Pressure fluctuations below this frequency are effectively communicated to the interior of the building.

n_c is given by Equation (6.29) (Vickery, 1994).

$$\frac{n_c V_o}{\bar{U} A_{W, \text{total}}} = \frac{\phi}{2\pi} \frac{k}{1 + (K_A/K_B)} \left(\frac{a_s}{\bar{U}}\right)^2 \frac{r^{1/2} \left(r + \dfrac{1}{r}\right)^{3/2}}{\sqrt{C_{pW} - C_{pL}}} \tag{6.29}$$

where

r is the ratio of total leeward wall surface area to windward wall surface area,
a_s is the speed of sound,

and the other parameters were defined previously. $A_{W, \text{total}}$ is the total surface area of the windward wall, and ϕ is the wall porosity. Equation (6.29) is essentially the same as Equation (6.28), with τ equal to $(1/2\pi n_c)$.

The peak internal pressure coefficient can be estimated by Equation (6.30).

$$\hat{C}_{pi} \cong \bar{C}_{pi} \left[1 + 2g \frac{\sigma_u'}{\bar{U}}\right] \tag{6.30}$$

where g is a suitable peak factor, and σ_u' is an effective, filtered standard deviation of velocity fluctuations that are capable of generating internal pressure fluctuation, given by:

$$\sigma_u'^2 = \int_0^\infty S_u(n) / \left[1 + (n/n_c)\right]^2 dn \tag{6.31}$$

Equation (6.31) has been evaluated using Equation (3.30) for the longitudinal turbulence spectrum, and σ_u'/σ_u is shown plotted against $\left(n_c \ell_u / \bar{U}\right)$ in

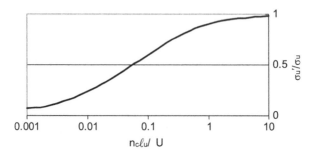

Figure 6.5 Reduction factor for fluctuating internal pressures, for a building with distributed porosity (Vickery, 1994).

Figure 6.5 (Vickery, 1994). g is a peak factor which lies between 3.0 and 3.5. The mean internal pressure coefficient in Equation (6.30) can be evaluated using Equation (6.26).

Evaluation of Equation (6.31) for a building with a wall porosity of 0.0005 (0.05%) gave a value of σ'_u/σ_u equal to 0.7, i.e. there is a 30% reduction in the effective velocity fluctuations resulting from the filtering effect of the porosity of the building (Vickery, 1994).

The effective areas of background leakage in buildings vary greatly depending on their usage, and on the climate of their location. Typical commercial buildings, such as offices, have effective leakage areas of 0.01–0.2 percent of the wall areas. Industrial warehouses and farm buildings may have permeabilities of up to 1.0% of the wall area. The latter value was measured in a closed building in a hot humid climate, and was contributed significantly by the leakage area around a closed roller door which amounted to about 2.4% of the door area (Humphreys, 2020).

6.5 WIND-TUNNEL MODELLING OF INTERNAL PRESSURES

To correctly model internal pressures in wind-tunnel tests, it is necessary to ensure that the frequencies associated with the internal pressure fluctuations are scaled correctly with respect to the frequencies in the external flow. The relevant internal pressure frequencies are the Helmholtz resonance frequency (Sections 6.2.3 and 6.2.7) and the 'characteristic frequency' (Section 6.4).

For correct scaling of internal pressure fluctuations at full-scale design wind speeds, it is usually necessary to increase the internal volume above that obtained from that obtained from normal geometric scaling. The details of the scaling rules for internal pressures are discussed in Chapter 7 (Section 7.4.2).

6.6 RISK-CONSISTENT INTERNAL PRESSURES

In the 'real world', it is very difficult to anticipate the number of large openings that are likely to be present at the time of an extreme wind event, particularly in the case of severe thunderstorm-generated winds such as downbursts, for which there is likely to be little warning. Thus, a large industrial warehouse may have many large doors which may or may not be left open in the extreme wind event. Similarly, a balcony door of an apartment building may be left open in an extreme wind, or not.

In the case of buildings in regions affected by tropical cyclones (hurricanes and typhoons), doors and windows often fail by direct wind pressure or by impact from flying debris; however, in a large building with many doors and windows, the number of failures of these producing large openings is uncertain. However, reasonable estimates of the probability of various potential opening scenarios can be made.

A probabilistic approach to the prediction of internal pressures is explored in this section. It involves the known extreme value distribution for wind speeds at a building site, and probabilities for all possible opening scenarios are required.

A rational approach to the determination of internal pressure coefficients, when a number of different possible opening scenarios need to be considered, should be based on assigning a probability of occurrence to each scenario. Thus, a particular scenario, denoted by j, would have a probability of occurrence of P_j. If all N possibilities have been considered, then we get

$$\sum_{j=1}^{N} P_j = 1 \tag{6.32}$$

For a known extreme wind climate, the risk of exceedance (or non-exceedance) of a given internal pressure p_i can be calculated for each scenario knowing the internal pressure coefficient, C_{pi}, for that scenario. Then, the combined risk of exceedance of p_i for all scenarios can be calculated.

Assuming that the extreme gust speed for an average recurrence interval, R_I, is given by the GEV model (Equation 2.6):

$$V_R = C - D \cdot R_I^{-n} \tag{6.33}$$

Equation (6.33) applies to the whole lifetime of the building – i.e. to all possible opening scenarios. If only the times when opening scenario, j, occurs are considered, then Equation (6.33) can be adjusted to give Equation (6.34):

$$V_{R,\,j} = C - D \cdot \left(R_I . P_j \right)^{-n} \tag{6.34}$$

For example, if P_j is equal to 0.2, the opening scenario j is applicable for only 20% of the life of the building, and the average recurrence interval for the wind speed V_R, *given the opening scenario j*, is effectively increased by a factor of 5.

Then, the probability of *non*-exceedance of the wind speed $V_{R,\,j}$, in any year, given the opening scenario, j, can be determined from Equation (6.34) as:

$$1 - \frac{1}{R_P} = exp\left[-\left(\frac{1}{R_I}\right)\right] = exp\left[-P_j\left(\frac{C - V_{R,\,j}}{D}\right)^{\left(\frac{1}{n}\right)}\right]$$ 6.35)

For all N possible scenarios, assuming statistical independence, the combined probability of non-exceedance is:

$$\prod_{j=1}^{N} exp\left[-P_j\left(\frac{C - V_{R,\,j}}{D}\right)^{\left(\frac{1}{n}\right)}\right] = exp\left[-\sum_{j=1}^{N} P_j\left(\frac{C - V_{R,j}}{D}\right)^{\left(\frac{1}{n}\right)}\right]$$

and the combined probability of exceedance

$$1 - exp\left[-\sum_{j=1}^{N} P_j\left(\frac{C - V_{R,\,j}}{D}\right)^{\left(\frac{1}{n}\right)}\right]$$ (6.36)

The wind gust speed producing the internal pressure, p_i for the opening scenario, j, is given by:

$$V_{R,\,j} = \sqrt{\frac{2\,p_i}{\rho\,C_{pi,\,j}}}$$ (6.37)

where $C_{pi,\,j}$ is the internal pressure coefficient.

Substituting for $V_{R,\,j}$ from Equation (6.37) into Equation (6.36), the combined probability of exceedance of the internal pressure p_i can be determined. By iteration (trial and error), the internal pressure, and the effective combined internal pressure coefficient, for a defined risk of exceedance, such as an annual risk of 1/500, can be determined.

To a very good approximation, Equation (6.36) can be written, combined probability of exceedance=

$$\sum_{j=1}^{N}\left[P_j\left(\frac{C - V_{R,j}}{D}\right)^{\left(\frac{1}{n}\right)}\right]$$ (6.38)

where Equation (6.38) is based on the approximation,

$$1 - exp(-x) \cong x, \quad for \ x \ll 1$$ (6.39)

Example 6.3

Consider a warehouse with several large doors on the western wall of a building, of which up to three could be open, if a severe thunderstorm wind event occurred during working hours – 12 hours per day. The building does not operate for the remaining 12 hours of nighttime.

Allowing for the background permeability and openings and vents elsewhere on the building, the internal pressure coefficients in Table 6.2 can be obtained using a mass-balance approach (Equation (6.23)), for ten different scenarios, for wind blowing from the west. The third column in Table 6.2 shows the pressure coefficients, with the estimated probability of occurrence of each opening scenario.

The building is assumed to be in a location where an extreme value analysis gives values of C and D of 67 m/s and 41 m/s in Equation 6.33. The nominal design wind speed at the top of the building is 40 m/s.

By trial and error, the internal pressure that satisfies Equation 6.38, for a combined annual risk of 0.02 (1/50), is *328.3 Pascals*. The calculations for this are shown in Table 6.3.

Note that whenever $V_{R,j}$ exceeds 67 m/s – i.e. C, the maximum value possible from Equation (6.33), the contribution to the risk defaults to 0.0

Then, an effective risk-consistent internal pressure coefficient can be calculated as follows:

$$C_{pi, \text{eff}} = \frac{328.3}{0.5 \times 1.2 \times 40^2} = 0.34$$

The effective internal pressure coefficient is about two-thirds of the maximum value of C_{pi} in Tables 6.2 and 6.3 (for Scenario 7).

An alternative approach to development of a risk-consistent internal pressure, based on the 'out-crossing' method (Section 9.11 and Appendix C5.3) was described by Irwin and Sifton (1998).

Table 6.2 Door-opening scenarios and internal pressure coefficients

Windward wall doors	Leeward wall doors	Probability P_k	C_{pi}
No doors open	No doors open	0.50	0.01
1	0	0.056	0.33
1	1	0.056	0.03
2	0	0.056	0.47
2	1	0.056	0.24
2	2	0.056	0.05
3	0	0.056	0.54
3	1	0.056	0.36
3	2	0.056	0.20
3	3	0.056	0.06
	total	1.0	

Table 6.3 Calculation of contributions to combined annual risk

Scenario	C_{pi}	$V_{R,j}$ (m/s) (Equation 6.37)	Risk (Equation 6.38)
1	0.01	233.9	0.00E+00
2	0.33	40.7	6.50E−04
3	0.03	135.1	0.00E+00
4	0.47	34.1	6.11E−03
5	0.24	47.7	2.89E−05
6	0.05	104.6	0.00E+00
7	0.54	31.8	1.20E−02
8	0.36	39.0	1.23E−03
9	0.20	52.3	1.94E−06
10	0.06	95.5	0.00E+00
		Total Σ	**0.02**

6.7 SUMMARY

The topic of internal pressures produced by wind has been covered in this chapter. The relevant non-dimensional parameters are introduced, and the response time of the interior of a building or a single room to a sudden increase in external pressure at an opening has been evaluated.

The dynamic response of an internal volume to excitation by a sudden generation of a windward wall opening, or by turbulence, using the Helmholtz resonator model, which includes inertial effects, has been considered. The effect of roof flexibility on the governing equation, and on the Helmholtz resonance frequency has been discussed. A formula has been derived for estimating reduction in peak internal pressure resulting from a large internal volume, and/or a small opening area. The effect of multiple windward and leeward openings on mean and fluctuating internal pressures has been introduced. The case of a nominally sealed building with distributed porosity was also considered.

The requirements for modelling of internal pressures in wind-tunnel studies have also been mentioned; however, the full details of this are given in Chapter 7.

The common situation in which varying numbers of openings may occur in a building envelope from time to time is discussed in Section 6.6. A 'risk-consistent' approach may be used to determine an effective internal pressure coefficient for this situation, and an example is given for the method.

Many of the results in this chapter have been validated by wind-tunnel studies, and, more importantly, by full-scale measurements (e.g. Ginger *et al.*, 1997; Humphreys, 2020).

REFERENCES

Euteneur, G.A. (1970) Druckansteig im Inneren von Gebauden bei Windeinfall. *Der Bauingenieur*, 45: 214–216 (in German).

Ginger, J.D., Mehta, K.C. and Yeatts, B.B. (1997) Internal pressures in a low-rise full-scale building. *Journal of Wind Engineering and Industrial Aerodynamics*, 72: 163–174.

Ginger, J.D., Holmes, J.D. and Kim, P.Y. (2010) Variation of internal pressure with varying sizes of dominant openings and volumes. *Journal of Structural Engineering, American Society of Civil Engineers*, 136: 1319–1326.

Harris, R.I. (1990) The propagation of internal pressures in buildings. *Journal of Wind Engineering and Industrial Aerodynamics*, 34: 169–184.

Holmes, J.D. (1979) Mean and fluctuating internal pressures induced by wind. *Proceedings, 5th Internal Conference on Wind Engineering*, Fort Collins, Colorado, USA, pp. 435–450, Pergamon Press.

Holmes, J.D. and Ginger, J.D. (2012) Internal pressures – the dominant windward opening case – a review. *Journal of Wind Engineering and Industrial Aerodynamics*, 100: 70–76.

Humphreys, M. (2020) Characteristics of wind-induced internal pressures in industrial buildings with wall openings. *Ph.D. thesis*, James Cook University, Townsville, Australia.

Irwin, P.A. and Sifton, V.L. (1998). Risk considerations for internal pressures. *Journal of Wind Engineering and Industrial Aerodynamics*, 77–78: 715–723.

Liu, H. (1982) Wind tunnel modeling of building internal pressure. *Transportation Engineering Journal (ASCE)*, 108: 691–696.

Liu, H. and Rhee, K.H. (1986) Helmholtz oscillation in building models. *Journal of Wind Engineering and Industrial Aerodynamics*, 24: 95–115.

Liu, H. and Saathoff, P.J. (1982) Internal pressure and building safety. *Journal of the Structural Division (ASCE)*, 108: 2223–2234.

Malecki, I. (1969) *Physical foundations of technical acoustics*. Pergamon Press, Oxford.

Rayleigh, L. (1896) *Theory of sound – Volume 2*. Macmillan, London. (reprinted by Dover Publications, 1945).

Rossiter, J.E. (1964) Wind-tunnel experiments on the flow over rectangular cavities at subsonic and transonic speeds. Aeronautical Research Council, R & M 3438, Ministry of Aviation, UK.

Saathoff, P.J. and Liu, H. (1983) Internal pressure of multi-room buildings. *Journal of the Engineering Mechanics Division, American Society of Civil Engineers*, 109: 908–919.

Sharma, R.N. (2008) Internal and net envelope pressures in a building having quasi-static flexibility and a dominant opening. *Journal of Wind Engineering and Industrial Aerodynamics*, 96: 1074–1083.

Vickery, B.J. (1986) Gust factors for internal pressures in low-rise buildings. *Journal of Wind Engineering and Industrial Aerodynamics*, 23: 259–271.

Vickery, B.J. (1994) Internal pressures and interaction with the building envelope. *Journal of Wind Engineering and Industrial Aerodynamics*, 53: 125–144.

Vickery, B.J. and Bloxham, C. (1992) Internal pressure dynamics with a dominant opening. *Journal of Wind Engineering and Industrial Aerodynamics*, 41: 193–204.

Xu, H., Yu, S. and Lou, W. (2017) Wind-induced internal pressure fluctuations of structure with single windward opening. *Wind and Structures*, 24: 79–93.

Yu, S., Lou, W. and Sun, B. (2006) Wind-induced internal pressure fluctuations of structure with single windward opening. *Journal of Zhejiang University Science B*, 7: 415–423.

Yu, S., Lou, W. and Sun, B. (2008) Wind-induced internal pressure response for structure with single windward opening and background leakage. *Journal of Zhejiang University Science A*, 9: 313–321.

Chapter 7

Laboratory simulation of strong winds and wind loads

7.1 INTRODUCTION

Practising structural engineers will not generally themselves operate wind tunnels, or other laboratory equipment, for simulation of strong wind effects on structures, but they may be clients of specialist groups who will provide wind loading information for new or existing structures, usually by means of model tests. For this reason, this chapter will not attempt to describe in detail wind tunnel, or other simulation, techniques. There are detailed references, guide books and manuals of practice available which perform this function (e.g. Cermak, 1977; Reinhold, 1982; Australasian Wind Engineering Society, 2019; American Society of Civil Engineers, 2012). However, sufficient detail is given here to enable the educated client to be able to 'ask the right questions' of their wind-tunnel contractors.

In the following sections, a brief description of wind tunnel layouts is given, and methods of simulation of natural wind flow and experimental measurement techniques are discussed.

7.2 WIND-TUNNEL HISTORY AND LAYOUTS

7.2.1 Historical

The first use of a wind tunnel to measure wind forces on buildings is believed to have been made by W. C. Kernot in Melbourne, Australia (1893). A sketch of the apparatus, which he called a 'blowing machine', is given in Figure 7.1 (Aynsley *et al.*, 1977). This would now be described as an "open-circuit, open test-section" arrangement. With this equipment, Kernot studied wind forces on a variety of bluff bodies – cubes, pyramids, cylinders, etc. – and on roofs of various pitches.

At about the same time, Irminger (1894) in Copenhagen, Denmark, used the flow in a flue of a chimney to study wind pressures on some basic shapes (Larose and Franck, 1997).

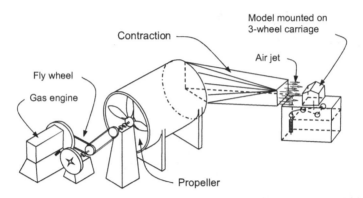

Figure 7.1 Sketch of W.C. Kernot's 'blowing machine' of 1893 (Aynsley et al., 1977).

Wind tunnels for aeronautical applications developed rapidly during the first half of the twentieth century, especially during and between the two World Wars. The two basic wind tunnel layouts: the open circuit, or 'N.P.L. (National Physical Laboratory)-type', and the closed circuit, or 'Göttingen-type' were developed during this period, named after the research establishments in the U.K. and Germany where they originated. These two types are outlined in Sections 7.2.2 and 7.2.3.

Simulation of the atmospheric boundary layer in a wind tunnel for a variety of problems, including wind loading of buildings, was pioneered by Jensen in the 1950s in Denmark (Jensen, 1958). Over the next 20 years, many attempts were made to simulate the boundary-layer flow of synoptic-scale wind events in wind tunnels. Hunt and Fernholtz (1975) surveyed 33 such facilities in Europe and the U.S.A.; many of these were conversions of aeronautical wind tunnels with short test sections (see Section 7.3.2), and most were developed for environmental problems such as the dispersion and diffusion of gaseous pollutants. At that time, there were probably a similar number of boundary-layer wind tunnels in operation in other parts of the world – particularly Australia, Canada and Japan. In the four decades since then, some older wind tunnels have been closed down, but numerous new ones have been constructed – many by private construction companies and consulting groups.

7.2.2 Open-circuit type

The simplest type of wind-tunnel layout is the open-circuit or N.P.L.-type. The main components are shown in Figure 7.2. The contraction, usually with a flow straightener and fine mesh screens, has the function of smoothing out mean flow variations, and reducing turbulence in the test section. For modelling atmospheric boundary layer flows, which are themselves very turbulent, as described in Chapter 3, it is not essential to include a contraction, although it is better to start with a reasonably uniform and smooth flow before commencing to simulate atmospheric profiles and turbulence.

The function of the diffuser, shown in Figure 7.2, is to conserve power by reducing the amount of kinetic energy that is lost with the discharging air. Again, this is not an essential item, but omission will be at the cost of higher power consumption.

Figure 7.2 shows an arrangement with an axial-flow fan downstream of the test section. This arrangement is conducive to better flow, but, since the function of the fan is to produce a pressure rise to overcome the losses in the wind tunnel, there will be a pressure drop across the walls and floor of the test section that can be a problem if leaks exist. An alternative is a 'blowing' arrangement in which the test section is downstream of the fan (see Figure 7.6). Usually, a centrifugal blower is used, and a contraction with screens is essential to eliminate the swirl downstream of the fan. However, in this arrangement the test section is at or near atmospheric pressure.

Both the arrangements described above have been used successfully in wind engineering applications.

7.2.3 Closed-circuit type

In the closed-circuit, or Göttingen-type, wind tunnel, the air is continually recirculated, instead of being expelled. The advantages of this arrangement are as follows:

- It is generally less noisy than the open-circuit type.
- It is usually more efficient. Although the longer circuit gives higher frictional losses, there is no discharge of kinetic energy at exit.
- More than one test section with different characteristics can be incorporated.

However, this type of wind tunnel has a higher capital cost, and the air heats up over a long period of operation before reaching a steady-state temperature. This can be a problem when operating temperature-sensitive

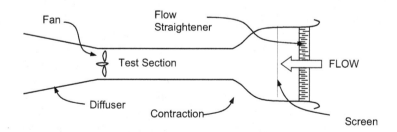

Figure 7.2 Layout of an open-circuit wind tunnel.

instruments, such as hot-wire or other types of thermal anemometers, which use a cooling effect of the moving air for their operation.

7.3 SIMULATION OF THE NATURAL WIND

In this section, methods of simulation of strong wind characteristics in a wind tunnel are reviewed. Primarily, the simulation of the atmospheric boundary layer in gale, or large-scale synoptic conditions, is discussed. This type of large-scale storm is dominant in the temperate climates, for latitudes greater than about 40°, as discussed in Chapter 1.

Even in large-scale synoptic windstorms, flows over sufficiently long homogeneous fetch lengths, so that the boundary layer is fully developed, are relatively uncommon. They will occur over open sea with consistent wave heights, and following large fetches of flat open country or desert terrain. Buildings or other structures, which are exposed to these conditions, are few in number, however. Urban sites, with flat homogeneous upwind roughness of sufficient length to produce full development of the boundary layer, are also relatively uncommon. However, there have been sufficient measurements in conditions that are close to ideal to produce generally accepted semi-theoretical models of the strong-wind atmospheric boundary layer for engineering purposes. These models have been validly used as the basis for wind-tunnel modelling of phenomena in the atmosphere, and the salient points have been discussed in Chapter 3.

In the case of the wind loading and response of structures, such as buildings, towers, bridges, etc., gales produced by large, mature, extra-tropical depressions are adequately described by these models, and they form a benchmark by which wind-tunnel flows are usually assessed. However, there are significant differences of opinion regarding some turbulence properties, such as length scales and spectra, which are important in determining wind forces and dynamic response. These uncertainties should be considered when assessing the reliability of wind-tunnel tests as a predictor of wind effects on real structures.

As outlined in Chapter 3, these models are also not good ones for storm winds produced by localized thermal mechanisms, namely tropical cyclones (hurricanes, typhoons), thunderstorms (including tornadoes) and monsoons. Winds produced by these storms are the dominant ones for design of structures in latitudes within about 40° from the Equator.

The following sections consider natural growth methods requiring long test sections, methods used for wind tunnels with short test sections, and methods developed for simulating only the inner or surface layer of the atmospheric boundary layer. Finally, some possibilities for simulations of strong winds in tropical cyclone and thunderstorm conditions are discussed. Laboratory modelling of these phenomena is still in an early stage of development, but some ideas on the subject are presented in Section 7.3.4.

7.3.1 Similarity criteria and natural growth methods

The 'ideal' neutral atmospheric boundary layer has two characteristic length scales – one for the outer part of the flow which depends on the rate of rotation of the earth and the latitude and on a velocity scale, and the other for the flow near the surface itself which depends on the size and density of the roughness on the surface. The region near the surface, which is regarded as being independent of the effects of the earth's rotation, has a depth of about 100 m, and is known as the inner or surface layer.

The first use of boundary-layer flow to study wind pressure on buildings was apparently by Flachsbart (1932). However, the work of Jensen in Denmark provided the foundation for modern boundary-layer wind tunnel testing techniques. Jensen (1958) suggested the use of the inner layer length scale, or roughness length z_o (see Section 3.2.1), as the important length scale in the atmospheric boundary-layer flow, so that for modelling phenomena in the natural wind, ratios such as building height to roughness length (h/z_o) – later known as the Jensen Number – are important. Jensen (1965) later described model experiments carried out in a small wind tunnel in Copenhagen, in which natural boundary layers were allowed to grow over a fetch of uniform roughness on the floor of the wind tunnel. In the 1960s, larger "boundary-layer" wind tunnels were constructed, and began to be used for wind engineering studies of tall buildings, bridges and other large structures (Cermak, 1971; Davenport and Isyumov, 1967). These tunnels are either of closed-circuit design (Section 7.2.3) or of open circuit of the 'sucking' type, with the axial flow fan mounted downstream of the test section (Section 7.2.2). In more recent years, several open-circuit wind tunnels of the 'blowing' type have been constructed with a centrifugal fan upstream of the test section, supplying it through a rapid diffuser, a settling chamber containing screens and a contraction. As discussed in Section 7.2.2, the latter system has the advantage of producing nearly zero static pressure difference across the wind tunnel walls at the end of the boundary-layer test section.

A naturally grown rough-wall boundary layer will continue to grow until it meets the boundary layer on the opposite wall or roof. In practical cases, this equilibrium situation is not usually reached, and tests of tall structures are carried out in boundary layers that are still developing, but are sufficient to envelop the model completely. In most cases of structural tests, more rapid boundary-layer growth must be promoted by a 'tripping' fence, grid or spires at the beginning of the test section. The test section of a boundary-layer wind tunnel used for studying wind loads on high-rise buildings is shown in Figure 7.3. Upstream can be seen as a barrier and triangular spires, which act in combination with the floor roughness formed by blocks, adjustable in height, to give a reasonable simulation, at a length scaling ratio of 1/400, of the boundary-layer flow at the city centre shown.

Figure 7.3 Simulation of an urban boundary-layer in a wind tunnel with a long test section. (Cermak, Peterka, Petersen, Sydney, Australia)

Dimensional analysis indicates that the full height of the atmospheric boundary layer depends on the wind speed and the latitude. However, the typical height is about 1,000 m. Assuming a geometric scaling ratio of 1/500, this means that a minimum wind-tunnel height of 2 m is required to model the full atmospheric boundary layer. Usually, a lower boundary layer height is accepted, but the turbulent boundary-layer flow should envelop completely any structure under test.

In the early days of boundary-layer wind tunnels, it was common to install a roof of adjustable height for the purpose of maintaining a constant pressure gradient in the along-wind direction. This allows for the increasing velocity deficit in the flow direction, and maintains the 'free-stream' velocity outside of the boundary layer approximately constant. This should also reduce the errors due to blockage for large models. For smaller models with lower blockage ratios, the errors in the measurements when the roof is maintained at a constant height, or with a fixed slope are quite small, and it has been found to be unnecessary to continually adjust the roof, in most situations. Blockage errors and corrections are discussed in Section 7.7.

As noted previously, the real atmospheric boundary layer is affected by the earth's rotation, and apparent forces of the Coriolis type must be included when considering the equations of motion of air flow in the atmosphere. One effect of this is to produce a mean velocity vector which is not constant in direction with height; it is parallel to the pressure gradient at the top of the boundary layer (or 'gradient' height), and rotates towards the lower static pressure side as the ground level is approached. This effect is known as the 'Ekman Spiral' (although the original solution by Ekman was obtained by assuming a shear stress in the flow proportional to the vertical velocity gradient – an assumption later shown to be unrealistic), and it has been shown to occur in full scale, with mean flow direction changes up to 30° having been measured. This effect cannot be achieved in conventional wind tunnels, and the direction change is usually regarded as unimportant over the heights of most structures.

7.3.2 Methods for short test sections

In the 1960s and 1970s, to avoid the costs of constructing new boundary-layer wind tunnels, several methods of simulating the atmospheric boundary layer in existing (aeronautical) wind tunnels with test sections of low aspect ratio, i.e. short with respect to their height and width, were investigated. These usually make use of tapered fins or spires, which produce an immediate velocity gradient downstream, and which develops into a mean velocity profile representative of that in the atmosphere within a short downstream distance. Other bluff devices, such as grids or barriers, are required upstream, together with roughness on the floor of the wind tunnel, to increase the turbulence intensities to full-scale values.

Flows produced by these methods are likely to be still in a process of rapid development at the end of the short test section, and the interaction of the vortex structures produced in the wakes of the various devices, may well result in unwanted characteristics in the turbulence at the measurement position. Unless detailed fluctuating velocity measurements, including spatial correlations, are made, such characteristics may never be detected. Fortunately, wind pressures and forces on structures appear to be dependent mainly on single point statistics, such as turbulence intensities, and integral length scales in the along-wind direction, and not on the detailed eddy structures within the turbulence, in the approach flow.

Of the several methods developed in the late 1960s and early 1970s, that of Counihan (1969) is perhaps the best documented. The upstream devices consisted of a castellated fence, or barrier, several elliptical 'sharks-fins', and a short fetch of surface roughness (Figure 7.4). Detailed measurements of mean velocity and turbulence intensity profiles at various spanwise stations, and of cross-correlations and spectra were made.

7.3.3 Simulation of the surface-boundary layer

For simulation of wind forces and other wind effects on low-rise buildings, say less than 10 m in height, geometric scaling ratios of 1/400, or 1/500, result in extremely small models and do not allow any details on the building to be reproduced. The large differences in Reynolds Numbers between model and full scale may mean that the wind-tunnel test data is quite unreliable. For this type of structure, no attempt should be made to model the complete atmospheric boundary layer. Simulation of the inner or surface

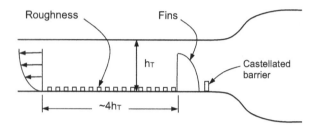

Figure 7.4 The Counihan method for short test sections.

layer, which is approximately 100 m thick in full scale, is sufficient for such tests. If this is done, larger and more practical, scaling ratios in the range from 1/50 to 1/200 can be used for the models.

Cook (1973) developed a method for simulation of the lower third of the atmospheric boundary layer. This system consists of a castellated barrier, a mixing grid and surface roughness. A simpler system consisting of a plain barrier, or wall, at the start of the test section, followed by several metres of uniform surface roughness has also been used (Figure 7.5) (Holmes and Osonphasop, 1983). This system has the advantage that simultaneous control of the longitudinal turbulence intensity and the longitudinal length scale of turbulence, to match the model scaling ratio, is obtained by adjustment of the height of the barrier. Larger scales of turbulence can be produced by this method than by other approaches – large horizontal vortices with their axes normal to the flow are generated in the wake of the barrier. Studies of the development of the flow in the wake of the barrier (Holmes and Osonphasop, 1983) showed that a fetch length of at least 30 times the barrier height is required to obtain a stable and monotonically increasing mean velocity profile. However, there is still a residual peak in the shear stress profile at the height of the barrier at this downstream position; this shows that the flow is still developing at the measurement position, but the effect of this on pressures on and flow around single buildings should not be significant.

7.3.4 Simulation of tropical cyclones and thunderstorm winds

As discussed in Chapter 1, strong winds produced by tropical cyclones and thunderstorms dominate the populations of extreme winds in most locations with latitudes less than 40°, including many sites in the U.S.A., Australia, India and South Africa. Unfortunately, full-scale measurements of such events are few in number, and there are no reliable analytical models for the surface wind structures in these storms. However, the few full-scale measurements, and some meso-scale numerical models, have enabled qualitative characteristics of the winds to be determined.

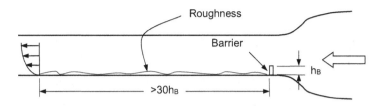

Figure 7.5 The barrier-roughness technique of boundary-layer simulation.

Tropical cyclones, also known as 'hurricanes' and 'typhoons' in some parts of the world, are circulating systems with a complex three-dimensional wind structure near their centre (Section 1.3.2) At the outer radii, where the wind speeds are lower, a boundary-layer structure should exist and conventional boundary-layer wind tunnels should be quite adequate for flow modelling. However, the region of maximum horizontal winds occurs just outside the eye wall. In that region, measurements with 'dropwindsondes' have indicated a maximum in the velocity profile at a height of about 300 m. Above that height, the mean wind velocity is approximately constant or falls slowly (Section 3.2.6). Measurements of turbulence intensities in typhoons have sometimes shown intermittent higher values than occur at the same site in non-cyclonic conditions (Section 3.3.1). However, generally, tropical cyclone wind flow can be represented by a boundary-layer flow similar to that used for simulating extra-tropical conditions.

The laboratory modelling of thunderstorm winds is a more difficult problem for a number of reasons. Firstly, there are a number of different types of local windstorms associated with thunderstorms, although some of these have similar characteristics. Secondly, these storms are individually transient, although a number of them may occur sequentially in the same day. The length of an individual storm rarely exceeds thirty minutes. Thirdly, thunderstorm winds are driven by thermodynamic processes which probably cannot be reproduced in a laboratory simulation.

The velocity profile in a thunderstorm microburst is quite similar to a wall jet. The latter has been proposed as a laboratory model of the flow in a microburst, and some studies have been conducted using the outlet jet from a wind tunnel impinging on a vertical board, as shown in Figure 7.6. Measurements can be carried out at various radial positions from the centre of the board. This system gives velocity profiles which are quite similar to those measured by radar in microbursts, but the transient characteristics of a real downburst flow are not reproduced, and the turbulence characteristics in the two flows could be quite different. To simulate the forward motion of such storms, the additional complexity of a moving impinging jet is required (Letchford and Chay, 2002).

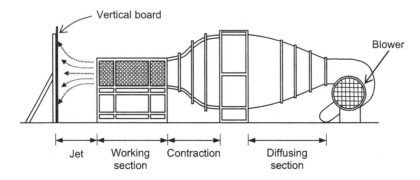

Figure 7.6 Simulation of thunderstorm downburst by impinging jet.

7.3.5 Laboratory simulation of tornadoes

Some characteristics of tornadoes and their effects on structures were discussed in Chapter 1 (Section 1.3.4) and Chapter 3 (Section 3.2.8). Davies-Jones (1976) gave a detailed review of the simulation of tornadoes, or 'tornado-like vortices' in laboratories. These have produced reasonable kinematic and dynamic similarity with full-scale tornadoes.

Chang (1971) and Ward (1972) used a ducted fan above a flat board, with rotary motion imparted to the air flowing into a convective chamber above the board by means of a rotating screen. In the Ward-type, the rising air exits the apparatus to an upper plenum, through a fine mesh honeycomb, which prevents fan-induced vorticity from entering the apparatus (Figure 7.7). In these systems, the rotational velocity is controlled by the rotational speed of the screen, and the core radius is controlled by the size of the opening to the upper plenum.

However, although methods of simulating tornadoes in laboratories are quite well developed, relatively few studies of wind pressures or forces on structures in laboratory simulations of tornadoes have been carried out, and few since the 1970s.

7.4 MODELLING OF STRUCTURES FOR WIND EFFECTS

7.4.1 General approach for structural response

The modelling of structures for wind effects, in wind tunnels, requires knowledge of dimensional analysis and the theory of modelling (e.g. Whitbread, 1963; Scanlan, 1974).

The general approach is as follows. It may be postulated that the response of a structure to wind loading, including resonant dynamic response, is

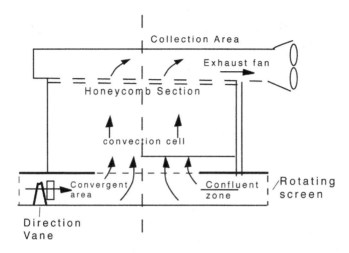

Figure 7.7 Laboratory simulation of tornado-like vortex (Ward, 1972).

dependent on a number of basic variables, such as the following (not necessarily exclusive):

$\bar{U}$ – the mean wind speed at some reference position

z_o – roughness length defining the approaching terrain and velocity profile (Section 3.2.1)

σ_u – standard deviation of longitudinal turbulence

σ_v – standard deviation of lateral turbulence

σ_w – standard deviation of vertical turbulence

ℓ_u – length scale of longitudinal turbulence (Section 3.3.4)

ℓ_v – length scale of lateral turbulence

ℓ_w – length scale of vertical turbulence

ρ_a – density of air

v – viscosity of air

g_0 – acceleration due to gravity

ρ_s – density of the structure

E – Young's modulus for the structural material

G – Shear modulus for the structural material

η – structural damping ratio

L – characteristic length of the structure

The above list has been simplified considerably. For example, for a bridge there will usually be different structural properties for the deck, the towers, the cables, etc. However, the above list will suffice to illustrate the principles of structural modelling.

The above 16 dimensioned variables can be reduced to 13 (16−3) independent dimensionless groups, according to the Buckingham-Pi theorem. A possible list of these is as follows:

L/z_0 – Jensen Number
$\sigma_u/\bar{U}$ – longitudinal turbulence intensity
$\sigma_v/\bar{U}$ – lateral turbulence intensity
$\sigma_w/\bar{U}$ – vertical turbulence intensity
ℓ_u/L – length ratio
ℓ_v/L – length ratio
ℓ_w/L – length ratio
$\dfrac{\bar{U}L}{v}$ – Reynolds Number (Section 4.2.4)

ρ_s/ρ_a – density ratio

$\dfrac{\bar{U}}{\sqrt{Lg_0}}$ – Froude Number (inertial forces (air)/gravity forces (structure))

$\dfrac{E}{\rho_a\bar{U}^2}$ – Cauchy Number (internal axial forces / inertial forces (air flow))

$\dfrac{G}{\rho_a\bar{U}^2}$ – Cauchy Number (internal shear forces / inertial forces (air flow))

η – critical damping ratio

For correct scaling, or similarity in behaviour between the model and full-scale structure, these non-dimensional groups should be numerically equal for the model (wind tunnel) and prototype situation.

The thirteen groups are not a unique set. Other non-dimensional groups can be formed from the sixteen basic variables, but there are only thirteen independent groups, and it will be found that the additional groups can be formed by taking products of the specified groups or their powers.

For example, it is often convenient to replace a Cauchy Number by a reduced frequency $(n_sL/\bar{U})$, where n_s is a structural frequency. For structures or structural members in bending, n_s is proportional to $\sqrt{(E/\rho_sL^2)}$.

Then the reduced frequency,

$$\frac{n_sL}{\bar{U}} = K\sqrt{\frac{E}{\rho_sL^2}}\cdot\frac{L}{\bar{U}} = K\sqrt{\frac{E}{\rho_a\bar{U}^2}}\cdot\sqrt{\frac{\rho_a}{\rho_s}} \tag{7.1}$$

where K is a constant.

Thus, the reduced frequency is proportional to the square root of the Cauchy Number divided by the square root of the density ratio.

7.4.2 Modelling of internal pressures

The phenomenon of Helmholtz resonance of internal pressures when the interior of a building is vented at a single opening was described in Chapter 6 (Sections 6.2.3 and 6.2.7) The 'characteristic' frequency of a building with distributed openings on windward and leeward walls was also discussed (Section 6.4). When simulating internal pressures in a wind-tunnel model of a building, these frequencies should be scaled correctly with respect to the frequencies in the external flow. The scaling requirements to ensure this are derived in the following equations.

For a single dominant opening (area A), the Helmholtz resonance frequency is given by Equation (6.10) in Chapter 6:

$$n_H = \frac{1}{2\pi} \sqrt{\frac{\gamma A p_o}{\rho_a \ell_e V_o}} \tag{6.10}$$

i.e. $\quad n_H \propto \sqrt{\dfrac{\sqrt{A} \cdot p_o}{\rho_a V_o}}$

$$n_H^2 \propto \frac{\sqrt{A} \cdot p_o}{\rho_a V_o}$$

Denoting the ratio of model to full-scale quantities by [ ]$_r$, the ratio of model to full-scale frequency is given by:

$$[n_H]_r^2 = \frac{[L]_r [p_o]_r}{[\rho_a]_r [V_o]_r} = \frac{[L]_r}{[V_o]_r}$$

since $[p_o]_r = [\rho_a]_r = 1.0$, for testing in air at normal atmospheric pressures.

However, for scaling with frequencies in the external flow:

$$[n_H]_r = \frac{[U]_r}{[L]_r}$$

Hence, for correct scaling, $[n_H]_r^2 = \dfrac{[L]_r}{[V_o]_r} = \dfrac{[U]_r^2}{[L]_r^2}$

i.e. $\quad [V_o]_r = \dfrac{[L]_r^3}{[U]_r^2}$ \hfill (7.2)

Thus, if the velocity ratio, $[U]_r$, is equal to 1.0, i.e. when the wind-tunnel speed is the same as full scale design speeds, then the internal volume should be scaled according to the geometrical scaling ratio,

$$[V_o]_r = [L]_r^3 \times 1.0$$

However, usually in wind-tunnel testing, the wind speed is considerably less than full-scale design wind speeds. Thus, $[U]_r$ is usually less than 1.0, and the internal volume should then be increased by a factor of $1/[U]_r^2$. For example, if the velocity ratio is 0.5, then the internal volume, V_o, should be increased by a factor of 4.

The characteristic response time for internal pressures in a building with distributed openings on the windward side A_W, and on the leeward side, A_L, is given by Equation (6.27) in Chapter 6 (neglecting inertial effects):

$$\tau = \frac{\rho_a V_o \bar{U} A_W A_L}{\gamma k p_o (A_W^2 + A_L^2)^{3/2}} \sqrt{C_{pW} - C_{pL}} \qquad (6.27)$$

For $A_W = A_L = A$ and fixed C_{pW} and C_{pL},

$$\tau \propto \frac{\rho_a V_o \bar{U}}{p_o A}$$

and the characteristic frequency,

$$n_C \propto \frac{p_o A}{\rho_a V_o \bar{U}}$$

Then, the ratio of model to full-scale frequency is given by:

$$\left[n_C\right]_r = \frac{[p_o]_r [L]_r^2}{[\rho_a]_r [V_o]_r [U]_r} = \frac{[L]_r^2}{[V_o]_r [U]_r}$$

For correct scaling with frequencies in the external flow,

$$\left[n_C\right]_r = \frac{[U]_r}{[L]_r} = \frac{[L]_r^2}{[V_o]_r [U]_r}$$

Hence $\quad \left[V_o\right]_r = \frac{[L]_r^3}{[U]_r^2} \qquad (7.2)$

Thus, the same scaling criterion applies, as for Helmholtz resonance frequency – i.e. the internal volume needs to be distorted if velocity ratio is not equal to 1.0.

It may be noted that agreement with Equation (7.2) amounts to ensuring equality of the non-dimensional parameter S^*, defined in Section 6.2.8, for the case when the speed of sound, a_s is the same in full and model scale, i.e. when the atmospheric pressure and air density are the same in full- and model-scale.

Liu (1982) discussed the modelling of internal pressures in buildings, and concluded that similarity of a parameter N, in model and full scale is required. N, as defined by Liu, is equal to $(\gamma/2S^*)$, where γ is the ratio of specific heats of air. Since γ is normally the same in wind tunnel and full scale, equality of N is equivalent to equality of S^*.

Pearce and Sykes (1999) also recognized that Equation (7.2) should be satisfied when modelling internal pressures in wind-tunnel tests.

The additional internal volume required by Equation (7.2) when the velocity ratio is less than 1.0, can be provided above (Pearce and Sykes, 1999), or beneath a wind-tunnel floor and connected to the interior of the model (Ginger *et al.*, 2010).

Failing to provide a sufficiently large volume will generally result in over-prediction of the fluctuating internal pressures, and incorrect frequency content (i.e. spectra) in the internal pressure fluctuations, but it is difficult to quantify the errors involved. Thus, it is advisable to correctly scale the internal volume, unless it is particularly difficult or inconvenient to do this.

7.4.3 Simulation requirements for structures in tornadoes

The similarity requirements in laboratory models of tornadoes, for simulating wind pressures on model structures, were discussed by Chang (1971) and Jischke and Light (1979). The latter proposed that the following non-dimensional parameters should be made the same in full- and model-scale for correct similarity:

$$\frac{h_i}{r_c}, \frac{\Gamma r_u}{Q}, \frac{z_o}{h_i}, \frac{L}{r_c}$$

where the dependent variables are as follows:

h_i – depth of the layer of horizontal inflow into the tornado
r_c – radius of the core
Γ – imposed circulation far from the axis of the tornado
r_u – radius of the updraft region
Q – volume flow rate
z_o – surface roughness length on the ground surface
L – characteristic length of the structure

7.4.4 Reynolds Numbers and roughening techniques

Reproducing full-scale flow conditions around circular cylinders on wind-tunnel models has been attempted on many occasions since the 1960s (e.g. Armitt, 1968). A rational approach to this was introduced by Szechenyi (1975); this involves similarity of the roughness Reynolds Number, Re_k, in which the average roughness height, k, is used as the length parameter.

In order for Re_k similarity to be achieved, it must also be checked that super-critical flow is obtained on the model. This can be checked by the use of Equation (4.12) or (4.13). The following example illustrates the use of this approach.

Example 7.1

Consider a reinforced concrete observation tower with a diameter of 13 m. The full-scale design mean wind speed at the top of the tower is 30 m/s, and the average roughness height of the concrete surface is 250 μm (250×10^{-6} m).

The equivalent wind speed for a 1/100 scale wind-tunnel model, which has been roughened with coarse sand with an average roughness height of 350 μm, is required.

The roughness Reynolds Number for full scale is as follows:

$$Re_k = \frac{30 \times 250 \times 10^{-6}}{14.5 \times 10^{-6}} \cong 517$$

To give the same value of Re_k in model scale,

$$\bar{U} = \frac{517 \times 14.5 \times 10^{-6}}{350 \times 10^{-6}} = 21.3 \text{ m/s}.$$

A check should now be made that super-critical flow is achieved on the model at this wind speed.

By Equation (4.12), the value of Re_b for minimum drag coefficient is as follows: $44{,}210 \left(\frac{k}{b} \right)^{-0.555} = 210 \left(\frac{350 \times 10^{-6}}{0.13} \right)^{-0.555} = 1.12 \times 10^5$

For the model test at 21.3 m/s, $Re_b = \dfrac{21.3 \times 0.13}{14.5 \times 10^{-6}} = 1.91 \times 10^5$.

Since this exceeds $Re_{b,minCd}$, super-critical flow will be obtained.

Note that if the above calculations are repeated for a model scale of 1/500, sub-critical flow occurs on the model, and correct similarity cannot be obtained by roughening.

7.4.5 Modelling of mullions on tall buildings

Mullions and other projections on side walls of tall buildings of all cross-sections, can have significant effects on local fluctuating and peak pressures. Modelling of these in wind-tunnel tests is discussed in Section 9.4.6.

7.5 MEASUREMENT OF LOCAL PRESSURES

Modern cheap sensitive solid-state pressure sensors, either as individual transducers or as part of a multi-channel electronic scanning system, enable near-simultaneous measurements of fluctuating wind pressures on wind-tunnel models of buildings and structures for up to several hundred measurement positions (Holmes, 1995).

For reasons of cost or geometric constraint, it is usually necessary to mount the pressure sensor or scanning unit remotely from the point where the pressure measurement is required. Then the fluctuating pressure must be transmitted by tubing between the measurement and sensing points. The dynamic frequency response of the complete pressure measurement system, including the sensor itself, the volume exposed to the diaphragm, and the tubing, is an important consideration.

Inadequate response can lead to significant errors especially when measuring peak pressures or suctions on building models (e.g. Durgin, 1982; Holmes, 1984; Irwin, 1988). As a rule of thumb, the equivalent full-scale upper frequency response limit should not be less than about 2 Hz. To convert this to model frequency, the frequency ratio is obtained by dividing the velocity ratio by the geometric length scaling ratio, e.g. for a typical velocity ratio of 1/3, and a geometric ratio of 1/300, the frequency ratio is 100, and the desirable upper limit is 200 Hz.

The transmission of pressure fluctuations is affected by the mass inertia, compressibility, and energy dissipation in the transmitting fluid (e.g. Bergh and Tijdeman, 1965). Standing waves can produce unwanted resonant peaks in the amplitude frequency response characteristics of the system, and a non-linear variation of phase lag with frequency.

An ideal system would have an amplitude response which is constant over the frequencies of interest, and a linear phase variation with frequency. The latter characteristic guarantees that there is no distortion of transient pressure "signatures" by the system.

The effect of bends in tubing systems on the dynamic response, with a radius of curvature of three or more times the internal radius of the tubing, appears to be negligible (Yoshida et al., 2001; Wang et al., 2018). However, sharp right-angled bends in tubing systems do have significant non-linear effects and should be avoided.

The availability of three-dimensional printing enables some part of the pressure-measurement paths to be 'printed' into building models. The effect on the dynamic response of the pressure-measurement systems ideally should be checked experimentally (e.g. Kay et al., 2020), as tubes with non-circular cross-sections, or changes in diameter, near the point of measurement, have the greatest effect on the response.

As well as pressure measurement at a single point, systems in which pressures from a number of points are connected to a common manifold or pneumatic averager have become widely used. In wind engineering, this arrangement has been used to obtain fluctuating and peak pressures appropriate to a finite area, or panel, on a building model in a turbulent wind-tunnel flow (e.g. Surry and Stathopoulos, 1977; Holmes and Rains, 1981; Gumley, 1984; Holmes, 1987; Kareem et al., 1989).

7.5.1 Single-point measurements

Three systems that have been used in the past are:

- "Short" tube systems.

  This system uses a relatively short length of tubing to connect the measurement point to the sensor. Typically, for wind-tunnel testing, this may consist of tubing 20–100 mm long, and 1–2 mm internal diameter. The short tube lengths will result in resonant frequencies that are high, hopefully well above the range of interest for the measurements. However, the short tube also results in low dissipation of energy, and the amplitude response rises to a high value at the peak.

- "Restricted" tube systems

  Restricted-tube systems may be defined as those involving one or more changes in internal diameter along the tube length. Such systems often allow location of pressure sensors at distances of 150–500 mm from the measurement point, with good amplitude and phase characteristics up to 200 Hz, or more. The simplest system of this type is the two-stage type, in which a section of narrower tube is inserted between the main tube section and the transducer. Restricted-tube systems are very effective in removing resonant peaks and giving linear phase response characteristics (e.g. Surry and Isyumov, 1975; Irwin et al., 1979; Holmes and Lewis, 1987a). An effective frequency range can be obtained which is better than that for a constant diameter tubing with a fraction of the length.

- "Leaked" tube systems.

  The leaked-tube system was proposed by Gerstoft and Hansen (1987). A theoretical model was developed by Holmes and Lewis (1989). A relatively flat amplitude frequency response to frequencies of 500 Hz, with 1 m of connecting tubing, is possible with a system of this type. This is achieved by inserting a controlled side leak part-way along the main connecting tube, usually close to the transducer. It has the effect of attenuating the amplitude response to low frequency fluctuations, and to steady pressures, to the level of a conventional closed system at higher frequencies. Thus, the leak effectively introduces a high-pass filter into the system. The amplitude ratio at frequencies approaching zero, is simply a function of the resistance to steady laminar flow of the main tube and leak tube. For multiple pressure tap measurements with this system, it is normally necessary to connect all the leaks to a common reference pressure, usually that inside a closed chamber, or plenum, to which the reference static pressure is also connected.

The general arrangement of the three types of single-point measurements are shown in Figure 7.8.

The systems described above were used in a period when computer-based data acquisition systems were somewhat slower than they are now. The

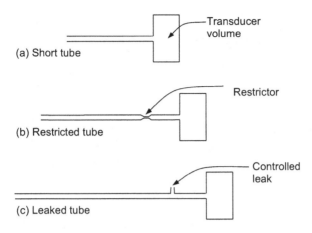

Figure 7.8 Tubing arrangements for measurement of point pressures. (a) short tube, (b) restricted tube, (c) leaked tube.

most common system in use today is digital correction, in which filtering, in the frequency domain, is used to correct the amplitude response to one that is near-flat over the frequency range of interest. Usually, the Fast Fourier Transform (FFT) is used for this purpose. The Inverse FFT is then used to recreate the time series (e.g. Irwin *et al.*, 1979). This method requires accurate knowledge of the frequency response characteristics of the tubing-transducer system before filtering. If the tubing dimensions and transducer volume are known accurately, the Bergh and Tijdeman (1965) theory can be used for this purpose; otherwise, experimental calibrations should be carried out.

7.5.2 Measurement of area-averaged pressures

Systems which average the pressure fluctuations from a number of measurement points, so that area-averaged wind loads on finite areas of a structure can be obtained, are now in common use. 'Pneumatic' averaging manifolds were first used in wind tunnels by Surry and Stathopoulos (1977). Gumley (1981, 1983) developed a theoretical model for their response.

Figure 7.9 shows the types of parallel tube and manifold arrangement that have been commonly used in wind engineering work. Provided that the inlet tubes are identical in length and diameter, such a system should provide a true average in the manifold, of the fluctuating pressures at the entry to the input tubes, assuming that laminar flow exists in them. Usually, flatter amplitude response curves to higher frequencies, can be obtained with the multi-tube-manifold systems, compared with single-point measurements using the same tube lengths, due to the reinforcement of the higher frequencies in the input tubes. However, once the number of input

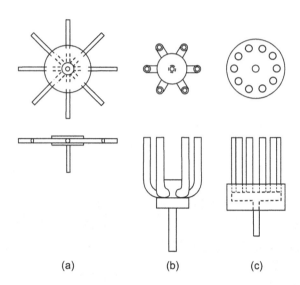

Figure 7.9 Manifolds for pressure averaging. (a) radial, (b) axial-radial, (c) axial.

tubes exceeds about five, there is little change to the response characteristics. The response is also not greatly sensitive to the volume of the averaging manifold.

With the reduction in cost of pressure sensors and digital processing in the last two decades, digital averaging of the outputs from individual pressure taps has largely replaced pneumatic averaging.

The assumption that the average of discrete fluctuating point pressures, sampled within a finite area of a surface, adequately approximates the continuous average aerodynamic load on the surface requires consideration (Surry and Stathopoulos, 1977; Holmes and Lewis, 1987b; Letchford, 1989). This is the case for both pneumatic averaging using manifolds, and for digital averaging.

Figure 7.10 shows the ratio of the variance of the averaged panel force to the variance of the point pressure, using firstly, the correct continuous averaging over the panel denoted by R_c, and secondly, the discrete averaging approximation performed using the ten pressure tappings within a panel, denoted by R_d. Calculations of these ratios were made, assuming a correlation coefficient for the fluctuating pressures of the form, $\exp(-Cr)$, where r is a separation distance, and C is a constant. The variance of the local pressure fluctuations across the panel of dimensions B by $B/2$, were assumed constant.

It can be seen that R_d exceeds R_c for all values of CB. This is due to the implied assumption, in the discrete averaging, that the pressure fluctuations are fully correlated in the tributary area around each pressure tap. Clearly, the errors increase with increasing C due to the lower correlation

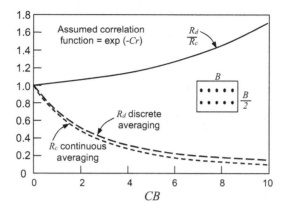

Figure 7.10 Discrete and continuous averaging of fluctuating pressures.

of the pressure fluctuations, and with increasing panel size, B. The errors can be decreased by increasing the number of pressure tappings within a panel of a certain size. However, it should be noted that the errors are larger at higher frequencies than at lower frequencies; a more detailed analysis of the errors requires knowledge of the coherence of the pressure fluctuations.

7.5.3 Equivalent time averaging

An alternative procedure for determining wind loads acting over finite surface areas from point pressures is known as 'equivalent time averaging'. In this approach, the time histories of fluctuating point pressures are filtered by means of a moving average filter. As originally proposed by Lawson (1976), the averaging time, τ, was estimated to be given by the following formula:

$$\tau \cong 4.5 \frac{L}{\bar{U}} \tag{7.3}$$

where L is usually taken as the length of the diagonal for the panel of interest.

However, a later analysis (Holmes, 1997) showed Equation (7.3) to be unconservative, and that a more correct relationship is:

$$\tau \cong 1.0 \frac{L}{\bar{U}} \tag{7.4}$$

However, the 'constants' in the above equations are likely to vary considerably depending on the location of the pressure measurement position on a building model – i.e. windward wall, roof, etc. This method is less accurate than the area-averaging technique by manifolding described in Section 7.5.2.

7.6 MODELLING OF OVERALL LOADS AND RESPONSE OF STRUCTURES

7.6.1 Base-pivoted model testing of tall buildings

This section describes the procedure for conducting of aeroelastic wind tunnel testing of high-rise buildings, using rigid models.

The use of rigid-body aeroelastic modelling of tall buildings is based on the following three basic assumptions:

- The resonant response of the building to wind loads in torsional (twisting modes) can be neglected.
- The response in sway modes higher than the first in each orthogonal direction, can be neglected.
- The mode shapes of the fundamental sway modes can be assumed to be linear.

With these assumptions, the motion of a rigid model of a building, pivotted at, or near, ground level, and located in a wind tunnel in which an acceptable model of the atmospheric boundary layer in strong winds has been set up, can be taken to represent the sway motion of the prototype building. The fact that a scaled reproduction of the building motion has been obtained, means that fluctuating aerodynamic forces that depend upon that motion have been reproduced in the wind tunnel. This is not the case when fixed models are used to measure the fluctuating wind pressures, or the 'base balance' technique is used.

Even buildings that have a non-linear mode shape can often be modelled by means of rigid-body rotation, but in these cases it may be appropriate to position the pivot point at a different level to ground level. For example, a building supported on stiff columns near ground level might be modelled by a rigid model pivotted at a height above ground level (e.g. Isyumov et al., 1975). The disadvantage of this approach is that the bending moment at ground level cannot be measured.

There is a direct analogy between the generalized mass of the prototype building, G_1, and the moment of inertia of the model building, including the contributions from the support shaft and any other moving parts.

Assuming that the mode shape of the building is given by Equation (7.5),

$$\phi_1(z) = \left(\frac{z}{h}\right) \tag{7.5}$$

and the generalized mass is given by Equation (7.6),

$$G_1 = \int_0^h m(z)\phi_1^2(z)dz \tag{7.6}$$

The equivalent prototype moment of inertia for rigid body rotation about ground level is then:

$$I_p = \int_0^h m(z)z^2\,dz = (1/h^2)G_1 \tag{7.7}$$

The equivalent model moment of inertia is then given by Equation (7.8),

$$I_m = M_r L_r^2 I_p = L_r^5(1/h^2)G_1 \tag{7.8}$$

where M_r and L_r and are the mass ratio and length ratio, respectively. In order to maintain a density ratio of unity in both model and full scale, assuming that air is the working fluid in both cases,

$$M_r = L_r^3 \tag{7.9}$$

Equation (7.8) can be used to establish the required model moment of inertia.

In order to obtain the correct moment of inertia, and at the same time to achieve a relatively rigid model, it is normally necessary to manufacture the model from a light material such as expanded foam, or balsa wood. A typical mounting is shown in Figure 7.11. The model is supported by gimbals

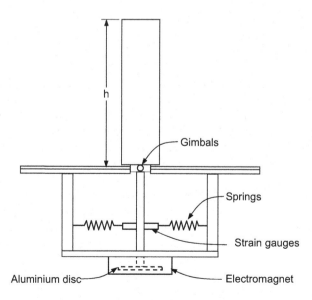

Figure 7.11 A base-pivoted tall building model.

of low friction, and rotation about any horizontal axis is permitted. Elastic support can be provided by springs whose position can be adjusted vertically. In the case of the system shown in Figure 7.11, damping is provided by an eddy current device, but vanes moving in a container of viscous liquid can also be used.

The moment of inertia of the model and the supporting rod and damper plates can be determined in one or more of the following three ways:

- By swinging the model, supporting rod and attachments, as a compound pendulum and measuring the period of oscillation,
- By measuring the frequency of vibration in the mounted position, and knowing the spring constants,
- By measuring the angular deflection of the supporting rod for known overturning moments applied to the model in position and using the measured frequencies.

The support system shown in Figure 7.11 is the most common arrangement, but a method of support based on a cantilever support has also been used. The vertical position of the model on the cantilever is adjusted to minimize the rotation at ground level. The advantage of this method is that base shear, as well as base bending moment, can be measured.

Testing of the model to determine either the base bending moment or the tip deflection over a range of reduced velocities should be carried out. The assumptions made to justify the rigid model aeroelastic testing, result in a relationship between the base bending moment, M_b, and the tip deflection, x, as follows:

$$M_b = \left(\omega_1^2 \, I_p/h\right)x = \left(\omega_1^2 \, G_1/h^3\right)x \cong \left(\omega_1^2 \, m \, h^2/3\right)x \tag{7.10}$$

where ω_1 is the natural circular frequency, and m is an average mass/unit height.

The relationship in Equation (7.10) implies that the mean and background wind loads are distributed over the height of the building in the same way as the resonant response, i.e. according to the distribution of inertial forces for first mode response. This is a consequence of the neglect of the higher modes of vibration.

The upper limit of reduced velocity should correspond to a mean wind speed which is larger than any design value for any wind direction. As it will be required to fit a relationship between response (either peak or r.m.s.) and mean wind speed, testing should be carried out for at least three reduced velocities.

It is wise to conduct aeroelastic tests for at least two different damping ratios – a value representative of that expected at perceptible accelerations for the height and construction type, and a higher value that may be achieved at ultimate conditions, or at serviceability design conditions when an auxiliary system is added. If the resonant response is dominant, values outside these conditions can be estimated by assuming that the r.m.s response varies as the inverse of the square root of the damping ratio.

The final stage of an aeroelastic investigation should be to provide the structural engineer with vertical distributions of loads which are compatible with the base bending moments obtained from the experiments and subsequent processing. As discussed in Chapter 5, there are different distributions for the mean component, background or sub-resonant fluctuating component, and the resonant component of the peak response, for any wind direction. If wind-tunnel pressure measurements are available, these can be used to determine the mean load distribution. Pressure measurements could, in principle, also be used to determine the background fluctuating loads, although this requires extensive correlation measurements; also, the loading distribution should be "tailored" to the particular load effect, such as a column load.

For tall buildings, a linear loading distribution with a maximum at the top, reducing to zero at the pivot point, is often assumed. Then the load per unit height at the top of the building, w_o, is given by Equation (7.11),

$$w_o = 3M_b / h^2 \tag{7.11}$$

For a linear mode of vibration, this is a realistic distribution for the inertial loading of the resonant part of the response (Section 5.4.4). However, this is not a realistic distribution for the mean (Section 5.4.2) or the background response (Section 5.4.3), when the loading is primarily along-wind.

7.6.2 The high-frequency base-balance

For most tall buildings, the 'high-frequency base-balance' (HFBB) technique (Tschanz and Davenport, 1983) has replaced aeroelastic model testing. This is alternatively known as the 'high-frequency force balance' (HFFB) method, or simply as the 'high-frequency balance' technique. In this method, there is no attempt to model the dynamic properties of the building – in fact the support system is made deliberately stiff to put the building model above the range of the exciting forces of the wind. A rigid model, which reproduces the building shape, is used. The model is supported at the base by a measurement system, which is capable of measuring the mean and fluctuating wind forces and moments to a high frequency, without significant amplification or attenuation. The spectral densities of the base forces and moments are measured, and the resonant response of the building, with appropriate dynamic properties incorporated, is computed using a spectral or random vibration approach, similar to that described in Section 5.3 of Chapter 5. A range of damping ratios and mean wind speeds can be simulated using this approach.

Note that the HFBB measures the mean and background fluctuating (quasi-static) base moments directly. Calculation is only required for the resonant components.

Figure 7.12 shows how the spectrum of wind force varies with different speeds in a wind tunnel. For a given design of balance there will be an upper limit to the wind force (proportional to wind speed squared) that is capable of being measured by the balance; this will be proportional to the stiffness of the balance for a particular force component. Thus, the maximum wind tunnel speed for which a balance can be used is proportional to the square root of the stiffness. Since the natural frequency of a model of given mass is also proportional to the square root of the stiffness, the ratio of maximum wind speed to maximum usable frequency will be a constant for a given design of balance.

When the prototype building does not have a linear sway mode shape, corrections are required to the computed resonant response, as they are for the base-pivoted aeroelastic model technique. Base torque can also be measured, and used to determine the response in torsional mode of vibration, although quite large mode shape corrections are required, as discussed following.

A variety of mode shape correction factors have been developed for the HFBB (e.g. Holmes, 1987; Boggs and Peterka, 1989). These depend on the assumptions made for the variation of the fluctuating wind forces (or torques) with height, and the correlation between the fluctuating sectional forces at different heights (Holmes *et al.*, 2003). There appear to be considerable differences between various commercial laboratories with regard to the corrections made, especially for the torsional, or twist, modes. Some laboratories make use of the measured base shears, as well as the base bending moments, available from a HFBB, and assume a linear variation of the instantaneous

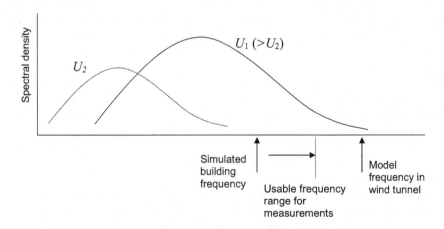

Figure 7.12 Frequency relationships for a high-frequency base balance.

wind force with height (Xie and Irwin, 1998). However, such methods do not eliminate the need for mode shape corrections (Chen and Kareem, 2005a).

The base moments $M_x(t)$, $M_y(t)$ and $M_z(t)$ measured by the HFBB must be converted into generalized forces for the two fundamental sway modes and twist mode. For example, using mode shape corrections proposed by Holmes (1987) and Holmes et al. (2003) are as follows:

$$Q_x(t) = \left(\frac{1}{h}\right)\sqrt{\frac{4}{1+3\beta_x}} \cdot M_y(t) \tag{7.12}$$

$$Q_y(t) = \left(\frac{1}{h}\right)\sqrt{\frac{4}{1+3\beta_y}} \cdot M_x(t) \tag{7.13}$$

$$Q_z(t) = \sqrt{\frac{1}{1+2\beta_z}} \cdot M_z(t) \tag{7.14}$$

where $Q_x(t)$, $Q_y(t)$, $Q_z(t)$ are the generalized forces in the x- and y-sway modes, and twist modes, respectively. h is the building height.

It has been assumed that the mode shapes can be represented by power functions of the form of Equations (7.15)–(7.17).

$$\phi_x = \left(\frac{z}{h}\right)^{\beta_x} \tag{7.15}$$

$$\phi_y = \left(\frac{z}{h}\right)^{\beta_y} \tag{7.16}$$

$$\phi_z = \left(\frac{z}{h}\right)^{\beta_z} \tag{7.17}$$

If the mode shape exponent for the twist mode, β_z is 1.0 (i.e. the dynamic twist varies linearly with height from the ground to the top of the building), then from Equation (7.14), the mode shape correction term is $\sqrt{(1/3)}$ or 0.58. This is significantly different from 1.0 because the HFBB measures the base torsional moment uniformly weighted with height, whereas the generalized force for the twist mode requires a linear weighting with height. On the other hand, the generalized forces for the sway modes usually have mode shape correction factors close to 1.0 (equal to 1.0 for linear mode shapes).

Many modern tall buildings have dynamic modes that involve coupled sway and twist motions. This often results from differences between the average positions of the centre of mass and centre of stiffness (shear centre) of the cross-sections of the building. It is extremely difficult (and expensive) to manufacture accurate aeroelastic wind-tunnel models of buildings with coupled modes. However, methods are available to make reasonable predictions of the resonant contributions from the coupled modes of tall buildings using the HFBB technique (Holmes *et al.*, 2003; Chen and Kareem, 2005b).

The HFBB technique requires relatively simple models, and clearly reduces the amount of wind-tunnel testing time by a large factor, at the expense of computing resources, which have rapidly become cheaper. There are methodologies to account for complex coupled sway and twist dynamic modes. Most tall buildings can adequately be studied using the HFBB technique – a very cost-effective method.

Recent developments in the HFBB technique, as applied to model studies of tall buildings, have been described in detail in several papers in a special issue of the journal *Wind and Structures* (Boggs, 2014; Ho, Jeong and Case, 2014; Xie and Garber, 2014; Chen *et al.*, 2014; Tse, *et al.*, 2014).

Comparisons of results (i.e. base moments and roof-top accelerations) from eight different laboratories for two 'benchmark' tall buildings, based on the HFBB method, have been compared in an international comparison reported by Holmes and Tse (2014). The dimensions and dynamic properties (i.e. frequencies, mode shapes and damping) of the two buildings were specified tightly, together with the flow properties in the approaching simulated atmospheric boundary layer. The agreement between the peak base moments (including base torque) was good, with coefficients of variation of less than 14% for the base bending moments. There was greater variation in the predictions of the peak accelerations at the corner of the roofs; these are more sensitive to assumptions and correction methods for mode shape, coupled and closely spaced modes, and combination of acceleration components. More details of these comparisons are given, in a graphical form, on the web site of the International Association of Wind Engineering (*www.iawe.org*).

7.6.3 Sectional and taut strip models of bridges

A common, and long-standing, technique to confirm the aerodynamic stability of the decks of long-span suspension or cable-stayed bridges is the section model test. This is another form of rigid body aeroelastic modelling. The technique dates back to the investigations following the failure of the first Tacoma Narrows bridge (Farquarson, Smith and Vincent, 1949–1954). A short section of the bridge deck is supported on springs and allowed to move in translation and rotation (Figure 7.13). By suitable adjustment of the springs, the model frequencies in rotation and vertical translation can be arranged to have the same ratio as those for the primary bending and torsional modes of the prototype bridge. Then, in order to achieve similarity between model, m, and prototype, p, the reduced frequencies (Section 7.4.1) should be kept equal as in Equation (7.18).

Figure 7.13 A sectional model of a footbridge. (Cermak, Peterka, Petersen, Sydney, Australia)

$$\left(\frac{n_s L}{U}\right)_m = \left(\frac{n_s L}{U}\right)_p \tag{7.18}$$

where n_s should be taken both as the lowest frequencies in vertical translation (bending), and in rotation (torsion).

The models are made as rigid as possible, but they are also required to satisfy the density scaling requirement that the ratio ρ_s/ρ_a should be the same in model and full scale, where ρ_s is the average density of the structure, and ρ_a is the air density. The details of the deck at the leading edge – such as edge beams and guard railings are usually modelled in some detail, as these have been found to affect the aeroelastic behaviour.

To achieve two-dimensional flow over a section model, and to minimize end effects, the ratio of span to chord should be at least 4 to 1 (Wardlaw, 1980).

Section models are primarily used to determine the critical flutter speeds of the section in both smooth and turbulent flow. The static aerodynamic coefficients can also be determined for use in calculations of turbulent buffeting of the section. Another use is for determination of the aeroelastic coefficients, or flutter derivatives (Sections 5.5.3 and 12.3.2), for subsequent use in more complete computational modelling of bridge behaviour; both free- (Scanlan and Tomko, 1971) and forced vibration (e.g. Matsumoto *et al.*, 1992) methods have been developed. More advanced methods, using system identification techniques, have been developed (Sarkar *et al.*, 1994; Gu *et al.*, 2001); these will produce the most accurate flutter derivatives, although complex to implement. If extraction of flutter derivatives is the only application for a section model, it is unnecessary to model the mass or density, and it is advantageous to add extra mass (Gu *et al.*, 2001)

Sectional models are primarily a two-dimensional simulation, and in turbulent flow, which of course is more representative of atmospheric flow and three-dimensional in nature, significant corrections are required to reproduce the behaviour of a full bridge in an atmospheric boundary layer. A more advanced test method for bridges, known as 'taut strip', involves the central span of the model bridge deck supported on two parallel wires, pulled into an appropriate tension, and separated by an appropriate distance, so that the bending and torsional modes are approximately matched. The deck is made in elements or short sections, so that no stiffness is provided. Such a model can be tested in full simulated boundary-layer flows, but is more economical than a full aeroelastic model test.

Scanlan (1983) and Tanaka (1990) have given useful reviews of the section model and taut-strip techniques for bridge decks, together with a discussion of full aeroelastic model testing of bridges. Scanlan *et al.* (1997) showed that, with additional computation and knowledge of properties of the wind-tunnel turbulence, the results from a taut-strip model could be reproduced by a sectional model test. However, to reproduce the aerodynamic and structural dynamic effects of the cables and supporting towers on a long-span bridge, a full aeroelastic model is necessary.

7.6.4 Multi-mode aeroelastic modelling

For the modelling of structures with non-linear mode shapes, or for structures which respond dynamically to wind in several of their natural resonant modes of vibration, such as tall towers and long-span bridges, the rigid body modelling technique is not sufficient. In the case of long-span bridges, the aerodynamic influences of the cables and the supporting towers, which are not included in section model or taut-strip testing (Section 7.6.3), may often be significant. More complete aeroelastic and structural modelling techniques are then required.

There are three different types of these multi-mode models:

- 'Replica' models – in which the construction of the model replicates that of the prototype structure.
- 'Spine' models which reproduce the stiffness properties of the prototype structure by means of smaller central members or 'spines'. Added sections reproduce the mass and aerodynamic shape of the prototype.
- 'Lumped mass' models, in which the mass of the model is divided into discrete 'lumps', connected together by flexible elements. The number of vibration modes that can be reproduced by this type of model is limited by the number of lumped masses.

The design of these models generally follows the scaling laws based on dimensional analysis, as outlined in Section 7.4. Full model testing of suspension bridges and cable suspended roofs, where stiffness is, at least partially, provided by gravitational forces, requires equality of Froude Number, $U/\sqrt{(L \cdot g_0)}$, (introduced in Section 7.4), between model and full scale.

$$\left(\frac{U}{\sqrt{Lg_0}} \right)_m = \left(\frac{U}{\sqrt{Lg_0}} \right)_p$$

Since the gravitational constant, g_0, is the same in model and full scale, this results in a velocity scaling given by Equation (7.19).

$$\frac{U_m}{U_p} = \sqrt{\frac{L_m}{L_p}} \tag{7.19}$$

Thus, the velocity ratio is fixed at the square root of the length ratio (or model scale). Thus, for a 1/100 scale suspension bridge model, the velocity in the wind tunnel is one tenth of the equivalent velocity in full scale.

For the majority of structures, in which the stiffness is provided by internal stresses (e.g. axial, bending, shear), Froude Number scaling is not required for aeroelastic models, and a free choice can be made of the velocity scaling when designing a model. Usually a fine adjustment of the velocity scaling is made after the model is built, to ensure equality of reduced frequency (see Equation 7.18).

Examples of aeroelastic models are shown in Figures 11.8 (observation tower) and 12.7 (a bridge under construction). These are both 'spine' models.

7.6.5 Modelling of tensioned and pneumatic structures

Large roofs comprising tensioned membranes made from fabric, present particular challenges if it is required to completely reproduce the dynamic and aeroelastic responses in a wind-tunnel model. However, in many cases it is not necessary to correctly scale the elastic forces in a membrane structure, since the geometric stiffness (i.e. resistance to deflection provided by changes in angle of the membrane material as it deforms) usually dominates over the elastic stiffness of the membrane material (Tryggvason, 1979).

In pneumatic, or air-supported, structures the tension in the membrane surface is determined by the geometry, and the internal pressurization, Δp_i (i.e. the difference between the internal and external pressures in no-wind conditions). In air-supported roofs, the pressurization is comparable to the weight per unit area of the membrane material. For this situation, Tryggvason (1979) proposed a scaling of the wind velocity that ensures the correct scaling of wind pressures, the internal pressure and the weight per unit area of the membrane:

$$[U]_r^2 = [\Delta p_i - \mu g]_r \qquad (7.20)$$

To correctly scale the fluctuating component of the internal pressure for pneumatic structures, it is necessary to scale the internal volume according to Equation (7.2). This will ensure that the 'acoustic stiffness' is correctly scaled. The additional volume can be provided beneath the floor of a wind tunnel (Tryggvason, 1979).

However, for large air-supported roofs, the 'pneumatic' damping, associated with the ventilation air flow within the building volume, is large, and sufficient to effectively damp any resonant response of a roof of this type to wind (see Kind, 1984, and Section 10.5). Hence, aeroelastic models are usually not necessary for roofs of this type.

7.6.6 Aeroelastic modelling of chimneys

Chimneys and other slender structures of circular cross-sections are vulnerable to cross-wind excitation by fluctuating pressures due to vortex shedding (Sections 4.6.3 and 11.5). In the 1950s and 1960s, it was quite common to investigate this behaviour with small-scale wind-tunnel models. However, the forces from vortex shedding are quite dependent on Reynolds Number (Section 4.2.4), and wind-tunnel tests will severely over-estimate the cross-wind response of prototype large chimneys (Vickery and Daly, 1984). The prediction of full-scale response of such structures is better undertaken by the use of mathematical models of the response (Section 11.5) with input parameters derived from full-scale measurements at high Reynolds Numbers.

7.6.7 Distorted 'dynamic' models

In many cases, the resonant response of a structure may be significant, but the prototype structure may be stiff enough such that aeroelastic forces (i.e. the motion-dependent forces) are not significant. Furthermore, the scaling requirements (Section 7.4.1), and properties of the available modelling materials may make it difficult, or even impossible, to simultaneously scale the mass, stiffness and aerodynamic shape of a structure.

In such cases, the mass and stiffness properties of the structure can both be distorted by the same factor (usually greater than 1.0). Then, the correct frequency relationship for the applied fluctuating wind forces and the structural frequencies is obtained. That is, it retains the correct value of reduced frequency (Section 7.4.1) and preserves the correct relationship between the frequencies associated with the flow (e.g. turbulence and vortex shedding), and those related to the structure. Internal forces and moments are correctly modelled (including resonant effects), but the deflections, accelerations and motion-induced forces, such as aerodynamic damping (Section 5.5.1), are not scaled correctly.

Such dynamic models have been used on some open-frame structures, where aeroelastic 'spine' models were not possible.

7.6.8 Structural loads through pressure measurements

For structures such as large roofs of sports stadiums, or large low-rise buildings, with structural systems that are well defined and for which resonant dynamic action is not dominant, or can be neglected, wind-tunnel pressure measurements on rigid models can be used effectively to determine load effects such as member forces and bending moments, or deflections. This method is often used in conjunction with the area-averaging pressure technique described in Section 7.5.2. Also required are influence coefficients, representing the values of a load effect under the action of a single uniformly distributed static 'patch load' acting on the area corresponding to a panel on the wind-tunnel model. Two methods are possible:

- Direct on-line weighting of the fluctuating panel pressures recorded in the wind-tunnel test with the structural influence coefficients, to determine directly fluctuating and peak values of the load effects (Surry and Stathopoulos, 1977; Holmes, 1988).
- Measurement of correlation coefficients between the fluctuating pressures on pairs of panels, and calculation of root-mean-square and peak load effects by integration (Holmes and Best, 1981; Holmes *et al.*, 1997).

The latter method has advantages that the influence coefficients are not required at the time of the wind-tunnel testing, and also that the information can be used to determine equivalent static load distributions, as discussed in Chapter 5. When resonant response is of significance, as may be the case for the largest stadium roofs, time histories of the fluctuating pressures can be used to generate a time history of generalized force for each mode of significance. From the spectral density of the generalized force, the mean square generalized displacement (modal coordinate), and effective inertial forces acting can be determined (Section 5.4.4). The application of pressure model studies to large roofs is discussed in Chapter 10.

7.6.9 High-frequency pressure-integration

The high-frequency pressure-integration (HFPI) technique (e.g. Aly, 2013) has become commonly used for commercial wind-tunnel testing of tall buildings in the last few years, as an alternative to the HFBB (–Section 7.6.2). In the HFPI method, fluctuating pressures from a dense array of pressure tappings distributed over all walls of a building are weighted by their respective tributary areas to form sectional forces at discrete height increments. The latter forces can then be weighted by structural dynamic mode shapes to reproduce modal (generalized) forces, without the need for mode shape corrections. The dynamic response of the building can then be calculated in a similar way to the HFBB technique.

The HFPI method has the advantages that the same model used for local cladding pressures can be used, and that response for complex mode shapes can be accurately calculated. A disadvantage is that large amounts of connecting tubing are required, and there may be insufficient space inside some models – particularly those of buildings of high aspect ratio – to accommodate this tubing.

7.7 BLOCKAGE EFFECTS AND CORRECTIONS

In a wind tunnel with a closed test section, the walls and roof of the wind tunnel provide a constraint on the flow around a model building or group of buildings, which depends on the blockage ratio. The blockage ratio is the maximum cross-sectional area of the model at any cross section, divided by the area of the wind-tunnel cross section. If this ratio is high enough, there may be significant increases in the flow velocities around, and pressures on, the model. In the case of an open test section, the errors are in the opposite direction; that is the velocities around the model are reduced. To deal with the blockage problem, several approaches are possible:

- Ensure that the blockage ratio is small enough that the errors introduced are small, and no corrections are required. The usual rule for this approach is that the blockage ratio should not exceed 5%.
- Accept a higher blockage ratio, and attempt to make corrections. The difficulty with this approach is that the appropriate correction factors may themselves be uncertain. Although there are well-documented correction methods for drag and base pressure on stalled airfoils, and other bluff bodies in the centre of a wind tunnel with uniform, or homogeneous turbulent flow, there is very little information for buildings or other structures, mounted on the floor of a wind tunnel in turbulent boundary-layer flow. McKeon and Melbourne (1971) provided corrections for mean windward and leeward pressures, and total drag force, on simple plates and blocks. However, no corrections are readily available for pressures, mean or fluctuating, in separated flow regions, such as occur on roofs or side walls of building models.

• Design the walls and/or roof of the working section in such a way as to minimize the blockage errors. The most promising method for doing this appears to be the slotted wall concept (Parkinson, 1984; Parkinson and Cook, 1992). In this system, the walls and roof of the tests section are composed of symmetrical aerofoil slats, backed with a plenum chamber. The optimum open area ratio is about 0.55, and it is claimed that blockage area ratios of up to 30% can be used without correction.

7.8 MODELLING OF TOPOGRAPHY

In some situations, the effect of large-scale topographic features on the local wind flow at the site of a structure is important. A way of approaching this is to use small-scale (1/2,000–1/10,000) topographic models in wind-tunnel tests to determine mean velocity ratios and profiles, which can then assist the larger scale modelling of the building or structure, in a wind-tunnel test in which the topography is not included (Figure 7.14).

A significant difficulty with the modelling of topography with shallow slopes is the sensitivity of the airflow over curved surfaces to Reynolds Number (Section 4.2.4), and the inability to match full-scale Reynolds Numbers at small model scales. One effect of this is that the high-frequency end of the spectral densities of turbulence cannot be reproduced, so that gust velocities generally cannot be reproduced at small modelling scales. The use of surface roughening or 'terracing', rather than smoothed contouring, is often used by wind-tunnel groups to improve the simulations. These techniques produce mixed results, with the terracing approach, in particular, greatly increasing local turbulence intensities (Neal *et al.*, 1981). In

Figure 7.14 Topographic model of Hong Kong (Hong Kong University of Science and Technology, Hong Kong, China).

general, small-scale topographic models should only be used to investigate local mean velocities.

7.9 LARGE-SCALE AND SPECIALIST TEST FACILITIES

As discussed in Sections 7.2 and 7.3, small-scale aerodynamic modelling in wind tunnels was refined and developed over the twentieth century, initially for the aeronautical industry, and in the second half of the century, for building studies through the development of the boundary-layer wind tunnel. However, there are some clear limitations to small-scale modelling. The final link between wind loading and structural load effects (internal forces and stresses) is not reproduced and must be made in other ways – usually by computational modelling. Also, for some structures, particularly those with circular cross sections, the inability to reproduce full-scale Reynolds Numbers in the wind-tunnel flow may lead to significant errors.

In the present century, the ability to test full-scale structures, or parts of them, has been achieved in a few facilities worldwide. For example, Figure 7.15 shows a test of a television antenna in a large wind-tunnel cross section, mainly used for tests on full-scale motor vehicles. The

Figure 7.15 Wind-tunnel testing of a television antenna at full scale (Monash University, Victoria, Australia).

Figure 7.16 Exterior of a test facility for full-scale housing in hurricane winds (IBHS Research Center, South Carolina, USA).

near-circular nature of the antenna cross section indicates sensitivity of the sectional forces to Reynolds Number (Section 4.2.4), but testing at full scale allows full-scale values of this important non-dimensional parameter to be achieved. In the case shown, only the lower half of the antenna was immersed in the flow, but this was sufficient to determine the required sectional force coefficients.

An impressive facility that allows testing (to destruction if necessary) of full-scale one- and two-storey houses, in winds that are closely representative, in both strength and structure, of those in hurricanes, is shown in Figure 7.16. This unique facility is equipped with 105 fans and a test section of about 13 m high by 40 m wide.

Earlier chapters have noted the growing recognition of importance of non-boundary layer, and non-synoptic, wind events. These include downbursts and tornados usually associated with severe thunderstorms. In many parts of the world, and for some structures, such as transmission lines (see Chapter 13), these are the dominant extreme wind events. However, the lack of a large-enough facility for reproducing the wind flow associated with these storms has been a major barrier to understanding the loading effects on structures at ground level. The WIND-EEE facility in London, Canada (Figure 7.17), opened in 2013, that allows for simulation for all types of severe wind events, including downbursts, tornados and synoptic-scale boundary-layer winds, is thus a significant development. Described as a 'hexagonal wind tunnel' it comprises a 25 m diameter inner 'dome and a 40 m diameter outer return dome. Equipped with 106 fans in various orientations, it can produce three-dimensional wind structure with reasonable model-scale reproductions of most types of severe windstorms.

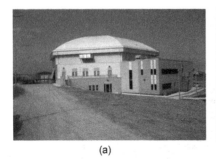
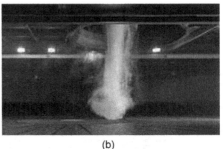

(a)   (b)

Figure 7.17 WIND-EEE facility at Western University, London, Ontario, Canada. (a) exterior (image by Professor K.C.S. Kwok). (b) interior with simulation of an impacting microburst.

7.10 OTHER SOURCES

Many different approaches to simulating the turbulent atmospheric boundary layer in a wind tunnel were published from the 1960s to the 1980s. Some of these were mentioned earlier in this chapter. Probably the best documented method is that of Armitt and Counihan (1968) and Counihan (1969, 1970, 1973) (see also Figure 7.4). This approach has been adopted by several wind tunnels since.

The principles of dimensional analysis are outlined by Douglas (1969), and the general theory of experimental modelling, based on dimensional analysis, is discussed by David and Nolle (1982). The latter reference includes some discussion of the modelling of wind loads on buildings.

'Wind tunnel modeling for civil engineering applications' (Reinhold ed., 1982) brought together chapters by several individual authors covering all aspects of the wind-tunnel modelling of wind effects on buildings and structures, and is still representative of much of current practice.

Minimum acceptable requirements for experimental wind-tunnel modelling for wind engineering studies of buildings are outlined in ASCE-49 (American Society of Civil Engineers, 2012) and AWES-QAM-1 (Australasian Wind Engineering Society, 2019).

The details of experimental techniques pressure, force and structural response at model scale are also covered in past papers in international journals, particularly the *Journal of Wind Engineering and Industrial Aerodynamics* (Elsevier).

7.11 SUMMARY

In this chapter, a review of methods of laboratory simulation of natural strong wind characteristics for the investigation of wind pressures, forces and structural response has been given. Early methods used natural growth of boundary layers on the floor of wind tunnels to simulate the mean flow and turbulence structure in the fully developed boundary layer in gale wind conditions. To make use of shorter test sections in aeronautical wind tunnels, rapid growth methods were developed and described. For investigations on smaller structures, such as low-rise buildings, methods of simulating only the lower part, or surface layer, of the atmospheric boundary layer were devised.

Laboratory methods of simulating tornadoes, which were quite advanced as early as the 1970s, are discussed. Methods of simulating strong winds in tropical cyclones and thunderstorms, which are the dominant types for structural design at locations in the tropics and sub-tropics at latitudes from 0° to 40°, are still at an early stage of development. A major problem is the lack of good full-scale data of the wind structure, on which the simulations can be based.

Experimental methods of measuring local pressures, and overall structural loads in wind-tunnel tests are described in Sections 7.5 and 7.6, and the problem of wind-tunnel blockage, and its correction is discussed in Section 7.7. The use of small-scale topographic models is discussed in Section 7.8, and some recently developed large-scale and specialist experimental facilities are discussed in Section 7.9.

REFERENCES

Aly, A.M. (2013) Pressure integration technique for predicting wind-induced response in high-rise buildings. *Alexandria Engineering Journal*, 52: 717–731.

American Society of Civil Engineers. (2012) *Wind tunnel testing for buildings and other structures*, ASCE Standard ASCE/SEI 49-12. ASCE, Reston, VA.

Armitt, J. (1968) The effect of surface roughness and free-stream turbulence on the flow around a model cooling tower at critical Reynolds Numbers. *Proceedings, Symposium on Wind effects on Buildings and Structures*, Loughborough, UK, 2–4 April.

Armitt, J. and Counihan, J. (1968) The simulation of an atmospheric boundary layer in a wind tunnel. *Atmospheric Environment*, 2: 49–71.

Australasian Wind Engineering Society. (2019) Wind-engineering studies of buildings. Quality Assurance Manual. AWES-QAM-1-2019.

Aynsley, R.D., Melbourne, W.H. and Vickery, B.J. (1977) *Architectural aerodynamics*. Applied Science Publishers, London.

Bergh, H. and Tijdeman, H. (1965) Theoretical and experimental results for the dynamic response of pressure measurement systems. National Aero- and Astronautical Research Institute (Netherlands), Report NLR-TR-F.238, 1965 January.

Boggs, D.W. (2014) The past, present and future of high-frequency balance testing. *Wind and Structures*, 18: 323–345.

Boggs, D.W. and Peterka, J.A. (1989) Aerodynamic model tests of tall buildings. *Journal of Engineering Mechanics*, 115: 618–635.

Cermak, J.E. (1971) Laboratory simulation of the atmospheric boundary layer. *AIAA Journal*, 9: 1746–1754.

Cermak, J.E. (1977) Wind-tunnel testing of structures. *ASCE Journal of the Engineering Mechanics Division*, 103: 1125–1140.

Chang, C.C. (1971) Tornado effects on buildings and structures by laboratory simulation. *Proceedings, 3rd. International Conference on Wind effects on Buildings and Structures*, Tokyo, Japan, 6–9 September, pp. 231–240.

Chen, X. and Kareem, A. (2005a) Validity of wind load distribution based on high frequency force balance measurements. *Journal of Structural Engineering*, 131: 984–987.

Chen, X. and Kareem, A. (2005b) Dynamic wind effects on buildings with 3D coupled modes: application of High Frequency Force Balance measurements. *Journal of Engineering Mechanics*, 131: 1115–1125.

Chen, X., Kwon, D-K. and Kareem, A. (2014) High-frequency force balance technique for tall buildings: a critical review and some new insights. *Wind and Structures*, 18: 391–422.

Counihan, J. (1969) An improved method of simulation of an atmospheric boundary layer in a wind tunnel. *Atmospheric Environment*, 3: 197–214.

Counihan, J. (1970) Further measurements in a simulated atmospheric boundary layer. *Atmospheric Environment*, 4: 259–275.

Counihan, J. (1973) Simulation of an urban adiabatic boundary layer in a wind tunnel. *Atmospheric Environment*, 7: 673–689.

Cook, N.J. (1973) On simulating the lower third of the urban adiabatic boundary layer in a wind tunnel. *Atmospheric Environment*, 7: 691–705.

Davenport, A.G., and Isyumov, N. (1967) The application of the boundary layer wind tunnel to the prediction of wind loading. *Proceedings, International Research Seminar on Wind effects on Buildings and Structures*, Ottawa, Canada, 11–15 September, pp. 201–230.

David, F.W., and Nolle, H. (1982) *Experimental modelling in engineering.* Butterworths-Heinemann, Oxford.

Davies-Jones, R.F. (1976) Laboratory simulation of tornadoes. *Symposium on Tornadoes: Assessment of Knowledge and Implications for Man*, Texas Tech University, Lubbock, Texas, USA, 22–24 June 1976, pp. 151–174.

Douglas, J.F. (1969). *Introduction to dimensional analysis*, Sir Isaac Pitman, London.

Durgin, F. (1982) Instrumentation requirements for measuring aerodynamic pressures and forces on buildings and structures. In *Wind tunnel modeling for civil engineering applications.* ed. T. Reinhold. Cambridge University Press, Cambridge, MA.

Farquarson, F.B., Smith, F.C. and Vincent, G.S. (1949–1954) Aerodynamic stability of suspension bridges with special reference to the Tacoma Narrows Bridge. *University of Washington Engineering Experiment Station. Bulletin No. 116*, Parts I to V.

Flachsbart, O. (1932) Winddruck auf geschlossene und offene Gebäude. In *Ergebnisse der aerodynamischen Versuchanstalt zu Göttingen, IV. Lieferung.* eds. L. Prandl and A. Betz. Verlag von R. Oldenbourg, Munich and Berlin.

Gerstoft, P. and Hansen, S.O. (1987) A new tubing system for the measurement of fluctuating pressures. *Journal of Wind Engineering and Industrial Aerodynamics*, 25: 335–354.

Ginger, J.D., Holmes, J.D. and Kim, P.Y. (2010) Variation of internal pressure with varying sizes of dominant openings and volumes. *Journal of Structural Engineering (ASCE)*, 136: 1319–1326.

Gu, M., Zhang, R. and Xiang, H. (2001) Parametric study on flutter derivatives of bridge decks. *Engineering Structures*, 23: 1607–1613.

Gumley, S.J. (1981) Tubing systems for the measurement of fluctuating pressures in wind engineering. *D.Phil. thesis*, University of Oxford.

Gumley, S.J. (1983) Tubing systems for pneumatic averaging of fluctuating pressures. *Journal of Wind Engineering and Industrial Aerodynamics*, 12: 189–228.

Gumley, S.J. (1984) A parametric study of extreme pressures for the static design of canopy structures. *Journal of Wind Engineering and Industrial Aerodynamics*, 16: 43–56.

Ho, T.C.E., Jeong, U.Y. and Case, P.C. (2014) Components of wind tunnel analysis using force balance test data. *Wind and Structures*, 18: 347–373.

Holmes, J.D. (1984) Effect of frequency response on peak pressure measurements. *Journal of Wind Engineering and Industrial Aerodynamics*, 17: 1–9.

Holmes, J.D. (1987) Mode shape corrections for dynamic response to wind. *Engineering Structures*, 9: 210–212.

Holmes, J.D. (1988) Distribution of peak wind loads on a low-rise building. *Journal of Wind Engineering and Industrial Aerodynamics*, 29: 59–67.

Holmes, J.D. (1995) *Methods of fluctuating pressure measurement in wind engineering*. In *A state of the art in wind engineering*. ed. P. Krishna. Wiley Eastern Limited, New Delhi.

Holmes, J.D. (1997) Equivalent time averaging in wind engineering. *Journal of Wind Engineering and Industrial Aerodynamics*, 72: 411–419.

Holmes, J.D. and Best, R.J. (1981) An approach to the determination of wind load effects for low-rise buildings. *Journal of Wind Engineering and Industrial Aerodynamics*, 7: 273–287.

Holmes, J.D. and Lewis, R.E. (1987a) Optimization of dynamic-pressure-measurement systems. I. Single point measurements. *Journal of Wind Engineering and Industrial Aerodynamics*, 25: 249–273.

Holmes, J.D. and Lewis, R.E. (1987b) Optimization of dynamic-pressure-measurement systems. II. Parallel tube-manifold systems. *Journal of Wind Engineering and Industrial Aerodynamics*, 25: 275–290.

Holmes, J.D. and Lewis, R.E. (1989) A re-examination of the leaked-tube dynamic pressure measurement system. *10th. Australasian Fluid Mechanics Conference*, Melbourne, Victoria, Australia, pp. 5.39–5.42, December.

Holmes, J.D. and Osonphasop, C. (1983) Flow behind two-dimensional barriers on a roughened ground plane, and applications for atmospheric boundary-layer modelling. *Proceedings, 8th Australasian Fluid Mechanics Conference*, Newcastle, N.S.W., Australia.

Holmes, J.D. and Rains, G.J. (1981) Wind Loads on flat and curved roof low rise buildings. *Colloque "Construire avec le Vent"*, Nantes, France, July 1981.

Holmes, J.D. and Tse, T.K.T. (2014) International high-frequency base balance benchmark study. *Wind and Structures*, 18: 457–471.

Holmes, J.D., Denoon, R.O., Kwok, K.C.S. and Glanville, M.J. (1997) Wind loading and response of large stadium roofs. *International Symposium on Shell and Spatial Structures*, Singapore, 10–14 November.

Holmes, J.D., Rofail, A. and Aurelius, L. (2003) High frequency base balance methodologies for tall buildings with torsional and coupled resonant modes. *Proceedings, 11th International Conference on Wind Engineering*, Lubbock, Texas, USA, 1–5 June.

Hunt, J.C.R. and Fernholz, H. (1995) Wind-tunnel simulation of the atmospheric boundary layer: a report on Euromech 50. *Journal of Fluid Mechanics*, 70: 543–559.

Irminger, J.O.V. (1894) Nogle forsog over trykforholdene paa planer og legemer paavirkede af luftstrominger. *Ingenioren*, 17.

Irwin, P.A. (1988) Pressure model techniques for cladding wind loads. *Journal of Wind engineering and Industrial Aerodynamics*, 29: 69–78.

Irwin, H.P.A.H., Cooper, K.R. and Girard, R. (1979) Correction of distortion effects caused by tubing systems in measurements of fluctuating pressures. *Journal of Industrial Aerodynamics*, 5: 93–107.

Isyumov, N., Holmes, J.D., Surry, D. and Davenport, A.G. (1975) A study of wind effects for the First National City Corporation Project, New York. University of Western Ontario, Boundary Layer Wind Tunnel Special Study Report, BLWT-SS1-75.

Jensen, M. (1958) The model law for phenomena in the natural wind. *Ingenioren (International edition)*, 2: 121–128.

Jensen, M. (1965) *Model scale tests in the natural wind. (Parts I and II)*. Danish Technical Press, København.

Jischke, M.C. and Light, B.D. (1979) Laboratory simulation of tornadic wind loads on a cylindrical structure. *Proceedings, 5th International Conference on Wind Engineering*, Fort Collins, Colorado, USA, July, pp. 1049–1059, Pergamon Press.

Kareem, A., Cheng, C-M. and Lu, P.C. (1989) Pressure and force fluctuations on isolated circular cylinders of finite height in boundary layer flows. *Journal of Fluids and Structures*, 3: 481–508.

Kay, N.J., Oo, N.L., Richards, P.J. and Sharma, R.N. (2020) Characteristics of fluctuating pressure measurement systems utilising lengths of 3D-Printed tubing. *Journal of Wind Engineering and Industrial Aerodynamics*, 199: 104121.

Kernot, W.C. (1893) Wind pressure. *Proceedings, Australasian Society for the Advancement of Science*, Adelaide, Australia, 25 September –3 October, V: 573–581.

Kind, R.J. (1984) Pneumatic stiffness and damping in air-supported structures. *Journal of Wind Engineering and Industrial Aerodynamics*, 17: 295–304.

Larose, G.L. and Franck, N. (1997) Early wind engineering experiments in Denmark. *Journal of Wind Engineering and Industrial Aerodynamics*, 72: 493–499.

Lawson, T.V. (1976) The design of cladding. *Building and Environment*, 11: 37–38.

Letchford, C.W. (1989) On the discrete approximation in pneumatic averaging. *Proceedings, 2nd Asia-Pacific Conference on Wind Engineering*, Beijing, China, June, pp. 1159–1167, Pergamon Press.

Letchford, C.W. and Chay, M.T. (2002) Pressure distributions on a cube in a simulated thunderstorm downburst. Part B. Moving downburst observations. *Journal of Wind Engineering and Industrial Aerodynamics*, 90: 733–753.

Liu, H. (1982) Wind tunnel modeling of building internal pressure. *Transportation Engineering Journal (ASCE)*, 108: 691–696.

Matsumoto, M., Shirato, H. and Hirai, S. (1992) Torsional flutter mechanism of 2-d H-shaped cylinders and effect of flow turbulence. *Journal of Wind Engineering and Industrial Aerodynamics*, 41: 687–698.

McKeon, R. and Melbourne, W.H. (1971) Wind-tunnel blockage effects and drag on bluff bodies in rough wall turbulent boundary layers. *3rd International Conference on Wind Effects on Buildings and Structures*, Tokyo, Japan, Saikon Shuppan Publishers.

Neal, D., Stevenson, D.C. and Lindley, D. (1981) A wind-tunnel boundary-layer simulation of wind flow over complex terrain: effect of terrain and model construction. *Boundary-Layer Meteorology*, 21: 217–293.

Parkinson, G.V. (1984) A tolerant wind tunnel for industrial aerodynamics. *Journal of Wind Engineering and Industrial Aerodynamics*, 16: 293–300.

Parkinson, G.V. and Cook, N.J. (1992). Blockage tolerance of a boundary-layer wind tunnel. *Journal of Wind Engineering and Industrial Aerodynamics*, 42: 873–884.

Pearce, W. and Sykes, D.M. (1999) Wind tunnel measurements of cavity pressure dynamics in a low-rise flexible roofed building. *Journal of Wind Engineering and Industrial Aerodynamics*, 82: 27–48.

Reinhold, T. (ed.) (1982) Wind tunnel modeling for civil engineering applications. *International Workshop on Wind Tunnel Modeling Criteria and Techniques in Civil Engineering Applications*, Gaithersburg, Maryland, USA, Cambridge University Press.

Sarkar, P.P., Jones, N.P., and Scanlan, R.H. (1994). Identification of aeroelastic parameters of flexible bridges. *Journal of Engineering Mechanics, ASCE*, 120: 1718–1741.

Scanlan, R.H. (1974). Scale models and modelling laws in fluidelasticity. *ASCE National Structural Engineering Meeting*, Cincinnati, Ohio, USA, April 22–26.

Scanlan, R.H. (1983). Aeroelastic simulation of bridges. *Journal of Structural Engineering, ASCE*, 109: 2829–2837.

Scanlan, R.H. and Tomko, J.J. (1971). Airfoil and bridge deck flutter derivatives. *Journal of the Engineering Mechanics Division, ASCE*, 97: 1717–1737.

Scanlan, R.H., Jones, N.P. and Lorendeaux, O. (1997). Comparison of taut-strip and section-model-based approaches to long-span bridge aerodynamics. *Journal of Wind Engineering and Industrial Aerodynamics*, 72: 275–287.

Surry, D. and Isyumov, N., (1975) Model studies of wind effects - a perspective on the problems of experimental technique and instrumentation. *6th International Congress on Aerospace Instrumentation*, Ottawa, Canada.

Surry, D. and Stathopoulos, T. (1977) An experimental approach to the economical measurement of spatially-averaged wind loads. *Journal of Industrial Aerodynamics*, 2: 385–397.

Szechenyi, E. (1975) Supercritical Reynolds number simulation for two-dimensional flow over circular cylinders, *Journal of Fluid Mechanics*, 70: 529–542.

Tanaka, H. (1990) Similitude and modelling in wind tunnel testing of bridges. *Journal of Wind Engineering and Industrial Aerodynamics*, 33: 283–300.

Tryggvason, B.V. (1979) Aeroelastic modelling of pneumatic and tensioned fabric structures. *Proceedings, 5th Internal Conference on Wind Engineering*, Fort Collins, Colorado, USA, pp. 1061–1072, Pergamon Press.

Tschanz, T. and Davenport, A.G. (1983) The base balance technique for the determination of dynamic wind loads. *Journal of Wind Engineering and Industrial Aerodynamics*, 13: 429–439.

Tse, K.T., Yu, X.J. and Hitchcock, P.A. (2014) Evaluation of mode shape linearization for HFBB analysis of real tall buildings. *Wind and Structures*, 18: 423–441.

Vickery, B.J. and Daly, A. (1984) Wind tunnel modelling as a means of predicting the response to vortex shedding. *Engineering Structures*, 6: 363–368.

Wang, X., Wang, X., Ren, X., Yin, X. and Wang, W. (2018) Effects of tube system and data correction for fluctuating pressure test in wind tunnel. *Chinese Journal of Aeronautics*, 31: 710–718.

Ward, N.B. (1972) The exploration of certain features of tornado dynamics using a laboratory model. *Journal of Atmospheric Sciences*, 29: 1194–1204.

Wardlaw, R.L. (1980) Sectional versus full model testing of bridge road decks. *Proceedings, Indian Academy of Sciences*, 3: 177–198.

Whitbread, R.E. (1963). Model simulation of wind effects on structures. *Proceedings, International Conference on Wind Effects on Buildings and Structures*, Teddington, UK, 26–28 June, pp. 284–302.

Xie, J. and Garber, J. (2014) HFFB technique and its validation studies. *Wind and Structures*, 18: 375–389.

Xie, J., and Irwin, P.A. (1998) Application of the force balance technique to a building complex. *Journal of Wind Engineering and Industrial Aerodynamics*, 77/78.: 579–590.

Yoshida,Y., Tamura, Y. and Kurita, T. (2001) Effects of bends in a tubing system for pressure measurement. Journal of Wind Engineering and Industrial Aerodynamics, 89: 1701–1716.

Chapter 8

Low-rise buildings

8.1 INTRODUCTION

For the purposes of this chapter, *low-rise buildings* are defined as roofed low-rise structures less than 15 m in height. Large roofs on major structures such as sports stadia, including arched roofs, are discussed in Chapter 10; free-standing roofs and canopies are covered in Chapter 14.

The assessment of wind loads for low-rise buildings is as equally difficult as it is for taller buildings and other larger structures due to the following factors:

- They are usually immersed within the layer of aerodynamic roughness on the earth's surface, where the turbulence intensities are high, and interference and shelter effects are important, but difficult to quantify.
- Roof loadings, with all the variations due to changes in geometry, are of critical importance for low-rise buildings. The highest wind loadings on the surface of a low-rise structure are generally the suctions on the roof, and many structural failures are initiated there.
- Low-rise buildings often have a single internal space, and internal pressures can be very significant, especially when a dominant opening occurs in a windward wall. The magnitude of internal pressure peaks, and their correlation with peaks in external pressure, must be assessed.

However, resonant dynamic effects can normally be neglected for smaller buildings. The majority of structural damage in windstorms is incurred by low-rise buildings, especially family dwellings, which are often non-engineered and lacking in maintenance.

The following sections will discuss the history of research on wind loads on low-rise buildings, the general characteristics of wind pressures and model scaling criteria, and a summary of the results of the many studies that were carried out in the 1970s–1990s.

8.2 HISTORICAL

8.2.1 Early wind-tunnel studies

Some of the earliest applications of wind tunnels were in the study of wind pressures on low-rise buildings. The two earliest investigations were by Kernot (1893) in Melbourne, Australia, and Irminger (1894) in Copenhagen, Denmark. Kernot used what would now be called an open-jet wind tunnel (see Section 7.2.1), as well as a whirling arm apparatus, and measured forces on a variety of building shapes. The effects of roof pitch, parapets and adjacent buildings were all examined.

Irminger used a small tunnel driven by the suction of a factory chimney, and measured pressures on a variety of models, including one of a house. He demonstrated the importance of roof suction, a poorly understood concept at the time.

Over the following thirty years, isolated studies were carried out in aeronautical wind tunnels at the National Physical Laboratory (NPL) in the United Kingdom, the DLR laboratories (Deutches Zentrum für Luft- und Raumfahrt) at Göttingen, Germany, the National Bureau of Standards in the United States, and the Central Aero-Hydrodynamical Institute of the USSR These early measurements showed some disagreement with each other, although they were all measurements of steady wind pressures in nominally steady flow conditions. This was probably due to small but different levels of turbulence in the various wind tunnels (Chapter 4 discusses the effect of turbulence on the mean flow and pressures on bluff bodies), and other effects, such as blockage.

In Denmark, Irminger and Nokkentved (1930) carried out further wind-tunnel studies on low-rise buildings. These tests were again carried out in steady, uniform flow conditions, but included some innovative work on models with porous walls, and the measurement of internal as well as external pressures. Dryden and Hill (1931) studied wind pressures on a mill building with a 20° roof pitch, with and without a ridge ventilator. Measurements on low-rise building models were also carried out by Richardson and Miller (1932) in Australia.

In 1936, the American Society of Civil Engineers (1936) surveyed the data available at that time on wind loads on steel buildings. This survey included consideration of "rounded and sloping roofs". These data consisted of a variety of early wind-tunnel measurements presumably carried out in smooth flow.

Flachsbart, at the Göttingen Laboratories in Germany, is well known for his extensive wind-tunnel measurements on lattice frames and bridge trusses, in the 1930s. Less well-known, however, is the work he did in comparing wind pressures on a low-rise building in smooth and boundary-layer flow (Flachsbart, 1932). This work – probably the first boundary-layer wind tunnel study – has been rediscovered, and reported, by Simiu and Scanlan (1996).

Recognition of the importance of boundary-layer flow was also made by Bailey and Vincent (1943) at the National Physical Laboratories in the United Kingdom. In doing so, they were able to make some progress in explaining differences, between wind-tunnel and full-scale measurements of pressures, on a low-rise shed.

However, it was not until the 1950s that Jensen (1958), at the Technical University of Denmark, explained satisfactorily the differences between full-scale and wind-tunnel model measurements of wind pressures. Figure 8.1 reproduces some of his measurements, which fully established the importance of using a turbulent boundary-layer flow to obtain pressure coefficients in agreement with full-scale values. The non-dimensional ratio of building height to roughness length, h/z_o, was later named, the *Jensen*

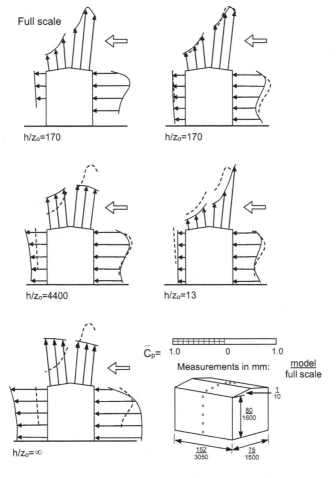

Figure 8.1 Pioneer boundary-layer measurements of Jensen (1958).

Number (see Section 4.4.5), in recognition of this work. Jensen and Franck (1965) later carried out extensive wind-tunnel measurements on a range of building shapes in a small boundary-layer wind tunnel.

The work of Jensen and Franck was the precursor to a series of generic, wind-tunnel studies of wind loads on low-rise buildings in the 1970s and 1980s, including those on industrial buildings by Davenport *et al.* (1977), and on houses by Holmes (1983, 1994). Results from these studies are discussed in later sections.

Important contributions to the understanding of the effect of large groupings of bluff bodies in turbulent boundary layers, representative of large groups of low-rise buildings, were made by Lee and Soliman (1977) and Hussain and Lee (1980). Three types of flow were established, depending on the building spacing: *skimming* flow (close spacing), *wake-interference* flow (medium spacing) and *isolated-roughness* flow (far spacing).

8.2.2 Full-scale studies

The last three decades of the twentieth century were notable for a number of full-scale studies of wind loads on low-rise buildings. In these studies, advantage was taken of the considerable developments that had taken place in electronic instrumentation, and computer-based statistical analysis techniques, and provided a vast body of data which challenged wind-tunnel modelling techniques.

In the early 1970s, the Building Research Establishment in the United Kingdom commenced a programme of full-scale measurements on a specially-constructed experimental building, representative of a two-storey low-rise building at Aylesbury, England. The building had the unique feature of a roof pitch which was adjustable between 5° and 45° (Figure 8.2).

The results obtained in the Aylesbury Experiment emphasized the highly fluctuating nature of wind pressures, and the high pressure peaks in separated flow regions near the roof eaves and ridge, and near the wall corners (Eaton and Mayne, 1975; Eaton *et al.*, 1975). Unfortunately the experiment was discontinued, and the experimental building was dismantled after only 2 years at the Aylesbury site. However, interest from wind-tunnel researchers in the Aylesbury data continued through the 1980s, when an International Aylesbury Comparative Experiment was established. Seventeen wind-tunnel laboratories around the world tested identical 1/100 scale models of the Aylesbury building, using various techniques for modelling the upwind terrain and approaching flow conditions. This unique experiment showed significant differences in the measured pressure coefficients – attributed mainly to different techniques used to obtain the reference static and dynamic pressures, and in modelling the hedges in the upwind terrain at the full-scale site (Sill *et al.*, 1989, 1992).

In the late 1980s, two new full-scale experiments on low-rise buildings were set up in Lubbock, Texas, U.S.A., and Silsoe, U.K. The Lubbock experiment,

Figure 8.2 AylesburyExperimental Building (United Kingdom 1970–1975).

Figure 8.3 Texas Tech Field Experiment (United States).

known as the Texas Tech Field Experiment, comprised a small steel shed of height, 4.0 m, and plan dimensions, 9.1 and 13.7 m; the building had a near-flat roof (Figure 8.3). The building had the unique capability of being mounted on a turntable, thus enabling control of the building orientation relative to the mean wind direction. Pressures were measured with high response pressure transducers mounted close to the pressure tappings on the roof and walls; the transducers were moved around to different positions at different times during the course of the experiments. A 50 m high mast, upwind of the building, in the prevailing wind direction, had several levels of anemometers, enabling the approaching wind properties to be well defined. The upwind terrain was quite flat and open. The reference static pressure was obtained from an underground box, 23 m away from the centre of the test building. (Levitan and Mehta, 1992a,b).

The Texas Tech experiment has produced a large amount of wind pressure data for a variety of wind directions. External and internal pressures, with and without dominant openings in the walls, were recorded. Very high extreme pressures at the windward corner of the roof for 'quartering' winds blowing directly on to the corner, at about 45° to the walls, were measured; these were considerably greater than those measured at equivalent positions on small (1/100) scale wind-tunnel models. The internal pressures, however, showed similar characteristics to those measured on wind-tunnel models, and predicted by theoretical models.

The Silsoe Structures Building was a larger steel portal-framed structure, 24 m long, 12.9 m span, and 4 m to the eaves, with a 10° roof pitch, located in open country. With seventy pressure tapping points on the building roof and walls, the building was equipped with twelve strain gauge positions on the central portal frame to enable measurements of structural response to be made (Robertson, 1992).

The building could be fitted with both curved and sharp eaves. The curved eaves were found to give lower mean negative pressures immediately downwind of the windward wall, than those produced by the sharp eaves. Measurements of strain in the portal frame were found to be predicted quite well by a structural analysis computer program when the correct column fixity was applied. Spectral densities of the strains were also measured – these showed the effects of Helmholtz resonance (Section 6.2.3) on the internal pressures, when there was an opening in the end wall of the building. Generally, these measurements justified a *quasi-steady* approach to wind loads on low-rise buildings (Section 4.6.2).

8.3 GENERAL CHARACTERISTICS OF WIND LOADS ON LOW-RISE BUILDINGS

Full-scale measurements of wind pressures on low-rise buildings, such as those described in Section 8.2.2, showed the highly fluctuating nature of wind pressures, area-averaged wind loads, and load effects, or responses, on these structures. The fluctuations with time can be attributed to two sources (see also Section 4.6.1):

- Pressure fluctuations induced by upwind turbulent velocity fluctuations (see Chapter 3). In an urban situation, the turbulence may arise from the wakes of upwind buildings.
- Unsteady pressures produced by local vortex shedding, and other unsteady flow phenomena, in the separated flow regions near sharp corners, roof eaves and ridges (see Chapter 4).

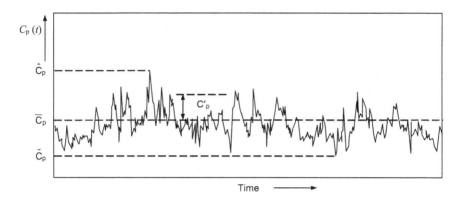

Figure 8.4 Typical variation of wind pressure and definition of pressure coefficients.

These two phenomena may interact with each other to further complicate the situation.

It should be noted that, as well as a variation with time, as shown for a single point on a building in Figure 8.4, there is a variation with space, i.e. the same pressure or response variation with time may not occur simultaneously at different points separated from each other on a building.

8.3.1 Pressure coefficients

The basic definition of a pressure coefficient for a bluff body was given in Section 4.2.1., and the root-mean-square fluctuating (standard deviation) pressure coefficient was defined in Section 4.6.4. A general time-varying pressure coefficient, $C_p(t)$, for buildings in stationary, or synoptic, wind storms is calculated as follows:

$$C_p(t) = \frac{p(t) - p_o}{\frac{1}{2}\rho_a \bar{U}^2} \qquad (8.1)$$

where p_o is a static reference pressure (normally atmospheric pressure measured at a convenient location near the building, but not affected by the flow around it), ρ_a is the density of air, and $\bar{U}$ is the mean (time-averaged) velocity measured at an appropriate defined reference height, as in the atmospheric boundary layer, there is a variation of mean wind speed with height (Section 3.2). In the case of a low-rise building, this is usually taken to be at roof height, either at eaves level, mid-height of the roof, or at the highest level of the roof; as for the static pressure, this must be away from the direct influence of the building.

Figure 8.4 shows a typical variation of $C_p(t)$ on a low-rise building, and four significant values of the pressure coefficient:

$\bar{C}_p$ – the mean or time-averaged pressure coefficient

C_p' ($= \sigma_{Cp}$) – the root-mean-squared (r.m.s.) fluctuating value, or standard deviation, representing the average departure from the mean

$\hat{C}_p$ (or $C_{\hat{p}}$) – the expected maximum value of the pressure coefficient in a given time period

$\check{C}_p$ (or $C_{\check{p}}$) – the expected minimum value of the pressure coefficient in a given time period

8.3.2 Dependence of pressure coefficients

The dependence of pressure coefficients on other non-dimensional quantities, such as Reynolds Number and Jensen Number, in the general context of bluff-body aerodynamics, was discussed in Section 4.2.3. This dependence is applicable to wind loads on low-rise buildings.

For bodies which are sharp-edged, and on which points of flow separation are generally fixed, the flow patterns and pressure coefficients are *relatively* insensitive to viscous effects and hence Reynolds Number. This means that, provided an adequate reproduction of the turbulent flow, characteristics in atmospheric boundary-layer flow is achieved, and the model is geometrically correct, wind-tunnel tests can be used to predict pressure and force coefficients on full-scale buildings. However, the full-scale studies from the Texas Tech Field Experiment have indicated that for certain wind directions, pressure peaks in some separated flow regions are not reproduced in wind-tunnel tests with small scale models, and some Reynolds Number dependency *is* indicated.

As discussed in Section 8.2.1, Jensen (1958) identified the ratio, h/z_o, the ratio of building height to the aerodynamic roughness length in the logarithmic law (Sections 3.2.1 and 4.4.5) as the most critical parameter in determining mean pressure coefficients on low-rise buildings. The Jensen Number clearly directly influences the mean pressure distributions on a building through the effect of the mean velocity profile with height. However, in a fully developed boundary-layer over a rough ground surface, the turbulence quantities such as intensities (Section 3.3.1), and spectra (Section 3.3.4) should also scale with the ratio z/z_o near the ground. There is an indirect influence of the turbulence properties on the mean pressure coefficients (Section 4.4.3), which would have been responsible for some of the differences observed by Jensen (1958), and seen in Figure 8.1. In wind-tunnel tests, the turbulence intensity similarity will only be achieved with h/z_o equality, if the turbulent inner surface layer in the atmospheric boundary layer has been correctly simulated in the boundary-layer in the wind tunnel. Many researchers prefer to treat parameters such as turbulence intensities and ratios of turbulence length scale to building dimension as independent non-dimensional quantities (See

Section 4.2.3), but unfortunately it is difficult to independently vary these parameters in wind-tunnel tests.

Fluctuating and peak external pressures on low-rise buildings which are most relevant to structural design, are highly dependent on the turbulence properties in the approach flow, especially turbulence intensities. Consequently peak load effects, such as bending moments in framing members, are also dependent on the upwind turbulence. For correctly simulated boundary layers, in which turbulence quantities near the ground scale as z/z_o, peak load effects can be reduced to a variation with Jensen Number (e.g. Holmes and Carpenter, 1990).

Finally, the question of the dependency of pressures and load effects on low-rise buildings in wind storms of the downdraft type (Section 1.3.5) arises. As discussed in Section 3.2.6, these winds have boundary layers which are not strongly dependent on the surface roughness on the ground – hence the Jensen Number may not be such an important parameter. Further research is required to identify non-dimensional parameters in the downdraft flow which are relevant to wind pressures on buildings in these types of storms.

8.3.3 Flow patterns and mean pressure distributions

Figure 8.5 shows the main features of flow over a building with a low-pitched roof, which has many of the features of flow around a two-dimensional bluff body described in Section 4.1. The flow separates at the top of the windward wall and *re-attaches* at a region further downwind on the roof, forming a separation zone or 'bubble'. However this bubble exists only as a time average. The separation zone is bounded by a free shear layer, a region of high velocity gradients, and high turbulence. This layer rolls up intermittently to form vortices; as these are shed downwind, they may produce high

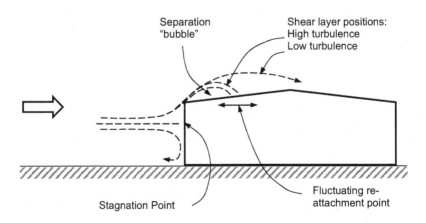

Figure 8.5 Wind flow around a low-rise building.

negative pressure peaks on the roof surface. The effect of turbulence in the approaching flow is to cause the vortices to roll up closer to the leading edge, and a shorter distance to the re-attachment zone results.

The longitudinal intensities of turbulence at typical roof heights of low-rise buildings, are 20% or greater, and separation zone lengths are shorter, compared to those in smooth, or low turbulence flow. Small separation zones with high shear layer curvatures, are associated with low pressures, that is high initial negative pressures, but rapid pressure recovery downwind.

Roof pitches up to about 10°, for wind normal to a ridge or gable end, are *aerodynamically flat*. When the mean wind direction is parallel to a ridge line, the roof is also seen as aerodynamically flat, for any roof pitch. For winds normal to the ridge line, and roof pitches between 10° and 20°, a second flow separation occurs at the ridge, producing regions of high negative pressures on both sides of the ridge. Downwind of the ridge, a second re-attachment of the flow occurs with an accompanying recovery in pressure. At roof pitches greater than about 20°, positive mean pressures occur on the upwind roof face, and fully separated flows, without re-attachment occur downwind of the ridge giving relatively uniform negative mean pressures on the downwind roof slope.

It should be noted that the above comments are applicable only to low-rise buildings with height /downwind depth (h/d) ratios less than about 0.5. As this ratio increases, roof pressures generally become more negative. This influence can be seen in Figure 8.6 which shows the mean pressure distribution along the centre line of low-rise buildings for various roof pitches and h/d ratios; the horizontal dimension across the wind (into the paper in Figure 8.6) is about twice the along-wind dimension ($b/d \approx 2$). For higher buildings with h/d ratios of 3 or greater, the roof pressure will be negative on both faces, even for roof slopes greater than 20°.

Similar flow separation and re-attachment, as described for roofs, occurs on the side walls of low-rise buildings, although the magnitude of the mean pressure coefficients is generally lower. The mean pressures on windward walls are positive with respect to the freestream static pressure. Leeward walls are influenced by the re-circulating wake, and generally experience negative pressures of lower magnitude; however, the values depend on the building dimensions, including the roof pitch angle.

When the wind blows obliquely on to the corner of a roof, a more complex flow pattern appears as shown in Figure 8.7. *Conical* vortices similar to those found on delta-wings of aircraft occur. Figure 8.8 shows these vortices visualised by smoke – their axes are inclined slightly to the adjacent walls forming the corner. The pressures underneath these are the largest to occur on the low-pitched roofs, square or rectangular in planform, although the areas over which they act are usually quite small, and are more significant for pressures on small areas of cladding than for the loads in major structural members.

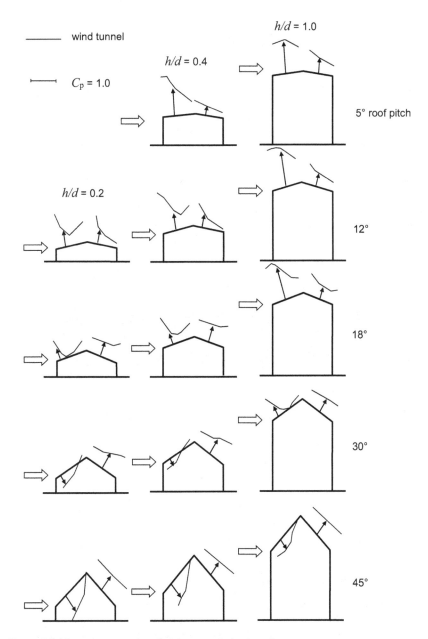

Figure 8.6 Mean pressure distributions on pitched roofs.

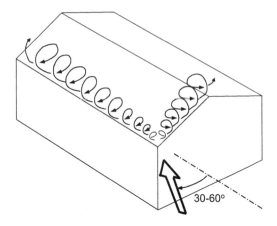

Figure 8.7 Conical vortices for oblique wind directions.

Figure 8.8 Corner vortices generated by quartering winds (from the Texas Tech Field Experiment).

8.3.4 Fluctuating pressures

The root-mean-squared fluctuating, or standard deviation, pressure coefficient, defined in Sections 4.6.4 and 8.3.1, is a measure of the general level of pressure fluctuations at a point on a building. As discussed in Section 8.3.2, the values obtained on a particular building are generally dependent on the turbulence intensities in the approaching flow, which in turn are dependent on the Jensen Number. In boundary-layer winds over open country terrain, for which longitudinal turbulence intensities are typically around 20%, at heights typical of eaves heights on low-rise buildings, the values of r.m.s.

pressure coefficients (based on a dynamic pressure calculated from the mean wind speed at eaves height) on windward walls, are typically in the range 0.3–0.4. In separated–reattaching flow regions on side walls, values of C_p' of 0.6 or greater can occur. Even higher values can occur at critical points on roofs, with values greater than 1.0 being not uncommon.

High instantaneous peak pressures tend to occur at the same locations as high r.m.s. fluctuating pressures. The highest negative peak pressures are associated with the conical vortices generated at the roof corners of low-pitch buildings, for quartering winds blowing on to the corner in question (Figures 8.7 and 8.8). Figure 8.9 shows a short sample of pressure – time history, from a pressure measurement position near the formation point of one of these vortices, on the Texas Tech building (Mehta *et al.*, 1992). This shows that high pressure peaks occur as 'spikes' over very short time periods. Values of negative peak pressure coefficients as high as –10 often occur, and magnitudes of –20 have also occasionally been measured.

The probability density function (p.d.f.) and cumulative distribution function (c.d.f.) are measures of the amplitude variations in pressure fluctuations at a point. Even though the upwind velocity fluctuations in boundary-layer winds are nearly Gaussian (Section 3.3.2 and Appendix C, Section C3.1), this is not the case for pressure fluctuations on buildings. Figure 8.10 shows a wind-tunnel measurement of the c.d.f. for pressure fluctuations on the windward wall of a low-rise building model (Holmes, 1981, 1983). On this graph, a straight line indicates a Gaussian distribution. Clearly, the measurements showed an upward curvature, or positive skewness (Appendix C, Figure C3). This can, in part, be explained by the square-law relationship between pressure and velocity (see Equation (4.14))

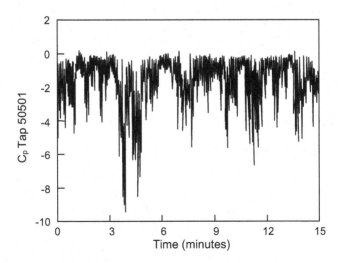

Figure 8.9 Pressure coefficient versus time from a corner pressure tap (Mehta et al., 1992).

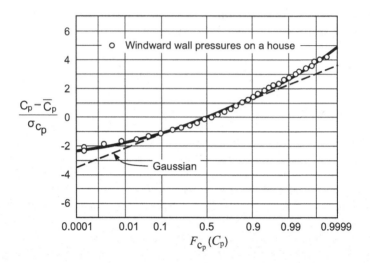

Figure 8.10 Cumulative probability distribution for pressure fluctuations on the windward wall of a house.

(Holmes, 1981 and Appendix C3.3). Negative skewness occurs for pressure fluctuations in separated flow regions of a building.

The spatial structure of fluctuating pressures on low-rise buildings has been investigated in detail, using a technique known as 'proper orthogonal decomposition' (e.g Best and Holmes, 1983; Holmes, 1990a; Letchford and Mehta, 1993; Bienkiewicz *et al.*, 1993; Ho *et al.*, 1995; Holmes *et al.*, 1997; Baker, 1999; Ginger, 2004). This method allows the complexity of the space-time structure of the pressure fluctuations on a complete roof, building, or tributary area, to be simplified into a series of 'modes', each with its own spatial form. Surprisingly, few of these modes are required to describe the complexity of the variations. Invariably, for low-rise buildings, the first, and strongest mode is driven by the quasi-steady mechanism associated with upwind turbulence fluctuations.

8.4 BUILDINGS WITH PITCHED ROOFS

8.4.1 Cladding loads

Figures 8.11 and 8.12 show contours of the worst minimum pressure coefficients, for any wind direction, measured in wind-tunnel tests on models of single storey houses with gable roofs of various pitches (Holmes, 1994). The simulated approach terrain in the approach boundary-layer flow, was representative of open country, and the wind direction was varied at 10° intervals during the tests. The coefficients are all defined with respect to the mean wind speed at *eaves* height.

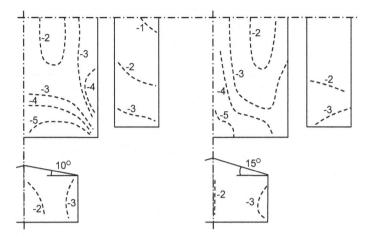

Figure 8.11 Largest minimum pressure coefficients, $\check{C}_p$, for houses with roofs of 10° and 15° pitch (for any wind direction) (Holmes, 1994).

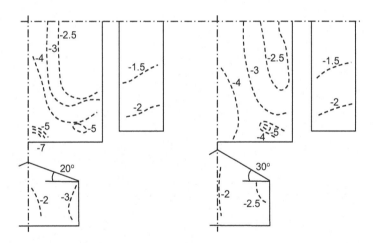

Figure 8.12 Largest minimum pressure coefficients, $\check{C}_p$, for houses with roofs of 20° and 30° pitch (for any wind direction) (Holmes, 1994).

The highest magnitude coefficients occur on the roof. At the lowest pitch (10°) the contours of highest negative pressures converge towards the corner of the roof; the effect of increasing the roof pitch is to emphasize the gable end as the worst loaded region. The worst local negative peak pressures occur on the 20° pitch roof in this area. The highest magnitude minima on the walls occur near a corner.

Similar plots for shapes representative of industrial buildings with roof pitches of 5°, 18° and 45° pitch, are shown in Figures 8.13, 8.14, and 8.15

(Davenport *et al.*, 1977). In these figures, contours of maximum pressure coefficients, as well as minimum pressure coefficients, are plotted. Plots are given for three different eaves heights, for each roof pitch. Results from building models located in simulated urban terrain are shown.

For any given roof pitch, there is not a large variation in the magnitudes of the minimum and maximum pressure coefficients with eaves height – however the pressure coefficients are defined with respect to the mean dynamic pressure at eaves height in each case. Since the mean velocity, and hence dynamic pressure, in a boundary layer increases with increasing height, the pressures themselves will generally increase with height of the building. Since the fluctuating pressure coefficients are closely related to the turbulence intensities in the approach flow, lower magnitudes might be expected at greater eaves heights, where the turbulence intensities are lower, as shown in Figures 8.13–8.15. However, the local pressure peaks are also influenced by local flow separations, and hence by the relative building dimensions.

The worst minimum pressure coefficients for the 18° pitch roofs (Figure 8.14) occur near the ridge at the gable end (compare also the house

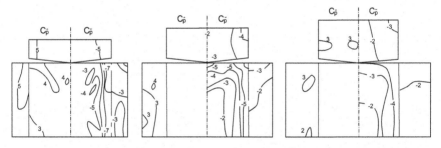

Figure 8.13 Largest maximum and minimum pressure coefficients, $\hat{C}_p$ and $\check{C}_p$, for industrial buildings with roofs of 5° pitch (for any wind direction) (Davenport et al., 1977).

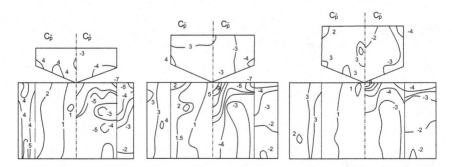

Figure 8.14 Largest maximum and minimum pressure coefficients, $\hat{C}_p$ and $\check{C}_p$, for industrial buildings with roofs of 18° pitch (for any wind direction) (Davenport et al., 1977).

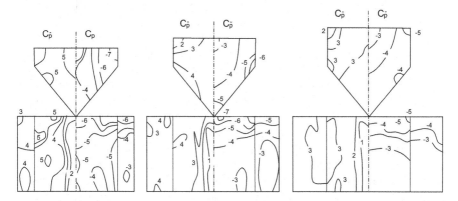

Figure 8.15 Largest maximum and minimum pressure coefficients, $\hat{C}_p$ and $\check{C}_p$, for industrial buildings with roofs of 45° pitch (for any wind direction) (Davenport et al., 1977).

with the 20° pitch roof in Figure 8.12). For the 5° pitch case (Figure 8.13), there is a more even distribution of the largest minimum (negative) pressure coefficients around the edge of the roof. For the 45° pitch, the corner regions of the roof generally experience the largest minima; the maximum pressure coefficients are also significant in magnitude on the 45° pitch roof.

Plots such as those in Figures 8.11–8.15 can be used as a guide to the specification of wind loads for the design of cladding. However, it should be noted that if the design wind speeds are non-uniform with direction, as they normally will be, the contours of maximum and minimum *pressures* (as opposed to pressure coefficients) will be different, and will depend on the site and the building orientation.

8.4.2 Structural loads and equivalent static load distributions

The effective peak wind loads acting on a major structural element such as the portal frame of a low-rise building are dependent on two factors:

- the correlation or statistical relationship between the fluctuating pressures on different parts of the tributary surface area 'seen' by the frame; this can be regarded as an area-averaging effect
- the influence coefficients which relates pressures at points or panels on the surface to particular load effects, such as bending moments or reactions.

Chapter 5 described methods for determining 'effective static loading distributions', which represent the wind loads that are equivalent in their structural effect to fluctuating (background) wind pressures, and to the resonant (inertial) loads when they are significant. For the low-rise buildings under discussion in this chapter, resonant effects can be ignored, but the fluctuating, or background loading is quite significant because of the high turbulence intensities near the ground. Some examples of the application of the methods discussed in Chapter 5, will be given in this chapter.

To illustrate the problem, consider Figure 8.16. This shows instantaneous external pressure distributions occurring at three different times during a wind storm around a portal frame supporting a low-rise building. These pressure distributions are clearly different from each other in both shape and magnitude. The value of a load effect such as the bending moment at the knee of the frame, will respond to these pressures in a way that might produce the time history of bending moment versus time given in Figure 8.17. Over a given time period, a maximum bending moment will

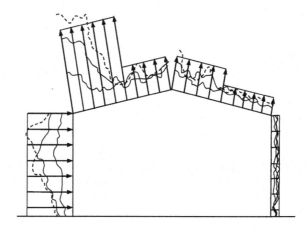

Figure 8.16 Instantaneous external pressure distributions on the frame of a low-rise building, and simplified code distributions (Holmes and Syme, 1994).

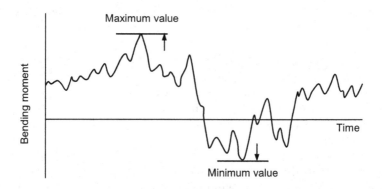

Figure 8.17 Time history of a bending moment (Holmes and Syme, 1994).

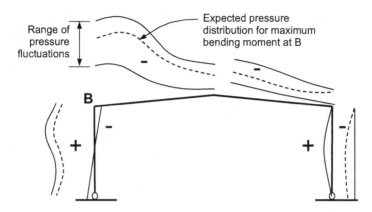

Figure 8.18 Effective static load distribution for a corner bending moment (Holmes and Syme, 1994).

occur. A minimum bending moment will also occur. Depending on the sign of the bending moment produced by the dead loads acting on the structure, one of these extremes will be the critical one for design of the structure. Methods for determination of the *expected* pressure distribution which corresponds to the maximum or minimum wind-induced bending moment have been discussed in Chapter 5. The effective static pressure distribution so determined, must lie between the extreme point pressure limits of the pressures around the frame, as shown in Figure 8.18.

In wind codes and standards, usually an 'envelope' loading is specified with pressures uniformly distributed in length along the columns and rafters, as shown in Figure 8.16. These are usually, but not always, conservative loadings which will give over-estimates of load effects such as bending moments.

8.4.3 Hipped-roof buildings

Damage investigations following severe windstorms, have sometimes noted that hipped-roof buildings have generally suffered lesser damage. Meecham *et al.* (1991) studied wind pressures on hipped and gable-roof buildings of 18.4° pitch in a boundary-layer wind tunnel. Although there is little difference in the largest peak total lift force, or overturning moment on the two roofs, the gable end region of the gable roof experiences around 50% greater peak negative local pressures, than does the corresponding region on the hipped roof. Furthermore the largest area-averaged full-span truss load was about twice as high on the gable roof.

However, Xu and Reardon (1998), who studied pressures on hipped roofs with three different roof pitches (15°, 20°, and 30°), found that the benefits of a hipped configuration, compared with a gable-roof type, reduces as the roof pitch increases. Figure 8.19 shows contours of worst minimum

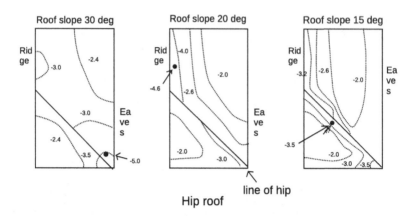

Figure 8.19 Largest minimum pressure coefficients for hipped roofs of 15°, 20° and 30° pitch, for any wind direction. (Xu and Reardon, 1998).

(negative) peak pressure coefficients, referenced to mean dynamic pressure at eaves height, and can be compared with the equivalent gable-roof values in Figures 8.11 and 8.12. Note that, at 30° pitch, the worst negative pressure coefficients of about −5.0 are similar for the two roof types.

8.4.4 Effect of surrounding buildings – shelter and interference

Most low-rise buildings are in an urban situation, and are often surrounded by buildings of similar size. The shelter and aerodynamic interference effect of upstream buildings can be very significant on the wind loads. This aspect was the motivation for the studies by Lee and Soliman (1977) and Hussain and Lee (1980) on grouped buildings, as discussed in Section 8.2.1. Three flow regimes were identified that depended upon the building spacing. A study on tropical houses (Holmes, 1994) included a large number of grouped building situations for buildings with roofs of 10° pitch. This study showed that upstream buildings of the same height reduced the wall pressures and the pressures at the leading edge of the roof significantly, but had less effect on pressures on other parts of the roof. The building height/spacing ratio was the major parameter, with the number of shielding rows being of lesser importance.

A series of wind-tunnel pressure measurements, for both structural loads and local cladding loads, on a flat-roofed building, situated in a variety of 'random city' environments was carried out by Ho *et al.* (1990, 1991). It was found that the mean component of the wind loads decreased, and the fluctuating component increased, resulting in a less-distinct variation in peak wind load with direction. The expected peak loads in the urban environment were much lower than those on the isolated building. It was also found that a high coefficient of variation (60%–80%) of wind loads occurred on the building in the urban environment due to the variation in *location* of the building. For the isolated building, similar coefficients of variation occurred, but in this case, they resulted from variation due to *wind direction*.

8.5 MULTI-SPAN BUILDINGS

The arrangement of industrial low-rise buildings as a series of connected spans is common practice for reasons of structural efficiency, lighting and ventilation. Such configurations also allow for expansion in stages of a factory or warehouse.

Wind-tunnel studies of wind pressures on multi-span buildings of the 'saw-tooth' type with 20° pitch were reported by Holmes (1990b), and by Saathoff and Stathopoulos (1992) on 15° pitch buildings of this type. Multi-span gable-roof buildings were studied by Holmes (1990b) (5° pitch), and by Stathopoulos and Saathoff (1994), (18° and 45° pitch). The main focus of these studies was to determine the difference in wind loads for multi-span buildings, and the corresponding single-span monoslope and gable-roof buildings, respectively.

As for single-span buildings, the aerodynamic behaviour of multi-span buildings is quite dependent on the roof pitch. Multi-span buildings of low pitch (say less than 10°) are aerodynamically flat, as discussed in Section 8.3.3. Consequently, quite low mean and fluctuating pressures are obtained on the downwind spans, as illustrated in Figure 8.20. The pressures on the first windward span are generally similar to those on a single span building of the same geometry.

For gable-roof buildings, and for saw-tooth roofs with the roofs sloping downwards away from the wind, the downwind spans experience much lower magnitude negative mean pressures than the windward spans. For the opposite wind direction on the saw-tooth configuration, the highest magnitude mean pressure coefficients occur on the second span downwind, due to the separation bubble formed in the valley.

8.6 EFFECTS OF PARAPETS ON
LOW-RISE BUILDINGS

A detailed wind-tunnel study of the wind effects of parapets on the roofs of low-rise buildings was carried out by Kopp *et al.* (2005a,b). Earlier work was reviewed by Stathopoulos and Baskaran (1988).

It was found that tall parapets $(h_p/(h + h_p) > 0.2)$, where h_p is the parapet height, can reduce peak local negative pressures by up to 50%, in corner regions of a roof, when they are installed around the complete perimeter of a roof. Lower parapets, $(h_p/(h+h_p) < 0.2)$, increase the worst negative peak

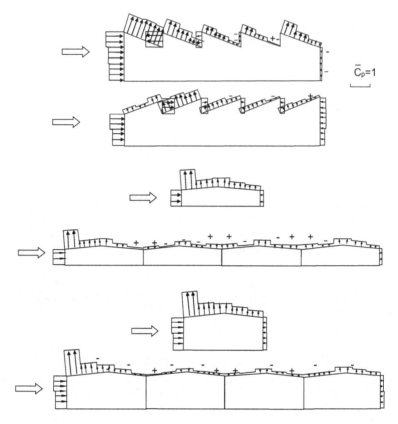

Figure 8.20 Mean pressure distributions on multi-span buildings and comparison with a single span (Holmes, 1990b).

pressure coefficients, apparently by stabilizing the corner conical vortices that occur on flat or near-flat roofs (see Figure 8.8). However, high parapets increase the positive (downwards) pressure peaks on the roof upwind of leeward parapets.

Isolated parapets. i.e. those installed adjacent to one wall of a building, always increase the corner roof loads, irrespective of their height (Kopp et al., 2005a).

Structural loads on roofs are affected somewhat differently than local loads. For a wind direction normal to a wall, a parapet will move the point of re-attachment (see Figure 8.5) further downwind, thus giving a larger region of separated flow. This increases the structural loads, presumably because of the increased correlations of roof pressures in the separated flow region. The loading on interior bays was increased by about 10% for low parapets ($h_p/(h + h_p) < 0.09$), with greater increases for higher parapets (Kopp *et al.*, 2005b).

8.7 EFFECT OF BUILDING LENGTH

In codes and standards for wind loading, the effect of the horizontal aspect ratio (*b/d*) on the specified wind pressure coefficients for low-rise buildings is often neglected. Coefficients are typically based on data obtained from wind-tunnel tests with *b/d* ratios of about two. However, several wind-tunnel studies (e.g. Ginger and Holmes, 2003; Ginger, 2004) have shown that horizontal aspect ratio (or building length) has a significant effect on roof pressures when it becomes large, and particularly for long gable-roofed buildings with very high roof pitches. Such buildings are often used for bulk storage of solids, such as mineral ore or sugar.

The windward ends of these buildings can experience very high loads for oblique wind directions. Very high negative pressures have been observed on the leeward roof surfaces in these situations. The flow and resulting pressures are, in fact, similar to those occurring at the ends of free-standing walls for oblique wind directions (see Section 14.2.1 and Figure 14.3).

A wind-tunnel study of buildings with roof pitches greater than 30° (Ginger and Holmes, 2003) showed that some major codes and standards underestimated load effects such as frame bending moments by up to 70%, for a building with a *b/d* ratio of 6. The Australian Standard of 1989 (Standards Australia, 1989) underestimated some bending moments for this case by 50%, but revisions included in later editions (e.g. Standards Australia, 2011) have resulted in much better agreement (Ginger and Holmes, 2003).

Another class of low-rise building is the new generation of regional warehouses with large horizontal dimensions, and with length/breadth/height ratios that are well outside the range used for shape factors in wind codes and standards. The lengths of some of these buildings may be as high as 400 m, although the roof heights are in the range of 10–15 m. Often the roof pitches are as low as 2°–3°. A detailed study of the aerodynamics and pressure distributions for buildings of these shapes is apparently yet to be undertaken, but it is known that re-attached flow covers large areas of roof. Both positive (downward acting) and negative pressures act in this region and both need to be accounted for in design together with internal pressures. Also, the re-attached flow is accompanied by large areas of frictional drag forces in the plane of the roof, which will combine to give large horizontal forces which must be resisted by structural bracing.

8.8 INTERNAL PRESSURES

In Chapter 6, the prediction of internal pressures in buildings in general is discussed in detail. For low-rise buildings in particular, the internal pressure loading may form a high proportion of the total wind loading for both major structural elements and cladding. In severe wind storms, such as hurricanes or typhoons, failures of roofs often occur following window failure on the windward wall, which generates high positive internal pressures acting together with negative external pressures.

8.9 A CASE STUDY – OPTIMUM SHAPING OF A LOW-RISE BUILDING

Low-rise buildings are not generally shaped to optimize wind loading, unlike some larger structures like long-span bridges, and some tall buildings. However, Figure 8.21 shows such an example – a standard design for public shelter buildings for towns threatened by tropical cyclones in the state of Queensland, Australia.

In order to satisfy onerous wind loading guidelines for public shelters, with respect to both direct wind loading, and windborne debris (Department of Public Works, Queensland, 2006), several features to achieve this are incorporated. These include: chamfering of the corners on the walls on the lower wall corners, architectural 'fins on the upper floor corners, venting of the corner eaves on the roof, and heavy-duty debris resistant screens (Figure 8.21).

Wind-tunnel tests were carried out to validate the features. The venting of the roof corners on the standard shelter produced reductions of 50%–70% in

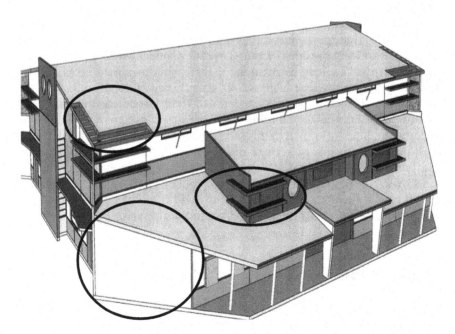

Figure 8.21 A low-rise building with design features to improve the aerodynamic characteristics in strong winds.

peak negative pressures at critical roof taps. Compared with calculations for a building of simple rectangular planform, as prescribed by the Australian and New Zealand Standard for Wind Actions (Standards Australia, 2011), reductions in peak roof and wall pressures of 13%–41% were achieved (Holmes, Ginger and Mullins, 2019).

Other options for mitigating local pressures on low-rise buildings were investigated by Bitsuamlak *et al.* (2013).

8.10 WIND-TUNNEL DATABASES

A significant development for wind engineering specialists and designers has been the availability of two very useful comprehensive sources of data on external shape factors and pressure coefficients on low-rise buildings on the internet. These are as follows:

- The National Institute of Standards and Technology (NIST) of the United States has provided data produced by the University of Western Ontario on 37 different configurations of low-rise gable-roof buildings (no eaves). These include both internal and external pressures. Two configurations included roof parapets and a shape representing the full-scale building of the Texas Tech Field Experiment (Figure 8.3). The data are presented as time histories of local pressures and are available for further processing.
- The Wind Engineering Group of the Tokyo Polytechnic University has provided data on 116 configurations of low-rise buildings with gable hipped and flat roofs, both with and without eaves. The data are presented in a summarized form for local and area-averaged external pressures, as contour plots and graphs. In some cases, time histories of fluctuating pressures are also provided as.MAT files. The URL for this database is: http://www.wind.arch.t-kougei.ac.jp/info_center/windpressure/lowrise/mainpage.html

8.11 OTHER SOURCES

For further information on the wind loading of low-rise buildings, the reader is directed to omprehensive reviews of wind loads on low-rise buildings by Holmes (1983), Stathopoulos (1984, 1995), Krishna (1995) and Surry (1999). More recent research can be found in papers in journals such as the *Journal of Wind Engineering and Industrial Aerodynamics* (Elsevier), *Wind and Structures* (Techno Press) and the *Journal of Structural Engineering* (ASCE).

8.12 SUMMARY

This chapter has discussed various aspects of the design of low buildings for wind loads. The long history of investigation into wind loads has been discussed, and the use of the modern boundary-layer wind tunnel for determination of design loading coefficients is covered. The characteristics of loads for major structural members and foundations, and for local cladding have been considered for buildings with flat, and pitched roofs. The effect of shelter and interference from surrounding buildings has been considered. Multi-span building configurations and parapets have also been discussed.

An example of a building in which architectural features have been used to minimize the aerodynamic effects of severe winds has been shown, and two web-based databases providing comprehensive information on aerodynamic shape factors and pressure coefficients for generic low-rise building shapes have also been highlighted.

REFERENCES

American Society of Civil Engineers. (1936) Wind-bracing in steel buildings. Fifth Progress Report of Sub-Committee No. 31. *Proceedings ASCE*, March 1936, pp. 397–412.

Baker, C.J. (1999) Aspects of the use of the technique of orthogonal decomposition of surface pressure fields. *10th International Conference on Wind Engineering*, Copenhagen, 21–24 June, pp. 393–400, A.A. Balkema, Rotterdam.

Bailey, A. and Vincent, N.D.G. (1943) Wind pressure on buildings including the effects of adjacent buildings. *Journal of the Institution of Civil Engineers*, 2: 243–275.

Best, R.J. and Holmes, J.D. (1983) Use of eigenvalues in the covariance integration method for determination of wind load effects. *Journal of Wind Engineering and Industrial Aerodynamics*, 13: 359–370.

Bitsuamlak, G.T., Warsido, W., Ledesma, E., and Chowdhury, A.G. (2013) Aerodynamic mitigation of roof and wall corner suctions. *Journal of Engineering Mechanics, ASCE*, 139: 396–408.

Bienkiewicz, B., Ham, H.J. and Sun, Y. (1993) Proper orthogonal decomposition of roof pressure. *Journal of Wind Engineering and Industrial Aerodynamics*, 50: 193–202.

Davenport, A.G. Surry, D. and Stathopoulos, T. (1977) Wind loads on low-rise buildings. Final report of Phases I and II. University of Western Ontario, Boundary Layer Wind tunnel Report BLWT-SS8-1977.

Department of Public Works, Queensland. (2006) Design guidelines for Queensland public cyclone shelters. Prepared by Mullins Consulting for the Queensland Government, September.

Dryden, H.L. and Hill, G.C. (1931) Wind pressures on a model of a mill building. *Journal of Research of the National Bureau of Standards*, 6: 735–755 (Research Paper RP301).

Eaton, K.J. and Mayne, J.R. (1975) The measurement of wind pressures on two-storey houses at Aylesbury. *Journal of Industrial Aerodynamics*, 1: 67–109.

Eaton, K.J., Mayne, J.R. and Cook, N.J. (1975) Wind loads on low-rise buildings – effects of roof geometry. *Fourth International Conference on Wind Effects on Buildings and Structures*, London, September.

Flachsbart, O. (1932) Winddrucke auf geschlossene und offene Gebaude. In *Ergebnisse der Aerodynamischen Versuchanstalt zu Gottingen, IV. Lieferung.* eds. L. Prandtl and A. Betz. Verlag von R. Oldenbourg, Munich and Berlin.

Ginger, J.D. (2004) Fluctuating wind loads across gable-end buildings with planar and curved roofs. *Wind and Structures*, 7: 359–372.

Ginger, J.D. and Holmes, J.D. (2003) Effect of building length on wind loads on low-rise buildings with a steep roof pitch. *Journal of Wind Engineering and Industrial Aerodynamics*, 91: 1377–1400.

Ho, T.C.E., Surry, D. and Davenport, A.G. (1990) The variability of low building wind loads due to surrounding obstructions. *Journal of Wind Engineering and Industrial Aerodynamics*, 36: 161–170.

Ho, T.C.E., Surry, D. and Davenport, A.G. (1991) Variability of low building wind loads due to surroundings. *Journal of Wind Engineering and Industrial Aerodynamics*, 38: 297–310.

Ho, T.C.E., Davenport, A.G. and Surry, D. (1995) Characteristic pressure distribution shapes and load repetitions for the wind loading of low building roof panels. *Journal of Wind Engineering and Industrial Aerodynamics*, 57: 261–279.

Holmes, J.D. (1981) Non-Gaussian characteristics of wind pressure fluctuations. *Journal of Wind Engineering and Industrial Aerodynamics*, 7: 103–108.

Holmes, J.D. (1983) *Wind loads on low rise buildings - A review.* CSIRO, Division of Building Research, Highett, Victoria.

Holmes, J.D. (1990a) Analysis and synthesis of pressure fluctuations on bluff bodies using eigenvectors. *Journal of Wind Engineering and Industrial Aerodynamics*, 33: 219–230.

Holmes, J.D. (1990b) Wind loading of multi-span buildings. *Civil Engineering Transactions, Institution of Engineers, Australia*, CE32: 93–98.

Holmes, J.D. (1994) Wind pressures on tropical housing. *Journal of Wind Engineering and Industrial Aerodynamics*, 53: 105–123.

Holmes, J.D., and Carpenter, P. (1990) The effect of Jensen Number variations on the wind loads on a low-rise building. *Journal of Wind Engineering and Industrial Aerodynamics*, 36: 1279–1288.

Holmes, J.D., and Syme, M.J. (1994) Wind loads on steel-framed low-rise buildings. *Steel Construction (Australian Institute of Steel Construction)*, 28: 2–12.

Holmes, J.D., Sankaran, R., Kwok, K.C.S. and Syme, M.J. (1997) Eigenvector modes of fluctuating pressures on low-rise building models. *Journal of Wind Engineering and Industrial Aerodynamics*, 69–71: 697–707.

Holmes, J.D., Ginger, J.D. and Mullins, P. (2019) Aerodynamically-shaped cyclone shelters for Queensland. *Australian Journal of Civil Engineering*, 17: 188–194. doi: 10.1080/14488353.2019.1664227.

Hussain, M. and Lee, B.E. (1980) A wind tunnel study of the mean pressures acting on large groups of low-rise buildings. *Journal of Wind Engineering and Industrial Aerodynamics*, 6: 207–225.

Irminger, J.O.V. (1894) Nogle forsog over trykforholdene paa planer og legemer paavirkede af luftstrominger. *Ingenioren*, 17.

Irminger, J.O.V. and Nokkentved, C. (1930) Wind pressures on buildings. *Ingeniorvidenskabelige Skrifter*, A23.

Jensen, M. (1958) The model law for phenomena in the natural wind. *Ingenioren*, 2: 121–128.

Jensen, M, and Franck, N. (1965) *Model-scale tests in turbulent wind. Part II.* Danish Technical Press, Copenhagen.

Kernot, W.C. (1893) Wind pressure. *Proceedings, Australasian Association for the Advancement of Science*, Adelaide, Australia, V: 573–581, and VI: 741–745.

Krishna, P. (1995) Wind loads on low rise buildings - A review. *Journal of Wind Engineering and Industrial Aerodynamics*, 55: 383–396

Kopp, G.A., Surry, D. and Mans, C. (2005a) Wind effects of parapets on low buildings: part 1. Basic aerodynamics and local loads. *Journal of Wind Engineering and Industrial Aerodynamics*, 93: 817–841.

Kopp, G.A., Mans, C. and Surry, D. (2005b) Wind effects of parapets on low buildings: part 2. Structural loads. *Journal of Wind Engineering and Industrial Aerodynamics*, 93: 843–855.

Lee, B.E. and Soliman, B.F. (1977) An investigation of the forces on three-dimensional bluff bodies in rough wall turbulent boundary layers. *Journal of Fluids Engineering*, 99: 503–510.

Letchford, C.W. and Mehta, K.C. (1993) The distribution and correlation of fluctuating pressures on the Texas Tech Building. *Journal of Wind Engineering and Industrial Aerodynamics*, 50: 225–234.

Levitan, M.L. and Mehta, K.C. (1992a) Texas Tech field experiments for wind loads. Part I. Building and pressure measuring system. *Journal of Wind Engineering and Industrial Aerodynamics*, 43: 1565–1576.

Levitan, M.L. and Mehta, K.C. (1992b) Texas Tech field experiments for wind loads. Part II. Meteorological instrumentation and terrain parameters. *Journal of Wind Engineering and Industrial Aerodynamics*, 43: 1577–1588.

Meecham, D., Surry, D. and Davenport, A.G. (1991) The magnitude and distribution of wind-induced pressures on hip and gable roofs. *Journal of Wind Engineering and Industrial Aerodynamics*, 38: 257–272.

Mehta, K.C., Levitan, M.L., Iverson, R.E. and Macdonald, J.R. (1992) Roof corner pressures measured in the field on a low-rise building. *Journal of Wind Engineering and Industrial Aerodynamics*, 41: 181–192.

Richardson, E.B. and Miller, B.H. (1932) The experimental determination of the pressures and distribution of pressures of an airstream on model buildings. *Journal of the Institution of Engineers Australia*, 4: 277–282.

Robertson, A.P. (1992) The wind-induced response of a full-scale portal framed building. *Journal of Wind Engineering and Industrial Aerodynamics*, 43: 1677–1688.

Saathoff, P. and Stathopoulos, T. (1992) Wind loads on buildings with sawtooth roofs. *ASCE Journal of Structural Engineering*, 118: 429–446.

Sill, B.L., Cook, N.J. and Blackmore, P.A. (1989) IAWE Aylesbury Comparative Experiment – preliminary results of wind-tunnel comparisons. *Journal of Wind Engineering and Industrial Aerodynamics*, 32: 285–302.

Sill, B.L., Cook, N.J. and Fang, C. (1992) The Aylesbury Comparative Experiment - a final report. *Journal of Wind Engineering and Industrial Aerodynamics*, 43: 1553–1564.

Simiu, E. and Scanlan, R.H. (1996) *Wind effects on structures – an introduction to wind engineering*, 3rd Edition. John Wiley, New York

Standards Australia (1989) *SAA loading code. Part 2: wind loads*. Standards Australia, North Sydney. Australian Standard, AS1170.2-1989.

Standards Australia (2011) *Structural design actions. Part 2: wind actions*. Standards Australia, Sydney. Australian/New Zealand Standard, AS/NZS1170.2: 2011.

Stathopoulos, T. (1984) Wind loads on low-rise buildings: a review of the state of the art. *Engineering Structures*, 6: 119–135.

Stathopoulos, T. (1995) Evaluation of wind loads on low buildings – a brief historical review. In *A state of the art in wind engineering*. ed P. Krishna. Wiley Eastern Limited, New Delhi.

Stathopoulos, T. and Baskaran, A. (1988) Turbulent wind loading on roofs with parapet configurations. *Canadian Journal of Civil Engineering*, 29: 570–578.

Stathopoulos, T. and Saathoff, P. (1994) Codification of wind-pressure coefficients for multispan gable roofs. *ASCE Journal of Structural Engineering*, 120: 2495–2519.

Surry, D. (1999) Wind loads on low-rise buildings : past, present and future. *10th International Conference on Wind Engineering*, Copenhagen, Denmark, 21–24 June, pp. 105–114, A.A. Balkema, Rotterdam.

Xu, Y.L. and Reardon, G.F. (1998) Variations of wind pressure on hip roofs with roof pitch. *Journal of Wind Engineering and Industrial Aerodynamics*, 73: 267–284.

Chapter 9

Tall buildings

9.1 INTRODUCTION

Tall buildings, now up to 830 m in height, project well into the atmospheric boundary layer, and their upper levels may experience the highest winds of large-scale windstorms, such as tropical cyclones or the winter gales of the temperate regions. Resonant dynamic response in along-wind, cross-wind and torsional modes are a feature of the overall structural loads experienced by these structures. Extreme local cladding pressures may be experienced on their side walls.

The post-World War II generation of high-rise buildings were the stimulus for the development of the boundary-layer wind tunnel, which remains the most important tool for the establishment of design wind loads on major building projects in many countries.

In this chapter, the history of investigations into wind loading of tall buildings, the major response mechanisms and phenomena, and the available analytical and semi-analytical techniques, will be discussed.

9.2 HISTORICAL

Tall buildings, or 'skyscrapers', are amongst the more wind-sensitive of structures, and it was inevitable that their response to wind would be of concern to structural engineers, and attract the interest of early experimenters, both in the wind tunnel and in full scale.

Dryden and Hill (1926) conducted pioneering wind-tunnel study of pressures on a prismatic shape representative of a tall building with a square planform (height to breadth ratio of about 3); the tests were carried out in smooth uniform flow, but some comment was made on the possible effects of turbulence.

The Empire State Building, at 380 m, was the tallest building in the world for 40 years, and was the subject of three significant studies in the 1930s (Coyle, 1931; Dryden and Hill, 1933; Rathbun, 1940). These studies have been re-appraised in some detail by Davenport (1975).

Coyle (1931) used a portable horizontal pendulum to record the motion of the building. This clearly revealed resonant dynamic response with a period of around 8 seconds. Rathbun's (1940) extensive full-scale measurements were described by Davenport as: "a monumental piece of full-scale experimentation" (Davenport, 1975, p. 35). Wind pressures on three floors of the building were measured with 30 manometers and 28 flash cameras. The pressure coefficients showed considerable scatter, but were clearly much lower than those obtained by Dryden and Hill (1933) on a wind-tunnel model in a uniform flow some years earlier. Rathbun also performed deflection measurements on the Empire State Building using a plumb bob extending from the 86th floor to the 6th floor. These results (as re-analysed by Davenport) indicated the significantly different stiffness of the building in the east-west direction in comparison with the north-south direction (Figure 9.1)

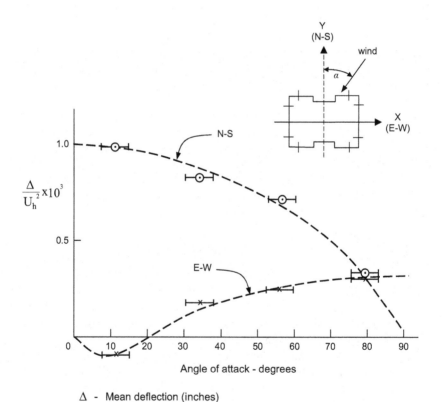

Δ - Mean deflection (inches)
U_h - Mean wind speed at 1250 feet MPH (uncorrected)

Figure 9.1 Full-scale measurements of mean deflection on the Empire State Building by Rathbun (1940) – reanalysed by Davenport (1975).

In the 1960s and 1970s, a resurgence in the building of skyscrapers occurred – particularly in North America, Japan and Australia. There was great interest in wind loads on tall buildings at this time – this has continued to the end of the twentieth century. The two main problem areas to emerge were:

- The vulnerability of glazed cladding to both direct wind pressures, and flying debris in windstorms,
- Serviceability problems arising from excessive motion near the top of tall buildings.

From the early 1970s, many new building proposals were tested in the new boundary-layer wind tunnels (see Chapter 7), and quite a few full-scale monitoring programmes were commenced.

One of the most comprehensive and well-documented full-scale measurement studies, with several aspects to it, which lasted for most of the 1970s, was that on the 239 m tall Commerce Court building in Toronto, Canada (Dalgleish, 1975; Dalgleish *et al.*, 1979; Dalgleish *et al.*, 1983). The full-scale studies were supplemented with wind-tunnel studies, both in the design stage (Davenport *et al.*, 1969) and later on a pressure model (Dalgleish *et al.*, 1979), and a multi-degree-of-freedom aeroelastic model, in parallel with the full-scale studies (Templin and Cooper, 1981; Dalgleish *et al.*, 1983).

The early full-scale pressure measurements on the Commerce Court building showed good agreement with the wind-tunnel study (at 1/400 scale) for mean pressure coefficients, and for the mean base shear and overturning moment coefficients. Not as good agreement with the 1/400 scale wind-tunnel tests, was found for the r.m.s. fluctuating pressure coefficients for some wind directions (Dalgleish, 1975). The later reported pressure measurements (Dalgleish *et al.*, 1979) showed better agreement for the fluctuating pressure and peak measurements on a larger (1/200) scale wind-tunnel model, with accurately calibrated tubing and pressure measurement system. The full-scale pressure study on Commerce Court highlighted the importance of short duration peak pressures in separated flow regions (at around this time similar observations were being made from the roof of the low-rise building at Aylesbury – Section 8.2.2). Subsequently, detailed statistical studies of these were carried out for application to glass loading (see Section 9.4.5). Although the Commerce Court pressure measurements were of high quality, they suffered from the lack of an independent reference pressure for the pressure coefficients – an internal pressure reading from the building was used. For comparison of mean pressure coefficients with the wind-tunnel results, it was necessary to force agreement at one pressure tapping – usually in wake region.

The full-scale study of acceleration response (Dalgleish et al., 1983) showed the following features:

- the significance of the torsional (twisting) motions superimposed on the sway motions for one direction (E–W) – i.e. a 'coupled' mode. This was explained by an eccentricity in the north-south direction between the centre of mass, and the elastic axis,
- generally good agreement between the final aeroelastic model, which included torsional motions, and the full-scale data, for winds from a range of directions,
- reasonable agreement between the full-scale data and predictions of the National Building Code of Canada for along- and cross-wind accelerations.

The agreements observed occurred despite some uncertainties in the reference velocity measured at the top of the building, and in the dynamic properties (frequency and damping) of the building. An interesting observation, not clearly explained, was a clear decrease in observed building frequency as the mean speed increased.

Another important full-scale study, significant for its influence on the development of the British Code of Practice for Wind Loads, was carried out on the 18-storey Royex House in London (Newberry *et al.*, 1967). This study revealed aspects of the transient and fluctuating pressures on the windward and side walls.

The first major boundary-layer wind-tunnel study of a tall building was carried out for the twin towers of the World Trade Center, New York, in the mid-1960s, at Colorado State University. This was the first of many commercial studies, now numbering in the thousands, in boundary-layer wind tunnels.

9.3 FLOW AROUND TALL BUILDINGS

Tall buildings are bluff bodies of medium to high aspect ratio, and the basic characteristics of flow around this type of body were covered in some detail in Chapter 4.

Figure 9.2 shows the general characteristics of boundary-layer wind flow around a tall building. On the windward face, there is a strong downward flow below the stagnation point, which occurs at a height of 70%–80% of the overall building height. The down flow can often cause problems at the base, as high velocity air from upper levels is brought down to street level. Separation and re-attachment at the side walls are associated with high local pressures. The rear face is a negative pressure region of lower magnitude mean pressures, and a low level of fluctuating pressures.

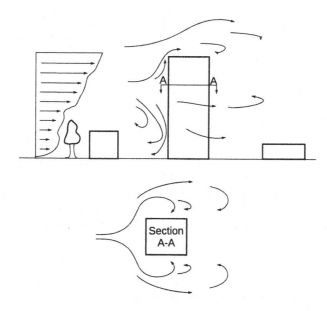

Figure 9.2 Wind flow around a tall building.

In a mixed extreme wind climate of thunderstorm microbursts (Section 1.3.5) and synoptic winds, the dominant wind for wind loading of tall buildings will normally be the latter, as the microburst profile has a maximum at a height of 50–100 m (Figure 3.6).

9.4 CLADDING PRESSURES

9.4.1 Pressure coefficients

As in previous chapters, pressure coefficients in this chapter will be defined with respect to a mean wind speed at the top of the building, denoted by $\bar{U}_h$. Thus, the mean, root-mean-square fluctuating (standard deviation), maximum and minimum pressure coefficients are defined according to Equations (9.1), (9.2), (9.3) and (9.4), respectively.

$$\bar{C}_p = \frac{\bar{p} - p_o}{\frac{1}{2}\rho_a \bar{U}_h^2} \tag{9.1}$$

$$C_p' = \sigma_{Cp} = \frac{\sqrt{\overline{p'^2}}}{\frac{1}{2}\rho_a \bar{U}_h^2} \tag{9.2}$$

$$\hat{C}_p = \frac{\hat{p} - p_o}{\frac{1}{2}\rho_a \bar{U}_h^2} \tag{9.3}$$

$$\check{C}_p = \frac{\check{p} - p_o}{\frac{1}{2}\rho_a \bar{U}_h^2} \tag{9.4}$$

In Equations (9.3) and (9.4), the maximum and minimum pressures, $\hat{p}$ and $\check{p}$, are normally defined as the average or expected peak pressure at a point in a given averaging time, which may be taken as a period between 10 minutes and 3 hours in full scale. It is not usually convenient, or economic, to measure such average peaks directly in wind-tunnel tests, and various alternative statistical procedures have been proposed. These are discussed in Section 9.4.4.

9.4.2 Pressure distributions on buildings of rectangular prismatic shape

The local pressures on the wall of a tall building can be used directly for the design of cladding, which is generally supported over small tributary areas.

Figure 4.15 shows the distribution of mean pressure coefficient on the faces of tall prismatic shape, representative of a very tall building, with aspect ratio (height/width) of 8, in a boundary-layer flow.

Figures 9.3–9.5 show the variation in mean, maximum and minimum pressure coefficients on the windward, side and leeward faces, for a lower

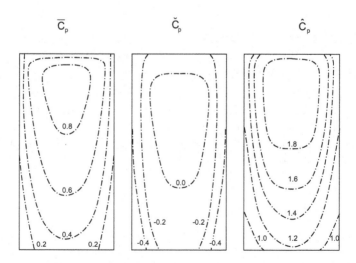

Figure 9.3 Mean, maximum and minimum pressure coefficients – windward wall of a building with square cross-section – height/width=2.1 (Cheung, 1984).

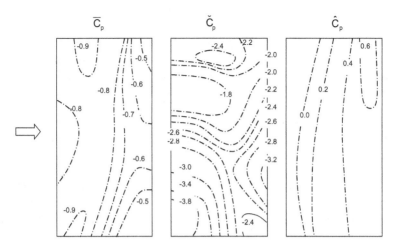

Figure 9.4 Mean, maximum and minimum pressure coefficients – side wall of a building with square cross-section – height/width=2.1 (Cheung, 1984).

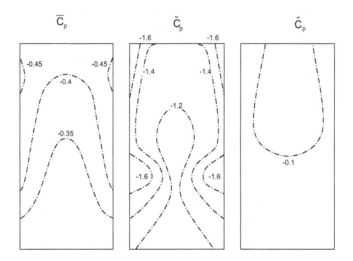

Figure 9.5 Mean, maximum and minimum pressure coefficients – leeward wall of a building with square cross-section – height/width=2.1 (Cheung, 1984).

building of square cross-section, with aspect ratio equal to 2.1 (Cheung, 1984). The pressures were measured on a wind-tunnel models which represented a building of 85 m height; the building is isolated, that is there is no shielding from buildings of comparable height, and the approaching flow was boundary-layer flow over suburban terrain. The value of Jensen Number, h/z_o (see Section 4.4.5), was then approximately 40.

Figure 9.3 shows a stagnation point on the windward face, where the value of $\bar{C}_p$ reaches a maximum, at about $0.8h$. The heights for largest maximum pressure coefficient are slightly lower than this.

The side walls (Figure 9.4) are adjacent to a flow which is separating from the front wall, and generating strong vortices (see Figures 4.1 and 9.2). The mean pressure coefficients are generally in the range from -0.6 to -0.8, and not dissimilar to the values on the much taller building in Figure 4.15. The largest magnitude minimum pressure coefficients of about -3.8 occur near the base of the buildings, unlike the windward wall pressures. A wind direction parallel to the side wall produces the largest magnitude negative pressures in this case.

The mean and largest peak pressures on the leeward wall (Figure 9.5) is also negative, but are typically half the magnitude of the side wall pressures. This wall is of course sheltered, and exposed to relatively slowly moving air in the near wake of the building.

9.4.3 The nature of fluctuating local pressures and probability distributions

As discussed in Section 9.2, in the 1970s, full-scale and wind-tunnel measurements of wind pressures on tall buildings, highlighted the local peak negative pressures, that can occur, for some wind directions, on the walls of tall buildings, particularly on side walls at locations near windward corners, and on leeward walls. These high pressures generally only occur for quite short periods of time, and may be very intermittent in nature. An example of the intermittent nature of these pressure fluctuations is shown in Figure 9.6 (from Dalgleish, 1971).

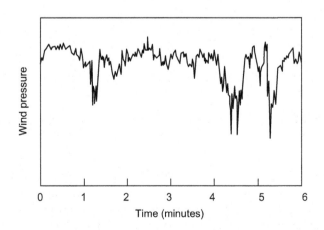

Figure 9.6 Record of fluctuating pressure from the leeward wall of a full-scale office building (Dalgleish, 1971).

Several studies (e.g. Dalgleish, 1971; Peterka and Cermak, 1975) indicated that the probability densities of pressure fluctuations in separated flow regions on tall buildings were not well-fitted by the normal or Gaussian probability distribution (Appendix C3.1). This is the case, even though the latter is a good fit to the turbulent velocity fluctuations in the wind (see Section 3.3.2). The 'spiky' nature of local pressure fluctuations (Figure 9.6) results in probability densities of peaks of five standard deviations, or greater, below the mean pressure, being several times greater than that predicted by the Gaussian distribution. This is illustrated in Figure 9.7, derived from wind-tunnel tests of two tall buildings (Peterka and Cermak, 1975).

A consequence of the intermittency and non-Gaussian nature of pressure fluctuations on tall buildings, is that the maximum pressure coefficient measured at a particular location on a building in a defined time period – say 10 minutes in full scale – may vary considerably from one time period to the next. Therefore, they cannot be predicted by knowing the mean and standard deviation, as is the case with a Gaussian random process (Davenport, 1964). This has led to a number of different statistical techniques being adopted to produce more consistent definitions of peak pressures for design – these are discussed in Section 9.4.4. A related matter is the response characteristics of glass cladding to short duration peak loads. The latter aspect is discussed in the Section 9.4.5.

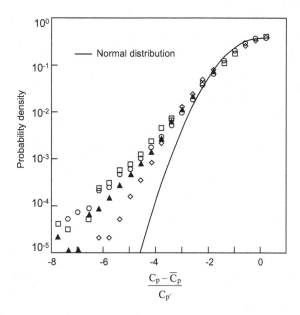

Figure 9.7 Probability densities of pressure fluctuations from regions in separated flow on tall buildings (Peterka and Cermak, 1975).

A detailed study (Surry and Djakovich, 1995) of local negative peak pressures on generic tall building models of constant cross-section, with four different corner geometries, indicated that the details of the corner geometry do not affect the general magnitude of the minimum pressure coefficients, but rather the wind direction at which they occur. The highest peaks were associated with vortex shedding.

9.4.4 Statistical methods for determination of peak local pressures

A simple approach, originally proposed by Lawson (1976), uses the parent probability distribution of the pressure fluctuations, from which a pressure coefficient, with a designated (low) probability of exceedance is extracted. The probability of exceedance is normally in the range from 1×10^{-4} to 5×10^{-4}, with the latter being suggested by Lawson. This method can be programmed 'on the run' in wind-tunnel tests, relatively easily; sometimes a standard probability distribution, such as the Weibull type (see Appendix C3.4) is used to fit the measured data and interpolate, or extrapolate, to the desired probability level.

Cook and Mayne (1979) proposed a method in which the total averaging time, T, is divided into 16 equal parts and the measured peak pressure coefficient (maximum or minimum) within each reduced time period, t, is retained. A Type I Extreme Value (Gumbel) distribution (see Section 2.2.1 and Appendix C.4) is fitted to the measured data, giving a mode, c_t, and slope, $(1/a_t)$. These can then be used to calculate the parameters of the Extreme Value Type I distribution appropriate to the maxima (or minima) for the original time period, T, as follows:

$$c_T = c_t + a_t \log_e 16 \tag{9.5}$$

$$a_T = a_t \tag{9.6}$$

Knowing the distribution of the extreme pressure coefficients, the expected peak, or any other percentile, can then be easily determined. The method proposed by Cook and Mayne (1979), in fact, proposes an effective peak pressure coefficient C_p^* given by Equation (9.7).

$$C_p^* = c_T + 1.4a_t \tag{9.7}$$

Equation (9.7) can be rewritten in terms of the mean and standard deviation of the extremes (Kasperski, 2003). For the Extreme Value Type I distribution, the mean and standard deviation are related to the mode, c_T, and scale factor, a_T, by Equations (9.8) and (9.9).

$$\text{mean} = m = u_T + 0.577a_T \tag{9.8}$$

standard deviation $= \sigma = (\pi/\sqrt{6})a_T = 1.282a_T$ $\qquad$ (9.9)

Hence, Equation (9.7) can be written as:

$C_p^* = u + 0.577a + 0.823a = m + 0.64(1.282a)$

$C_p^* = m + 0.64\sigma$ $\qquad$ (9.10)

Equation (9.10) can be used to calculate C_p^* by estimating m and σ from the measured extremes without having to fit a distribution. In fact, it does not require an assumption about the distribution of the extremes. An alternative, more conservative, form is Equation (9.11); this corresponds to Equation (9.7) with the 1.4 replaced by 1.5.

$C_p^* = m + 0.7\sigma$ $\qquad$ (9.11)

Peterka (1983) proposed the use of the probability distribution of 100 independent maxima within a time period equivalent to 1 hour, to determine C_p^*.

Another approach is to make use of level crossing statistics. Melbourne (1977) proposed the use of a normalized rate of crossing of levels of pressure (or structural response). A nominal rate of crossing (e.g. 10^{-4} per hour) was chosen to determine a nominal level of 'peak' pressure.

The parameters of the (Type I) extreme value distribution for the extreme pressure in a given time period can also be derived from level crossing rates as follows. The level crossings are assumed to be uncorrelated events which can be modelled by a Poisson distribution (Appendix C.3.5).

The Poisson distribution gives the probability for the number of events, n, in a given time period, T, when the average rate of occurrence of the events is v:

$$P(n, v) = \frac{(vT)^n}{n!} \exp(-vT) \qquad (9.12)$$

The 'event' in this case can be taken as an upcrossing of a particular level, e.g. the exceedence a particular pressure level. The probability of getting no crossings of a pressure level, p, during the time period, T, is also the probability that the largest value of the process $p(t)$, during the time period, is less than that level, i.e. the cumulative probability distribution of the largest value in the time period, T.

Thus,

$$F(p) = P(0, v) = \frac{(vT)^0}{0!} \exp(-vT) = \exp(-vT) \qquad (9.13)$$

If we assume that the average number of crossings of level x in time T, is given by:

$$\upsilon T = \exp\left[-\frac{1}{a}(p-u)\right] \qquad (9.14)$$

where a and u are constants, then,

$$F(p) = \exp\left\{-\exp\left[-\frac{1}{a}(p-u)\right]\right\} \qquad (9.15)$$

This is the Type I (Gumbel) extreme value distribution with a mode of u and a scale factor of a.

From Equation (9.14), taking natural logarithms of both sides,

$$\log_e(\upsilon T) = -\frac{1}{a}(p-u) \qquad (9.16)$$

The mode and scale factor of the Type I extreme value distribution of the process $p(t)$ can be estimated by the following procedure:

- Plot the natural logarithm of the number of up-crossings against the level, p
- Fit a straight line. From Equation (9.16), the slope is $(-1/a)$, and the intercept $(p=0)$ is (u/a)
- From these values, estimate u and a, the mode and scale factor of the Type I extreme value distribution of p.

The above method, and that of Cook and Mayne (1979), are based on the Type I (Gumbel) extreme value distribution. However, for local pressures on buildings, there is evidence (e.g. Holmes and Cochran, 2003) that the generalized extreme value distribution (Appendix C.4.1), with a small positive shape factor, may be more applicable to wind-tunnel data. In that case, the methods of Appendix G, based on 'peaks-over-threshold' data can be applied.

9.4.5 Strength characteristics of glass in relation to wind loads

Direct wind loading is a major design consideration in the design of glass and its fixing in tall buildings. However, the need to design for wind-generated flying debris (Section 1.5) – particularly roof gravel – in some cities, also needs to be considered (Minor, 1994).

As has been discussed, wind pressures on the surfaces of buildings fluctuate greatly with time, and it is known that the strength of glass is quite dependent on the duration of the loading. The interaction of these two phenomena results in a complex design problem.

The surfaces of glass panels are covered with flaws of various sizes and orientations. When these are exposed to tensile stresses, they grow at a rate dependent on the magnitude of the stress field, as well as relative humidity and temperature. The result is a strength reduction which is dependent on the magnitude and duration of the tensile stress. Drawing on earlier studies of this phenomenon, known as 'static fatigue', Brown (1972) proposed a formula for damage accumulation which has the form of Equation (9.17), at constant humidity and temperature.

$$D = \int_0^T \left[s(t) \right]^n dt \qquad (9.17)$$

where

D is the accumulated damage,
$s(t)$ is the time varying stress,
T is the time over which the glass is stressed,
n is a higher power (in the range of 12–20).

The expected damage, in time T, under a fluctuating wind pressure $p(t)$, in the vicinity of a critical flaw can be written as Equation (9.18).

$$E(D) = K \int_0^T E\left\{ \left[p(t) \right]^m \right\} dt \qquad (9.18)$$

where K is a constant, and m is a different power, usually lower than n, but dependent on the size and aspect ratio of the glass, which allows for the non-linear relationship between load and stress for glass plates due to membrane stresses (Calderone and Melbourne, 1993). $E\{\}$ is the expectation or averaging operation.

Calderone (1999), after extensive glass tests, found a power law relationship between maximum stress anywhere in a plate, and the applied pressure, for any given plate; this may be used to determine the value of m for that plate. Values fall in the range of 5–20.

The integral on the right-hand side of Equation (9.18) is T times the mth moment of the pressure fluctuation, so that:

$$E\{D\} = KT \left(\frac{1}{2} \rho \bar{U}^2 \right)^m \int_0^\infty C_p^m f_{Cp} \left(C_p \right) dC_p \qquad (9.19)$$

where $C_p(t)$ is the time-varying pressure coefficient, and $f_{Cp}(C_p)$ is the probability density function for C_p.

The integral in Equation (9.19) is proportional to the rate at which damage is accumulated in the glass panel. It can be evaluated from known or expected probability distributions (e.g. Holmes, 1985), or directly from wind-tunnel or full-scale pressure-time histories (Calderone and Melbourne, 1993).

The high weighting given to the pressure coefficient by the power, m, in Equation (9.19) means that the main contribution to glass damage comes from isolated peak pressures, which typically occur intermittently on the walls of tall buildings (see Figure 9.6).

An equivalent static pressure coefficient, C_{ps}, which corresponds to a constant pressure which gives the same rate of damage accumulation as a fluctuating pressure-time history, can be defined as:

$$C_{ps} = \left[\int_0^\infty C_p^m f_{Cp}\left(C_p\right) dC_p \right]^{1/m} \qquad (9.20)$$

For the structural design of glazing, it is necessary to relate the computed damage caused by wind action, to failure loads obtained in laboratory tests of glass panels. The damage integral (Equations (9.17) or (9.18)) can be used to compute the damage sustained by a glass panel under the ramp loading (i.e. increasing linearly with time), commonly used in laboratory testing. In these tests, failure typically occurs in about 1 minute.

Dalgleish (1979) defined an equivalent glass design coefficient, C_k, which, when multiplied by the reference dynamic pressure, $\left(\frac{1}{2}\rho_a\bar{U}^2\right)$, gives a pressure which produces the same damage in a 60-second ramp increase, as in a wind storm of specified duration.

Making use of Equations (9.19) and (9.20), it can be easily shown that for a statistically stationary (synoptic) windstorm of 1-hour duration:

$$C_k = \left[60(1+m) \right]^{1/m} C_{ps} \qquad (9.21)$$

Using typical values of m and typical probability distributions, it can be shown (Dalgleish, 1979; Holmes, 1985) that C_k is approximately equal to the expected peak pressure coefficient occurring during the hour of storm wind. This fortuitous result, which is insensitive to both the value of m and the probability distribution, means that measured peak pressure coefficients from wind-tunnel tests are valid for use in calculation of design loads, for comparison with 1-minute loads in glass design charts.

9.4.6 Effect of mullions and other elements on local wind pressures

Local elements that protrude from the walls of tall buildings such as mullions, and other architectural elements such as sunshades, have the potential to affect local wind pressures by disrupting local boundary layers on the building surface, or by initiating flow separations.

Standen *et al.* (1971) suggested that the height of surface features should be scaled in the ratio of $(Re_{model}/ Re_{full\ scale})^{4/5}$ to obtain the same ratio of surface element height to local surface boundary-layer thickness (Re is Reynolds Number – see Section 4.2.4). An alternative scaling approach would be to match the 'roughness' Reynolds Numbers for the elements, as often used for measurements on circular cylinders (see Section 7.4.4). These scaling methods usually result in distorted (i.e. greater) projection heights of the modelled mullions from those dictated by the overall model scaling.

This modelling approach of Standen *et al.* was used when comparing mean and root-mean-square fluctuating pressure coefficients on the windward wall, the model of a 43 m tall building (Holmes, 1973). In that case, the differences between the pressures on a smooth wall model and the model with ribs were insignificant.

However, for side-wall pressures significant differences are found, with greatly increased r.m.s. and peak negative pressure coefficients adjacent to the leading edges where flow separations occur (e.g. Stathopoulos and Zhu, 1990, based on tests on a 60 m tall building).

Hence, protruding elements such as mullions, if they are significant projections from a wall, should be modelled when conducting wind-tunnel model tests of tall buildings.

9.5 OVERALL LOADING AND DYNAMIC RESPONSE

In Chapter 5, the random, or spectral, approach to the along-wind response of tall structures was discussed. This approach is widely used for the prediction of the response of tall buildings in simplified forms in codes and standards (see Chapter 15). Dynamic response of a tall building in the along-wind direction is primarily produced by the buffeting by turbulent velocity fluctuations in the natural wind (Section 3.3). In the cross-wind direction, loading and dynamic response is generated by random vortex shedding (Section 4.6.3) – that is, it is a result of unsteady separating flow generated by the building itself, with a smaller contribution from cross-wind turbulence.

9.5.1 Correlation of along-wind pressures

Along-wind force fluctuations on a tall building are primarily generated by pressure fluctuations on the windward face. The magnitude of the overall

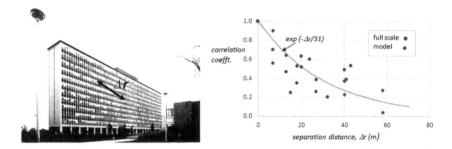

Figure 9.8 Correlation coefficients of fluctuating pressures as a function of separation distance, □r, on the windward face of a tall building.

along-wind buffeting forces depends upon the correlations of the pressure fluctuations across the windward face, as well as their magnitude.

Figure 9.8 shows the correlation coefficients (Section 3.3.5) for windward-wall pressure fluctuations on a large building, as a function of separation distance. Values from a full-scale building (shown in the figure), and from a wind-tunnel model of the building are shown. Shown are values of cross-correlation coefficients for zero-time lag. An exponential decay curve has been fitted to the data, with a 'best-fit' length scale of 31 m. This can be compared with the lateral length scale of about 40 m in the approaching turbulent flow (see Figure 3.8). These scales are similar, although there is some distortion of the turbulence as it approaches a bluff body like a large building.

9.5.2 General response characteristics

In this section, some general characteristics of the dynamic response of tall buildings to wind will be outlined.

By a dimensional analysis, or by application of the theory given in Section 5.3.1, it can be demonstrated (Davenport, 1966, 1971) that the root-mean-square fluctuating deflection at the top of a tall building of given geometry in a stationary (synoptic) wind, is given to a good approximation for the along-wind response by:

$$\frac{\sigma_x}{h} = A_x \left(\frac{\rho_a}{\rho_b} \right) \left(\frac{\bar{U}_h}{n_1 b} \right)^{k_x} \frac{1}{\sqrt{\eta}} \qquad (9.22)$$

and for the cross-wind response:

$$\frac{\sigma_y}{h} = A_y \left(\frac{\rho_a}{\rho_b} \right) \left(\frac{\bar{U}_h}{n_1 b} \right)^{k_y} \frac{1}{\sqrt{\eta}} \qquad (9.23)$$

where

h is the building height,
A_x, A_y are constants for a particular building shape,
ρ_a is the density of air,
ρ_b is an average building density,
$\bar{U}_h$ is the mean wind speed at the top of the building,
b is the building breadth,
k_x, k_y are exponents,
n_1 is the first mode natural frequency,
η is the critical damping ratio in the first mode of vibration.

Equations (9.22) and (9.23) are based on the assumption that the responses are dominated by the resonant components. For along-wind response, the background component is independent of the natural frequency. In the case of the cross-wind response, there is no mean component but some background contribution due to cross-wind turbulence. The assumption of dominance of resonance is valid for slender tall buildings with first mode natural frequencies less than about 0.5 Hz, and damping ratios less than about 0.02.

The equations illustrate that the fluctuating building deflection can be reduced by either increasing the building density or the damping. The damping term, η, includes aerodynamic damping as well as structural damping; however, this is normally small for tall buildings.

The term $\left(\bar{U}_h \, / \, n_1 b\right)$ is a non-dimensional mean wind speed, known as the reduced velocity. The exponent, k_x, for the fluctuating along-wind deflection is greater than 2, since the spectral density of the wind speed near the natural frequency, n_1, increases at a greater power than 2, as does the aerodynamic admittance function (Section 5.3.1 and Figure 5.4) at that frequency. The exponent for cross-wind deflection, k_y, is typically about 3, but can be as high as 4.

Figure 9.9 shows the variation of (σ_x/h) and (σ_x/h) with reduced velocity for a building of circular cross section (as well as the variation of $\bar{X}$).

9.5.3 Effect of building cross-section and corner modifications

In a study used to develop an optimum building shape for the U.S. Steel building, Pittsburgh, the response of six buildings of identical height and dynamic properties, but with different cross sections were investigated in a boundary-layer wind tunnel (Davenport, 1971). The probability distributions of the extreme responses in a typical synoptic wind climate was determined, and are shown plotted in Figure 9.10. The figure shows a range of 3:1 in the responses with a circular cross-section producing the lowest response,

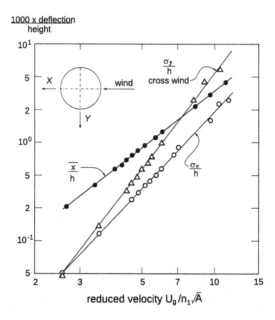

Figure 9.9 The mean and fluctuating response of a tall building of circular cross-section (Davenport, 1971).

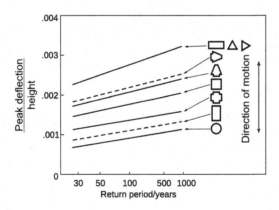

Figure 9.10 Effect of cross-sectional shape on maximum deflections of six buildings (Davenport, 1971).

and an equilateral triangular cross-section the highest. Deflection across the shortest (weakest) axis of a 2:1 rectangular cross-section was also large.

Slotted and chamfered corners on rectangular building cross-sections have significant effects on both along-wind and cross-wind dynamic responses to wind (Kwok and Bailey, 1987; Kwok, Wilhelm and Wilkie, 1988; Kwok, 1995). Chamfers of the order of 10% of the building width produce up to

40% reduction in the along-wind response and 30% reduction in the cross-wind response.

9.5.4 Prediction of cross-wind response

Along-wind response of isolated tall buildings can be predicted reasonably well from the turbulence properties in the approaching flow by applying the random vibration theory methods discussed in Section 5.3.1. Cross-wind response, however, is more difficult to predict, since vortex shedding plays a dominant role in the exciting forces in the cross-wind direction. However, an approach which has been quite successful, is the use of the high-frequency base balance technique to measure the spectral density of the generalized force in wind-tunnel tests (Section 7.6.2). Multiplication by the mechanical admittance and integration over frequency can then be performed to predict the building response.

Examples of generalized force spectra for buildings of square cross-section are shown in Figure 9.11. Non-dimensional spectra for three different height/breadth ratios are shown, and the approach flow is typical of suburban terrain. The mode shapes are assumed to be linear with height. The abscissa of this graph is reduced frequency – the reciprocal of reduced velocity.

For reduced velocities of practical importance (2–8), the non-dimensional spectra vary with reduced velocity to a power of 3–5, or with reduced frequency to a power of −3 to −5 (represented by the slope on the log-log plot). Such data has been incorporated in some standards and codes for design purposes (see Section 15.8).

9.5.5 Database for tall building loading and response

Most tall buildings of 100 m height, or greater, are the subject of special wind-tunnel tests at the later design stages; the techniques for this are well-developed and many of these are discussed in Section 7.6. However, the designer usually needs preliminary estimates of overall wind loading early in the design stage. As discussed in Chapter 15, several wind loading codes and standards contain methods for prediction of along-wind and cross-wind response. A more comprehensive alternative to codes and standards is provided by internet databases such as the one compiled by the Natural Hazards Modeling Laboratory of the University of Notre Dame (www.nd.edu/~nathaz/database). It provides information on the spectral densities of three components of base moment for 27 different building shapes, in two different terrain types. This information has been obtained from a high-frequency base balance (Section 7.6.2). By application of the random vibration, or spectral approach, described in Section 5.3, reasonable preliminary predictions of basic building responses, such as base bending moments and accelerations at the top of the building can be obtained. Subsequently,

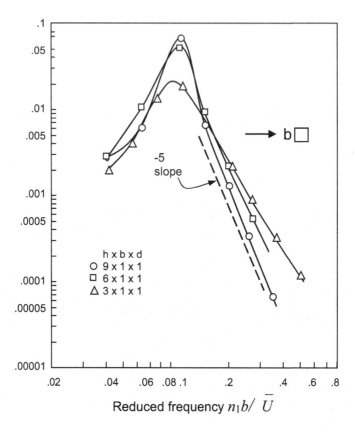

$$\text{Reduced frequency } n_1 b / \ \overline{U}$$

Figure 9.11 Cross-wind generalized force spectra for buildings of square cross-section (Saunders, 1974).

more accurate predictions can be obtained from specific wind-tunnel tests, allowing for accurate reproduction of the building shape, the surroundings, including shielding and interference effects from other buildings, and the mode shapes, including sway-twist coupling effects.

9.6 COMBINATION OF ALONG- AND CROSS-WIND RESPONSE

When dealing with the response of tall buildings to wind loading, the question arises: how should the responses in the along- and cross-wind directions be combined statistically? Since clearly the along-wind and cross-wind responses are occurring simultaneously on a structure it would be unconservative (and potentially dangerous!) to treat these as separate load cases. The question arises when applying those wind loading codes and standards which provide methods for calculating both along-wind and cross-wind dynamic response for tall buildings (see Chapter 15). It also arises when wind-tunnel tests are carried out using either aeroelastic (Section 7.6.1), or base-balance methods (Section 7.6.2). In these cases, predictions are usually provided for each wind direction, with respect to body- or building- axes rather than wind axes (see Section 4.2.2 and Figure 4.2). These axes are usually the two principal axes for sway of the building.

Two cases can be identified:

- 'scalar' combination rules for load effects
- 'vector' combination of responses

The former case is the more relevant case for structural load effects being designed for strength, as in most cases structural elements will 'feel' internal forces and stresses from both response directions, and will be developed in the following. The second case is relevant when axi-symmetric structures are under consideration, i.e. structures of circular cross-section such as chimneys.

Load effects (i.e. member forces and internal stresses) resulting from overall building response in two orthogonal directions (x- and y-) can very accurately be combined by use of Equation (9.24).

$$\hat{\varepsilon}_t = \overline{\varepsilon}_x + \overline{\varepsilon}_y + \sqrt{\left(\hat{\varepsilon}_x - \left| \overline{\varepsilon}_x \right| \right)^2 + \left(\hat{\varepsilon}_y - \left| \overline{\varepsilon}_y \right| \right)^2}$$

(9.24)

where

$\overline{\varepsilon}_t$ is total combined maximum peak load effect (e.g. the axial load in a column),

$\overline{\varepsilon}_x$ is the load effect derived from the mean response in the x-direction (usually derived from the mean base bending moment in that direction),

$\overline{\varepsilon}_y$ is the load effect derived from the mean response in the y-direction,

$\hat{\varepsilon}_x$ is the peak load effect derived from the response in the x-direction,

$\hat{\varepsilon}_y$ is the peak load effect derived from the response in the x-direction.

Equation (9.24) is quite an accurate one, as it is based on the combination of uncorrelated Gaussian random processes, for which it is exact. Most responses dominated by resonant contributions to wind, have been found to be very close to Gaussian, and if the two orthogonal sway frequencies are well-separated, the dynamic responses will be poorly correlated.

As an alternative approximation, the following load cases can be studied:

a. [Mean x-load+0.75(peak−mean)$_x$] with [mean y-load+0.75(peak−mean)$_y$]

b. [Mean x-load+(peak−mean)$_x$] with [mean y-load]
c. [Mean x-load] with [mean y-load+(peak−mean)$_y$]

The case (a) corresponds to the following approximation to Equation (9.24) for peak load effect:

$$\varepsilon_t = \bar{\varepsilon}_x + \bar{\varepsilon}_y + 0.75\left(\left(\hat{\varepsilon}_x - |\bar{\varepsilon}_x|\right) + \left(\hat{\varepsilon}_y - |\bar{\varepsilon}_y|\right)\right) \tag{9.25}$$

Equation (9.25) is a good approximation to Equation (9.24) for the range:

$$1/3 < \left(\hat{\varepsilon}_x - |\bar{\varepsilon}_x|\right)\Big/\left(\hat{\varepsilon}_y - |\bar{\varepsilon}_y|\right) \rightleftharpoons < 3$$

The other two cases (b) and (c) are intended to cover the cases outside this range, i.e. when $\left(\hat{\varepsilon}_x - |\bar{\varepsilon}_x|\right)$ is much larger than $\left(\hat{\varepsilon}_y - |\bar{\varepsilon}_y|\right)$, and vice-versa.

9.7 TORSIONAL LOADING AND RESPONSE

The significance of torsional components in the dynamic response of tall buildings was highlighted by the Commerce Court study of the 1970s (Section 9.2), when a building of a uniform rectangular cross-section experienced significant and measurable dynamic twist due to an eccentricity between the elastic and mass centres. Such a possibility had been overlooked in the original wind-tunnel testing. Now, when considering accelerations at the top of tall building, the possibility of torsional motions increasing the perceptible motions at the periphery of the cross-section may need to be considered.

There are two mechanisms for producing dynamic torque and torsional motions in tall buildings:

- Mean torque and torsional excitation resulting from non-uniform pressure distributions, or from non-symmetric cross-sectional geometries, and
- Torsional response resulting from sway motions through coupled mode shapes and/or eccentricities between elastic (shear) and geometric centres.

The first aspect was studied by Isyumov and Poole (1983), Lythe and Surry (1990), and Cheung and Melbourne (1992). Torsional response of tall buildings has been investigated both computationally making use of experimentally obtained dynamic pressure or force data from wind-tunnel models (Tallin and Ellingwood, 1985; Kareem, 1985), and experimentally on aeroelastic models with torsional degrees of freedom (Xu et al., 1992a; Beneke and Kwok, 1993; Zhang et al., 1993).

A mean torque coefficient, $\bar{C}_{Mz}$, can be defined as:

$$\bar{C}_{Mz} = \frac{\bar{M}_z}{\frac{1}{2}\rho_a \bar{U}_h^2 b_{max}^2 h} \tag{9.26}$$

where

$\bar{M}_z$ is the mean torque,
b_{max} is the maximum projected width of the cross-section,
h is the height of the building.

Lythe and Surry (1990), from wind-tunnel tests on 62 buildings, ranging from those with simple cross-sections to complex shapes, found an average value of $\bar{C}_{Mz}$, as defined above, of 0.085, with a standard deviation of 0.04. The highest values appear to be a function of the ratio of the minimum projected width, b_{min} to the maximum projected width, b_{max}, with a maximum value of $\bar{C}_{Mz}$ approaching 0.2, when (b_{min}/b_{max}) is equal to around 0.45 (Figure 9.12 from Cheung and Melbourne, 1992). The highest value of $\bar{C}_{Mz}$ for any section generally occurs when the mean wind direction is about 60°–80° from the normal to the widest building face.

Isyumov and Poole (1983) used simultaneous fluctuating pressures and pneumatic averaging (Section 7.5.2) on building models with a square or 2:1 rectangular cross-section in a wind tunnel, to determine the contribution to the fluctuating torque coefficient from various height levels on the

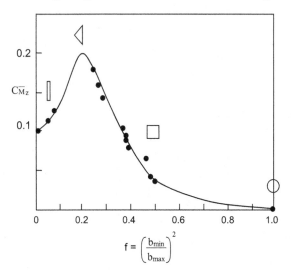

Figure 9.12 Mean torque coefficients on tall buildings of various cross-sections (Cheung and Melbourne, 1992).

buildings, and from the various building faces. The main contribution to the fluctuating torque on the square and rectangular section with the wind parallel to the long faces, came from pressures on the side faces, and could be predicted from the mean torque by quasi-steady assumptions (Section 4.6.2). On the other hand, for a mean wind direction parallel to the short walls of the rectangular cross-section, the main contribution was pressure fluctuations on the rear face, induced by vortex shedding.

A double peak in the torque spectra for the wind direction parallel to the long face of a 2:1 building has been attributed to buffeting by lateral turbulence, and by reattaching flow on to the side faces (Xu et al., 1992a). Measurements on an aeroelastic wind-tunnel tall building model designed only to respond torsionally (Xu et al., 1992a), indicated that aerodynamic damping effects (Section 5.5.1) for torsional motion of cross-section shapes characteristic of tall buildings are quite small in the range of design reduced velocities, in contrast to bridge decks. However, at higher reduced velocities, high torsional dynamic response and significant negative aerodynamic damping has been found for a triangular cross-section (Beneke and Kwok, 1993).

A small amount of eccentricity can increase both the mean twist angle and dynamic torsional response. For example, for a building with square cross-section, a shift of the elastic centre from the geometric and mass centre by 10% of the breadth of the cross-section, is sufficient to double the mean angle of twist and increase the dynamic twist by 40%–50% (Zhang et al., 1993).

9.8 INTERFERENCE EFFECTS

High-rise buildings are most commonly clustered together in groups – as office buildings grouped together in a city-centre business district, or in multiple building apartment developments, for example. The question of aerodynamic interference effects from other buildings of similar size on the structural loading and response of tall buildings arises.

9.8.1 Upwind building

A single similar upwind building on a building with square cross section and height/width (aspect) ratio of six produces increases of up to 30% in peak along-wind base moment, and 70% in cross-wind moment, at reduced velocities representative of design wind conditions in suburban approach terrain (Melbourne and Sharp, 1976). The maximum increases occur when the upwind building is 2–3 building widths to one side of a line taken upwind, and about 8 building widths upstream. Contours of percentage increase in peak cross-wind loading for square-section buildings with an aspect ratio of 4, are shown in Figure 9.13. It can be seen that reductions, i.e. shielding, occurs when the upstream building is within 4 building heights upstream and ±2 building heights to one side of the downstream building.

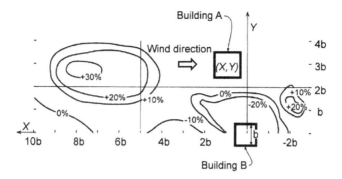

Figure 9.13 Percentage change in cross-wind response of a building (B) due to a similar building (A) at (*X*, *Y*) (Standards Australia, 1989).

The effect of increasing turbulence in the approach flow, i.e. increasing roughness lengths in the approach terrain, is to reduce the increases produced by interference.

The effect of increasing aspect ratio is to further increase the interference effects of upstream buildings, with increases of up to 80% being obtained, although this was for buildings with an aspect ratio of 9, and in relatively low turbulence conditions (Bailey and Kwok, 1985).

9.8.2 Downwind building

As shown in Figure 9.13, downwind buildings can also increase cross-wind loads on buildings if they are located in particular critical positions. In the case of the buildings of 4:1 aspect ratio of Figure 9.13, this is about one building width to the side, and two widths downwind.

More detailed reviews of interference effects on wind loads on tall buildings are given by Kwok (1995) and Khanduri *et al.* (1998). For a complex of tall buildings in the centres of large cities, wind-tunnel model tests (Chapter 7) will usually be carried out, and these should reveal any significant interference effects on new buildings such as those described in the previous paragraphs. Anticipated new construction should be included in the models when carrying out such tests. However, existing buildings may be subjected to unpredicted higher loads produced by new buildings of similar size at any time during their future life, and this should be considered by designers, when considering load factors.

9.8.3 Interference effects on local pressures

Adjacent buildings can also have dramatic effects on local cladding loads on tall buildings. An interesting, unusual example, based on wind-tunnel tests of a commercial high-rise development is shown in Figure 9.14 (Surry and

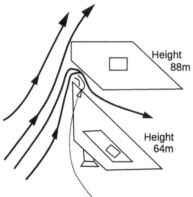

Wall region with very high local negative pressures.

Figure 9.14 Effect of a downwind building on local cladding pressures (Surry and Mallais, 1983). (Reproduced by permission of ASCE.)

Mallais, 1983). In this case, a downwind taller building resulted in an increase in the design local pressures by a factor of about three. It was explained by the presence of the adjacent building inducing re-attachment of the separated shear layers (see Section 4.1) on to the wall of the upwind building.

The large increases in local pressures caused by interference from an adjacent building in Figure 9.14 would have been accentuated by the non-rectangular cross sections of both buildings. For buildings of similar square cross sections and equal heights, spaced with a gap of half a building width, the maximum increase in the magnitude of the minimum pressure coefficients anywhere on the principal building, $\check{C}_p$, is about 30%. When the interfering building is twice the height of the principal building, the increase can reach 60%. However, maximum pressures on windward walls are invariably reduced by an adjacent building, and when the buildings are spaced apart by six times their width, the effects on both maximum and minimum pressures is negligible (Kim *et al.*, 2011).

9.9 DAMPING

The dynamic response of a tall building or other structure, to along-wind or cross-wind forces, depends on its ability to dissipate energy, known as 'damping'. Structural damping is derived from energy dissipation mechanisms within the material of the structure itself (i.e. steel, concrete etc.), or from friction at joints or from movement of partitions, etc.

For some large structures, the structural damping alone has been insufficient to limit the resonant dynamic motions to acceptable levels for serviceability considerations, and auxiliary dampers have been added. Three types of auxiliary damping devices will be discussed in this chapter: viscoelastic dampers (Section 9.9.2), tuned mass dampers (TMD, Section 9.9.3) and tuned liquid dampers (TLD, Section 9.9.4).

9.9.1 Structural damping

An extensive database of free vibration measurements from tall buildings in Japan has been collected (Tamura *et al.*, 2000). This database includes data on frequency as well as damping. More than 200 buildings were studied, although there is a shortage of values at larger heights – the tallest (steel encased) reinforced concrete building was about 170 m in height, and the highest steel-framed building was 280 m.

For reinforced concrete buildings, the Japanese study proposed the following empirical formula for the critical damping ratio in the first mode of vibration, for buildings less than 100 m in height, and for low-amplitude vibrations (drift ratio, (x_t/h) less than 2×10^{-5}):

$$\eta_1 \cong 0.014 n_1 + 470 \left(\frac{x_t}{h} \right) - 0.0018 \qquad (9.27)$$

where

n_1 is the first mode natural frequency,
x_t is the amplitude of vibration at the top of the building $(z=h)$.

The corresponding relationship for steel-framed buildings is:

$$\eta_1 \cong 0.013 n_1 + 400 \left(\frac{x_t}{h} \right) + 0.0029 \qquad (9.28)$$

The range of application for Equation (9.28) is stated to be: $h < 200$ m, and (x_t/h) less than 2×10^{-5}.

Equations (9.27) and (9.28) may be applied to tall buildings for serviceability limits states criteria (i.e. for the assessment of acceleration limits). Higher values are applicable for the high amplitudes appropriate to strength (ultimate) limits states, but unfortunately little, or no, measured data is available.

9.9.2 Visco-elastic dampers

Visco-elastic dampers incorporate visco-elastic material which dissipates energy as heat through shear stresses in the material. A typical damper, as shown in Figure 9.15a, consists of two visco-elastic layers bonded between three parallel plates (Mahmoodi, 1969). The force versus displacement

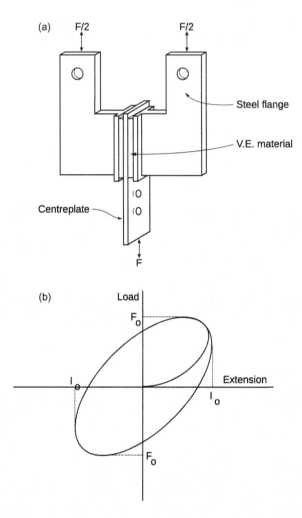

Figure 9.15 (a) A viscoelastic damper; (b) Hysteresis loop (Mahmoodi, 1969).

characteristic of such a damper forms a hysteresis loop as shown in Figure 9.15b. The enclosed area of the loop is a measure of the energy dissipated per cycle, and for a given damper, is dependent on the operating temperature and heat transfer to the adjacent structure (Mahmoodi and Keel, 1986).

The original World Trade Center buildings in New York City were the first major structures to utilize visco-elastic dampers (Mahmoodi, 1969). Approximately 10,000 dampers were installed in each 110-storey tower, with about 100 dampers at the ends of the floor trusses at each floor from the 7th to the 107th. More recently visco-elastic dampers have been installed in the 76-storey Columbia Seafirst Center Building, in Seattle, U.S.A. The dampers used in this building were significantly larger than those used at the World Trade Center, and only 260 were required to effectively reduce accelerations in the structure to acceptable levels (Skilling *et al.*, 1986; Keel and Mahmoodi, 1986).

A detailed review of the use of visco-elastic dampers in tall buildings has been given by Samali and Kwok (1995).

9.9.3 Tuned mass dampers

A relatively popular method of mitigating vibrations has been the tuned mass damper (TMD) or vibration absorber. Vibration energy is absorbed through the motion of an auxiliary or secondary mass connected to the main system by viscous dampers. The characteristics of a vibrating system with a TMD can be investigated by studying the two-degree-of-freedom system shown in Figure 9.16 (e.g. den Hartog, 1956; Vickery and Davenport, 1970).

TMD systems have successfully been installed in the Sydney Tower in Australia, the Citycorp Center, New York (275 m), the John Hancock Building, Boston, U.S.A. (60 storeys), and in the Chiba Port Tower in Japan (125 m). In the first and last of these, extensive full-scale measurements have been made to verify the effectiveness of the systems.

For the Sydney Tower, a 180-tonne doughnut-shaped water tank, located near the top of the Tower, and required by law for fire protection, was incorporated into the design of the TMD. The tank is 2.1 m deep and 2.1 m from inner to outer radius, weighs about 200 tonnes, and is suspended from the top radial members of the turret. Energy is dissipated in 8 shock absorbers attached tangentially to the tank and anchored to the turret wall. A 40-tonne secondary damper is installed lower down on the tower to further increase the damping, particularly in the second mode of vibration (Vickery and Davenport, 1970; Kwok, 1984).

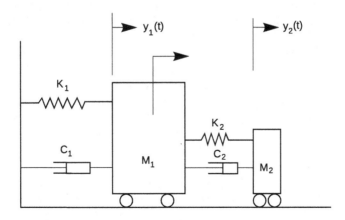

Figure 9.16 Two degree-of-freedom representation of a tuned mass damper.

The system installed in the Citycorp Center building, New York (McNamara, 1977), consists of a 400-tonne concrete mass riding on a thin oil film. The damper stiffness is provided by pneumatic springs, whose rate can be adjusted to match the building frequency. The energy absorption is provided by pneumatic shock absorbers, as for the Sydney Tower. The building was extensively wind-tunnel tested (Isyumov et al., 1975). The aeroelastic model tests included the evaluation of the TMD. The TMD was found to significantly reduce the wind-induced dynamic accelerations to acceptable levels. The effective damping of the model damper was found to be consistent with theoretical estimates of effective viscous damping based on the two-degree-of-freedom model (Vickery and Davenport, 1970).

TMD systems similar to those in the CityCorp Center have been installed in both the John Hancock Building, Boston, and in the Chiba Port Tower. In the case of the latter structure, the system has been installed to mitigate vibrations due to both wind (typhoon) and earthquake. Adjustable coil springs are used to restrain the moving mass, which is supported on frames sliding on rails in two orthogonal directions.

Taipei 101 in Taiwan, 508 m tall, has a pendulum type TMD, consisting of a 660-tonne spherical steel mass, suspended at a length to tune it to the building frequency of 0.14 Hz. It is claimed to reduce the accelerations at the top of the building by 30%–40%.

The performance of tuned mass dampers in tall buildings and towers under wind loading has been reviewed by Kwok and Samali (1995).

9.9.4 Tuned liquid dampers (TLD)

Tuned liquid dampers are relatively new devices in building and structures applications, although similar devices have been used in marine and aerospace applications for many years. They are similar in principle to the TMD, in that they provide a heavily damped auxiliary vibrating system attached to the main system. However, the mass, stiffness and damping components of the auxiliary system are all provided by moving liquid. The stiffness is in fact gravitational; the energy absorption comes from mechanisms such as viscous boundary layers, turbulence or wave breaking, depending on the type of system. Two categories of TLD will be discussed briefly here: tuned sloshing dampers (TSD) and tuned-liquid-column dampers (TLCD).

The TSD type (Figure 9.17) relies on the motion of shallow liquid in a rigid container for absorbing and dissipating vibrational energy (Fujino et al., 1988; Sun et al., 1989). Devices of this type have already been installed in at least two structures in Japan (Fujii et al., 1990), and on a television broadcasting tower in Australia.

Although a very simple system in concept, the physical mechanisms behind this type of damper are in fact quite complicated. Parametric studies of dampers with circular containers were carried out by Fujino et al. (1988). Some of their conclusions can be summarized as follows:

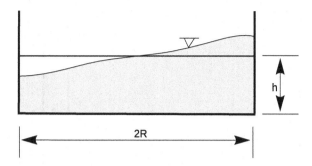

Figure 9.17 Tuned sloshing damper.

- Wave breaking is dominant mechanism for energy dissipation but not the only one.
- The additional damping produced by the damper is highly dependent on the amplitude of vibration.
- At small to moderate amplitudes, the damping achieved is sensitive to the frequency of sloshing of liquid in the container. For dampers with circular containers, the fundamental sloshing frequency is given by Equation (9.29):

$$n_s = (1 / 2\pi)\sqrt{[(1.84g_0 / R)\tanh(1.84h / R)]} \qquad (9.29)$$

where g_0 is the acceleration due to gravity, h is the height of the liquid and R is the radius of the container, as shown in Figure 9.17.

This formula is derived from linear potential theory of shallow waves.

- High-viscosity sloshing liquid is not necessarily desirable at high amplitudes of vibration, as wave breaking is inhibited. However, at low amplitudes, at which energy is dissipated in the boundary layers on the bottom and side walls of the container, there is an optimum viscosity for maximum effectiveness (Sun *et al.*, 1989).
- Roughening the container bottom does not improve the effectiveness because it has little effect on wave breaking.

The above conclusions were based on a limited number of free vibration tests with only two diameters of container. Further investigations are required, including the optimal size of T.S.D. for a given mass of sloshing liquid. However, the simplicity and low cost of this type of damper makes them very suitable for many types of structure.

Variations in the geometrical form are possible; for example, Modi *et al.* (1990) have examined T.S.D.s with torus ('doughnut') shaped containers.

The TLCD damper (Figure 9.18) comprises an auxiliary vibrating system consisting of a column of liquid moving in a tube-like container. The restoring

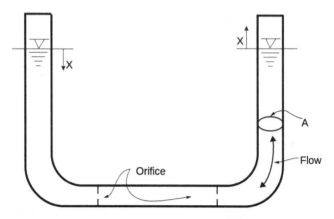

Figure 9.18 Tuned liquid-column damper.

force is provided by gravity, and energy dissipation is achieved at orifices installed in the container, (Sakai *et al.*, 1989, Hitchcock *et al.*, 1997a,b). The same principle has been utilized in anti-rolling tanks used in ships.

The TLCD, like the TSD, is simple and cheap to implement. Unlike the TSD, the theory of its operation is relatively simple and accurate. Sakai *et al.* (1989) have designed a TLCD system for the CityCorp Building, New York as a feasibility study; he found that the resulting damper was simpler, lighter and presumably cheaper than the TMD system actually used in this building (Section 9.9.3). Xu *et al.* (1992b) have examined theoretically the along-wind response of tall, multi-degree-of-freedom structures, with TMDs, TLCDs, and a hybrid damper – the tuned-liquid-column-mass damper (TLCMD). They found that the TMD and TLCD, with the same amount of added mass, achieved similar response reductions. The TLCMD, in which the mass of the container, as well as the liquid, is used as part of the auxiliary vibrating system, is less effective when the liquid column frequency is tuned to the same frequency as the whole damper frequency (with the water assumed to remain still). The performance of the latter is improved when the liquid column frequency is set higher than the whole damper frequency.

The effectiveness of tuned liquid structures in several tall structures in Japan has been reviewed by Tamura *et al.* (1995).

9.10 MOTION PERCEPTION AND ACCELERATION CRITERIA

The limiting of the wind-induced sway and twist motion of tall buildings to acceptable levels is an important part of the assessment of a tall building response to dynamic wind forces. In some cases, this assessment may result in the need for installation of auxiliary damping systems such as those described in Section 9.9.

The response of occupants to building vibration is a complex mixture of physiological and psychological factors. Excessive vibrations may affect the ability of an individual to perform manual tasks (e.g. Burton *et al.*, 2011). In extreme cases, some occupants may suffer motion sickness (Walton *et al.*, 2011).

Studies have generally accepted that humans respond primarily to fluctuating and peak acceleration of a building, and the most generally-accepted criterion is given in Annex D of ISO 10137 (ISO, 2009). This provides acceptability limits of peak horizontal wind-induced accelerations for a 1-year return period. Separate limits are given for residences and offices, with the latter being higher by 50% – i.e. working occupants of a building are deemed to be accepting higher motion than those of a residential building. The criteria also reduce with frequency for the typical range of first-mode building frequencies of 0.06–1.0 Hz. For example, for a frequency of 0.4 Hz, the limit on peak acceleration for office buildings is about 0.09 m/s^2 (approximately 9 milli-gs), and for residences, about 0.06 m/s^2 (about 6 milli-gs). At 1 Hz, the limits are 0.06 m/s^2 for offices, and 0.04 m/s^2 for residences. Note that these limits are defined for buildings only, and higher limits might be expected to apply to other structures such as observation towers, in which the occupancy is transient.

9.11 DIRECTIONALITY

When, on occasions, the predicted responses of a tall building to wind action from two or more different wind consultants have been compared, the principal source of differences has often turned out to be the treatment of the wind climate, and its interaction with the measured response to wind of the building in wind-tunnel tests. This involves the directional variability of the wind climate at the building site and its probabilistic treatment. Several methods have been developed to account for the direction effects – however, these give results that are not necessarily consistent with each other.

This topic has been discussed in detail elsewhere – particularly by Simiu and Miyata (2006), Isyumov *et al.* (2013), Holmes and Bekele (2015), Holmes (2020), and in Section 2.2.7 in this book; hence, a brief summary will only be given here. The methods for dealing with directionality can be classified into four categories:

- The 'out-crossing' approach (Davenport, 1977; Lepage and Irwin, 1985). This is based on the parent distributions of wind speed (see Section 2.5) by direction, and assumes that the extreme wind speeds, and hence extreme structural responses, can be predicted from probability distributions fitted to the parent population of wind speeds in each direction sector. Since these approaches are based on parent distributions of hourly, or ten-minute, mean wind speeds, it is clear that they cannot be applied to the many locations in the world that are dominated by thunderstorm-generated winds, such as downbursts (see Sections 1.3.5 and 3.2.7) that typically last only a few minutes. There are also many mathematical assumptions required to derive usable expressions for the rate of crossing of response boundaries. Recent comparisons with the multi-sector method and direct calculation (see following descriptions) suggests that, in many cases, the outcrossing method underpredicts the extreme responses, even when both methods are applied to a 'well-behaved' climate, subject only to synoptic extreme wind events (Holmes, 2020).

- Combined probability based on extreme wind speeds measured by direction sector. Two subsets of this approach are:
  a. The 'multi-sector' approach. Fitting of extreme value distributions to the responses in each direction sector and then computing a combined probability of non-exceedance by Equation (2.8) (Holmes, 1990, 2020; Holmes and Bekele, 2015). That is, Equation (2.8) is applied to a building response, instead of the wind speed itself. This method requires an assumption of independence between extreme wind speeds and responses in different direction sectors, and hence preferably angular increments between sectors that are not too small. However, use of smaller angular increments will lead to conservative predictions.
  b. Direct calculation. Forming a time series of annual extreme responses from the directional wind speed data, and directional response coefficients, and fitting a single extreme value probability distribution to the resulting 'one-dimensional' series of structural responses (e.g. Simiu and Filliben (1981), Simiu and Miyata (2006)).

- 'Sector-by-sector' approaches. These approaches attempt to produce directional wind speeds, or directional wind speed multipliers, for a discrete number of defined wind directions. Directional wind speeds, or multipliers, derived by these methods, are discussed in more detail in Section 2.2.6. When the directional wind speeds are combined with the measured structural response coefficients, the largest resulting response from any direction is deemed to be an appropriate design value. This approach is particularly appropriate for codes and standards, due to the requirement for a reduced level of computation; however, variants of it have also been adopted by several wind-tunnel testing groups. Within this category, at least three variants have been developed (e.g. Cook, (1983), Melbourne (1984), and Simiu and Filliben (2005)). These are referred to as Methods (b), (c) and (a), respectively, in Section 2.2.6. The method by Melbourne (1984) has been used since 1989 in the Australian and Australian/New Zealand

Standards for wind loading (Standards Australia, 1989 and 2011). The method of Cook (1983) was used in the British Standard (British Standards Institution, 1997).

The variations of this method were compared for a generic building at various orientations to a typical directional wind climate by Holmes and Bekele (2015).

- The 'storm passage' approach. This approach simulates the passage of wind storms across a building site using methods discussed in Section 2.2.4. This allows extreme structural responses from each storm to be extracted and fitted with appropriate distributions to allow predictions of extreme building responses to be made for any defined risk level. This method was described by Isyumov *et al.* (2003). According to this reference, the out-crossing approach, under-predicts peak load effects produced by tropical cyclones, such as hurricanes and typhoons. The storm passage approach appears to be an accurate one, but it is not yet available for less structured high-wind events such as extra-tropical cyclones or thunderstorms.

With such a variety of methods available, designers of tall buildings undergoing wind-tunnel testing, might be wise to request a sensitivity analysis of the results to the method used to account for directional variability when making extreme response predictions.

9.12 CASE STUDIES

Very many tall buildings have been studied in wind tunnels over several decades. These studies include the determination of the overall loading and response, cladding pressures, and other wind effects, such as environmental wind conditions at ground level. However, these studies are usually proprietary in nature, and not generally available. However, Willford (1985) has described a response study for the Hong Kong and Shanghai Bank Building, Hong Kong. A detailed wind engineering study for a building of intermediate height, including wind loading aspects, was presented by Surry et al. (1977).

Baskaran (1993) reviewed several early wind engineering studies of prominent tall buildings such as the original twin towers of the World Trade Center, New York City, the Sears Tower, Chicago (late renamed as 'Willis Tower'), the Bank of China, Hong Kong and the First National City Corporation (CityCorp) building, New York. The latter building required extensive strengthening of the bracing connections to resist winds from oblique wind directions, and is an example of the need to incorporate wind-tunnel testing (Isyumov *et al.*, 1975) early in the structural design process.

Despite the many hundreds of tall buildings that have been studied in boundary-layer wind tunnels over more than 40 years, very little 'benchmarking' has been carried out to promote confidence in structural designers that consistent and repeatable assessments of wind loads can be obtained for the same building by different groups in different facilities. An early attempt at this was the CAARC Standard Tall Building model that was studied in the 1970s – the results were summarized by Melbourne (1980). However, the measurements made, and techniques used, were relatively primitive compared to those currently available.

A more recent study, in which eight wind-tunnel laboratories participated, involved two 'benchmark' buildings (Holmes and Tse, 2014). This study was intended to compare the results of practitioners of the high-frequency base balance technique (Section 7.6.2), and base moments and accelerations. One building was 180 m tall and similar in dimensions to the CAARC building used in the 1970s, and mentioned in the previous paragraph. The second building, 240 m tall, was a modified version of an earlier building used as a benchmark for vibration control (Tse *et al.*, 2007). Both buildings had simple rectangular crosssections. The mean and peak base moments, including base torque, from the participating groups show quite good agreement with coefficients of variation about the averages of 8%–20% for the lower building and 20%–30% for the taller one. The peak corner accelerations from five groups show greater variation, especially for the lowest wind speed. The greater scatter in the acceleration predictions could be attributed to the dominance of the resonant components, which rely on accurate measurement of spectra of generalized forces at relatively high reduced frequencies requiring higher bandwidth in the instrumentation, a known difficulty with the high-frequency base balance technique.

Relatively few tall buildings have been studied in full scale for wind loads, although many have been studied for their basic dynamic properties (e.g. Tamura *et al.*, 2000). Case studies of wind-induced accelerations on medium height buildings were described by Wyatt and Best (1984) and Snaebjornsson and Reed (1991).

Numerous auxiliary damping systems have been installed in tall structures. A listing of over 50 cases up to the mid-1990s (Holmes, 1995), included installations in chimneys, airport control towers and bridge pylons, as well as tall buildings.

'Super-tall' buildings, above 400 m in height, have particular problems from a wind engineering perspective. Not the least of these is the lack of knowledge of wind and turbulence properties at these heights, especially as these structures are often being constructed in locations where 'well-behaved' large scale boundary-layer winds are not necessarily dominant (Irwin, 2007). The 830 m Burj Khalifa (formerly 'Burj Dubai'), completed in 2009 (Figure 9.19), required extensive wind engineering in its design process with extensive wind-tunnel testing at two different scales (Irwin and Baker, 2005), including an aeroelastic model test using the 'spine' technique (Section 7.6.4). The tapering and multiple changes of cross-section over its height are

Figure 9.19 The 830 m tall Burj Khalifa building, Dubai, UAE.

entirely a result of the need to limit the cross-wind, vortex-induced vibrations of the building. This building had extensive monitoring of its behaviour during construction, and this has continued since. The instrumentation includes the permanent installation of accelerometers, a weather station, and a GPS displacement measurement to continuously monitor the response of the structure to wind and seismic excitation (Abdelrazaq, 2011).

9.13 SUMMARY AND OTHER SOURCES

This chapter has discussed various aspects of the design of tall buildings for wind loads. The general characteristics of wind pressures on tall buildings, and local cladding loads have been considered. The special response characteristics of glass have been discussed. The overall response of tall buildings in along-wind and cross-wind directions, and in twist (torsion) has been covered. Aerodynamic interference effects, and the application of auxiliary damping systems to mitigate wind-induced vibration have been discussed.

Human perception of wind-induced tall building motions, and the several methods used to treat the varying response of building elements with wind direction, have also been discussed.

Discussion of several case studies includes the new generation of 'super-tall' buildings, whose design is dominated by wind engineering considerations.

Numerous studies of the aerodynamics of tall buildings have appeared in the international journals, *Journal of Wind Engineering and Industrial Aerodynamics* (Elsevier) and *Wind and Structures* (Techno Press). A review of the rapid development of wind engineering into the design of tall buildings up to 1990 was given by Baskaran (1993).

There also have been many publications and conference proceedings sponsored by the Council of Tall Buildings and Urban Habitat (CTBUH). A recent publication of the CTBUH focused on wind-tunnel testing techniques for tall buildings (Irwin *et al.*, 2013).

REFERENCES

Abdelrazaq, A. (2011) Validating the dynamics of the Burj Khalifa. CTBUH Technical Paper, Council of Tall Buildings and Urban Habitat, Chicago, Illinois, USA.

Bailey, P.A., and Kwok, K.C.S. (1985) Interference excitation of twin tall buildings. *Journal of Wind Engineering and Industrial Aerodynamics*, 21: 323–338.

Baskaran, A. (1993) Wind engineering studies on tall buildings – transitions in research. *Building and Environment*, 28: 1–19.

Beneke, D.L., and Kwok, K.C.S. (1993) Aerodynamic effect of wind induced torsion on tall buildings. *Journal of Wind Engineering and Industrial Aerodynamics*, 50: 271–280.

British Standards Institution. (1997) *Loading for buildings. Part 2: Code of practice for wind loads.* BS 6399: Part 2:1997, BSI, London.

Brown, W.G. (1972) A load duration theory for glass design. Division of Building Research. National Research Council of Canada. Research paper 508.

Burton, M.D., Kwok, K.C.S. and Hitchcock, P.A. (2011) Effect of low-frequency motion on the performance of a dynamic manual tracking task. *Wind and Structures*, 14: 517–536.

Calderone, I.J. (1999) The equivalent wind load for window glass design. *Ph.D. thesis*, Monash University.

Calderone, I. and Melbourne, W.H. (1993) The behaviour of glass under wind loading. *Journal of Wind Engineering and Industrial Aerodynamics*, 48: 81–94.

Cheung, J.C.K. (1984) Effect of tall building edge configurations on local surface wind pressures. *3rd International Conference on Tall Buildings*, Hong Kong and Guangzhou, China, 10–15 December.

Cheung, J.C.K. and Melbourne, W.H. (1992) Torsional moments of tall buildings. *Journal of Wind Engineering and Industrial Aerodynamics*, 41–44: 1125–1126.

Cook, N.J. (1983) Note on directional and seasonal assessment of extreme wind speeds for design. *Journal of Wind Engineering and Industrial Aerodynamics*, 12: 365–372.

Cook, N.J. and Mayne, J.R. (1979) A novel working approach to the assessment of wind loads for equivalent static design. *Journal of Industrial Aerodynamics*, 4: 149–164.

Coyle, D.C. (1931) Measuring the behaviour of tall buildings. *Engineering News-Record*, 1931: 310–313.

Dalgleish, W.A. (1971) Statistical treatment of peak gusts on cladding. *ASCE Journal of the Structural Division*, 97: 2173–2187.

Dalgleish, W.A. (1975) Comparison of model/full-scale wind pressures on a high-rise building. *Journal of Industrial Aerodynamics*, 1: 55–66.

Dalgleish, W.A. (1979) Assessment of wind loads for glazing design. *IAHR/IUTAM Symposium on Flow-Induced Vibrations*, Karslruhe, Germany, September.

Dalgleish, W.A., Templin, J.T. and Cooper, K.R. (1979) Comparison of wind tunnel and full-scale building surface pressures with emphasis on peaks. *Proceedings, 5th International Conference on Wind Engineering*, Fort Collins, Colorado, pp. 553–565, Pergamon Press, Oxford, UK.

Dalgleish, W.A., Cooper, K.R. and Templin, J.T. (1983) Comparison of model and full-scale accelerations of a high-rise building. *Journal of Wind Engineering and Industrial Aerodynamics*, 13: 217–228.

Davenport, A.G. (1964) Note on the distribution of the largest value of a random function with application to gust loading. *Proceedings, Institution of Civil Engineers*, 28: 187–196.

Davenport, A.G. (1966) The treatment of wind loading on tall buildings. *Proceedings Symposium on Tall Buildings*, Southampton, UK, April, pp. 3–44.

Davenport, A.G. (1971) The response of six building shapes to turbulent wind. *Philosophical Transactions, Royal Society, A*, 269: 385–394.

Davenport, A.G. (1975) Perspectives on the full-scale measurements of wind effects. *Journal of Industrial Aerodynamics*, 1: 23–54.

Davenport, A.G. (1977) The prediction of risk under wind loading. *2nd International Conference on Structural Safety and Reliability*, Munich, Germany, 19–21 September.

Davenport, A.G., Hogan, M. and Isyumov, N. (1969) A study of wind effects on the Commerce Court Tower, Part I. University of Western Ontario, Boundary layer Wind Tunnel Report, BLWT-7-69.

den Hartog, J.P. (1956). *Mechanical vibrations*. McGraw-Hill, New York.

Dryden, H.L. and Hill, G.C. (1926) Wind pressures on structures. *Scientific papers of the Bureau of Standards*, 20: 697–723 (Paper S523).

Dryden, H.L. and Hill, G.C. (1933) Wind pressure on a model of the Empire State Building. *Journal of Research of the National Bureau of Standards*, 10: 493–523.

Fujii, K., Tamura, Y., Sato, T. and Wakahara, T. (1990) Wind-induced vibration of tower and practical applications of tuned sloshing damper. *Journal of Wind Engineering and Industrial Aerodynamics*, 33: 263–272.

Fujino, Y., Pacheco, B.M., Chaiseri, P. and Sun, L-M. (1988) Parametric studies on tuned liquid damper (TLD) using circular containers by free-oscillation experiments. *Structural Engineering/Earthquake Engineering, Japan Society of Civil Engineers*, 5: 381s–391s.

Hitchcock, P.A., Kwok, K.C.S., Watkins, R.D. and Samali, B. (1997a) Characteristics of liquid column vibration absorbers I. *Engineering Structures*, 19: 126–134.

Hitchcock, P.A., Kwok, K.C.S., Watkins, R.D. and Samali, B. (1997b) Characteristics of liquid column vibration absorbers II. *Engineering Structures*, 19: 135–144.

Holmes, J.D. (1973) Wind pressure fluctuations on a large building. *Ph.D thesis*, Monash University, Victoria, Australia.

Holmes, J.D. (1985) Wind action on glass and Brown's integral. *Engineering Structures*, 7: 226–230.

Holmes, J.D. (1990) Directional effects on extreme wind loads. *Civil Engineering Transactions, Institution of Engineers, Australia*, 32: 45–50.

Holmes, J.D. (1995) Listing of installations. *Engineering Structures*, 17: 676–678.

Holmes, J.D. (2020) Comparison of probabilistic methods for the effect of wind direction on structural response. Structural Safety, 87, 101983.

Holmes, J.D. and Cochran, L.S. (2003). Probability distributions of extreme pressure coefficients. Journal of Wind Engineering and Industrial Aerodynamics, 91: 893–901.

Holmes, J.D. and Bekele, S.A. (2015) Directionality and wind-induced response – calculation by sector methods. *14th International Conference on Wind Engineering*, Porto Alegre, Brazil, 21–26 June.

Holmes, J.D. and Tse, T.K.T. (2014) International high-frequency base balance benchmark study. *Wind and Structures*, 18: 457–471.

International Organization for Standardization. (2009) Bases for design of structures – serviceability of buildings and walkways against vibration. Annex D. Guidance for human response to wind-induced motions in buildings. International Standard ISO 10137:2007(E).

Irwin, P.A. (2007) Wind engineering challenges of the new generation of supertall buildings. *12th International Conference on Wind Engineering*, Cairns, Queensland, Australia, 1–6 July.

Irwin, P.A. and Baker, W.F. (2005) The wind engineering of the Burj Dubai Tower. *7th CTBUH World Congress 'Renewing the urban landscape'*, New York, USA, 16–19 October.

Irwin, P.A., Denoon, R. and Scott, D. (2013) *Wind-tunnel testing of high-rise buildings – an output of the CTBUH Wind Engineering Working Group.* Council of Tall Buildings and Urban Habitat, Chicago, IL.

Isyumov, N. and Poole, M. (1983) Wind induced torque on square and rectangular building shapes. *Journal of Wind Engineering and Industrial Aerodynamics*, 13: 183–196.

Isyumov, N., Holmes, J.D., Surry, D. and Davenport, A.G. (1975) A study of wind effects for the First National City Corporation Project - New York, USA. University of Western Ontario, Boundary Layer Wind Tunnel Laboratory, Special Study Report, BLWT-SS1-75.

Isyumov, N., Mikitiuk, M.J., Case, P.C., Lythe, G.R. and Welburn, A. (2003) Prediction of wind loads and responses from simulated tropical storm passages. *11th International Conference on Wind Engineering*, Lubbock, Texas, USA, 2–5 June.

Isyumov, N., Ho, T.C.E. and Case, P.C. (2013) Influence of wind directionality on wind loads and responses. *12th Americas Conference on Wind Engineering*, Seattle, Washington, USA, 16–20 June.

Kareem, A. (1985) Lateral-torsional motion of tall buildings to wind loads. *Journal of the Structural Division, ASCE*, 111: 2479–2496.

Kasperski, M. (2003) Specification of the design wind load based on wind-tunnel experiments. *Journal of Wind Engineering and Industrial Aerodynamics*, 91: 527–541.

Keel, C.J. and Mahmoodi, P. (1986) Design of viscoelastic dampers for Columbia Center Building. In *Building motion in wind*. eds. N. Isyumov and T. Tschanz. ASCE, New York.

Khanduri, A.C., Stathopoulos, T., and Bedard, C. (1998) Wind-induced interference effects on buildings – a review of the state-of-the-art. *Engineering Structures*, 20: 617–630.

Kim, W., Tamura, Y. and Yoshida, A. (2011) Interference effects on local peak pressures between two buildings. *Journal of Wind Engineering and Industrial Aerodynamics*, 99: 584–600.

Kwok, K.C.S. (1984). Damping increase in building with tuned mass damper. *ASCE, Journal of Engineering Mechanics*, 110: 1645–1649.

Kwok, K.C.S. (1995) Aerodynamics of tall buildings. In *A state of the art in wind engineering*. ed. P. Krishna. Wiley Eastern Limited, New Delhi.

Kwok, K.C.S. and Bailey, P.A. (1987) Aerodynamic devices for tall buildings and structures. *ASCE, Journal of Engineering Mechanics*, 113: 349–365.

Kwok, K.C.S. and Samali, B. (1995) Performance of tuned mass dampers under wind loads. *Engineering Structures*, 17: 655–667.

Kwok, K.C.S., Wilhelm, P.A. and Wilkie, B.G. (1988) Effect of edge configuration on wind-induced response of tall buildings. *Engineering Structures*, 10: 135–140.

Lawson, T.V. (1976) The design of cladding. *Building and Environment*, 11: 37–38.

Lepage, M.F. and Irwin, P.A. (1985) A technique for combining historical wind data with wind tunnel tests to predict extreme wind loads. *5th US National Conference on Wind Engineering*, Lubbock, Texas, USA, 6–8 November.

Lythe, G.R. and Surry, D. (1990) Wind induced torsional loads on tall buildings. *Journal of Wind Engineering and Industrial Aerodynamics*, 36: 225–234.

Mahmoodi, P. (1969) Structural dampers. *ASCE, Journal of the Structural Division*, 95: 1661–1672.

Mahmoodi, P. and Keel, C.J. (1986). Performance of viscoelastic dampers for Columbia Center Building. In *Building motion in wind*. eds. N. Isyumov and T. Tschanz, ASCE, New York.

McNamara, R.J. (1977) Tuned mass dampers for buildings. *ASCE, Journal of the Structural Division*, 103: 1785–1798.

Melbourne, W.H. (1977) Probability distributions associated with the wind loading of structures. *Civil Engineering Transactions, Institution of Engineers, Australia*, 19: 58–67.

Melbourne, W.H. (1980) Comparisons of measurements on the CAARC Standard tall building model in simulated model wind flows. *Journal of Wind Engineering and Industrial Aerodynamics*, 6: 73–88.

Melbourne, W.H. (1984) Designing for directionality. *1st Workshop on Wind Engineering and Industrial Aerodynamics*, Highett, Victoria, Australia, July.

Melbourne, W.H. and Sharp, D.B. (1976) Effects of upwind buildings on the response of tall buildings. *Proceedings Regional Conference on Tall Buildings*, Hong Kong, China, September, pp. 174–191.

Minor, J.E. (1994) Windborne debris and the building envelope. *Journal of Wind Engineering and Industrial Aerodynamics*, 53: 207–227.

Modi, V.J., Welt, P. and Irani, P. (1990) On the suppression of vibrations using nutation dampers. *Journal of Wind Engineering and Industrial Aerodynamics*, 33: 273–282.

Newberry, C.W., Eaton, K.J. and Mayne, J.R. (1967) The nature of gust loading on tall buildings. *Proceedings, International Research Seminar on Wind effects on Buildings and Structures*, Ottawa, Canada, 11–15 September, pp. 399–428, University of Toronto Press, Toronto, Canada.

Peterka, J.A. (1983) Selection of local peak pressure coefficients for wind tunnel studies of buildings. *Journal of Wind Engineering and Industrial Aerodynamics*, 13: 477–488.

Peterka, J.A. and Cermak, J.E. (1975) Wind pressures on buildings – probability densities. *ASCE Journal of the Structural Division*, 101: 1255–1267.

Rathbun, J.C. (1940) Wind forces on a tall building. *Transactions, American Society of Civil Engineers*, 105: 1–41.

Sakai, F., Takeda, S. and Tamaki, T. (1989) Tuned liquid column damper - new type device for suppression of building vibrations. *International Conference on High-rise Buildings*, Nanjing, China, 25–27 March.

Saunders, J.W. (1974) Wind excitation of tall buildings. *Ph.D thesis*, Monash University, Victoria, Australia.

Samali, B. and Kwok, K.C.S. (1995) Use of viscoelastic dampers in reducing wind- and earthquake-induced motion of building structures. *Engineering Structures*, 17: 639–654.

Simiu, E. and Filliben, J.J. (1981) Wind direction effects on cladding and structural loads. *Engineering Structures*, 3: 181–186.

Simiu, E. and Filliben, J.J. (2005) Wind tunnel testing and the sector-by sector approach. *Journal of Structural Engineering, ASCE*, 131: 1143–1145.

Simiu, E. and Miyata, T. (2006) *Design of buildings and bridges for wind*. John Wiley, New York

Skilling, J.B., Tschanz, T., Isyumov, N., Loh, P. and Davenport, A.G. (1986) Experimental studies, structural design and full-scale measurements for the Columbia Seafirst Center. In *Building motion in wind*. eds. N. Isyumov and T. Tschanz, ASCE, New York.

Snaebjornsson, J. and Reed, D.A. (1991) Wind-induced accelerations of a building: a case study. *Engineering Structures*, 13: 268–280.

Standards Australia (1989) *Minimum design loads on structures. Part 2: wind loads*. Standards Australia, Sydney. Australian Standard AS1170.2-1989.

Standards Australia (2011) *Structural design actions. Part 2: wind actions*, Standards Australia, Sydney. Australian/New Zealand Standard, AS/NZS1170.2: 2011.

Standen, N.M., Dalgleish, W.A. and Templin, R.J. (1971) A wind-tunnel and full-scale study of turbulent wind pressures on a tall building. *Proceedings, 3rd International Conference on Wind effects on Buildings and Structures*, Tokyo, Japan, 11–15 September, Paper II.3, Saikon Shuppan Publishers, Tokyo.

Stathopoulos, T. and Zhu, X. (1990) Wind pressures on buildings with mullions. *Journal of Structural Engineering, ASCE*, 116: 2272–2291.

Sun, L-M., Chaiseri, P., Pacheco, B.M., Fujino, Y. and Isobe, M. (1989) Tuned liquid damper (TLD) for suppressing wind-induced vibration of structures. *2nd Asia-Pacific Symposium on Wind Engineering*, Beijing, China, 26–29 June.

Surry, D. and Djakovich, D. (1995) Fluctuating pressures on tall buildings. *Journal of Wind Engineering and Industrial Aerodynamics*, 58: 81–112.

Surry, D. and Mallais, W. (1983) Adverse local wind loads induced by adjacent building. *ASCE Journal of Structural Engineering*, 109: 816–820.

Surry, D., Kitchen, R.B. and Davenport, A.G. (1977) Design effectiveness of wind tunnel studies for buildings of intermediate height. *Canadian Journal of Civil Engineering*, 4: 96–116.

Tallin, A. and Ellingwood, B. (1985) Wind induced lateral-torsional motion of buildings. *ASCE Journal of the Structural Division*, 111: 2197–2213.

Tamura, Y., Fujii, K., Ohtsuki, T., Wakahara, T. and Kohsaka, R. (1995) Effectiveness of tuned liquid dampers under wind excitation. *Engineering Structures*, 17: 609–621.

Tamura, Y., Suda, K. and Sasaki, A. (2000) Damping in buildings for wind-resistant design. *1st International Symposium on Wind and Structures for the 21st Century*, Cheju, Korea, 26–28 January.

Templin, J.T. and Cooper, K.R. (1981) Design and performance of a multi-degree-of-freedom aeroelastic building model. *Journal of Wind Engineering and Industrial Aerodynamics*, 8, 157–175.

Tse, K.T., Kwok, K.C.S., Hitchcock, P.A., Samali, B. and Huang, M.F. (2007) Vibration control of a wind-excited benchmark tall building with complex lateral-torsional modes of vibration. *Advances in Structural Engineering*, 10: 283–304.

Vickery, B.J. and Davenport, A.G. (1970) An investigation of the behaviour in wind of the proposed Centrepoint Tower, in Sydney, Australia. University of Western Ontario, Boundary Layer Wind Tunnel Laboratory, Research Report, BLWT-1-70.

Walton, D., Lamb, S. and Kwok, K.C.S. (2011) A review of two theories of motion sickness and their implications for tall building motion sway. *Wind and Structures*, 14: 499–515.

Willford, M.R. (1985) The prediction of wind-induced responses of the Hong Kong and Shanghai Banking Corporation headquarters, Hong Kong. *Engineering Structures*, 7: 35–45.

Wyatt, T.A. and Best, G. (1984) Case study of the dynamic response of a medium height building to wind-gust loading. *Engineering Structures*, 6: 256–261.

Xu, Y.L., Kwok, K.C.S and Samali, B. (1992a) Torsion response and vibration suppression of wind-excited buildings. *Journal of Wind Engineering and Industrial Aerodynamics*, 43: 1997–2008.

Xu, Y.L., Samali, B. and Kwok, K.C.S. (1992b) Control of along-wind response of structures by mass and liquid dampers. *ASCE, Journal of Engineering Mechanics*, 118: 20–39.

Zhang, W.J., Xu, Y.L. and Kwok, K.C.S. (1993) Torsional vibration and stability of wind-excited tall buildings with eccentricity. *Journal of Wind Engineering and Industrial Aerodynamics*, 50: 299–309.

Chapter 10

Large roofs and sports stadiums

10.1 INTRODUCTION

Wind loading is usually the dominant structural loading on the roofs of large buildings, such as entertainment or exhibition centres, closed or partially closed sports buildings, aircraft hangars, etc. The wind loads on these structures have some significant differences in comparison with those on the roofs of smaller, low-rise buildings that justify separate treatment:

- The quasi-steady approach (Section 4.6.2), although appropriate for small buildings, is not applicable for large roofs,
- Resonant effects, although not dominant, can be significant.

These roofs are commonly of low pitch, and experience large areas of attached flow, with low correlations between the pressure fluctuations acting on different parts. Downward as well as upward external pressures can be significant. These roofs are often arched or domed structures, which are sensitive to the distributions of wind loads, and the possibility of critical 'unbalanced' pressure distributions should be considered.

This chapter will first consider the aerodynamic aspects of wind flow over large roofs, which will facilitate an understanding of the steady and fluctuating components of wind pressures that act on these structures. Then, methods of obtaining design wind loads are described, with emphasis on the method of effective static wind-load distributions, introduced in earlier chapters. The incorporation of resonant contributions is also discussed.

10.2 WIND FLOW OVER LARGE ROOFS

Figure 8.5 in Chapter 8 shows the main features of the flow over a low-pitched roof, with the wind blowing normal to one wall. At the top of the windward wall, the flow 'separates' and 're-attaches' further along the roof, forming a separation 'bubble'. The turbulence in the wind flow plays an important role in determining the length of the separation bubble - high turbulence gives a shorter bubble length, low turbulence produces a longer bubble. Even in open country turbulence, intensities in windstorms are equal to 10%–20% of the mean or slowly varying wind speed, and in this situation mean separation bubble lengths are equal to 2–3 wall heights.

The separation bubble region is very important for large roofs because the upwards (negative) pressures are the greatest in this region. In the reattached-flow region, the pressures are quite small. Thus, for very large flat or near-flat roofs, only the edge regions within two to three wall heights from the edge will experience large pressures, whereas large areas of the roof will experience quite low pressures. The variations of mean uplift pressures, measured in some wind-tunnel tests (Davenport and Surry, 1974) for flat roofs, are shown in Figure 10.1. It should be noted that fluctuations in pressures occur, so that downwards as well as upwards pressures can occur for short time periods. These positive pressure coefficients are not specified by all codes or standards on wind loads.

As the roof pitch increases, the point of flow separation moves away from the leading edge of the roof and, in the case of a curved or arched roof, separation usually occurs downstream of the apex (Figure 10.2, from Blessmann, 1991). Upwind of the separation point, the pressures may be downwards (positive) or upwards (negative) near the leading edge, depending on the rise to span ratio, but are always upwards at the apex. Downwind of the separation point they are upwards with small magnitudes.

The form of the net mean pressure coefficient distribution (i.e. the top surface pressure minus the bottom surface pressure) on a large cantilevered stadium roof is shown in Figure 10.3. Negative values indicate net upwards pressure differences. The largest uplift occurs at the leading edge, and reduces to a small pressure difference at the rear. The top surface experiences flow separation, so that the characteristic pressure distribution peaks at the leading edge, and reduces quite rapidly downstream. The flow stagnates at the back of the grandstand if there is no gap present, and reaches

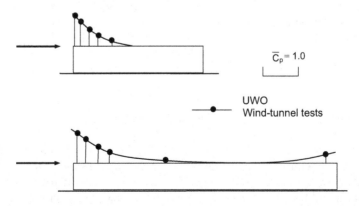

$\bar{C}_p = 1.0$

UWO
Wind-tunnel tests

Figure 10.1 Mean pressure distributions on flat roofs.

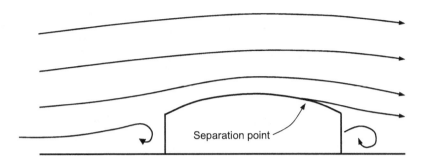

Figure 10.2 Flow separations over arched roofs (Blessmann, 1991).

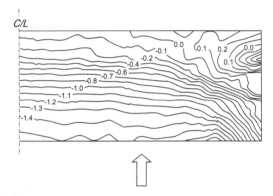

Figure 10.3 Mean net pressure distributions on a cantilevered stadium roof (Lam and To, 1995).

a pressure approaching the dynamic pressure of the freestream. However, the underside pressure will reduce in magnitude with increasing vent gaps at the back of the grandstand.

10.3 ARCHED AND DOMED ROOFS

10.3.1 Arched roofs

Arched roofs are structurally efficient, and are popular for structures like aircraft hangars, and enclosed sports arenas, which require large clear spans. Figure 10.4 shows the geometric variables that are relevant to the wind loading of arched-roof buildings. The variables are: the span, S; the length, L; the rise, R; and the height of the walls to the eaves level, h_e.

Some very early studies on arched roofs were carried out in an aeronautical wind tunnel in the Soviet Union in the 1920s (Bounkin and Tcheremoukhin,

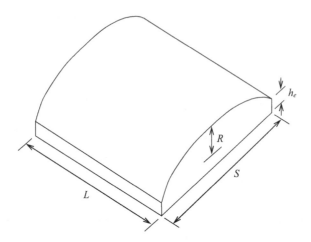

Figure 10.4 Geometric parameters for arched roof buildings.

1928). These data found their way into a number of national codes and standards on wind loading, and were used for many years after reference by the American Society of Civil Engineers (1936). Some early full-scale measurements on the Akron Airship Hangar, which had an arched roof of high rise-to-span ratio, were described by Arnstein and Klemperer (1936).

Arched roofs were apparently given very little attention by researchers after 1936 until the 1980s. Grillaud (1981) conducted a full-scale study of wind loads on an inflatable structure. Hoxey and Richardson (1983) also measured full-scale loads on film plastic greenhouses. Both structures had rise/span ratios of 0.5.

Johnson *et al.* (1985) reviewed existing model and full-scale data, and described some new wind-tunnel results from the University of Western Ontario. They found significant Reynolds Number effects in their wind-tunnel data for models with a rise/span ratio of 0.5.

Toy and Tahouri (1988) carried out measurements on models of semi-cylindrical structures ($R/S=0.5$; $h_e/R=0$). These wind-tunnel measurements were carried out with a smooth-wall boundary flow (very high Jensen Number – see Section 4.4.5), as well as low Reynolds Number (6.6×10^4, based on model height), and so the results are questionable in terms of applicability to full-scale structures. However, the data is useful in illustrating the strong effect of length/span ratio (L/S) on the mean pressures near the crest of the roofs. In this study the effects of lengthening the cross-section to produce a "flat-top", and shortening it to produce a "ridge", were also investigated. The latter modification has a particularly strong effect in modifying the mean pressure distribution over the roof.

Cook (1990) considered the measurements of Toy and Tahouri and also discussed the work of Blessmann (1987a,b) on arched roofs mounted on flat

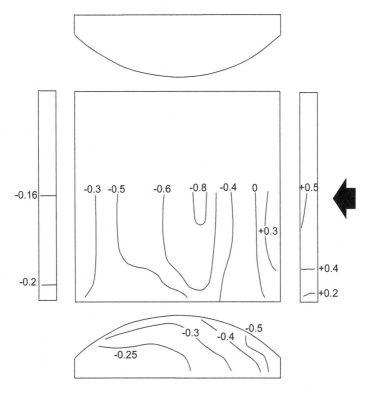

Figure 10.5 Mean pressure coefficients on an arched roof building – 0° (rise/span=0.2) (Paterson and Holmes, 1993).

vertical walls. It is suggested that flow separations occur at the eave when the roof pitch angle is less than 30°.

In a computational study of mean wind pressures on arched roofs (Paterson and Holmes, 1993), eleven separate geometrical configurations were examined. Figures 10.5 and 10.6 show the computed mean of external pressure coefficients on a building with a rise/span ratio (R/S) of 0.2, a length/span ratio (L/S) of 1.0, and a height to eaves/rise ratio (h_e/R) of 0.45, for wind directions of 0° and 45° from the normal to the axis of the arch. Because of symmetry, values on one half are shown for the 0° case.

For the 0° direction, positive pressure coefficients occur on the windward wall and the windward edge of the roof, with negative values over the rest of the structure. The highest magnitude negative values occur just upwind of the apex to the roof.

At wind direction of 45°, positive pressures occur near the windward corner of the building. The negative pressures on the roof and walls are generally higher than those obtained for the 0° case, with particularly high suctions occurring along the windward end of the arch roof.

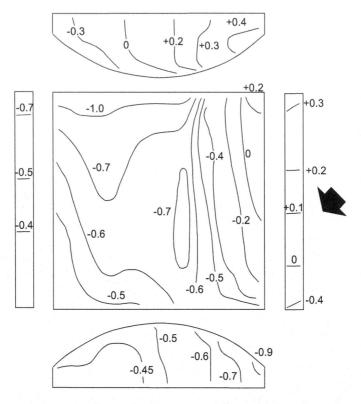

Figure 10.6 Mean pressure coefficients on an arched roof building – 45° (rise/span=0.2)
(Paterson and Holmes, 1993).

The effect of rise/span ratio is illustrated in Figure 10.7. The rise/span
ratio of the building in this figure is 0.50, compared with the building in
Figures 10.5 and 10.6, which has a rise/span of 0.20. It should be noted that
the reference dynamic pressure $(\frac{1}{2}\rho_a \bar{U}_h^2)$ is taken as the apex height of the
structure in both cases, so that for a fixed span and wall height, the reference
dynamic pressure will increase with increasing rise/span ratio.

As for high pitch gable roofs (Figure 8.6), there is a region of positive
pressure on the windward side of the roof.

The effect of increasing length/span ratio is to increase the magnitude of
both the positive and negative pressures in the central part of the building
as the flow becomes more two dimensional. Increasing wall height to rise
ratio (h_e/R) produces more negative values of external pressure coefficient
on the roof, side walls and leeward wall (Paterson and Holmes, 1993).

For wind directions parallel to the axis of the arch, arched roofs are aerodynamically flat, with similar pressure distributions as gable roofs, for the
same wind direction.

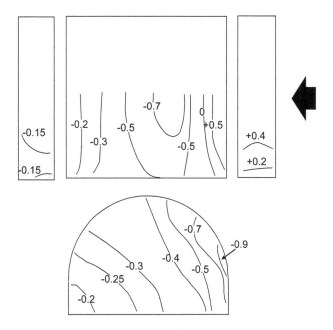

Figure 10.7 Mean pressure coefficients on an arched roof building (rise/span=0.5) (Paterson and Holmes, 1993).

10.3.2 Domed roofs

Domed roofs are less common than arched roofs, but are often used for air-supported, or 'pneumatic' roofs (see Section 10.5).

Two configurations of domed roofs on a circular planform were studied in a wind tunnel with boundary-layer flow, by Kawamura and Kiuchi (1986). Both configurations had a rise/span ratio of 0.15; eaves height-to-span ratios were 0.1 and 0.2, giving eaves height/rise ratios of 0.67 and 1.33, respectively. The mean pressure coefficients were negative over most of the roof, in both cases, with only a small positive pressure on the windward side. At the apex, the mean pressure coefficients (based on the mean wind speed upwind at the same height) were about −0.7 and −0.8 for the eaves-height-to span ratios of 0.1 and 0.2, respectively.

For design purposes, domed roofs can conservatively be assumed to have similar pressure distributions as those on arched roofs, of the same rise/span ratio, for a wind direction normal to the axis.

10.4 EFFECTIVE STATIC LOAD DISTRIBUTIONS

The statistical correlation between pressures separated by large distances can be quite small due to large fluctuating component in the wind loading on large roofs. Designers can make use of this, to the advantage of the cost of the structure, by determining effective static load distributions. This approach enables realistic and economical design wind load distributions to be obtained using wind-tunnel tests. Two possible methods can be used:

• A direct approach in which simultaneous time histories of fluctuating pressures from the whole roof are recorded and stored. These are subsequently weighted with structural influence coefficients to obtain time histories of load effects. The instantaneous pressure distributions coinciding with peak load effects are then identified and averaged.
• In the other approach, correlations between pressure fluctuations at different parts of the roof are measured, and expected pressure distributions corresponding to peak load effects are obtained using methods discussed in Chapter 5.

The effective static load distribution method, discussed in Section 5.4, tries to simplify the complex time and space variation of wind pressures on structures (produced by upwind turbulence and local building-induced effects) into a number of effective static pressure distributions for structural design. It is a particularly appropriate method for large roofs, over which the pressure fluctuations are not strongly correlated (or statistically related to each other). Significant reductions in design load effects, such as axial forces and bending moments in major structural members, can be obtained by this method, although normally wind-tunnel tests are required to obtain the necessary statistical data.

The principles behind the method as applied to large roofs are illustrated in Figure 10.8. In this figure, a section through a large arched roof is shown and the instantaneous external pressure distributions at three different points in time are shown. Clearly, there are considerable variations from time to time in these loadings. The variations are due to turbulence in the approaching wind flow, and local effects such as vortex shedding at the leading edge of the roof. The mean pressure distribution indicates only

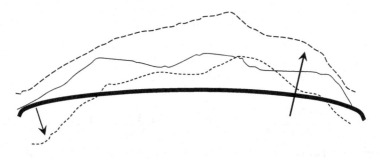

Figure 10.8 Instantaneous pressure distributions at three different times.

the average pressure at each point, but this distribution usually forms the basis for the design distributions of pressure found in codes. However, the instantaneous pressure distributions producing the largest load effects may be quite different in shape to the mean.

The question of interest to the structural engineer is: what are the critical instantaneous distributions which produce the largest structural load effects in the structure? The maximum and minimum values of each load effect will be produced by two particular expected instantaneous pressure distributions, which can be determined. The main factors determining these distributions are:

- the influence line for the load effect – an example of the influence line for a bending moment in an arch is given in Figure 5.8,
- the correlation properties of the wind pressures acting on the roof (both internal and external).

The influence lines can be calculated by the structural engineer, by applying point loads in a static structural analysis, and the correlation information can be obtained easily from wind-tunnel tests.

The effective static loading distributions for the various load effects of interest can be obtained by the formula developed by Kasperski and Niemann (1992) (see also Kasperski, 1992). Examples of two of these distributions are given in Figure 5.10 in Chapter 5. The distributions for a support reaction and a bending moment are shown. Clearly these two distributions differ considerably from each other, due to the different influence lines for the two load effects. They also differ from the mean pressure distribution. The shaded area in Figure 5.10 indicates the limits of the instantaneous maximum and minimum peak pressures around the arch, which form an 'envelope' within which the effective static loading distributions must fall.

When applying the effective static wind-load distribution approach to large roofs, usually a limited number of load effects are considered and effective static load distributions are computed for them. These are then 'enveloped' to give a smaller number of wind pressure distributions, which are then used by the structural engineers to design all the members of the structure. If required by structural designers, the peak values of critical load effects, such as forces in main members, or deflections can be directly computed.

10.4.1 Contributions from resonant components

When considering dynamic response of any structure to wind, it is necessary to distinguish between the resonant response at or near the natural frequencies of the structure, and the fluctuating response at frequencies below the first or lowest frequency, or 'background response', which is usually the largest contributor. As for all structures, the significance of resonant dynamic response to wind for large roofs depends on the natural frequencies of vibration, which are in turn dependent on the mass (inertia) and stiffness properties, and the damping. For roofs which are supported on two or four sides in the case of a rectangular plan, or all the way round in the case of a circular plan, the stiffness is usually high enough that resonant response is very small and can be ignored. For totally enclosed buildings, additional stiffness may be provided by compression of the air inside the building. Also, there may be additional positive aerodynamic damping which further acts to mitigate any resonant dynamic response.

Extra large stadium roofs may have several natural frequencies below 1 Hz, although these can be expected to have quite high damping.

Roofs supported on one side only, i.e. cantilevered roofs, however, are more prone to significant resonant response due to the lower stiffness. Figure 10.9 shows some resonant response in the time history of vertical deflection at the leading edge of a model cantilevered roof in wind-tunnel tests. The use of stiffening cables often increases the stiffness sufficiently to reduce the resonant contribution to minor proportions.

Most codes and standards do not include the effects of resonant response on large roofs, although the Australian Standard AS/NZS1170.2 (Standards Australia, 2011) is an exception as it contains a design load distribution which is dependent on natural frequency.

Wind-tunnel testing with an aeroelastic model is recommended when resonant response is anticipated to be substantial on a large roof. These can be very useful, but they have limitations, in that accurate effective load distributions cannot be determined from them, and the structural stiffness cannot be altered to accommodate design changes once the model has been built. As discussed earlier, for important structures, a rigid pressure model test is advisable to obtain the distributions in pressure for the mean and background components. The resonant response can also be computed from the spectra and cross-spectra of the fluctuating pressures at the natural frequency, or from the time histories of the

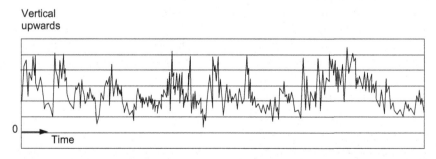

Vertical
upwards

0 ⟶ Time

Figure 10.9 Vertical displacement of the leading edge of a cantilevered roof showing some resonant contributions to the response (Melbourne, 1995).

generalized forces in the contributing modes of vibration. Both methods are computationally complex and require simultaneous pressure measurement over the entire roof (including the underside pressure for an open stadium roof). Both methods have been used for large projects at wind-tunnel laboratories.

Usually the resonant response will comprise no more than 10%–20% of the peak values of critical load effects (Holmes et al., 1997), and this contribution can be calculated separately and added to the fluctuating background response using a 'root-sum-of squares approach'. The effective static load distribution corresponding to each peak load effect can then scale up to match the recalculated peak load effect.

For very large roofs, several resonant modes can contribute, and the evaluation of effective static loads becomes more difficult. In general, it is necessary to adopt the approach of Section 5.3.7 in which the background response is separated from the resonant components, as these components all have different loading distributions. The magnitude of the contribution of each resonant mode depends on the load effect through its influence line. Section 12.3.6 describes the application of the equivalent static load approach to long-span bridges, when more than one resonant mode contributes. This approach can also be applied to very large roofs; in this case, the background contribution is treated as an additional 'mode', and the effective load distribution is calculated separately.

Thus, the effective static load distribution for the combined background and resonant contributions is:

$$p'_{\text{eff}}(x) = W'_B \cdot p_{\text{eff, back}}(x) + \sum_{j}^{N} W'_j m(x) \phi_j(x) \tag{10.1}$$

where the weighting factors are given by:

$$W'_B = \frac{\sigma_{r,B}}{\left[\left\{ \sigma^2_{r,B} + \sum_{j=1}^{N} \alpha_j^2 \omega_j^4 \overline{a_j^2} \right\} \right]^{1/2}} \tag{10.2}$$

$$W'_j = \frac{\alpha_j \omega_j^4 \overline{a_j^2}}{\left[\left\{ \sigma^2_{r,B} + \sum_{j=1}^{N} \alpha_j^2 \omega_j^4 \overline{a_j^2} \right\} \right]^{1/2}} \tag{10.3}$$

where $\sigma_{r,B}$ is the background component of the load effect, and the other terms are defined in Section 12.3.4. The derivation of the background effective static load distribution, $p_{\text{eff, back}}(x)$, is described in Chapter 5.

10.5 MEMBRANE AND AIR-SUPPORTED ROOFS

Lightweight roofs, made from fabric forming a tension membrane, are an architecturally attractive, low-cost option for large spans, particularly for sporting and music venues that require a large column-free space (Figure 10.10). These structures are potentially sensitive to wind loading, but this can be effectively mitigated by pre-tension; the latter can be achieved through a combination of cables and poles, or by air pressure for completely enclosed spaces. The largest example of a membrane roof is that covering the Hajj Terminal at Jeddah Airport, Saudi Arabia; it consists of 210 separate modules, each of 45 m², covering a total area of about 46 hectares.

Information on wind loads on generic forms of membrane roofs is sparse. However, some useful data are available for 'horn-shape' roofs, a shape often used as a basic module for large spans. Nagai *et al.* (2011) describe wind-tunnel tests on modules with a rise-to-span ratio of 0.2 on a square planform. A free roof (i.e. a roof with an open space underneath) and a roof above an enclosed building were studied. The mean external pressure coefficients on the roof in the latter case are all negative, whereas for the free roof, net pressures are positive (downwards) on the windward third. Some testing was carried out with multi-bay models. Surprisingly, there is little reduction in net pressures on the downwind bays in the case of the open-under free roof. However, there is significant reduction in external mean pressure coefficients up to the third bay for the enclosed roof.

Although membrane roofs with long spans may have quite low natural frequencies, significant resonant response to wind is unlikely, as they have high positive aerodynamic damping (see Section 5.5.1) for vertical motion at high wind speeds. For example, Sun and Gu (2014) in a numerical study of a 'saddle' membrane roof found aerodynamic damping of up to 10% of critical – about five times the assumed structural damping.

Figure 10.10 Example of a large tensioned membrane roof – Myer Music Bowl, Melbourne.

Air-supported roofs form a special class of membrane structures that require special treatment when considering the action of wind pressures (e.g. Tryggvason, 1979; Kind, 1984; Kawamura and Kiuchi, 1986). The resistance to wind forces is provided by the internal pressurization and by geometric and elastic stiffness of the membrane. The latter component is usually small in comparison to the first two, and is usually neglected in the scaling of aeroelastic wind-tunnel models (Tryggvason, 1979, and Section 7.6.5).

Air changes are required within the volume of a building of this type for ventilation purposes, and must be provided by fans and controlled leakage. Kind (1984) clearly showed that fluctuations in this air flow provided large pneumatic damping of the roof, particularly in large structures. This damping effectively suppressed any possibility of significant dynamic response to wind of these roofs. Designers should rather concentrate on providing resistance to overall uplift at the supports of the roof, and on maintaining sufficient internal air pressure to avoid local distortions, or total collapse. For these purposes, consideration should be given to the fluctuations in wind pressure, and spatial variations in instantaneous distributions in wind pressure, as discussed earlier in this chapter.

10.6 WIND-TUNNEL METHODS

Large roofs are usually dominated by the mean wind pressures and the background fluctuating components. Resonant contributions to the wind-induced structural load effects are usually small, even though natural frequencies as low as 0.5 Hz can occur for the largest roofs. The main reason for this behaviour is the nature of the separating-reattaching flow over large roofs of low pitch, and the consequent very low correlations between fluctuating pressures acting on different parts of the roof. Excitation of a dynamic mode requires pressure 'modes' which are coincident with the mode shape at the modal frequency in question. Usually, the excitation energy satisfying these conditions is small. Another reason for low resonant response is high damping with significant positive contributions from aerodynamic damping (Section 5.5.1).

For the reasons given above, modern wind-tunnel testing of large roofs for sports stadiums or arenas is usually carried out with rigid models on which detailed pressure measurements are made. The techniques used are described in Section 7.6.8. Using recorded time histories of fluctuating pressures, computations can be made of the resonant contributions, and added to the mean and background fluctuating contributions.

Full aeroelastic models of large roofs, although used quite frequently in the past, are now much less common. They are quite expensive to design and build, are structure-dependent, and do not lend themselves to changes in the underlying structure during the design process. Also, they can only normally be used for deflection measurements. However, for very flexible cantilevered roofs, the use of aeroelastic models may be required in conjunction with tests on rigid models.

10.7 CASE STUDIES

Holmes (1984) carried out wind-tunnel model measurements for an arched-roof aircraft hangar building with a rise/span ratio (R/S) of 0.20. Although the tests were carried out at low Reynolds Numbers, the curved roof surface was roughened. The effect of a ridge ventilator on the apex of the roof was also investigated and found to be significant. A significant aspect of this work was an early attempt to establish effective static load distributions for load effects such as axial forces and bending moments in the arch, and structural influence lines for the arch with both pinned and fixed supports were incorporated into the processing of the pressure data.

Wind-tunnel studies for a large cable-supported roof system with an elliptical planform of the Stadio delle Alpi, in Turin, Italy were described by Vickery and Majowiecki (1992). These consisted of pressure measurements and deflection measurements on an aeroelastic model. Due to the lack of multiple pressure-measuring instrumentation available at that time, unfortunately the pressure measurements covered only about 10% of the total roof area. However, the spectra of the fluctuating pressures indicated a dominant frequency which was attributed to vortex shedding from the upwind roof. This frequency was about one quarter of the first-mode natural frequency of the roof, and the resonant response was found to be relatively small, as it often is for large stadium roofs supported circumferentially (see Section 10.4.1). Cross-spectra (i.e. coherence functions) for panels separated on the roof indicated low correlations between fluctuating pressures on different parts of the roof. Vickery and Majowiecki identified the need to provide designers of structures with a number of effective wind pressure distributions, and suggested reducing the large amount of data generated from pressure-time histories by weighting with orthogonal functions (this suggests the use of 'proper orthogonal decomposition' as discussed in Section 8.3.4). Techniques for achieving these objectives were subsequently developed, and have been described earlier in this chapter.

A study for the re-roofing of the Olympic Stadium in Rome was described by Borri et al. (1992). This roof has a similar planform and cable-supported structure to the Turin stadium discussed in the previous paragraph. The limited fluctuating pressure measurements from a wind-tunnel study were supplemented with computer-generated time histories to cover the whole roof. These were applied as inputs to a non-linear finite element structural analysis program to compute the response of the structure to wind action. As for the roof of the Turin stadium, the response was primarily quasi-static with only small resonant contributions.

Wind-tunnel tests for the 206.5 m long roof of the Karaiskaki stadium in Piraeus (Greece) used for the Olympic Games in 2004, and later by the Olympiacos football team were described by Biagini *et al.* (2006). A rigid pressure model was used to simultaneously sample fluctuating pressures from 126 pressure taps, on both upper and lower surfaces of the roof. A numerical model was used with recorded time histories of the net pressure fluctuations to determine key structural load effects such as maximum and minimum tensile and compressive forces in the upper ties and lower struts of the lattice cantilevers supporting the roof. Equivalent uniform pressure distributions which gave similar values of the peak load effects were also derived.

Sykes (1994) described wind-tunnel measurements of mean net pressure distributions on two large tension roofs for EXPO'92 in Seville, Spain. The pressure coefficients and overall vertical and horizontal force coefficients were similar to those expected for 'rigid' free roofs with similar slope. The effect of 30% porosity of one roof resulted in an approximate halving of the overall force coefficients. However, these studies did not include measurement of wind pressure fluctuations, and the structural loads in large roofs of this type are sensitive to spatially varying wind load distributions as are large roofs generally, as discussed in Section 10.4. A difference with tensioned fabric roofs is their non-linear structural behaviour; this results in structural influence coefficients that vary with applied load. This can be handled by determining the influence coefficients by superimposing unit point or 'patch loads' on the design mean wind loading.

The effective static wind-load distribution method (Section 10.4), based on measurement of correlations between fluctuating pressures on panels on different parts of the roof, applied in conjunction with wind-tunnel tests, to two large stadium roofs in Australia is described by Holmes *et al.* (1997). This reference also discusses the effects of resonant load components. Some results from that study are given in Figure 10.11.

The alternative approach, based on the direct weighting of the recorded fluctuating pressures by influence coefficients, is described by Xie (2000). This is a case study of a stadium roof consisting of two large cantilevered panels with a complex curvature.

The *Journal of Wind Engineering and Industrial Aerodynamics* (Elsevier) contains a number of other descriptions of wind engineering studies of stadium roofs. There have been many other unpublished studies of the response to wind of large stadium roofs. One example worth mentioning briefly is the roof of the rebuilt Wembley Stadium in London. This is a very large roof with a number of resonant modes below 1 Hz that have the potential to be excited by wind action. However, it was found that the percentage of dynamic amplification of the internal forces in individual members of the roof structures varied greatly – between about 2% and 40%. This can be attributed to the similarity, or lack of, between the influence lines of the particular structural load effect associated with a member and the mode shapes for any of the resonant modes excited by the wind.

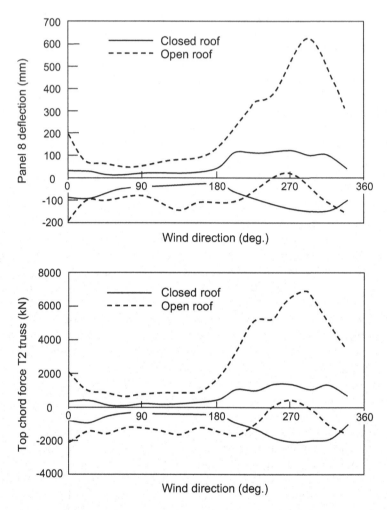

Figure 10.11 Variation of a deflection and a main truss force for a large stadium roof computed from a wind-tunnel pressure model test. (From Holmes *et al.*, 1997.)

10.8 SUMMARY

This chapter has attempted to cover the main aspects of wind loads on large roofs, including those used for sports stadiums. The characteristics of airflow and mean pressure distributions on flat, arched and domed roofs are discussed. There is some overlap with Chapter 8 'Low-rise buildings', but there are some significant differences, such as the large effects of reduced correlations between fluctuating pressures over large expanses of low-pitch roofs, and the possibility of some resonant response contributions.

The application of wind-tunnel methods, using pressure measurements on rigid models, to determine effective static wind-load distributions is discussed. Several case studies documented in the public domain have also been discussed.

REFERENCES

American Society of Civil Engineers. (1936) Wind-bracing in steel buildings. Fifth Progress Report of Sub-Committee No. 31. *Proceedings ASCE*, March 1936, pp. 397–412.

Arnstein, K. and Klemperer, W. (1936) Wind pressures on the Akron Airship-dock. *Journal of the Aeronautical Sciences*, 3: 88–90.

Biagini, P., Borri, C., Majowiecki, M., Orlando, M. and Procino, L. (2006). BLWT tests and design loads on the roof of the new Olympic stadium in Piraeus. *Journal of Wind Engineering and Industrial Aerodynamics*, 94: 293–307.

Blessmann, J. (1987a) Acao do vento em coberturas curvas, la Parte. Caderno Tecnico CT-86. Universidade Federale do Rio Grande do Sul (in Portugese).

Blessmann, J. (1987b) Vento em coberturas curvas - pavilhoes vizinhos. Caderno Tecnico CT-88. Universidade Federale do Rio Grande do Sul (in Portugese).

Blessmann, J. (1991) Acao do vento em telhados. SAGRA, Porto Alegre, Brazil.

Borri, C., Majowiecki, M. and Spinelli, P. (1992) Wind response of a large tensile structure: the new roof of the Olympic Stadium in Rome. *Journal of Wind Engineering and Industrial Aerodynamics*, 41–44: 1435–1446.

Bounkin, A. and Tcheremoukhin, A. (1928) Wind pressures on roofs of buildings. Transactions, Central Aero- and Hydrodynamical Institute, Moscow, No. 35.

Cook, N.J. (1990) *The designer's guide to wind loading of building structures. Part. 2. Static structures.* Building Research Establishment, UK.

Davenport, A.G. and Surry, D. (1974) The pressures on low-rise structures in turbulent wind. *Canadian Structural Engineering Conference*, Toronto.

Grillaud, G. (1981) Effet du vent sur une structure gonflable. Colloque, 'Construire avec le vent', Nantes, France, June.

Holmes, J.D. (1984) Determination of wind loads for an arch roof. *Civil Engineering Transactions, Institution of Engineers, Australia*, CE26: 247–253.

Holmes, J.D., Denoon, R.O., Kwok, K.C.S. and Glanville, M.J. (1997) Wind loading and response of large stadium roofs. *Proceedings, IASS International Symposium '97 on Shell and Spatial Structures*, Singapore, 10–14 November.

Hoxey, R. and Richardson, G.M. (1983) Wind loads on film plastic greenhouses. *Journal of Wind Engineering and Industrial Aerodynamics*, 11: 225–237.

Kasperski, M. (1992) Extreme wind load distributions for linear and nonlinear design. *Engineering Structures*, 14: 27–34.

Kasperski, M. and Niemann, H.-J. (1992) The L.R.C. (Load-Response-Correlation) method: A general method of estimating unfavourable wind load distributions for linear and non-linear structural behavior. *Journal of Wind Engineering & Industrial Aerodynamics*, 43: 1753–1763.

Johnson, G.L., Surry, D., and Ng, W.K. (1985) Turbulent wind loads on arch-roof structures: A review of model and full-scale results and the effect of Reynolds Number. *5th US National Conference on Wind Engineering*, Lubbock, Texas, USA, November 6–8.

Kawamura, S. and Kiuchi, T. (1986) An experimental study of one-membrane type pneumatic structure – wind load and response. *Journal of Wind Engineering and Industrial Aerodynamics*, 23: 127–140.

Kind, R.J. (1984) Pneumatic stiffness and damping in air-supported structures. *Journal of Wind Engineering and Industrial Aerodynamics*, 17: 295–304.

Lam, K.M. and To, A.P. (1995) Generation of wind loads on a horizontal grandstand roof of large aspect ratio, *Journal of Wind Engineering and Industrial Aerodynamics*, 54/55: 345–357.

Melbourne, W.H. (1995) The response of large roofs to wind action, *Journal of Wind Engineering and Industrial Aerodynamics*, 54/55: 325–335.

Nagai, Y., Okada, A., Miyasato, N. and Saitoh, M. (2011) Chapter 15: Wind-tunnel tests on the horn-shaped membrane roof. In *Wind tunnels and experimental fluid dynamics research*. ed. J.C. Lerner, Intech Open, Shanghai, China.

Paterson, D.A. and Holmes, J.D. (1993) Mean wind pressures on arched-roof buildings by computation. *Journal of Wind Engineering and Industrial Aerodynamics*, 50: 235–43.

Standards Australia. (2011) *Structural design actions. Part 2: wind actions*. Standards Australia, Sydney, Australian/New Zealand Standard AS/NZS1170.2: 2011.

Sun, F-J. and Gu M. (2014) A numerical solution to fluid-structure interaction of membrane structures under wind action. *Wind and Structures*, 19: 35–58.

Sykes, D.M. (1994) Wind loading tests on models of two tension structures for EXPO'92, Seville. *Journal of Wind Engineering and Industrial Aerodynamics*, 52: 371–383.

Toy, N. and Tahouri, B. (1988) Pressure distributions on semi-cylindrical structures of different geometrical cross-sections. *Journal of Wind Engineering and Industrial Aerodynamics*, 29: 263–272.

Tryggvason, B.V. (1979) Aeroelastic modelling of pneumatic amd tensioned fabric structures. *Proceedings, 5th Internal Conference on Wind Engineering*, Fort Collins, Colorado, USA, pp. 1061–1072, Pergamon Press, Oxford, UK.

Vickery, B.J. and Majowiecki, M. (1992) Wind induced response of a cable supported stadium roof. *Journal of Wind Engineering and Industrial Aerodynamics*, 41–44: 1447–1458.

Xie, J. (2000) Gust factors for wind loads on large roofs. *1st International Symposium on Wind and Structures*, Cheju, Korea, January.

Chapter 11

Towers, chimneys and masts

11.1 INTRODUCTION

In this chapter, the wind loading and wind-induced response of a variety of slender vertical structures will be considered; chimneys of circular cross-section, free-standing lattice towers, observation towers of varying cross-section, poles carrying lighting arrays or mobile telephone antennas, and guyed masts. Natural draft cooling towers, although not slender, are large wind-sensitive structures; the loading and response under wind action of these structures will be considered briefly in Section 11.6.

The methodology for determination of the loading and response of slender structures will first be described (making use of the general principles outlined in Chapters 1 – 7), followed by descriptions of several test case examples.

The dynamic response to wind of slender structures is quite similar in nature to that of tall buildings (described in Chapter 9). However, there are some significant differences:

- Fundamental mode shapes are generally non-linear.
- Higher modes are more likely to be significant in the resonant dynamic response.
- Since the aspect ratio is higher, i.e. the width is much less than the height, aerodynamic 'strip' theory can be applied, so that total aerodynamic coefficients for the cross-section can be used with the wind properties upstream at the same height.
- Aerodynamic damping (Section 5.5.1) will be significant when the mass per unit height is low.
- As for tall buildings, cross-wind response can be significant (except for lattice structures). However, due to the smaller cross-wind breadth, the velocity at which the vortex shedding frequency (or the maximum frequency of the cross-wind force spectrum) coincides with the first mode vibration frequency, is usually much lower than for tall buildings, and it often lies within the range of frequently occurring mean wind speeds.

11.2 HISTORICAL

11.2.1 Lattice towers

When the Eiffel Tower in Paris was completed in 1889, at 300 m it was easily the tallest structure in the world, and one of the first major towers of lattice construction. The designer Gustav Eiffel described the wind loading assumptions used in the design in an address to the Societe' des Ingenieurs Civils (Eiffel, 1885). He assumed a static horizontal pressure of 2 kPa at the base increasing to 4 kPa at the top. Over a large part of the top and base of the tower, he replaced the area of members in the lattice with solid surfaces with the same enclosed area. In the middle section where the tower solidity is lower, he assumed a frontal area equal to 'four times the actual area of iron'. These very conservative assumptions, of course, resulted in a very stiff structure with no serviceability problems in strong winds.

Eiffel constructed a laboratory at the top of the Tower, and carried out various scientific experiments, including measurements of the deflection of the tower, using a telescope aimed vertically at the target at the top. Some of these measurements were later analysed by Davenport (1975). This analysis indicated that the effective drag coefficient used in the design was approximately 3.5 times that required to produce the measured deflections, and well above that currently used in design for a tower with a solidity of about 0.3 (see Figure 11.1).

Since Eiffel was probably concerned about the over-conservatism of his designs, he carried out some experiments on wind forces on simple plates.

In the twentieth century, development of high-voltage power transmission, and radio and television broadcasting, from about the 1920s onwards, promoted the efficient use of steel for lattice tower construction.

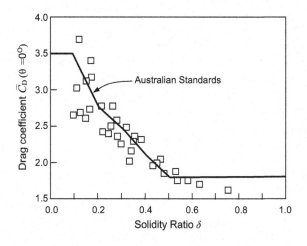

Figure 11.1 Drag coefficients for square towers with flat-sided members.

11.2.2 Tall chimneys

In the nineteenth and early twentieth centuries, most factory and power station chimneys were of masonry construction. With the known weakness of masonry joints to resist tension, these structures relied on dead load to resist the overturning effect of wind loads. Although, many of these failed in severe windstorms, W.C. Kernot commented in 1893 that "… there are thousands of such chimneys in existence, many in very open and exposed situations, which, apart from the adhesion of the mortar, would infallibly overturn with a pressure of not more than 15 pounds per square foot" (Kernot, 1893). Kernot concluded that the currently used design wind pressures were over-conservative (perhaps an early recognition of the effect of correlation), and proceeded to carry out some important early research in wind loads (see Section 7.2.1 and Figure 7.1).

The first full-scale wind pressure measurements on a cylindrical chimney were performed by H.L. Dryden and G.C. Hill on the newly erected masonry chimney of the power plant of the Bureau of Standards near Washington, DC (Dryden and Hill, 1930). These measurements were carried out together with full-scale measurements on another shorter cylinder (aspect ratio of 3) mounted on a roof, and wind tunnel measurements on circular cylinders. Through comparison of the resulting pressure distributions, this important study recognized, at an early stage, the effects of Reynolds Number, surface roughness (Section 4.5.1), and aspect ratio (Section 4.5.2) on the pressure distribution and drag coefficients of slender circular cylinders.

In the 1950s, extensive work on the cross-wind vibration of steel chimneys was carried out at the National Physical Laboratory (N.P.L.) in the U.K. under the direction of Christopher ('Kit') Scruton. This work (e.g. Scruton and Flint, 1964) included some important measurements on circular cylinders obtained in a compressed air wind tunnel, and the development of the now-ubiquitous helical strakes for the mitigation of vibration due to vortex shedding on tall chimneys (Section 4.6.3 and Figure 4.27).

11.3 BASIC DRAG COEFFICIENTS FOR TOWER SECTIONS

11.3.1 Drag coefficients for solid cross-sections

Many observation towers, communication towers, and chimneys have cross-sections which are circular or square. Drag coefficients for these cross-sections were discussed in Chapter 4. The effect of aspect ratios less than 20 is significant on the effective total drag coefficient (see Figures 4.10 and 4.19). Other cross sections may require wind-tunnel tests to determine drag coefficients.

The mean or time-averaged drag force per unit height, and hence bending moments, can be calculated using an appropriate sectional drag coefficient with a wind speed appropriate to the height, using an appropriate expression for mean wind speed profile (see Equation 5.32).

11.3.2 Drag coefficients for lattice towers

A basic formula for drag force for winds blowing at any angle to a face of a rectangular lattice tower is:

$$D = C_D A_z \cdot q_z \tag{11.1}$$

where

> D is the drag force on a complete tower panel section (i.e. all four sides of a square section tower),
> C_D is the drag coefficient for the complete tower section - it depends on the solidity of a face, and the wind direction,
> A_z is the projected area of tower members in one face of the tower,
> $q_z \left(= \frac{1}{2} \rho_a U_z^2 \right)$ is the dynamic wind pressure at the average height, z, of the panel under consideration.

Figure 11.1 shows the values of C_D specified in the Australian Standard for steel lattice Towers, AS3995 (Standards Australia, 1994) for square sections with flat-sided members, as a function of the solidity, compared with experimental values obtained from wind-tunnel tests for wind blowing normally to a face. For the range of solidity from 0.1 to 0.5, the following equations are appropriate (from Bayar, 1986).

$$C_D = 4.2 - 7\delta \text{ (for } \delta < 0.2) \tag{11.2}$$

$$C_D = 3.5 - 3.5\delta \text{ (for } 0.2 < \delta < 0.5) \tag{11.3}$$

The ASCE Guidelines (2020) and CSIR Recommendations (1990) for transmission line structures give equations for the wind drag force on a section of a lattice tower for any arbitrary wind direction, θ, with respect to the face of the tower. The CSIR equation may be written as follows:

$$D = q_z [C_{dn1} A_{n1} \cos^2 \theta + C_{dn2} A_{n2} \sin^2 \theta] K_\theta \tag{11.4}$$

where

> C_{dn1}, C_{dn2} are drag coefficients for wind normal to adjacent faces, 1 and 2, of the tower,

A_{n1}, A_{n2} are the total projected areas of faces 1 and 2, respectively, θ is the angle of incidence of the wind with respect to the normal to face 1 of the tower, K_θ is a wind incidence factor (derived empirically), given by:

$$K_\theta = 1 + 0.55\delta \sin^2(2\theta) \tag{11.5}$$

where δ is the solidity ratio (for $0.2 \leq \delta \leq 0.5$).

The ASCE Guidelines (2020) give a similar form to Equation (11.4), with a slightly different form for K_θ.

The drag of a lattice tower can also be computed by summing the contributions from every member. However, this is a complex calculation, as the effect of varying pitch and yaw angles on the various members, must be considered. This method also cannot easily account for interference and shielding effects between members and faces.

11.4 DYNAMIC ALONG-WIND RESPONSE OF TALL SLENDER TOWERS

The application of random vibration theory to the along-wind response of structures with distributed mass is discussed in Sections 5.3.6 and 5.3.7. The application of the equivalent static load distribution method to the along-wind response of tall structures is described in Section 5.4. These methods are applicable to all the structures covered in this chapter. However, a simple gust response factor (Section 5.3.2), in which a single multiplier, G, is applied to the mean pressure distribution, or a structural response derived from it, is generally not applicable in its simplest form to slender structures. Modifications are required to allow for a varying gust response factor, depending on the height, s, at which the load effect is required. A similar argument applies when a dynamic response factor approach is used (Section 5.3.4).

Two effects produce an increase in the gust response factor with height of load effect:

- the curved mode shape which gives an increasing contribution from the resonant component as the height, s, increases.
- since wind gusts of size equal to, or greater than, the distance $(h-s)$ between the height s and height of the top of the structure, h, are fully effective in producing stresses at the level s, the background contribution also increases as the height s increases.

An analysis for slender towers (Holmes, 1994) gives the following expressions for the gust response factors for shearing force, G_q, and bending moment, G_m, at any arbitrary height level, s, on a tower.

$$G_q = 1 + \frac{r\sqrt{g_B^2 B_s F_2 + g_R^2 \left(\dfrac{SE}{\eta_1}\right) F_3 F_4 F_5}}{F_1} \qquad (11.6)$$

$$G_m = 1 + \frac{r\sqrt{g_B^2 B_s F_7 + g_R^2 \left(\dfrac{SE}{\eta_1}\right) F_3 F_4 F_8}}{F_6} \qquad (11.7)$$

where

r is a roughness factor (=2 I_u), i.e. twice the longitudinal turbulence intensity at the top of the tower (Section 3.3.1),

B_s is a background factor reflecting the reduction in correlation of the fluctuating loads between the height level s and the top of the tower (Section 4.6.6),

g_B and g_R are peak factors (Section 5.3.3) separately calculated for the background and resonant components, respectively,

S is a size factor representing the aerodynamic admittance (Section 5.3.1) evaluated at the natural frequency of the tower,

$E = \dfrac{\pi n_1 S_u(n_1)}{4\sigma_u^2}$ is a non-dimensional form of the spectral density of longi-

tudinal turbulence (Section 3.3.4) evaluated at the natural frequency of the tower,

η_1 is the critical damping ratio for the first mode of vibration (this should also include aerodynamic damping contributions),

$F_1 \dots F_8$ are non-dimensional parameters depending on properties of the approaching wind and geometrical and dynamic properties of the tower, such as mean velocity profile, taper ratio, mode shape, and mass distribution. They also depend on the ratio (s/h), i.e. the ratio of the height level, s, at which the shearing force and bending moments are required, and the height of the top of the tower.

By evaluation of Equations (11.6) and (11.7) for a typical lattice tower (Holmes, 1994), it was shown that the increase in the value of gust response factor over the height of a structure will typically be in the range of 5%–15%.

A similar analysis for the deflection at the top of the tower, x, gives a similar expression to Equations (11.6) and (11.7) for the gust response factor for deflection, G_x (Holmes, 1996a):

$$G_x = 1 + \frac{r\sqrt{g_B^2 B_o F_{11} + g_R^2 \left(\dfrac{SE}{\eta_1}\right) F_3 F_4 F_{12}}}{F_{10}} \qquad (11.8)$$

where B_o is B_s evaluated at s equal to 0 (the reduction due to correlation over the whole height of the tower is important). F_{10}, F_{11}, and F_{12} are additional non-dimensional parameters; F_{12} is a non-dimensional stiffness for the tower.

It can be seen from Equations (11.6)–(11.8) that the gust response factor depends on the type of load effect under consideration, as well as the height on the tower at which it is evaluated.

An alternative approach, for the along-wind loading and response of slender towers and chimneys is the equivalent (or effective) static load distribution approach discussed in Section 5.4 (see also Holmes, 1996b). This approach allows variations in dimension shape and mass over the height of a tower of complex shape to be easily incorporated. Examples of effective static wind load distributions derived for a 160 m tower are given in Figures 5.11 and 5.12.

11.5 CROSS-WIND RESPONSE OF TALL SLENDER TOWERS

The strength of regular vortex shedding from a tower of uniform or slightly tapered cross-section is often strong enough to produce significant dynamic forces in the cross-wind direction. If the damping of a slender tower of a solid cross-section is low, high amplitude vibrations can occur when the frequency of vortex shedding coincides with a natural frequency of the structure. The velocity at which this coincidence occurs is known as the *critical velocity*. If the critical velocity is very high, i.e. outside the design range, no problems should arise, as the resonant condition will not occur. Conversely, if the critical velocity is very low, there will also not be a problem as the aerodynamic excitation forces will be low. However, significant vibration could occur if a critical velocity falls in the range 10–40 m/s.

Because of the higher rate of vortex shedding for a circular cross-section compared with that for a square or rectangular section of the same cross-wind breadth, the critical velocity is significantly lower.

Methods of calculation of cross-wind response of slender towers or chimneys fall into two classes:

i. those based on sinusoidal excitation
ii. those based on random excitation

The following sections describe the methods developed mainly for structures of circular cross-section. However, in principle they can be applied to structures of any (constant) cross-section.

11.5.1 Sinusoidal excitation models

The assumption that the vortex shedding phenomenon generates near-sinusoidal cross wind forces on circular cylinders can be linked to the work of

Scruton and his co-workers in the 1950s and 1960s (summarized in Scruton, 1981). In the original formulation, the excitation forces were treated solely as a form of negative aerodynamic damping, but this is equivalent to sinusoidal excitation by applied forces. Such models are good for situations in which large oscillations occur, and the shedding has 'locked-in' to the cross-wind motion of the structure (Section 5.5.4).

Sinusoidal excitation models were also proposed by Rumman (1970) and Ruscheweyh (1990).

Unlike other loading models in wind engineering, sinusoidal excitation models are *deterministic*, rather than random. The assumption of sinusoidal excitation leads to responses which are also sinusoidal.

To derive a simple formula for the maximum amplitude of vibration of a structure undergoing cross-wind vibration due to vortex shedding, the following assumptions will be made:

- Sinusoidal cross-wind force variation with time
- Full correlation of the forces over the height over which they act
- Constant amplitude of fluctuating cross-wind force coefficient.

None of these assumptions are very accurate for structures vibrating in the turbulent natural wind. However, they are useful for simple initial calculations to determine whether vortex-induced vibrations are a potential problem.

The structure is assumed to vibrate in the jth mode of vibration (in practice j will be equal to 1 or 2), so that Equation (5.17) applies:

$$G_j \ddot{a}_j + C_j \dot{a}_j + K_j a_j = Q_j(t) \qquad (5.17)$$

where

G_j is the generalized mass equal to $\int_0^h m(z)\phi_j^2(z)\, dz$,

$m(z)$ is the mass per unit length along the structure,

h is the height of the structure,

C_j is the modal damping,

K_j is the modal stiffness,

ω_j is the natural undamped circular frequency for the jth mode ($= 2\pi n_j = \sqrt{\dfrac{K_j}{G_j}}$),

$Q_j(t)$ is the generalized force, equal to $\int_{z1}^{z2} f(z,t)\phi_j(z)\, dz$, where $f(z, t)$ is the fluctuating force per unit height,

z_1 and z_2 are the lower and upper limits of the height range over which the vortex shedding forces act.

In this case, the applied force is assumed to be harmonic (sinusoidal) with a frequency equal to the vortex shedding frequency, n_s. The maximum amplitude of vibration will occur at resonance, when n_s is equal to the natural frequency of the structure, n_j.

Thus, the generalized force (Section 5.3.6) is given by:

$$Q_j(t) = \int_{z1}^{z2} f(z,t)\, \phi_j(z)\; dz = \left(\frac{1}{2}\right)\rho_a C_\ell b \sin\left(2\pi n_j t + \psi\right)\int_{z1}^{z2} \bar{U}^2(z)\phi_j(z)\; dz$$

$$= Q_{j,\max} \sin\left(2\pi n_j t + \psi\right)$$

where $Q_{j,\,\max}$ is the amplitude of the applied generalized force, given by:

$$Q_{j,\max} = \left(\frac{1}{2}\right)\rho_a C_\ell b \int_{z1}^{z2} \bar{U}^2(z)\phi_j(z)\; dz \tag{11.9}$$

where

C_ℓ is the amplitude of the sinusoidal lift (cross-wind force) per unit length coefficient,
ρ_a is the density of air.

The result for the maximum amplitude at resonance for a single-degree-of-freedom system can be applied:

$$a_{\max} = \frac{Q_{j,\max}}{2K_j \eta_j} = \frac{Q_{j,\max}}{8\pi^2 n_j^2 G_j \eta_j} \tag{11.10}$$

where η_j is the critical damping ratio for the jth mode, equal to $\dfrac{C_j}{2\sqrt{G_j K_j}}$.

Substituting for $Q_{j,\,\max}$ from Equation (11.9) in Equation (11.10),

$$a_{\max} = \frac{\left(\frac{1}{2}\right)\rho_a C_\ell\, b \int_{z1}^{z2} \bar{U}^2(z)\phi_j(z)\, dz}{8\pi^2 n_j^2 G_j \eta_j} \tag{11.11}$$

$$= \frac{\rho_a C_\ell\, b^3 \int_{z1}^{z2} \phi_j(z)\, dz}{16\pi^2 G_j \eta_j St^2}$$

where St is the *Strouhal Number* for vortex shedding (Section 4.6.3), which in this case can be written as:

$$St = \frac{n_s b}{\bar{U}(z_e)} = \frac{n_j b}{\bar{U}(z_e)}$$

where z_e is an average or effective height for the vortex shedding frequency. The maximum amplitude of deflection at any height on the structure is given by:

$$y_{max} = a_{max} \phi_j(z)$$

$$= \frac{\rho_a C_\ell \, b^3 \phi_j(z) \int_{z1}^{z2} \phi_j(z) \, dz}{16\pi^2 G_j \eta_j St^2} \qquad (11.12)$$

For a tower with a uniform mass per unit height, the maximum deflection at the tip ($z=h$), and where $\phi(h)$ is chosen as 1.0, is given by:

$$\frac{y_{max}(h)}{b} = \frac{\rho_a C_\ell \, b^2 \int_{z1}^{z2} \phi_j(z) \, dz}{16\pi^2 G_j \eta_j St^2} = \frac{C_\ell \int_{z1}^{z2} \phi_j(z) \, dz}{4\pi \, Sc \, St^2 \int_0^h \phi_j^2(z) \, dz} \qquad (11.13)$$

where Sc is the *Scruton Number*, or 'mass-damping parameter', previously introduced in Equation (5.48) and, in this case, defined as:

$$Sc = \frac{4\pi m \eta_j}{\rho_a b^2} \qquad (11.14)$$

where m is the average mass per unit length along the structure.

The ratio of vibration amplitude at the tip of a uniform cantilevered tower, to the tower breadth can thus be evaluated as:

$$\frac{y_{max}}{b} = \frac{k \cdot C_\ell}{4\pi \cdot Sc \cdot St^2} \qquad (11.15)$$

where $k \left(= \dfrac{\int_{z1}^{z2} \phi_j(z) \, dz}{\int_0^h \phi_j^2(z) \, dz} \right)$ is a parameter dependent weakly on the mode shape of vibration.

Ruscheweyh (1990) has modified the basic sinusoidal model by the use of a 'correlation length'. The term 'correlation length' is one that is normally applied to random processes or excitation (Section 4.6.6), and a better term would be 'excitation length'. The vortex shedding forces are applied over a height range less than the total height of the structure in this model.

A simple formula based on Equation (11.13) can be derived to estimate the maximum amplitude of vibration as a fraction of the diameter. The version in the Eurocode (BSI, 2005) is written as follows:

$$\frac{y_{max}}{b} = \frac{1}{St^2}\frac{1}{Sc} K K_w C_{lat} \qquad (11.16)$$

where

y_{max} is the maximum amplitude of vibration at the critical wind speed,
K_w is an effective correlation length factor,
K is a mode shape factor,
C_{lat} is a lateral (cross-wind) force coefficient (= C_ℓ).

11.5.2 Random excitation – Vickery-Basu model

A random excitation model, for vortex shedding response prediction, was developed by Vickery and Basu (1983). The peak deflection at the tip for a uniform cantilever, as a ratio of diameter, can be written in the following form:

$$\frac{\hat{y}}{b} = g\frac{\left[n_1 S_{C\ell}(n_1)\right]^{1/2}\left(\rho_a b^2/m\right)}{16\pi^{3/2}\eta^{1/2}St^2}f(\phi) \qquad (11.17)$$

where

$S_{C\ell}(n)$ is the spectral density of the generalised crosswind force coefficient,
$f(\phi)$ is a function of mode shape,
g is a peak factor which depends on the resonant frequency, but is usually taken as 3.5–4,
η is the critical damping ratio, comprising both structural and aerodynamic components.

Equation (11.17) has some similarities with Equation (11.13), but it should be noted that in the case of random vibration, the response is inversely proportional to the *square root* of the damping, whereas in the case of sinusoidal excitation, the peak response is inversely proportional to the damping. The peak factor (ratio between peak and root-mean-square response) is also much greater than the value of $\sqrt{2}$ in the sinusoidal model. The spectral density includes the effect of correlation length on the fluctuating forces.

In Vickery and Basu's procedure, the spectral density of the local lift force per unit length is represented by a Gaussian function, as follows:

$$\frac{n \cdot S_\ell(n)}{\sigma_\ell^2} = \frac{(n/n_s)}{B\sqrt{\pi}}\exp\left[-\left(\frac{1-n/n_s}{B}\right)^2\right] \qquad (11.18)$$

where B is a *bandwidth parameter*.

This function is based on the assumption of a constant Strouhal Number and the shedding frequency varying with wind speed, as the large-scale turbulence generates a Gaussian variation in wind speed about the mean value (Vickery and Basu, 1983).

Lock-in (Sections 4.6.3 and 5.5.4), in which the vortex-shedding frequency 'locks-in' to the natural frequency of the structure, results in an increase in the magnitude of the fluctuating cross-wind forces, and an increase in their correlation along the length of the structure. It is dealt in the Vickery and Basu model with a non-linear, amplitude-dependent, aerodynamic damping, within the random excitation model.

Equation (11.17) can be written in the form:

$$\frac{\hat{y}}{b} = \frac{A}{\left[(Sc/4\pi) - K_{ao}\left(1 - y^2/y_L^2\right) \right]^{1/2}} \tag{11.19}$$

where

A incorporates all parameters not associated with damping,

y is the root-mean-square fluctuating amplitude,

y_L is a limiting root-mean-square amplitude,

K_{ao} is a non-dimensional parameter associated with the negative aerodynamic damping.

Equation (11.19) can be used to define three response regimes:

- A randomly 'forced' vibration regime, at high values of Scruton Number,
- A 'lock-in' regime for low values of Scruton Number, in which the response is driven by the negative aerodynamic damping, and is largely independent of A,
- A transition regime between the above two regimes.

These three regimes, with an empirical fit based on Equation (11.19), are shown in Figure 11.2 (from Vickery and Basu, 1983), and compared with experimental data from a model chimney (Wooton, 1969).

With appropriate input parameters, the Vickery/Basu method is applicable to any full-scale structure of constant, or slightly tapered cross-section, but it has been calibrated to the vortex-induced response of large concrete chimneys.

When making predictions on real towers, or chimneys in atmospheric turbulence, it is necessary to include the effect of lateral turbulence. Referring to Figure 11.3, the effect of lateral (horizontal) turbulence is for the instantaneous flow direction to be at an angle to the mean flow direction of θ, where

$$\sin\theta \cong \frac{v}{\overline{U}}$$

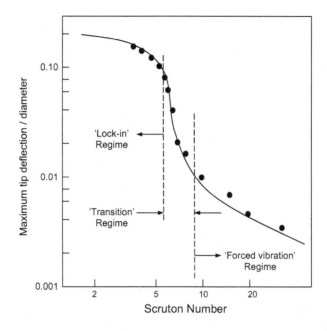

Figure 11.2 Response regimes for cross-wind vibration of circular towers and chimneys (Vickery and Basu, 1983).

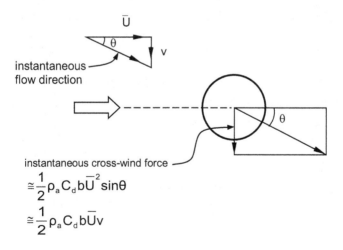

Figure 11.3 Cross-wind force due to lateral turbulence.

Thus, for a circular cross section, the instantaneous lateral force per unit length based on quasi-steady assumptions can be written as:

$$f_t(z,t) = \frac{1}{2}\rho_a b C_d \bar{U}^2 \sin\theta = \frac{1}{2}\rho_a b C_d \bar{U} v(z,t) \tag{11.20}$$

Basu and Vickery (1983), in developing a method suitable for prediction of the combined cross-wind response of real structures in the atmospheric boundary layer, used the following expression for the mean square modal coordinate in the jth mode:

$$\overline{a_j^2} = \frac{\pi n_j \left[S_{\ell,v}(n_j) + S_{\ell,t}(n_j) \right]}{4K_j^2 (\eta_s + \eta_a)} = \frac{S_{\ell,v}(n_j) + S_{\ell,t}(n_j)}{(4\pi n_j)^3 G_j^2 (\eta_s + \eta_a)} \tag{11.21}$$

where $S_{\ell,v}(n_j), S_{\ell,t}(n_j)$ are, respectively, the spectral densities, evaluated at the natural frequency, n_j, of the cross-wind forces due to vortex shedding and lateral turbulence. Equation (11.21) is based on the assumption that the spectral density is constant over the resonant peak, as previously used to derive Equation (5.13).

A comparison of the peak-to-peak cross-wind deflection at the top of the 330 m high Emley Moor television tower computed by the random vibration approach of Vickery and Basu, and compared with measurements is shown in Figure 11.4. Calculations were made for the first four modes of vibration. There was some uncertainty in the appropriate structural damping for this tower, but generally good agreement was obtained.

Comparisons were also made with full-scale response measurements from several reinforced concrete chimneys (Vickery and Basu, 1984). The average agreement was quite good, but some scatter was shown.

11.5.3 Random excitation – Hansen model

The Vickery and Basu model for vortex-induced cross-wind vibration was developed into a more usable closed-form model by Hansen (2007). This version has been adopted in the design codes of the International Committee on Industrial Chimneys (CICIND, 2002), and by the Eurocode on wind actions (BSI, 2005).

The equations for the ratio of the maximum standard deviation of the cross-wind deflection to the diameter are as follows (Hansen, 2007):

$$\left(\frac{\sigma_{max}}{b}\right)^2 = c_1 + \sqrt{c_1^2 + c_2} \tag{11.22}$$

$$c_1 = \frac{a_L^2}{2}\left(1 - \frac{Sc}{4\pi K_a}\right) \tag{11.23}$$

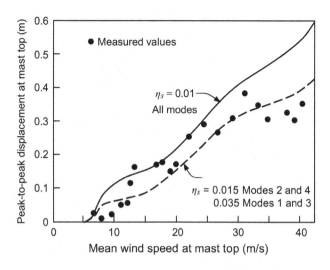

Figure 11.4 Comparison of measured and computed cross-wind response on the Emley Moor television tower (Basu and Vickery, 1983).

$$c_2 = \frac{a_L^2}{K_a} \frac{\rho b^2}{m} \frac{C_c^2}{St^4} \frac{b}{h} \qquad (11.24)$$

where

$a_L = \gamma_L \cdot a_{L,\,ref}$ is a limiting amplitude;
$C_c = \gamma_C \cdot C_{c,\,ref}$ is an aerodynamic excitation parameter;
K_a = aerodynamic damping parameter;
h = length of structure exposed to vortex-shedding forces (normally the height of a vertical structure;
ρ is the density of air;
b is the breadth (diameter);
m is the mass per unit length;
Sc is the Scruton Number (Section 11.5.1);
St is the Strouhal Number (Section 4.6.3).

Equation (11.22) is the solution of a quadratic equation for $(\sigma_{max}/b)^2$ similar to Equation (11.19) (Hansen, 2007). As a closed form solution, iterations are avoided.

For a circular cylinder, values of $a_{L,\,ref}$, $C_{c,\,ref}$ and K_a suggested by Hansen (2007) for various ranges of Reynolds Number are given in Table 11.1. Correction factors for mode shape, γ_L and γ_C, are given for various shapes (normalized to a maximum value of 1.0) in Table 11.2.

The standard deviation of deflection $\sigma_{y,\,max}$ given by Equation (11.22) must be multiplied by a peak factor to give the peak deflection y_{max}.

Table 11.1 Aerodynamic parameter for cross-wind response of circular sections (Hansen, 2007)

Parameter	$Re \leq 10^5$	$Re = 5 \times 10^5$	$Re \geq 10^6$
$a_{L,ref}$	0.4	0.4	0.4
$C_{c,ref}$	0.02	0.005	0.01
K_a	2	0.5	1

Table 11.2 Correction factors for mode shape (Hansen, 2007)

Mode shape	γ_L	γ_C
Uniform, $\phi(z) = 1$	1	1
Linear, $\phi(z) = z/h$	1.73	1.29
Parabolic, $\phi(z) = (z/h)^2$	2.24	1.34
Cantilever	1.41	1.16

Hansen (2007) recommended the following expression for peak factor (from Ruscheweyh and Sedlacek, 1988):

$$g = \sqrt{2} \left[1 + 1.2 \arctan\left(0.75 \left(\frac{Sc}{4\pi K_a} \right)^4 \right) \right] \tag{11.25}$$

For large amplitudes of vibration (i.e. 'lock-in' conditions), Equation (11.25) gives a peak factor which approaches $\sqrt{2}$ which is the value for sinusoidal vibrations. At small amplitudes the value approaches 3.5–4.

11.5.4 Hybrid model of ESDU

Item 96030 of the Engineering Sciences Data Unit (ESDU, 1996) covers the response of structures of circular and polygonal cross-section to vortex shedding. A computer program and spreadsheet is provided to implement the methods. ESDU 96030 covers uniform, tapered and stepped cylindrical or polygonal structures, and also yawed flow situations.

The method used in ESDU 96030 appears to be a hybrid of the two previously described approaches. For low amplitudes of vibration, a random excitation model similar to that of Vickery and Basu, has been adopted. At high amplitudes, i.e. in lock-in situations, a sinusoidal excitation model has been adopted, with a cross-wind force coefficient that is non-linearly dependent on the vibration amplitude. The response is assumed to switch intermittently between a random wide-band response and a constant amplitude of sinusoidal type, as lock-in occurs.

The effect of cross-wind turbulence excitation is also included in this method. This contribution becomes more significant with increasing wind speed, and thus is more important for larger cylinders (e.g. large diameter reinforced concrete chimneys with high critical wind speeds).

The ESDU method has the disadvantage of a discontinuity between the two response regimes.

11.5.5 Comparison of predictions of cross-wind response

In this section, a comparison of the computed response to vortex shedding for three representative slender structures with circular cross-section is made:

- a 100 m steel chimney,
- a 250 m reinforced concrete chimney,
- a 25 m thin-walled, steel lighting pole.

The relevant details of the three structures are given in Table 11.3.

These represent a wide range of structural types for which the cross-wind response needs to be assessed. In all three cases, the structures were assumed to be located in open country terrain, with relevant velocity profile and turbulence properties. In this comparison, only the first mode of vibration was considered.

The maximum root-mean-square tip deflection/mean diameters for the three structures have been calculated by the following methods and tabulated in Table 11.4: (a) the sinusoidal excitation method given in the Eurocode (BSI, 2005), (b) Vickery and Basu's random excitation approach (Structures 1 and 2 only) and (c) the hybrid approach of ESDU (ESDU, 1996).

The three methods compared in Table 11.4 clearly give significant variations in estimated response to vortex shedding, for all three structures. In the case of structure (1), all methods predict large amplitudes that

Table 11.3 Structural properties

Property	Structure 1	Structure 2	Structure 3
Height (m)	100	250	25
Diameter (m)	4.9	20	0.55–0.20 (tapered)
Surface roughness (mm)	0.1	1	0.15
Natural frequency (Hz)	0.5	0.3	0.5
Mode shape exponent	2	1.6	2
Mass/unit height (kg/m) (top third)	1,700	50,000	30
Critical damping ratio	0.005	0.01	0.005

Table 11.4 Calculated values of maximum root-mean-square tip deflection/diameter (at or near critical velocity)

Method	Structure 1	Structure 2	Structure 3
(a)	0.080	0.032	0.016
(b)	0.214	0.0045	n.a.
(c)	0.308	0.0054	0.014

are characteristic of lock-in, although Methods (b) and (c) predict higher amplitudes. Method (a), based on sinusoidal excitation overestimates the response of structure 2 (a large reinforced concrete chimney), which is subject to wide-band excitation with low amplitudes. Methods (b) and (c) predict similar maximum response for structure 2.

Vickery and Basu's model has generally been used for high Reynolds Numbers only, and has not been applied to structure 3, which is clearly in the sub-critical regime. The other methods predict a low response amplitude for structure 3 which has a very low critical velocity in the first mode, although this type of low-mass pole, or mast, has a history of occasional large vortex shedding responses, sometimes in higher modes, and often producing fatigue problems. One of the main problems in predicting their behaviour is in predicting the structural damping ratio, which is often very amplitude dependent.

Verboom and van Koten (2009) applied three methods for prediction of the cross-wind response to vortex shedding of thirteen steel chimneys in Europe, the operational history of which was known in some detail. The three methods used were:

Approach 1 in Eurocode 1 (British Standards Institution, 2005). This is a variant of the sinusoidal excitation model discussed in Section 11.5.1.

Approach 2 in Eurocode 1. This is the Hansen model (Section 11.5.3), and is essentially the same as that in the CICIND (2002) code for steel chimneys.

Approach 3 was described as 'an accurate implementation' of the Vickery-Basu model. In particular, this included the effect of atmospheric turbulence on the bandwidth parameter, B, and on the aerodynamic damping parameter, K_{ao} (Section 11.5.2).

Using the expected fatigue life as a criterion, Verboom and van Koten concluded that Approach 1 'seriously underestimated' the stresses caused by cross-wind vibrations, and hence would have overestimated the fatigue life for five of the 13 chimneys. Approach 2 mostly over-estimated the stresses, and hence unjustly rejected six out of the 13 chimneys as being 'unsafe'.

Approach 3, which accurately took account of the effects of turbulence, was stated as giving a 'good indication' of the stresses due to cross-wind vibrations, and accurately predicted the fatigue life for three out of the four chimneys, for which the operational life is known.

Hence, it appears that neither approach given in Eurocode 1, for the prediction of the cross-wind response of steel chimneys is completely satisfactory, and that the effects of turbulence intensity on the aerodynamic parameters are significant. However, it will be difficult to accurately incorporate the latter into a code-based approach, as the turbulence intensity can often vary considerably at the same site, depending on the wind speed, and other factors such as the atmospheric temperature, and the occurrence of temperature inversions, as well as the roughness of the surrounding terrain, for various directions.

11.6 COOLING TOWERS

The vulnerability of large hyperbolic natural draught cooling towers to wind action was emphasised in the 1960s due to collapse of the Ferrybridge towers in the U.K. (Figure 1.11). This event provoked research work on the wind loading and response of these large structures, especially in Europe. The sensitivity of wind pressures on circular cross-sections to Reynolds Number means that like chimneys, there are some questions about the validity of wind-tunnel tests to produce reliable results.

The main factors affecting wind loading of large cooling towers are:

• The partially correlated nature of fluctuating wind pressures acting on such large bluff structures, which means that quasi-steady design wind pressures are inadequate,
• The non-linear nature of the thin reinforced concrete,
• Aerodynamic interference effects from adjacent similar structures (as illustrated by the Ferrybridge failures).

Since the lowest natural frequency in the uncracked state usually exceeds 1 Hz, these structures are not particularly dynamically sensitive to wind, although after cracking of the concrete, the frequencies can apparently reduce significantly, with significant resonant contributions to the response (Zahlten and Borri, 1998).

A detailed discussion of the wind loading of these special structures will not be given in this text, although they are covered in some detail by Simiu and Scanlan (1996). There are a number of specialist design codes for cooling towers which include specification of wind loads (e.g. VGB, 1990; BSI, 1992).

Other useful references are by Shu and Wenda (1991) for soil interaction effects, Niemann and Köpper (1998) for aerodynamic interference, Zahlten and Borri (1998) for resonant amplification effects, and Niemann and Ruhwedel (1980) for wind-tunnel modelling.

11.7 GUYED MASTS

Since most guyed masts are lattice structures (usually with triangular cross-sections), wind-tunnel testing is neither appropriate nor required for this type of structure. Analytical methods are usually used for tall guyed masts.

However, guyed masts are complex structures to analyse for wind loading for a number of reasons.

- Their structural behaviour is non-linear,
- The influence lines for load effects such as bending moments and guy tensions are complex, and,
- When resonant dynamic response is important (for masts greater than about 150 m in height) many modes participate, and they are often coupled.

Generally, the dynamic response to wind may be analysed using the methods of random vibration outlined in Chapter 5. However, simple gust response factor approaches are not appropriate, because of the complex influence lines, with alternating positive and negative portions. The non-linear nature of the structure may be readily dealt with by computing the free-vibration frequencies and mode shapes, about the deflected position under the mean wind loading, rather than the 'no wind' condition. The effective static load methods outlined in Section 5.4 are very useful to derive effective static load distributions for both the background and resonant response of these structures.

A simplified approach to the dynamic response of tall guyed masts, in which the responses due to 'patch loads' are scaled to match the response calculated more rigorously from random vibration theory, is described by Davenport and Sparling (1992) and Sparling *et al.* (1996). The patch loads are applied on each span of the mast between adjacent guy levels, and from midpoint to midpoint of adjacent spans. The magnitude of the patch loads is taken as equal to the r.m.s. fluctuating drag force per unit height, at each height level, z:

$$d(z) = \rho_a C_d(z) b(z) \, \overline{U}(z) \sigma_u(z) \tag{11.26}$$

To simulate the lack of correlation of the fluctuating wind loads, the responses (bending moments, shear, deflections) due to the individual patch loads are combined by a root-sum-of-squares as in Equation (11.27).

$$\tilde{r}_{Pl} = \sqrt{\sum_{i=1}^{N} r_i^2} \tag{11.27}$$

where $\tilde{r}_{Pl}$ is the resultant patch load response, r_i is the response due to the ith patch load, and N is the total number of patch loads.

The design peak response is then determined from Equation (11.28).

$$\hat{r}_{Pl} = \tilde{r}_{Pl} \cdot \lambda_B \cdot \lambda_R \cdot \lambda_{TL} \cdot g \qquad (11.28)$$

where g is a peak factor, and λ_B, λ_R, λ_{TL} are a 'background scaling factor', a 'resonant magnification factor', and a 'turbulent length scale factor', respectively. These factors were determined by calibrating the method against the results of a full dynamic (random vibration) analysis for eight guyed masts ranging in height from 123 to 622 m. Expressions for these factors resulting from this calibration are given by Sparling *et al.* (1996).

This patch method has been adopted by the British Standard for lattice towers and masts (BSI, 1994). The results from the analysis of a 295 m guyed mast are shown in Figure 11.5. This shows that good agreement is achieved between the patch load method and the full dynamic analysis. The results from a conventional gust response factor approach (Section 5.3.2) are also shown. In this method, the mast is analysed under the mean wind loading, and the resulting responses are factored up by a constant factor (in this case, 2.0 was used). Clearly this method grossly underestimates the peak bending moments between the guy levels.

11.8 WIND TURBINE TOWERS

With the development of wind energy farms in many parts of the world, during the last thirty years, there are now many large wind turbine towers in existence, with their heights now exceeding 100 m. These towers are

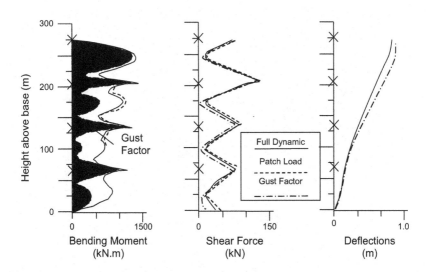

Figure 11.5 Comparison of peak responses for a 295 m guyed mast (Sparling *et al.*, 1996).

relatively flexible and carry a large mass at the top, due to the nacelles and turbine blades. Hence, their frequencies can be quite low (i.e. less than 0.5 Hz) and the towers, which are normally in exposed locations, may be subjected to significant turbulent buffeting (Section 5.3) at high wind speeds. There have been a number of failures of these towers in extreme winds (Figure 11.6).

The along-wind response of a typical wind turbine tower was investigated numerically by Murtagh *et al.* (2005) using simulated turbulent wind forces. The tower and the blades were modelled dynamically as multi-degree of freedom systems. The responses of the blades themselves were calculated using drag force time histories derived from rotationally-sampled wind spectra. The responses of the blades and the tower were coupled using compatibility of displacement at the top of the tower. It was found that neglect of the blade-tower interaction can significantly underestimate the response at the top of the tower, especially if the fundamental frequencies of the tower and turbine blades are close to each other.

11.9 INCLINED TOWERS

Towers with main axes inclined to the vertical are sometimes used to create 'sculptures' (Figure 11.7), or to support cable-stayed bridge decks. Inclined lattice towers are also often used on offshore platforms to ensure the gas-burner flames are located well away from the platform decks. However, there is a rather limited knowledge of wind loads and dynamic response of these structures.

Figure 11.6 A failed wind turbine tower following a typhoon (picture by Dr. M. Matsui, Tokyo Polytechnic University).

Figure 11.7 Slender inclined towers (as architectural features).

However, the galloping (Section 5.5.2) response of inclined slender towers with square cross sections has been studied in some detail. Towers inclined backwards – i.e. towards the downwind side by up to 10°, in a boundary-layer wind flow, have been found to exhibit greater cross-wind galloping amplitudes of vibration than the equivalent vertical tower. On the other hand, forward-inclined towers show reduced amplitudes at every angle of inclination (Hu *et al.*, 2015).

There appears to be no readily accessible generic information on the along-wind forces and buffeting behaviour of inclined towers of 'solid', square or rectangular, cross-sections.

Wyatt (1992) attempted an analysis of the dynamic response to wind buffeting of inclined *open-lattice* towers, representative of flare towers on offshore platforms. This required complex coordinate transformations of the components of turbulence in the approach flow, as well as the force coefficients for the tower members. Experimental studies to validate the approach appear to be lacking so far.

11.10 CASE STUDIES

An overview of a comprehensive wind-tunnel study was carried out and presented by Isyumov *et al.* (1984) comparing the 555 m high CN Tower in Toronto, Canada, with the full-scale observations. The wind-induced response of the 309 m Sydney Tower was described by Kwok and Macdonald (1990); the response was found to decrease markedly after a tuned mass damper system (Section 9.9.3) was installed. Numerical and wind-tunnel simulations of the wind-induced response of the 310 m Nanjing Tower are described by Kareem *et al.* (1998).

A case study of the wind loading and response study of the 338 m tall Macau Tower, which incorporates both wind-tunnel studies, and calculations, is described by Holmes (2000). The full aeroelastic model (1/150

Figure 11.8 Aeroelastic wind-tunnel model of a large free-standing tower.

scale) of the Macau Tower, used for the wind-tunnel testing, is shown in Figure 11.8.

There have also been a number of full-scale studies on the dynamic response of large reinforced concrete chimneys. Notable amongst these are studies by Muller and Nieser (1975), Hansen (1981), Melbourne *et al.* (1983) and Waldeck (1992). Ruscheweyh (1990) reported on some measurements on a number of steel stacks of cross-wind vibration, and made comparisons with predictions based on the sinusoidal model (Section 11.5.1).

Measurements on two tall guyed masts have been made by Peil *et al.* (1996) for comparison with theoretical predictions. One of these studies entailed the detailed measurement of turbulent wind speed at 17 height levels up to 340 m height (Peil and Nölle, 1992).

The wind-induced acceleration response of an air-traffic control tower was investigated by Park *et al.* (2006), using both a high-frequency base balance (Section 7.6.2), and an aeroelastic (Section 7.6.4) test in a wind tunnel. The excessive acceleration response at the top of the tower was controlled using a hybrid active passive tuned mass damper (Section 9.9.3). The effectiveness of the damper system was demonstrated by free vibration tests.

11.11 SUMMARY

In this chapter, the wind loading of slender towers, chimneys and masts of various types has been discussed. These structures are usually dynamically sensitive to wind, and response in both along-wind and cross-wind directions may need to be considered. Theoretical methods for calculating dynamic response, in both directions, are discussed.

The wind loading of hyperbolic cooling towers, guyed masts and wind turbine towers, are complex due to their structural behaviour. The main features of the wind loading and response of these structures are discussed.

REFERENCES

American Society of Civil Engineers. (2020) *Guidelines for electrical transmission line structural loading*, 4th Edition. American Society of Civil Engineers, Reston, VA.

Basu, R.I. and Vickery, B.J. (1983) Across-wind vibrations of structures of circular cross-section. Part II. Development of a mathematical model for full-scale applications. *Journal of Wind Engineering and Industrial Aerodynamics*, 12: 75–97.

Bayar, D.C. (1986) Drag coefficients of latticed towers. *ASCE, Journal of Structural Engineering*, 112: 417–430.

British Standards Institution. (1992) *Water cooling towers. Part 4. Code of practice for structural design and construction*. British Standard, BS 4485: Part 4: 1992. BSI, London.

British Standards Institution. (1994) *Lattice towers and masts. Part 4. Code of practice for lattice masts*. British Standard, BS 8100: Part 4: 1994. BSI, London.

British Standards Institution. (2005) *Eurocode 1: actions on structures - Part 1–4: general actions - wind actions*. BS EN 1991-1-4.6. BSI, London.

CICIND (International Committee on Industrial Chimneys). (2002) *Model code for steel chimneys. Revision 1- December 1999, Amendment A, March 2002*, CICIND. Zurich, Switzerland.

CSIR. (1990) *Transmission line loading. Part I: recommendations and commentary. Part II: appendices*. Engineering Structures Programme, CSIR Building Technology, Pretoria.

Davenport, A.G. (1975) Perspectives on the full-scale measurement of wind effects. *Journal of Industrial Aerodynamics*, 1: 23–54.

Davenport, A.G. and Sparling, B.F. (1992) Dynamic gust response factors for guyed masts. *Journal of Wind Engineering and Industrial Aerodynamics*, 44: 2237–2248.

Dryden, H.L. and Hill, G.C. (1930) Wind pressure on circular cylinders and chimneys. *Journal of Research of the National Bureau of Standards*, 5: 653–693.

Eiffel, G. (1885) Projet d'une tour en fer de 300m de hauteur destinees à L'Exposition de 1889. *Memoires de la Societe' des Ingenieurs Civils I*, 38: 345–370.

ESDU. (1996) *Response of structures to vortex shedding: structures of circular or polygonal cross-section*. Engineering Sciences Data Unit, ESDU Data Item 96030, ESDU International, London.

Hansen, S.O. (1981) Cross-wind vibrations of a 130 metre tapered concrete chimney. *Journal of Wind Engineering and Industrial Aerodynamics*, 8: 145–156.

Hansen, S.O. (2007) Vortex-induced vibrations of structures. *Proceedings, 3rd Structural Engineers World Congress*, Bangalore, India, 2–7 November.

Holmes, J.D. (1994) Along-wind response of lattice towers: Part I – derivation of expressions for gust response factors. *Engineering Structures*, 16: 287–292.

Holmes, J.D. (1996a) Along-wind response of lattice towers: Part II – aerodynamic damping and deflections. *Engineering Structures*, 18: 483–488.

Holmes, J.D. (1996b) Along-wind response of lattice towers: Part III – effective load distributions. *Engineering Structures*, 18: 489–494.

Holmes, J.D. (2000) Wind loading of the Macau Tower – application of the effective static load approach. *Proceedings, 1st International Symposium on Wind and Structures for the 21st Century*, Cheju, Korea, 26–28 January, pp. 81–90.

Hu, G., Tse, K.T. and Kwok, K.C.S. (2015) Galloping of forward and backward inclined slender square cylinders. *Journal of Wind Engineering and Industrial Aerodynamics*, 142: 232–245.

Isyumov, N., Davenport, A.G., and Monbaliu, J. (1984) CN Tower, Toronto: model and full-scale response to wind. *Proceedings, 12th Congress, International Association for Bridge and Structural Engineering*, Vancouver, Canada, 3–7 September, pp. 737–746.

Kareem, A., Kabat, S. and Haan, F.L. (1998) Aerodynamics of Nanjing Tower : a case study. *Journal of Wind Engineering and Industrial Aerodynamics*, 77–78: 725–739.

Kernot, W.C. (1893) Wind pressure. *Proceedings, Australasian Society for the Advancement of Science*, V: 573–581.

Kwok, K.C.S. and Macdonald, P.A. (1990) Full-scale measurements of wind-induced acceleration response of Sydney Tower. *Engineering Structures*, 12: 153–162.

Melbourne, W.H., Cheung, J.C.K. and Goddard, C. (1983) Response to wind action of 265-m Mount Isa stack. *ASCE, Journal of Structural Engineering*, 109: 2561–2577.

Muller, F.P. and Nieser, H. (1975) Measurements of wind-induced vibrations on a concrete chimney. *Journal of Industrial Aerodynamics*, 1: 239–248.

Murtagh, P.J., Basu, B. and Broderick, B.M. (2005) Along-wind response of a wind turbine tower with blade coupling subjected to rotationally sampled wind loading. *Engineering Structures*, 27: 1209–1219.

Niemann, H.-J. and Köpper, H.-D. (1998) Influence of adjacent buildings on wind effects on cooling towers. *Engineering Structures*, 20: 874–880.

Niemann, H.-J. and Ruhwedel, J. (1980) Full-scale and model tests on wind-induced, static and dynamic stresses in cooling tower shells. *Engineering Structures*, 2: 81–89.

Park, W., Park, K-S., Koh, H-M.. and Ha, D-H. (2006) Wind-induced response control and serviceability improvement of an air traffic control tower. *Engineering Structures*, 28: 1060–1070.

Peil, U. and Nölle, H. (1992) Guyed masts under wind load. *Journal of Wind Engineering and Industrial Aerodynamics*, 41–44: 2129–2140.

Peil, U., Nölle, H., and Wang, Z.H. (1996) Nonlinear dynamic behaviour of guys and guyed masts under turbulent wind load. *Journal of the International Association for Shell and Spatial Structures*, 37: 77–88.

Rumman, W.S. (1970) Basic structural design of concrete chimneys. *ASCE, Journal of the Power Division*, 96: 309–318.

Ruscheweyh, H. (1990) Practical experiences with wind-induced vibrations. *Journal of Wind Engineering and Industrial Aerodynamics*, 33: 211–218.

Ruscheweyh, H. and Sedlacek, G. (1988) Cross-wind vibrations of steel stacks – critical comparisons between some recently-proposed codes. *Journal of Wind Engineering and Industrial Aerodynamics*, 33: 173–183.

Scruton, C. (1981) *An introduction to wind effects on structures*. Oxford University Press, Oxford, UK.

Scruton, C. and Flint, A.R. (1964) Wind-excited oscillations of structures. *Proceedings, Institution of Civil Engineers (UK)*, 27: 673–702.

Shu, W. and Wenda L. (1991) Gust factors for hyperbolic cooling towers on soils. *Engineering Structures*, 13: 21–26.

Simiu, E. and Scanlan, R.H. (1996) *Wind effects on structures – fundamentals and applications to design*, 3rd Edition. John Wiley, New York.

Sparling, B.F., Smith, B.W. and Davenport, A.G. (1996) Simplified dynamic analysis methods for guyed masts in turbulent winds. *Journal of the International Association for Shell and Spatial Structures*, 37: 89–106.

Standards Australia. (1994) *Design of Steel Lattice Towers and Masts*. Standards Australia, North Sydney, AS3995-1994.

Verboom, V.K. and van Koten, H. (2009) Vortex excitation: Three design rules tested on 13 industrial chimneys. *Journal of Wind Engineering and Industrial Aerodynamics*, 98: 145–154.

VGB. (1990) *VGB: BTR Bautechnik bei Kültürmen*. (Construction guidelines for cooling towers). VGB Association of Large Powerplant Operators, Essen, Germany.

Vickery, B.J. and Basu, R I. (1983) Across-wind vibrations of structures of circular cross-section. Part I. Development of a mathematical model for two-dimensional conditions. *Journal of Wind Engineering and Industrial Aerodynamics*, 12: 49–73.

Vickery, B.J. and Basu, R I. (1984) The response of reinforced concrete chimneys to vortex shedding. *Engineering Structures*, 6: 324–333.

Waldeck, J.L. (1992) The measured and predicted response of a 300 m concrete chimney. *Journal of Wind Engineering and Industrial Aerodynamics*, 41: 229–240.

Wootton, L.R. (1969) The oscillations of large circular stacks in wind. *Proceedings of the Institution of Civil Engineers (UK)*, 43: 573–98.

Wyatt, T.A. (1992) Dynamic response of inclined towers. *Journal of Wind Engineering and Industrial Aerodynamics*, 43: 2153–2163.

Zahlten, W. and Borri, C. (1998) Time-domain simulation of the non-linear response of cooling tower shells subjected to stochastic wind loading. *Engineering Structures*, 20: 881–889.

Chapter 12

Bridges

12.1 INTRODUCTION

As discussed in Chapter 1, bridges have suffered some spectacular failures during wind storms (Figure 1.10. The history of the dynamically wind-sensitive suspension bridge from the nineteenth century onwards, including the periodic failures that have occurred, have been well documented (e.g. Steinman and Watson, 1957; Billington, 1977; Petroski, 1996).

Most of the early interest was in the drag or along-wind forces, and Baker (1884), Kernot (1893) and others, noted that peak wind forces acting on large areas, such as a complete bridge girder, were considerably less than those on a small plate or board. However, the great American builder of suspension bridges, John Roebling, was aware of the dynamic effects of wind as early as 1855. In commenting on the failure of the Wheeling Bridge, Ohio, in the previous year, he wrote:

> That bridge was destroyed by the momentum acquired by its own dead weight, when swayed up and down by the force of the wind.... A high wind, acting upon a suspended floor, devoid of inherent stiffness, will produce a series of undulations, which will be corresponding from the center each way.
>
> *Steinman and Watson (1957)*

However, it took several years for the dramatic failure of the first Tacoma Narrows suspension bridge in 1940 (Section 1.4) to direct serious attention to the dynamic actions of the wind, and other wind actions on bridge decks, such as vertical cross-wind forces and torsional moments.

The cable-stayed bridge emerged in the 1950s in Germany, as an efficient method of spanning intermediate length crossings. Gimsing (1983) and Virlogeux (1999) have reviewed recent developments in the design of bridges of this type.

At the start of the twenty-first century, the spans of the long-span suspension and cable stayed bridges have been extended to new limits. The longest bridge in the world in 2020 is the suspension bridge across the

Figure 12.1 Akashi-Kaikyo Bridge, Japan.

Akashi-Kaikyo Straits in Japan, which has an overall length of nearly 4 km, with a main span of 1,990 m (Figure 12.1). The design of this bridge was dominated by its aerodynamic characteristics.

The cable-stayed bridge with the longest span, in 2020, is the Russky Bridge, near Vladivostok, with an overall length of 3,100 m, and a main span of 1,104 m.

As the spans increase, wind actions become more critical in bridge design, and extensive wind studies are normally undertaken when designing the longest suspension or cable-stayed bridges. The dynamic wind forces will excite resonant response, often in several modes, and *aeroelastic* forces, which are generated by the motion of the structure itself, are important. Long-span bridges are usually crossings of large expanses of water, and may be exposed to relatively low-turbulence flow, at least at low wind speeds. This has contributed to a number of cases of vibrations of bridge decks induced by vortex shedding (Section 4.6.3). Recently the spans of cable-stayed bridges have been limited by problems with cable vibrations, sometimes involving rain, as well as wind (Section 12.5).

In the following sections, a review of the main aspects of wind forces and the wind-induced excitation of long-span bridges and their supporting cables is given. The aerodynamics of bridges is a large and specialist topic, and an in-depth treatment will not be given in this book.

12.2 BASIC FORCE COEFFICIENTS FOR BRIDGE DECKS

As for other structures, all bridges are subjected to mean and fluctuating wind forces. These may be estimated by the use of mean, or steady state, force coefficients, usually determined from wind-tunnel tests. Such coefficients are also required to determine dynamic response from turbulent buffeting.

Many wind-tunnel section tests of decks for long-span bridges (Section 7.6.3) have been carried out, primarily to determine their aerodynamic stability (Section 12.3.2). Determination of the basic section force coefficients, as a function of wind angle of attack, is also routinely done during the tests.

Most nineteenth-century suspension bridges were built with open lattice truss sections. This use has continued, as this type of section has some benefits from the point of view of dynamic response. The open structure prevents the formation of vortices, and dynamic excitation from vortex shedding (Section 4.6.3) is not usually a problem. Provided the torsional stiffness can be made high enough, the critical speed for flutter instability (Sections 5.5.3 and 12.3.2) will be high. However, the drag coefficients for open truss sections are high in comparison with other sections. For example, the drag coefficients for two cross-sections considered for the Little Belt suspension bridge completed in the 1960s in Denmark are shown in Figure 12.2 (Ostenfeld and Larsen, 1992). The drag coefficient for the trussed cross-section is more than three times that of the streamlined box girder section; the latter was eventually used for the bridge. However, after extensive aerodynamic testing (Miyata *et al.*, 1992), a truss girder, 11m deep, was chosen for the Akashi-Kaikyo suspension bridge – the world's longest (Figure 12.1).

Note that the along-wind *chord* dimension, d, rather than the cross-wind dimension, b, has been used to define the drag coefficients. This is usually the convention for bridges.

Very slender deck cross-sections, such as the box girder section shown in Figure 12.2, although they have low drag coefficients, will have high lift

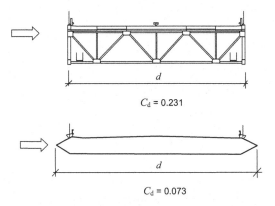

$C_d = 0.231$

$C_d = 0.073$

Figure 12.2 Comparison of drag coefficients for two bridge deck cross-sections (Ostenfeld and Larsen, 1992). (reproduced by permission from 'Aerodynamics of large bridges' – *Proceedings of the First International Symposium*, Copenhagen, Denmark, 19–21 February 1992, Larsen, Allan (ed).)

(cross-wind) force coefficients (Section 4.2.2) when the wind has a significant angle of attack, as does an airfoil. This situation will occur instantaneously in turbulent flow. This characteristic makes deck sections of this type prone to buffeting by vertical turbulence (Section 12.3.3).

Examples of the variation of static horizontal and vertical force coefficients, and moment coefficient about the mass centre of a bridge-deck section, with angle of attack, are given in Figure 12.3.

The conventional definition of section force and pitching moment coefficients for bridges is as follows:

$$C_X = \frac{F_x}{\frac{1}{2}\rho_a U^2 d} \quad C_Z = \frac{F_z}{\frac{1}{2}\rho_a U^2 d} \quad C_M = \frac{M}{\frac{1}{2}\rho_a U^2 d^2} \quad (12.1)$$

12.3 THE NATURE OF DYNAMIC RESPONSE OF LONG-SPAN BRIDGES

There are several mechanisms, in various wind speed ranges, which can excite resonant dynamic response in the decks of long-span bridges, as follows.

- Vortex shedding excitation (Section 4.6.3) occurs in low wind speeds and low turbulence conditions (e.g. Frandsen, 2001).

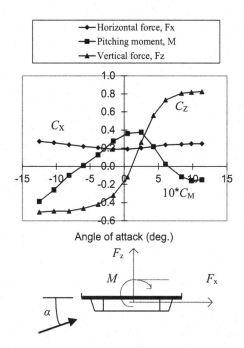

Figure 12.3 Static force coefficients for a typical bridge deck section.

- Flutter instabilities (Section 5.5.3) of several types occur at very high wind speeds for *aerodynamically stable* decks due to dominance of self-excited aerodynamic forces (Sabzevari and Scanlan, 1968). These always involve torsional (rotational) motions, and may also involve vertical bending motions.
- Buffeting excitation (Section 4.6.1) is caused by the fluctuating forces induced by turbulence (Davenport, 1962; Scanlan and Gade, 1977). This occurs over a wide range of wind speeds, and normally increases monotonically with increasing wind speed.

The nature of these mechanisms is discussed in the following sections.

12.3.1 Vortex-induced excitation

Under certain conditions, vortex-shedding excitation can induce significant, but limited, amplitudes of vibration. The conditions required for this to occur are most, or all, of the following:

- Wind direction normal to the longitudinal axis of the bridge
- Low turbulence conditions (typically I_u less than 0.05)
- A wind speed in a narrow critical range (5–12 m/s)
- Low damping (1% of critical or less).

The above conditions can be satisfied for both short to medium span cable-stayed bridges across water, and longer span suspension bridges. With Strouhal Numbers in the range of 0.1–0.2 (based on the depth of the deck cross-section), and natural frequencies in the range of 0.1–0.6 Hz, critical velocities of 6–15 m/s can produce significant amplitudes. Low turbulence conditions can occur in 'stable' atmospheric conditions, often in the early morning or evening. Some recorded examples of this behaviour are listed in Table 12.1.

Section tests carried out in smooth flow in wind tunnels can provide reasonably good predictions of the full-scale behaviour (Wardlaw, 1971; van

Table 12.1 Some recorded cases of vortex-shedding induced vibrations of bridges

Name	Natural frequency (Hz)	Critical velocity (m/s)	Maximum amplitude (mm)	References
Long's Creek Bridge	0.6	12	100–170	Wardlaw (1971)
Wye Bridge	0.46	7.5	35	Smith (1980)
Waal River	0.44	9–12	50	van Nunen and Persoon (1982)
Great Belt East	0.13–0.21	4.5–9	320	Larsen et al. (1999) Frandsen (2001)

Nunen and Persoon, 1982). In the case of the Long's Creek Bridge, Canada, where the vibrations were large enough to require remedial action, triangular fairings on the ends, and a soffit plate underneath the deck were added to the prototype structure, giving satisfactory results (Wardlaw, 1971). Guide vanes were used at the lower corners of the box girder of the Great Belt East suspension bridge, a method known to be successful in suppressing vortex shedding vibrations, which occurred at four different frequencies and a corresponding wide range of wind speeds. Lock-in effects (Sections 4.6.3 and 5.5.4) were also observed in the vortex-induced vibration on this bridge (Frandsen, 2001).

12.3.2 Flutter instabilities and prediction of flutter speeds

The coupled motion (rotation and vertical displacement) of a suspended bluff body was discussed in Section 5.5.3. Equations (5.49) and (5.50) – the coupled equations of motion – are repeated as follows:

$$\ddot{z} + 2\eta_z \omega_z \dot{z} + \omega_z^2 z = \frac{F_z(t)}{m} + H_1 \dot{z} + H_2 \dot{\theta} + H_3 \theta \tag{5.49}$$

$$\ddot{\theta} + 2\eta_\theta \omega_\theta \dot{\theta} + \omega_\theta^2 \theta = \frac{M(t)}{I} + A_1 \dot{z} + A_2 \dot{\theta} + A_3 \theta \tag{5.50}$$

Equations (5.49) and (5.50) are simplified forms of the full equations of motion, which include the horizontal motions of the deck, and eighteen different aeroelastic derivatives, corresponding to all possible motion-induced forces. However, many of these terms are small. The propensity of a bridge deck to flutter instability depends on the magnitudes and signs of some of the aeroelastic derivatives, or *flutter derivatives*, of the particular deck cross-section as a function of the wind speed. For example, a positive value of the derivative, A_2 is an indication of flutter in a pure rotational motion – sometimes known as 'stall flutter' in aeronautical terminology. This can be seen from Equation (5.50) when the term $A_2 \dot{\theta}$ is transposed to the left-hand-side of the equation – it takes the form of a negative damping term, with the ability to *extract* energy from the flow. If the magnitude of the negative aerodynamic damping is greater than the structural damping, then vibrations will grow in amplitude – i.e. an aeroelastic instability will occur.

The most commonly-understood use of the term 'flutter' however is to describe the coupled translational-rotational form of instability which is largely governed by the signs of the derivatives H_2 and A_1 (see Table 5.1).

Data on the flutter derivatives A_i to H_i are usually obtained experimentally from section tests in wind tunnels (see Section 7.6.3). Tests are usually done in smooth (low turbulence) flow; it has been found that the effects of turbulence on the derivatives are generally small (Scanlan and Lin, 1978; Gu *et al.*, 2001). The derivatives are a function of reduced velocity, $\left(\dfrac{U}{n\,d}\right)$ which incorporates the variation with frequency of vibration, n, as well as the wind speed, U. The following non-dimensional forms are usually used for the derivatives (Scanlan and Tomko, 1971).

$$H_1^* = \frac{mH_1}{\rho_a d^2 \omega}; \quad A_1^* = \frac{IA_1}{\rho_a d^3 \omega}$$

$$H_2^* = \frac{mH_2}{\rho_a d^3 \omega}; \quad A_2^* = \frac{IA_2}{\rho_a d^4 \omega}$$

$$H_3^* = \frac{mH_3}{\rho_a d^3 \omega^2}; \quad A_3^* = \frac{IA_3}{\rho_a d^4 \omega^2} \tag{12.2}$$

where

m and I are the mass and moment of inertia per unit length (spanwise), d is the width (chord) of the deck, ρ_a is the air density, and ω is the circular frequency (= $2\pi n$).

Examples of aeroelastic derivatives determined for two common types of bridge deck – an open truss and a box girder – are shown in Figure 12.4.

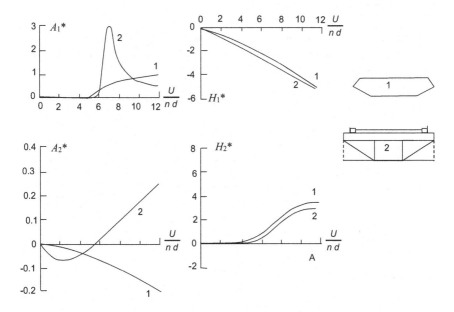

Figure 12.4 Aeroelastic derivatives for two types of bridge deck (Scanlan and Tomko, 1971).

Although the magnitude and sign of the derivatives give some indications of the tendency of a particular section in the design stage of important long-span bridges to aerodynamic instability, it is usual to attempt to determine a 'critical flutter speed' for the deck cross-section. If this wind speed does not exceed by a substantial margin, the design wind speed of the site at the deck height (suitably factored for ultimate limit states), then modifications to the deck cross-section are usually made.

Stall, or pure torsional, flutter is the most common type for bridge decks of bluff cross-section. This phenomenon was the primary cause of the failure of the Tacoma Narrows Bridge in Washington state, USA in 1940. The 'H' cross-section used for the deck of that bridge was subsequently found to have positive values of the derivative, A_2 (i.e. negative aerodynamic damping for torsional modes of vibration) for low values of reduced velocity. More streamlined sections of the box-girder type, such as Section 1 in Figure 12.4, typically have negative values of A_2 for all reduced velocities and thus, are not prone to torsional flutter. However, coupled flexural-torsional flutter should be investigated for bridges with those sections.

Several methods may be used to determine the critical wind speed for coupled flexural-torsional flutter:

- empirical formulae (e.g. Selberg, 1963)
- experimental determination by use of section model testing
- theoretical stability analysis of the equations of motion (Equations (5.46) and (5.47), with values of A_i and H_i obtained experimentally (e.g. Simiu and Scanlan, 1996).

Selberg (1961, 1963) proposed an empirical equation for critical flutter speed, U_F, which, in its simplest form, can be written as Equation (12.3).

$$U_F = 0.44d \sqrt{\left(\omega_T^2 - \omega_V^2\right)\frac{\sqrt{\upsilon}}{\mu}} \qquad (12.3)$$

where $\upsilon = 8\left(\dfrac{r}{d}\right)^2$ and $\mu = \dfrac{\pi \rho_a d^2}{2m}$; r is the radius of gyration of the cross-section ($I=mr^2$); and m is the mass per unit length.

ω_T ($=2\pi n_T$) and ω_V ($=2\pi n_V$) are the circular frequencies in the first torsional (twisting) mode and first vertical bending modes, respectively.

Alternative ways of expressing the Selberg formula are as follows:

$$U_F = 0.417\omega_T(d/2) \sqrt{\left(1 - \frac{\omega_V^2}{\omega_T^2}\right)\frac{mr}{\rho_a(d/2)^3}} \qquad (12.4)$$

where the half chord of the bridge deck ($d/2$) has been used.

$$\frac{U_F}{n_T d} = 3.72\sqrt{\left(\frac{mr}{\rho_a d^3}\right)\left(1 - \frac{n_V^2}{n_T^2}\right)}$$ (12.5)

Figure 12.5 shows measured flutter speeds for several bridge deck sections compared with predictions from the Selberg formula. Reasonable agreement is obtained although there is an overestimation at low angles of attack. It would appear to be unwise to rely on a prediction based on an empirical formula alone.

The analytical estimation of flutter speeds is a specialist function of bridge aerodynamicists, but Ge and Tanaka (2000) have given a useful summary of the techniques available, amongst many others.

12.3.3 Buffeting of long-span bridges

A bridge that is otherwise stable in flutter up to a high wind speed, and does not suffer from vortex induced vibrations at low wind speeds, will still experience dynamic response to atmospheric turbulence, known as *buffeting* over a wide range of wind speeds. This response will normally determine the size of the structural members and require evaluation at the design stage.

Davenport (1962) was the first to apply random vibration methods to the buffeting of a long-span suspension bridge. These methods were later validated by comparison of them with model studies in turbulent boundary

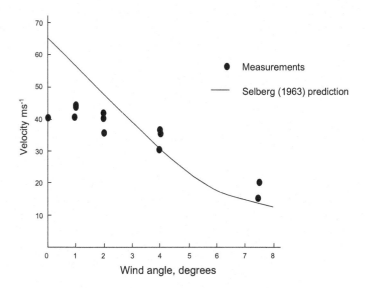

Figure 12.5 Measured critical flutter speeds and comparisons with the Selberg prediction formula (Wardlaw, 1971).

layer flow in the 1970s (e.g. Holmes, 1975, 1979; Irwin, 1977; Irwin and Schuyler, 1978).

The methodology described in Section 5.3.6 for the along-wind response of distributed mass structures can be adapted to the *cross*-wind response of bridge decks excited by vertical turbulence components.

Applying a 'strip' assumption, the sectional cross-wind force per unit span can be written as:

$$f'(z,t) = \rho_a \bar{U} d \left[C_{Z_0} u'(t) + \frac{1}{2} \frac{dC_Z}{d\alpha} w'(t) \right]$$ (12.6)

where

C_{Z_0} is the vertical force coefficient at zero angle of attack,

$\dfrac{dC_Z}{d\alpha}$ is the slope of the vertical force coefficient versus angle of attack, α,

$u'(t)$ and $w'(t)$ are the horizontal and vertical velocity fluctuations upstream of the deck section in question.

C_{Z_0} and $\dfrac{dC_Z}{d\alpha}$ can be obtained from static section tests of the deck cross-section (Section 12.2). If there is significant angular rotation of the bridge deck under the mean wind load (as is often the case with suspension bridges), then C_{Z_0} may need to be replaced by the value of C_Z at the mean (non-zero) angle of attack under the mean wind loading.

Following an argument similar to that used in Section 5.3.6, the spectral density of the generalised force for the jth mode of vibration can be obtained.

$$S_{Qj}(n) = \left(\rho_a \bar{U} d\right)^2 \left[C_{Z_0}^2 S_u(n) + \frac{1}{4} \left(\frac{dC_Z}{d\alpha}\right)^2 S_w \phi(n) \right] X^2(n)$$

$$\times \int_0^L \int_0^L \rho(y_1, y_2, n) \phi_j(y_1) \phi_j(y_2) \ dy_1 \ dy_2$$ (12.7)

In Equation (12.7), $X^2(n)$ is an *aerodynamic admittance*, allowing for the fact that smaller gusts (higher frequencies) do not completely envelope the bridge cross-section. Konishi *et al.* (1975), Shiraishi and Matsumoto (1977), Jancauskas (1986), Li *et al.* (2018) and others have directly measured this function for bridge deck sections and other streamlined and bluff shapes. Note that this aerodynamic admittance which applies to vertical (cross-wind) aerodynamic forces is similar, but not identical, to that discussed in Section 5.3.1, which relates to along-wind forces, and response.

Analysis using Equation (12.7) and the methods of random vibration analysis outlined in Section 5.3.6, have given good agreement with the measured response on full aeroelastic wind tunnel models (e.g. Holmes, 1975) and full-scale measurements (Melbourne, 1979). However, for large span bridges, the towers and cables play important parts in the overall bridge response, and it is the practice to carry out full aeroelastic model studies in simulated turbulent boundary-layer flow, as described in Section 12.4.

12.3.4 Effect of wind barriers on dynamic response

Semi-porous barriers are often used on exposed bridge decks to protect traffic from the effects of cross winds. These can affect the dynamic response to wind of the bridge itself. Clearly, the increased drag forces due to the barrier will increase the horizontal buffeting forces. Barriers of 30% and 50% porosities have been found to increase the propensity of bridges to torsional (stall) flutter, by causing the A_2 derivative (Section 12.3.2) to change sign from negative to positive at a lower reduced velocity (Buljac et al., 2017). However, the effects also depend on the shape of the bridge deck itself.

12.3.5 Use of fairings to mitigate bridge deck vibrations

As discussed in Section 12.3.1, triangular leading-edge fairings were found to be effective in reducing vortex-induced vibrations of suspended bridge decks at low wind speeds. Such modifications can also be effective in reducing the propensity for stall or torsional flutter – i.e. to increase the critical wind speed for the instability, if not reducing it entirely (Huston et al., 1988).

Fairings of different geometries may be more acceptable architecturally. Some examples of possible shapes as applied to a footbridge are shown in Figure 12.6.

Other shapes are possible. For example, Fumoto et al. (2007) found that an asymmetric trapezoidal fairing was more effective than a triangular fairing in mitigating flutter for the deck of a design of a very-long span suspension bridge.

12.3.6 Effective static load distributions

The method of equivalent static load distributions discussed in Section 5.4 can be applied to the response of bridges. In many cases of long span bridges, the background response can be neglected in comparison to the resonant contributions. However, it is often the case that several modes

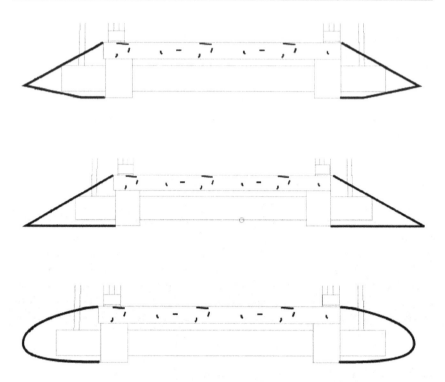

Figure 12.6 Some possible leading-edge fairing configurations for minimizing bridge-deck vibrations due to vortex shedding and flutter.

are significant. The following approach (Holmes, 1999) gives the correct method of combining inertial force distributions from more than one resonant mode of vibration. This approach is consistent with the weighting factor method discussed in Section 5.4.5.

The mean square fluctuating value of a load effect, r, resulting from the resonant response in mode j, can be written as:

$$\sigma_{rj}^2 = \alpha_j^2 \omega_j^4 \overline{a_j^2} \tag{12.8}$$

where the displacement response of the deck is written as:

$$y(x,t) = \phi_j(x) \cdot a_j(t)$$

$\phi_j(x)$ is the mode shape, and $a_j(t)$ is the modal coordinate for the jth mode, ω_j is the circular frequency in mode j $(= 2\pi n_j)$

$$\alpha_j \text{ is the integral}: \int_0^L m(x) \cdot \phi_j(x) i(x) dx \tag{12.9}$$

$m(x)$ is the mass per unit length,
$i(x)$ is the influence line for the load effect in question.

The contribution of the load at each spanwise position to the load effect, is the product of the inertial load on a small increment of span, centred at that position, $m(x) \cdot \omega_j^2\, y(x, t)\, \delta x$, is multiplied by the influence function, $i(x)$. Equation (12.8) is then obtained by integration of the contributions over the span, L, squaring and taking the mean value.

The *total* mean square fluctuating response is then obtained by summing the contributions from the N contributing modes.

$$\sigma_r^2 = \sum_{j}^{N} \alpha_j^2 \omega_j^4 \overline{a_j^2} \tag{12.10}$$

To obtain Equation (12.10), we have assumed that the modes are well separated, and hence the resulting responses can be assumed to be uncorrelated with each other.

The *envelope* of the combined dynamic loadings at each point along the span of a bridge can be obtained by taking the root sum of squares of the inertial loads from the contributing modes along the span, and adding to the mean loading, as in Equation (12.11):

$$f_{\mathrm{env}}(x) = \overline{f}(x) \pm \left[\sum_{j}^{N} \left(m(x) \cdot \omega_j^2 \phi_j(x) \right)^2 \overline{a_j^2} \right]^{1/2} \tag{12.11}$$

where $\overline{f}(x)$ is the mean wind loading at x.

Note that the envelope is independent of the influence line $I(x)$ of the load effect. It represents the limits within which the effective static load distributions for all load effects must lie.

The contribution of each mode to the total static equivalent load, corresponding to a peak load effect (e.g. a bending moment at any point along the span) depends on the shape of the influence line for that load effect. Thus, there is not a single static equivalent load. Equation (12.12) gives the weighting factor to be applied to obtain the contribution from mode j to the combined inertial load for a *root-mean-square* value of a given load effect, when a total of N modes contributes to it.

$$W_j' = \frac{\alpha_j \omega_j^4 \overline{a_j^2}}{\left\{ \sum_{j}^{N} \alpha_j^2 \omega_j^4 \overline{a_j^2} \right\}^{1/2}} \tag{12.12}$$

It can be demonstrated that Equation (12.12) will result in the correct mean square fluctuating response, as given by Equation (12.10).

The effective loading distribution for the root mean square fluctuating response, σ_r, obtained by summing over all modes is:

$$p'_{\text{eff}}(x) = m(x) \sum_j^N W'_j \phi_j(x) \tag{12.13}$$

The total root mean square fluctuating response is then,

$$\sigma_r = \int_0^L p'_{\text{eff}}(x)\, i\,(x)\,dx = \int_0^L i(x) \cdot m(x) \sum_j^N W'_j \phi_j(x)\, dx$$

$$= \frac{\sum_j^N \alpha_j^2 \omega_j^4 \overline{a_j^2}}{\left\{ \sum_j^N \alpha_j^2 \omega_j^4 \overline{a_j^2} \right\}^{1/2}} = \left\{ \sum_j^N \alpha_j^2 \omega_j^4 \overline{a_j^2} \right\}^{1/2}$$

which agrees with Equation (12.10).

The weighting factor for the contribution from mode j to the effective static loading for the *peak* (maximum or minimum) load effect, r, in a specified time period, T, can be closely approximated as:

$$W_i = \frac{\left\{ \sum_j^N \alpha_j^2 g_j^2 \omega_j^4 \overline{a_j^2} \right\}^{1/2} \alpha_j \omega_j^{\,4} \overline{a_j^{\,2}}}{\sum_j^N \alpha_j^2 \omega_j^4 \overline{a_j^2}} \tag{12.14}$$

where g_j is an expected peak factor for the response in mode j

Equation (12.14) can be obtained from Equation (12.12), as shown below:

$$W_j = g_r \cdot W'_j$$

where, g_r is the peak factor for the response, which can be approximated quite accurately by Equation (12.15).

$$g_r \cong \frac{\left\{ \sum_j^N \alpha_j^2 g_j^2 \omega_j^4 \overline{a_j^2} \right\}^{1/2}}{\left\{ \sum_j^N \alpha_j^2 \omega_j^4 \overline{a_j^2} \right\}^{1/2}} \tag{12.15}$$

This is a weighted average of the peak factors for the various modes. When only one mode is significant, Equation (12.14) reduces to Equation (12.16).

$$W_j = g_j \omega_j^2 \left(\overline{a_j^2} \right)^{1/2}$$ (12.16)

i.e. simply the peak inertial force in the mode, j.

Note that Equation (12.16) is independent of a_j, and hence of the influence line $i(x)$.

The contribution to the total inertial loading from mode j at a given spanwise position is then given by the product of W_j with the mass/unit length, $m(x)$, and the mode shape at that position. The total effective static loading for the peak load effect, r, is then given by Equation (12.17).

$$f_{\text{eff}}(x) = \overline{f}(x) + m(x) \sum_{j}^{N} W_j \phi_j(x)$$ (12.17)

The effective static loading depends on the influence line for r through the parameter a_j. Thus, the effective static loading will be different for load effects, such as bending moments at different spanwise positions. If the influence line is symmetrical about the centre of the bridge, as for example that for the bending moment at centre span, then a_j will be zero for anti-symmetric modes, i.e. only symmetric modes will contribute.

It should also be noted that since g_r from Equation (12.15) can be either positive or negative, the second term on the right-hand side of Equation (12.17) can also be either positive or negative, i.e. it may add or subtract from the mean loading.

12.4 WIND-TUNNEL TECHNIQUES

The verification of aerodynamic stability and determination of response to wind of long-span bridges, for structural design, is still largely an experimental process, making use of modern wind-tunnel techniques. Some of the experimental techniques were discussed in Chapter 7 (Sections 7.6.3 and 7.6.4).

A full wind-tunnel test series for a major long-span bridge might consist of all, or some, of the following phases.

- Section model tests to determine basic static aerodynamic force and moment coefficients (Section 12.2) for the deck section.
- Section model free or forced vibration tests to determine the aerodynamic or flutter derivatives (Sections 5.5.3 and 12.3.2).

- Section model tests in which the natural frequencies in vertical translation and rotation are scaled to match those of the prototype bridge, and critical flutter speeds are thence determined by slowly increasing the wind-tunnel speed (Section 7.6.3). This may be done in both smooth (low turbulence) and turbulent flow. (An alternative method which better reproduces the mode shapes of the prototype bridge is the 'taut strip' method described in Section 7.6.3).
- Scaled aeroelastic models of the completed bridge, i.e. deck, towers, and cables, tested in turbulent boundary-layer flow (Section 7.6.4). The multi-mode aeroelastic modelling scales the various parts of the bridge for elastic properties, mass (inertial), as well as geometric properties. Such tests are quite expensive, with much of the cost involved in the model designing and manufacturing.
- Scaled aeroelastic partial models of the bridge in various stages of erection. In most cases, the erection stages find a bridge in its most vulnerable state with respect to wind loading, with lower frequencies making them more prone to turbulent buffeting (Section 12.3.3) and lower flutter speeds, since flutter instabilities tend to occur at constant *reduced* velocity. The erection stage tests may include separate aeroelastic tests of the bridge towers as free-standing structures.

A complete series of tests as outlined above may require two or three different wind tunnels. The wind-tunnel testing of bridges tends to be a specialist activity for wind-tunnel laboratories, with few facilities being capable of carrying out all the above-listed tests. Some facilities restrict their involvement to section testing for bridge decks; others only carry out boundary-layer wind-tunnel tests. However, it should be noted that, to satisfactorily carry out aeroelastic tests on full models of the largest suspension bridges, a test section of at least 10 m width (e.g. Figure 12.7) is required. Few boundary layer wind tunnels are of this size.

Figure 12.7 A full aeroelastic model of a cable-stayed bridge in a construction stage.

12.5 VIBRATION OF BRIDGE CABLES

As the spans of cable-stayed bridges have increased and the cables themselves have become longer, cable vibration has become more of a problem. One of the more interesting excitation mechanisms, and until fairly
recently, least-understood ones, is the so-called "rain-wind" vibration. In
the following sections, the history of occurrences of this phenomenon,
suggested excitation mechanisms, and methods of mitigating the vibrations are reviewed.

12.5.1 Rain-wind vibration

The first, clearly-defined occurrence of wind-induced cable vibration, during which the presence of rain was an essential feature, was observed during the construction of Meiko-Nishi Bridge at Nagoya Harbour, Japan, in
1984. Low frequency (1–3 Hz) vibrations of some cables, with double amplitudes up to 300 mm were observed, over a 5-month period. This bridge has
a main span of 405 m with cables up to 165 mm in diameter, and lengths
varying from 65 to 200 m. The vibrations occurred in wind speeds between
7 and 14 m/s; these speeds greatly exceeded the critical wind speeds for
vortex shedding at the low frequencies observed. Using a section of polyethylene pipe casing from the prototype structure, wind-tunnel tests were
conducted, with and without simulated rain, and it was clearly established
that the rain was necessary to induce vibration over a defined range of wind
speeds (Hikami and Shiraishi, 1988).

Later, it was found that rain-wind induced vibration had occurred on
six bridges in Japan. A common feature was that the vibrating cables
were usually sloping downwards in a downwind direction, with the wind
approaching obliquely to the plane of the cable (Figure 12.8). Vibrations
were apparently observed only for cables encased in polyethylene.

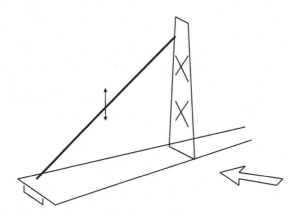

Figure 12.8 Typical cable/wind orientation for rain-wind vibration (Matsumoto *et al.*, 1993).

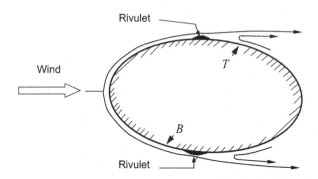

Figure 12.9 Flow separations produced by rivulets of rain water.

Outside Japan, rain-wind vibration of bridge cables have been observed on the Faroe Bridge (Denmark), Bretonne Bridge (France), Koehlbrand Bridge (Germany), Normandie Bridge (France), Fred Hartman Bridge (USA) and the Anzac Bridge (Sydney, Australia). Many other bridges have experienced cable vibrations – some from different mechanisms such as high-frequency vortex-shedding excitation, or from unknown or undefined mechanisms.

12.5.2 Excitation mechanisms

The wind-tunnel studies were carried out as the vibrations were observed on the Meiko-Nishi Bridge, indicating that the motion was induced by the presence of two water "rivulets", that oscillated in circumferential position with the cable motion. At low wind speeds, a single rivulet formed on the underside. Motion commenced at higher wind speeds when a second rivulet formed on the upper surface. The rivulets act as trigger points to promote flow separation on the vibrating cable, as shown in Figure 12.9. In this figure, the effective cross-wind shape is postulated to be elliptical. Other observations have suggested that the circumferential motion was not two-dimensional, and that the width and depth of the rivulet on the upper surface was less than that on the lower surface.

Wind-tunnel tests in France for the Normandie Bridge (Flamand, 1994) showed that carbon combustion products deposited on the surface of the casing were necessary for aerodynamic instability to occur, indicating the role played by surface tension in allowing the water rivulet to be maintained.

Fundamental wind-tunnel model studies of inclined cable aerodynamics, with and without rain, have been made at various angles of pitch (inclination), yaw, and rivulet position (Matsumoto *et al.*, 1993). Severe vibrations usually commenced at a reduced wind velocity of $(U/n_c b)$ (where U is the wind velocity, n_c is the cable frequency and b is the diameter) of about 40.

Matsumoto (2012) summarized more than 25 years of Japanese research into the excitation mechanisms for stay cable vibration. He concluded that inclined cable vibration is a form of galloping instability, but not one that can be explained by the 'quasi-steady' theory of Section 5.5.2. He suggested that rivulet formation, axial flow along the cable in the wake, and Reynolds Number variations were all contributing factors to the 'unsteady galloping' mechanism.

The criterion of the Federal Highway Administration (FHWA, 2007) suggests that if a cable has sufficient damping so that it has a Scruton Number (defined according to Equation 5.48 or 11.14) of 125 or greater, then rain-wind vibrations will not occur. If this is not possible to achieve, then other mitigation options may be adopted, as discussed in the following section.

12.5.3 Solutions

The solutions that have been successful in eliminating, or mitigating rain-wind induced vibration of bridge cables can be divided into the following categories:

- Aerodynamic treatments – i.e. geometrical modifications of the outer cable casing.
- Auxiliary cable ties.
- Auxiliary dampers.

Model measurements were carried out by Miyata *et al.* (1994), on sections of cable models with the same diameter as full-size cables, with a variety of roughened surface treatments (Figure 12.10). Discrete roughness of

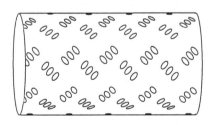

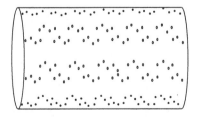

Figure 12.10 Surface roughness treatments for cable vibration mitigation (Miyata *et al.*, 1994).

about 1% of the diameter was found to be effective in suppressing rain-wind induced vibration. The explanation was that supercritical flow was promoted at lower Reynolds Numbers than would occur on cables with smooth surface finish.

Wind-tunnel tests in France (Flamand, 1994) found that parallel surface projections did stabilize a cable model, but produced a high drag coefficient in the super-critical Reynolds Number range. An alternative solution which minimized the drag increase, was adopted, namely the use of a double helix spiral, 1.3 mm high, 2 mm wide, and with a pitch of 0.6 m. This configuration was adopted for the Normandie Bridge.

Usually only one or two stay cables from a harp or fan array, will experience rain-wind vibration in particular atmospheric conditions. This observation led to a solution that has been used on several bridges, especially with cable cross-ties. They have also been used on the Normandie Bridge, where they are known as "aiguilles". They have been adopted for the Dane Point Bridge, Florida, USA, the Fred Hartman Bridge, Texas, and the Tatara Bridge, Japan (Figure 12.11).

A fundamental study of damping in stay cables, and of the effectiveness of cross ties, was carried out by Yamaguchi and Fujino (1994). Measurements on cables of a typical cable-stayed bridge indicated a range of critical damping ratios, from about 0.001 to 0.003, for the first mode, with lower values occurring for the low sag ratios, i.e. a higher pre-stress. A laboratory experiment on cross-ties showed that a 'stiff' cross-tie performed a function of transferring vibration energy from a vibrating cable to its neighbours. By use of 'soft' cross- ties, energy could also be dissipated in the cross-ties, making this system more effective.

Energy dissipation can also be provided by auxiliary damping devices mounted between the cable and the bridge girder, near the connection

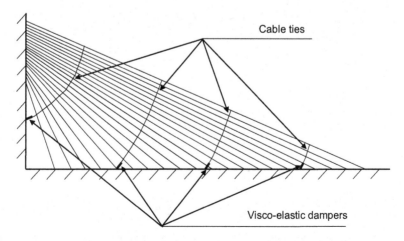

Cable ties

Visco-elastic dampers

Figure 12.11 Cable ties for vibration mitigation used on the Tatara Bridge.

points. This solution is more expensive than the cross-tie method, but more aesthetically pleasing. Oil dampers and visco-elastic dampers (Section 9.9.2) have been used for this purpose.

12.6 CASE STUDIES AND OTHER TOPICS

The literature on the aerodynamics of long-span bridges is extensive, and many papers on this subject contain references to particular bridges for illustration purposes. Sections 12.3.1 and 12.5 contain several examples in relation to vortex-shedding, induced vibrations and cable vibrations, respectively. Holmes (1999) has described the application of the equivalent static load method (Section 12.3.4) to generate design loadings for the Baram River (Malaysia) having a cable-stayed bridge.

The extensive wind engineering studies carried out for two of the longest bridges in the world were described by Miyata *et al.* (1992) for the Akashi Kaikyo Bridge, and by Reinhold *et al.* (1992) and Larsen and Jacobsen (1992) for the Great Belt East Bridge (Denmark). The determination of flutter derivatives for the Great Belt Bridge was described by Poulsen *et al.* (1992), and Jensen *et al.* (1999) using full-scale measurements on the completed bridge to estimate structural and aerodynamic damping.

The extensive wind-tunnel model testing, including those for various construction stages of the Izmit Bay Bridge (Turkey), a suspension bridge with a main span of 1,550 m completed in 2016, was described by Diana *et al.* (2013). The wind design of the Normandie Bridge in France (cable-stayed with a main span of 856 m), from the designer's point of view, was well covered by Virlogeux (1992).

Buljac *et al.* measured the static and aeroelastic force coefficients on wind-tunnel models of sections of the decks of three well-known bridges: Golden Gate Bridge (USA – suspension bridge), Kao-Pin Hsi Bridge (Taiwan – cable-stayed) and Great Belt Bridge (Denmark – suspension bridge). This work included investigations of the effects of wind barriers of various porosities (see Section 12.3.4).

Full-scale observation data of wind-induced behaviour of the Golden Gate Bridge, crossing the San Francisco Bay (USA) were analysed by Tanaka and Davenport (1983), using spectral analysis. An experimental investigation on the same bridge was also conducted using a taut-strip model (see Section 7.6.3), with simulation of natural wind turbulence at laboratory scale. In a similar approach, Nikitas *et al.* (2011) investigated vibrations of the historic Clifton suspension bridge (Bristol, U.K.), in hourly-mean wind speeds up to 16 m/s, and used modern system identification techniques to determine the flutter derivatives for the deck.

Many long-span bridges have been constructed in China in recent years, and for a number of these the wind engineering studies have been documented in publicly-available sources. For example, Gu *et al.* (2001) describe the extraction of flutter derivatives for several deck cross sections considered for the Jiangyin Bridge, a suspension bridge of 1385m over the Yangtze River. Wind-loading codes and design standards are the subject of Chapter 15. However, it is worth noting here the general design standard, BD49/01, for aerodynamic effects on road bridges in the U.K., produced by the Highways Agency of England and the corresponding agencies in Wales, Scotland and Northern Ireland. This is a useful set of guidelines for preliminary assessment of possible aeroelastic effects such as flutter and vortex-induced vibrations. It also includes provisions for wind-tunnel testing (Highways Agency, 2001).

The work described in this chapter primarily relates to the aerodynamics of *road* bridges, which have been the subjects of the majority of research and experience in wind engineering. Most footbridges have spans of less than 100 m and are not dynamically wind-sensitive. However, there have also been several cable-supported *footbridges* built with main spans of 400 m or greater. Based on criteria developed for road bridges, these should have become aerodynamically unstable at relatively low wind speeds. This appears not to have occurred; this may well be due to the low aspect ratio of the deck cross-section – i.e. the narrow deck chord in comparison to the height of the side railings. However, apparently few long-span footbridge decks have been studied in wind-tunnel tests to verify this.

12.7 SUMMARY AND OTHER SOURCES

In this chapter, the aerodynamics of bridges has been presented in a summary form. Long-span road bridges are probably the most dynamically 'wind-sensitive' of all structures, with complex aerodynamics and dynamics, and are generally within the sphere of specialist bridge aerodynamicists. However, the main phenomena of vortex shedding, flutter and buffeting have been discussed in this chapter.

The vibration of the cables on cable-stayed bridges has become the limiting factor on their ultimate spans, and this topic, with alleviation measures, has been discussed in some detail.

This chapter has attempted to give an 'overview' of the main features of the aerodynamics of long-span, cable-supported bridges. For the specialist topic of long-span bridge, aerodynamics has been reviewed in several other review papers and textbooks. For example, Larsen (1992), Simiu and Scanlan (1996), Xu (2013) and Cai *et al.* (2014) have treated the subject in more detail. Xu (2013), in particular, covers the whole subject comprehensively, including several topics not discussed in this chapter, such as wind-bridge-vehicle interaction.

There are also many journal papers on bridge aerodynamics, and related bluff-body aerodynamics, in international journals such as the *Journal of Wind Engineering and Industrial Aerodynamics* (Elsevier), the *Journal of Fluids and Structures* (Elsevier) and *Wind and Structures* (Techno Press).

REFERENCES

Baker, B. (1884) The Forth Bridge. *Engineering*, 38: 213.

Billington, D.P. (1977) History and esthetics in suspension bridges. *ASCE, Journal of the Structural Division*, 103: 1655–1672.

Buljac, A., Kozmar, H., Pospíšil, S. and Macháček, M. (2017) Flutter and galloping of cable-supported bridges with porous wind barriers. *Journal of Wind Engineering and Industrial Aerodynamics*, 171: 304–318.

Cai, S.C.S., Zhang, W, and Montens, S. (2014) Chapter 22: Wind effects on long-span bridges. In *Bridge engineering handbook*, 2nd Edition. ed W-F. Chan and L. Duan. CRC Press, Boca Raton, FL.

Davenport, A.G. (1962) Buffeting of a suspension bridge by storm winds. *ASCE, Journal of the Structural Division*, 88: 233–268.

Diana, G., Yamasaki, Y., Larsen, A., Rocchi, D., Giapinno, S., Argentini, T., Pagani, A., Villiani, M., Somaschino, C. and Portentoso, C. (2013) Construction stages of the long span suspension Izmit Bay Bridge: wind tunnel test assessment. *Journal of Wind Engineering and Industrial Aerodynamics*, 123: 300–310.

Federal Highway Administration (FHWA). (2007) Wind-induced vibration of stay cables. TechBrief. FHWA-HRT-05-084, July.

Flamand, O. (1994) Rain-wind induced vibrations of cables. *International Conference on Cable-stayed and Suspension Bridges*, Deauville, France, 12–15 October.

Frandsen, J.B. (2001) Simultaneous pressures and accelerations measured full-scale on the Great Belt East suspension bridge. *Journal of Wind Engineering and Industrial Aerodynamics*, 89: 95–129.

Fumoto, K., Hata, K. and Miyasaki, M. (2007) Basic research on the feasibility of super long suspension bridges. *12th International Conference on Wind Engineering*, Cairns, Queensland, Australia, 1–6 July.

Ge, Y.J. and Tanaka, H. (2000) Aerodynamic flutter analysis of cable-supported bridges by multi-mode and full-mode approaches. *Journal of Wind Engineering and Industrial Aerodynamics*, 86: 123–153.

Gu, M., Zhang, R. and Xiang, H. (2001) Parametric study on flutter derivatives of bridge decks. *Engineering Structures*, 23: 1607–1613.

Gimsing, N.J. (1983) *Cable-supported bridges*. John Wiley, New York.

Highways Agency (UK) and others (2001) Design rules for aerodynamic effects on bridges, BD 49/01.

Hikami, Y. and Shiraishi, N. (1988) Rain-wind induced vibrations of cables in cable-stayed bridges. *Journal of Wind Engineering and Industrial Aerodynamics*, 29: 409–418.

Holmes, J.D. (1975) Prediction of the response of a cable-stayed bridge to turbulence. *Proceedings, 4th International Conference on Wind Effects on Buildings and Structures*, London, UK, 8–12 September, pp. 187–197, Cambridge University Press, Cambridge, UK.

Holmes, J.D. (1979) Monte Carlo simulation of the wind-induced response of a cable-stayed bridge. *Proceedings, 3rd. International Conference on Applications of Statistics and Probability in Soil and Structural Engineering (ICASP-3)*, Sydney, Australia, pp. 551–565, University of New South Wales.

Holmes, J.D. (1999) Equivalent static load distributions for resonant dynamic response of bridges. *Proceedings, 10th International Conference on Wind Engineering,* Copenhagen, Denmark, 21–24 June, pp. 907–911, A.A. Balkema, Rotterdam.

Huston, D.R., Bosch, H.R., Scanlan, R.H. (1988) The effect of fairings and of turbulence on the flutter derivatives of a notably unstable bridge deck. *Journal of Wind Engineering and Industrial Aerodynamics,* 29: 339–349.

Irwin, H.P.A.H. (1977) Wind tunnel and analytical investigations of the response of Lions' Gate Bridge to a turbulent wind. National Aeronautical Establishment, Canada. Laboratory Technical Report, LTR-LA-210, June 1977.

Irwin. H.P.A.H. and Schuyler, G.D. (1978) Wind effects on a full aeroelastic bridge model. *ASCE Spring Convention,* Pittsburgh, Pennsylvania, 24-28 April.

Jancauskas, E.D. (1986) The aerodynamic admittance of two-dimensional rectangular section cylinders in smooth flow. *Journal of Wind Engineering & Industrial Aerodynamics,* 23: 395–408.

Jensen, J.L., Larsen, A., Andersen, J.E. and Vejrum, T. (1999) Estimation of structural damping of Great Belt suspension bridge. *Proceedings of the 4th European Conference on Structural Dynamics (Eurodyn '99),* Prague, Czech Republic, 7–10 June, 1999, A.A. Balkema, Rotterdam.

Kernot, W.C. (1893) Wind pressure. *Proceedings, Australasian Society for the Advancement of Science,* V: 573–81.

Konishi, I., Shiraishi, N. and Matsumoto, M. (1975) Aerodynamic response characteristics of bridge structures. *Proceedings, 4th Internal Conference on Wind Effects on Buildings and Structures,* London, UK, 8–12 September, pp. 199–208, Cambridge University Press, Cambridge, UK.

Larsen, A. (ed.) (1992) Aerodynamics of large bridges. *Proceedings of the 1st International Symposium on Aerodynamics of Large Bridges,* Copenhagen, Denmark, 19–21 February, 1992, A.A. Balkema, Rotterdam.

Larsen, A. and Jacobsen, A.S. (1992) Aerodynamic design of the Great Belt East Bridge. *Proceedings of the 1st International Symposium on Aerodynamics of Large Bridges,* Copenhagen, Denmark, 19–21 February, 1992, pp. 269–283, A.A. Balkema, Rotterdam.

Larsen, A., Esdahl, S., Andersen, J.E. and Vejrum, T. (1999) Vortex shedding excitation of the Great Belt suspension bridge. Proceedings, *10th International Conference on Wind Engineering,* Copenhagen, Denmark, 21–24 June, pp. 947–954, A.A. Balkema, Rotterdam.

Li, M., Li, M. and Yang, Y. (2018) A statistical approach to the identification of the two-dimensional aerodynamic admittance of streamlined bridge decks. *Journal of Fluids and Structures,* 83: 372–385.

Matsumoto, M. (2012) On galloping instability of stay cables of cable-stayed bridges. *2012 World Congress on Advances in Civil, Environmental and Materials Research,* Seoul, Korea, 26–30 August.

Matsumoto, M., Shirato, H., Saito, H., Kitizawa, H. and Nishizaki, T. (1993) Response characteristics of rain-wind induced vibration of stay cables of cable-stayed bridges. *1st. European-African Regional Congress on Wind Engineering,* Guernsey, 20–24 September.

Melbourne, W.H. (1979) Model and full-scale response to wind action of the cable-stayed, box-girder, West Gate Bridge. *IAHR/IUTAM Symposium on Flow-Induced Vibrations,* Karlsruhe, Germany, September 3–8.

Miyata, T., Yokoyama, Y., Yasuda, M. and Hikami, Y. (1992) Akashi Kaikyo Bridge: wind effects and full model tests. *Proceedings of the First International Symposium on Aerodynamics of Large Bridges*, Copenhagen, Denmark, 19–21 February, 1992, pp. 217–236, A.A. Balkema, Rotterdam.

Miyata, T., Yamada, H. and Hojo, T. (1994) Aerodynamic response of PE stay cables with pattern indented surface. *International Conference on Cable-stayed and Suspension Bridges*, Deauville, France, 12–15 October.

Nikitas, N., Macdonald, J.H.G. and Jakobsen, J.B. (2011) Identification of flutter derivatives from full-scale ambient vibration measurements of the Clifton Suspension Bridge. *Wind and Structures*, 14: 221–238.

Ostenfeld, K.H. and Larsen, A. (1992) Bridge engineering and aerodynamics. *Proceedings of the 1st International Symposium on Aerodynamics of Large Bridges*, Copenhagen, Denmark, 19–21 February, 1992, pp. 3–22, A.A. Balkema, Rotterdam.

Poulsen, N.K., Damsgaard, A. and Reinhold, TA. (1992) Determination of flutter derivatives for the Great Belt bridge. *Journal of Wind Engineering & Industrial Aerodynamics*, 41–44: 153–164.

Petroski, H. (1996) *Engineers of dreams*. Vintage Books, New York.

Reinhold, T.A., Brinch, M. and Damsgaard, A. (1992) Wind-tunnel tests for the Great Belt link. *Proceedings of the First International Symposium on Aerodynamics of Large Bridges*, Copenhagen, Denmark, 19–21 February, 1992, pp. 255–267, A.A. Balkema, Rotterdam.

Sabzevari, A. and Scanlan, R.H. (1968) Aerodynamic instability of suspension bridges. *ASCE, Journal of the Engineering Mechanics Division*, 94: 489–519.

Selberg, A. (1961) Aerodynamic effects on suspension bridges. Acta *Polytechnica Scandinavica*, Engineering and Building Construction Series No. 13, Trondheim, Norway.

Selberg, A. (1963) Aerodynamic effects on suspension bridges. *Proceedings, International Conference on Wind Effects on Buildings and Structures*, Teddington, UK, 26–28 June, pp. 462–486.

Scanlan, R.H. and Gade, R.H. (1977) Motion of suspended bridge spans under gusty winds. *Journal of the Structural Division ASCE*, 103: 1867–1883.

Scanlan, R.H. and Lin, W-H. (1978) Effects of turbulence on bridge flutter derivatives. *Journal of the Engineering Mechanics Division ASCE*, 104: 719–733.

Scanlan, R.H. and Tomko, J.J. (1971) Airfoil and bridge flutter derivatives. *Journal of the Engineering Mechanics Division ASCE*, 97: 1717–1737.

Shiraishi, N. and Matsumoto, M. (1977) Aerodynamic responses of bridge structures subjected to strong winds. *Symposium on Engineering for Natural Hazards*, Manila, September.

Simiu, E. and Scanlan, R.H. (1996) *Wind effects on structures – fundamentals and applications to design*, 3rd Edition. John Wiley, New York.

Smith, I.J. (1980) Wind induced dynamic response of the Wye Bridge. *Engineering Structures*, 2: 202–208.

Steinman, D.B. and Watson, S.R. (1957) *Bridges and their builders*. Dover Publications Inc., New York.

Tanaka, H. and Davenport, A.G. (1983) Wind-induced response of Golden Gate Bridge. *Journal of the Engineering Mechanics Division, ASCE*, 109: 296–312.

van Nunen, J.W.G. and Persoon, A.J. (1982) Investigation of the vibrational behaviour of a cable-stayed bridge under wind loads. *Engineering Structures*, 4: 99–105.

Virlogeux, M. (1992) Wind design and analysis for the Normandy Bridge. *Proceedings of the 1st International Symposium on Aerodynamics of Large Bridges*, Copenhagen, Denmark, 19–21 February, 1992, pp. 183–216, A.A. Balkema, Rotterdam.

Virlogeux, M. (1999) Recent evolution of cable-stayed bridges. *Engineering Structures*, 21: 737–755.

Wardlaw, R.L. (1971) Some approaches for improving the aerodynamic stability of bridge road decks. *3rd International Conference on Wind Effects on Buildings and Structures*, Tokyo, Japan, 6–9 September, Saikon Shuppan Co., Tokyo.

Xu, Y-L. (2013) *Wind effects on cable-supported bridges*. John Wiley, New York, ISBN 978-1-118-18828-6.

Yamaguchi, H. and Fujino, T. (1994) Damping in cables in cable-stayed bridges with and without damping control measures. *International Conference on Cable-stayed and Suspension Bridges*, Deauville, France, 12–15 October.

Chapter 13

Transmission lines

13.1 INTRODUCTION

Electrical transmission lines and their supporting towers are, like other structures, subjected to severe wind storms of various types, and their safe and economic design for wind loading is of concern to the power utilities. There are significant differences between the response of high-voltage transmission towers and other structures to wind:

- They are structurally designed with generally lower safety margins against collapse than other structures.
- The overall length of a transmission line system is relevant when considering probability and risk of a complete line receiving strong winds from localised wind storms such as thunderstorm downbursts and tornadoes.

This chapter deals with the wind loading of the suspended lines, and risk issues associated with a long transmission line as a system. The wind loading of the supporting towers and poles is covered in Chapter 11.

13.2 STRUCTURAL RESPONSE AND CALCULATION OF WIND LOADS

Basic design data for wind loads on transmission line conductors in temperate synoptic winds has been compiled by the American Society for Civil Engineers (ASCE, 2020) and CSIR in South Africa (CSIR, 1990).

13.2.1 Nature of the response

Fortunately, resonant dynamic response does not appear to be a major problem with transmission line systems. Although the suspended lines themselves usually have natural frequencies less than 1 Hz, the resonant response is largely damped out because of the very large aerodynamic damping (Section 5.5.1) (e.g. Matheson and Holmes, 1981).

The natural frequencies of supporting towers up to 50 m in height are normally greater than 1 Hz, and hence the resonant response is also negligible. Thus, except for extremely tall supporting towers and long line spans, we can safely compute the peak response of a transmission line system, neglecting the resonant dynamic response. Then, the peak response is directly related to the instantaneous gusts upwind, and hence transmission line structures can be designed using gust wind speeds. However, because of the non-uniform spatial gust structure, assumption of the same peak gust along the full span is conservative; this leads to the concept of a *span reduction factor*.

For those cases, where resonant response is significant, i.e. very high supporting towers, and very long spans, a simplified random response model of the tower-line combination, based on the gust response factor concept is available (Davenport, 1979).

13.2.2 Wind forces on conductors

The nominal wind force acting on a single conductor perpendicular to the span can be taken to be:

$$F_c = q_{zc} \cdot C_d \cdot A_c \sin^2 \theta \cdot \alpha \tag{13.1}$$

where q_{zc} is the free-stream dynamic wind pressure (= $\frac{1}{2}\rho_a \hat{U}_{zc}^2$) at a suitable mean conductor height, z_c. A suitable value for z_c is shown in Figure 13.1, taken from the South African Recommendations for transmission line loading (CSIR, 1990).

C_d is the drag force coefficient for the conductor,
A_c is the reference area, which may be taken as $s \times b$, where s is the wind span (see Figure 13.1), and b is the conductor diameter,
θ is the horizontal angle of incidence of the wind in relation to the direction of the line,
α is a span reduction factor.

The ASCE Guidelines show experimental data for the drag force coefficient as a function of Reynolds Number, Re (Section 4.2.4) for several conductor types, based on wind tunnel tests. This data is reproduced in Figure 13.2. The Reynolds Number can be calculated by:

$$Re = \frac{U_{zc}b}{15 \times 10^{-6}} \tag{13.2}$$

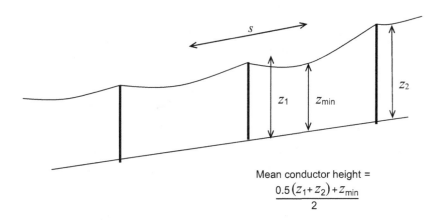

Mean conductor height =
$$\frac{0.5\left(z_1+z_2\right)+z_{min}}{2}$$

Figure 13.1 Mean conductor height for calculation of wind loads (CSIR, 1990).

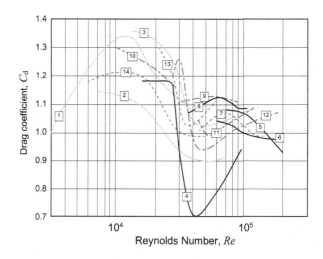

Figure 13.2 Drag force coefficients for conductors (ASCE, 2020).

where U_{zc} is the design gust wind speed in metres per second at the mean conductor height, z_c. The conductor diameter, b, is in metres.

The South African design recommendations (CSIR, 1990) have simplified the data to give the design line shown in Figure 13.3.

13.2.3 Span reduction factors

The span reduction factor, α, allows for the reduction in peak wind along the span of a conductor, due to non-simultaneous action of the gusts. Since it is determined by the structure of turbulence in the approaching wind

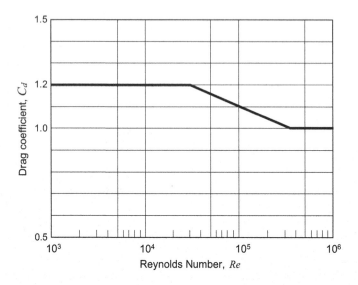

Figure 13.3 Design values of conductor drag coefficient recommended by CSIR (1990).

flow, the span reduction factor is a function of the approach terrain, the mean conductor height and the span. This factor has a direct relationship with the gust response factor, G (Section 5.3.2). The relationship is in Equation (13.3).

$$\alpha = G \left(\frac{\bar{U}_z}{\hat{U}_z} \right)^2$$

(13.3)

where $\bar{U}_z$ is the mean wind speed at height z, and $\hat{U}_z$ is the gust speed at the same height.

Span reduction factors are insensitive to the conductor height, and the following equations can be used to predict values of α.

$$\alpha = 0.58 + 0.42 \exp\left(\frac{-s}{180} \right) \quad \text{for rural terrain}$$

(13.4)

$$\alpha = 0.50 + 0.50 \exp\left(\frac{-s}{140} \right) \quad \text{for urban terrain}$$

(13.5)

where s is the span in metres.

In Table 13.1, values of span reduction factor for various spans have been calculated using Equations (13.4) and (13.5). Clearly, the span reduction factor reduces with increasing span, and with increasing terrain roughness.

Table 13.1 Span reduction factors for transmission line conductors

Conductor span (m)	Rural terrain ($z_o \cong 0.02\,m$)	Urban terrain ($z_o \cong 0.2\,m$)
200	0.72	0.62
300	0.66	0.56
400	0.63	0.53
500	0.61	0.51

In the latter case, the reduction occurs because of the increased fluctuating component in the peak load on the line.

Note that the above values are applicable only to the case of large-scale synoptic winds. Span reduction factors are greater in thunderstorm down-bursts, an intense local storm type that are a major source of failure of transmission lines in many countries.

13.2.4 Conductor shielding

In both the ASCE Guidelines (ASCE, 2010) and the CSIR Recommendations (CSIR, 1990), no allowance for shielding for individual conductors in a bundle is permitted. Such shielding effects would be small, and would not be present for every angle of attack of the instantaneous wind to the line.

13.2.5 Wind forces on lattice supporting towers

The calculation of wind forces on lattice towers typical of those used in high-voltage transmission line system is discussed in Section 11.3.2. The overall drag coefficients for lattice towers depends upon the solidity of the towers. Higher solidity results in greater mutual interference and shielding, and a reduction in drag coefficient, based on the projected area of members.

13.3 RISK MODELS FOR TRANSMISSION LINE SYSTEMS

Transmission line systems often extend for several hundred kilometres, and are prone to impact by small intense local windstorms, such as tornadoes (Section 1.3.4) and downbursts (Section 1.3.5). There has been a history of failures of transmission line systems from these events, especially in large continental countries like Australia, Brazil and Argentina (e.g. Hawes and Dempsey, 1993). Figure 13.4 shows the result of one such event. The risk of failure of *any one tower* along a line is much greater than that for a single isolated structure. Design of the supporting structures requires knowledge of the total risk to complete lines that are susceptible to these small intense windstorms. Knowledge of the risk of failures enables a balance to be made

Figure 13.4 Failure of a high-voltage transmission tower following a local downburst event.

between the cost of failures, and the cost of replacement towers. This may vary from country to country, as in some cases there are alternative routes for power transmission.

13.3.1 Tornado-risk model

Twisdale and Dunn (1983) describe several tornado risk models for point and 'lifeline' targets. Milford and Goliger (1997) developed a tornado risk model for transmission line design which considered normal intersection of a tornado with the line direction. A simplified probabilistic model is discussed in this section, considering both normal and oblique intersections of tornado tracks with the line target, representing a transmission line.

Since the width of tornado tracks (usually less than 100 m) is almost always less than the span length between towers, the critical factor in line failure is intersection of a tornado with a tower. Thus, the rate of intersection with a tower is required, rather than with the conductors.

Consider a region specified by its area, A_R (square kilometres), in which there is an average tornado occurrence of n events per year, so that the per square kilometre rate for the region is

$$v = n/A_R \qquad (13.6)$$

Normal intersection of a tornado path of length ℓ, and width, w, with a line of overall length L occurs only for those tracks whose centre falls within the zone of area, $L \times \ell$, adjacent to the line (see Figure 13.5), giving a rate of intersection, r,

$$r = vL\ell \qquad (13.7)$$

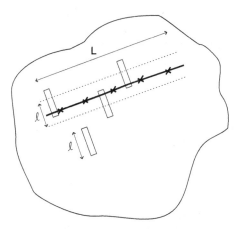

Figure 13.5 Normal intersection of a tornado with a transmission line system.

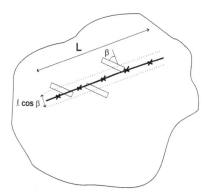

Figure 13.6 Oblique intersection of a tornado with a transmission line system.

This model can be extended to variable intersection angle as follows.

For a tornado path intersecting the transmission line at an angle β to normal (Figure 13.6), the width of the zone of intersection reduces to $\ell \cos \beta$ and the rate of intersection (with the line) per annum is now given by:

$$r = \upsilon L \ell \cos \beta \qquad (13.8)$$

Now the width of the intersection zone along the line is given by $w/\cos \beta$ and the probability of a *given single point* on the line which may represent a single tower falling within this zone is $w/(L \cos \beta)$. Thus, the number of intersections of tornadoes with this tower per year is given by:

$$r = \upsilon L \ell \cos \beta \cdot w / (L \cos \beta)$$

$$= \upsilon w \ell \qquad (13.9)$$

If the span between towers is s, the number of towers along a line of length, L, is equal to L/s, and assuming that intersections are independent (i.e. only one tower is intersected by any tornado), then the total number of intersections with *any* tower along the line per year is given by:

$$r_t = (vw\ell)(L / s) \tag{13.10}$$

It should be noted that the rate of intersection is independent of the intersection angle, β. Equation (13.10) may also be written as:

$$r_t = n(a/A_R)N = vaN \tag{13.11}$$

where n is the number of events per year in an area A_R, a is the area of tornado 'footprint' and N is the number of towers in the area.

Example 13.1

Assume : $L=500\,km$; $s=0.5\,km$; $\ell=5\,km$; $w=0.1\,km$; $v=10^{-4}/km^2/year$.
    Then, from Equation (13.10), the number of intersections with this line per year$=10^{-4}\times0.1\times5\times(500/0.5)=0.05$.
    i.e. average of 1 intersection every 20 years.

13.3.2 Downburst-risk model

Damage 'footprints' produced by severe thunderstorm downbursts (Section 1.3.5) are usually wider than those produced by tornadoes. The lengths of the damaged areas produced by downbursts, are generally shorter than those of tornadoes. The increased width usually results in several transmission line spans being enveloped by damaging winds, and several adjacent towers often fail as a group. The direct wind load on the conductors themselves is therefore a significant component of the overall wind load in downburst events. This must be incorporated into a risk model.

Oliver *et al.* (2000) describe a downburst risk model for transmission lines, which allows the prediction of an event frequency, where an event is the intersection of a region of wind above a given wind speed with a line of some defined length. The probability of such an event is dependent on:

- the overall length of the line, L;
- the relative angle, $\theta\text{–}\phi$, between the direction of the downburst path, θ, and the line orientation, ϕ;
- the probability of exceedence of the threshold wind speed of interest, U, at any point in the surrounding region derived from the anemometer records,
- the width of the path of winds above the threshold, w_u

The average recurrence interval, $R_{U,L}$, of the event was shown (Oliver *et al.*, 2000) to be given by:

$$R_{U,L} = (w_u / L) \Bigg/ \left\{ \sum_{i=1}^{N} \Pr\left(u > U/\left|\sin(\theta_i - \phi)\right|\right) \Pr(\theta_i)\left|\sin(\theta_i - \phi)\right| \right\} \quad (13.12)$$

where it is assumed that:

- there is an average or characteristic downburst damage footprint width associated with each wind speed U, given by w_u;
- for each direction, all downburst tracks can be represented in discrete directional ranges, centred on a characteristic direction θ_i and the summation is over each of these directions;
- the relative probability that the downburst should lie along each of these directions is directly related to the directional frequency of measured gusts;
- the distribution of wind speed, given a direction, is independent of the directional sector.

The presence of the overall line length, L, in the denominator of Equation (13.12) indicates that as the overall transmission line length increases, the average recurrence interval for damaging intersections decreases. Thus, in the case of very long lines orientated at right angles to the prevailing directions of severe thunderstorm winds, the risk of failure may be very high, if the above parameters are not taken into account. This is the experience in large continental countries such as Australia and Argentina, where many failures have occurred (e.g. Hawes and Dempsey, 1993).

An alternative model of downburst risk for transmission line systems has been developed for Argentina by Schwarzkopf and Rosso (2001).

Situations have occurred where multiple storm cells have produced line failures on several transmission lines. Unfortunately this can produce widespread 'blackouts' as occurred in South Australia on September 28, 2016. Modelling of such events with a relatively simple 'closed-form' equations, such as that of Equation (13.12), is difficult and a better approach is to use Monte Carlo-type simulation approach, similar to that used to predict the effects of tropical cyclones and hurricanes (Section 2.2.4).

For practical design purposes, the variation in overload risk to individual lines of various lengths and orientation can be incorporated approximately through multipliers for both variables and applied to the design wind speed. This is a suitable approach for a design code or standard.

A suitable form for a 'line-length' multiplier, M_L, has been found to be given by Equation (13.13).

$$M_L = 1 + (a - b \log_e R_L) \log_e L \quad (13.13)$$

where L is the line length in kilometres, and R_L is the specified value of average-recurrence interval for impact anywhere along the line of the design wind gust speed. a and b are parameters that may be determined by 'calibration' against a more rigorous evaluation of risk by use of Equation (13.12).

13.4 WIND-INDUCED VIBRATIONS OF TRANSMISSION LINES

Wind can induce vibrations in transmission lines through several different mechanisms, and over a range of frequencies and wind speeds. This topic is covered in specialist publications, usually originating within the electrical transmission line industry, and coverage in this book is restricted to summaries in the following sections.

13.4.1 Vortex-induced aeolian vibration

High-frequency cross-wind vibrations occur on transmission lines under certain atmospheric conditions, and were first observed as early as the 1920s. Early experiments (e.g. Varney, 1926) identified the exciting mechanism to be the regular shedding of vortices (see Section 4.6.3). It was observed that the vibrations consisted of travelling waves which are reflected at the points of support. Fatigue failures usually occurred at or near the point of support, as a result of local flexural stresses. It was also observed that vibrations occurred in relatively low wind speeds, and in conditions of low turbulence. Hence, significant vibrations are only observed on lines located in flat, open country, where the turbulence intensities (Section 3.3.1) are low. Several types of damping devices to mitigate this vibration have been devised over the years, and are widely used in the industry.

'Aeolian' vibration of transmission lines is covered in detail in many other publications (e.g EPRI, 1979 and 2017). Only the main features of the phenomenon, and mitigation measures will be discussed here.

The main conditions for significant aeolian vibrations of transmission lines are as follows:

- wind direction near normal to the line,
- mean wind speeds less than 7 m/s,
- low turbulence intensities – this may occur in thermally stable atmospheric conditions in which turbulence in the wind flow is likely to be suppressed.

Some characteristics of the vibration are:

- Excitation over short lengths of span by vortex shedding at high frequency. The Strouhal Number for the vortex shedding (Section 4.6.3) for typical transmission line cross sections is in the range of 0.18–0.19.

- 'Lock-in' (Section 5.5.4) may occur resulting in the vortex-shedding synchronizing with the vibration frequency,
- Amplitudes of vibrations are usually less than one diameter,
- Frequencies of vibrations are typically in the range of 10–150 Hz,
- Fatigue damage is typically initiated near supports by fretting or rubbing between strands of a conductor,
- Vibration amplitudes are enhanced by increasing line temperatures and reduced air temperatures.

Various types of damper have been devised to mitigate aeolian vibrations of transmission lines. These can broadly be categorized into impact dampers, and tuned mass dampers. An example of the former is the spiral type, which consists of a helically-wound spiral bar of length 1.2–1.5 m. The damper is attached to a conductor by a short 'gripping' section close to one end of a span. The cross-wind motion of the line is mitigated by impacts on the loosely wound part of the damper which has a diameter of about twice the diameter of the conductor (Figure 13.7).

Mass dampers for transmission lines come in several forms – such as the 'Stockbridge' or 'dog-bone' type (Figure 13.8). These are nominally 'tuned' mass dampers (Section 9.9.3), but in fact operate over a wide range of frequencies. Their performance is generally determined by experiment rather than by theoretical means.

13.4.2 Galloping vibrations

A common phenomenon in cold climates is the 'galloping' of iced-up conductors that involves cross-wind, vibrations of large amplitude and low frequency, produced by the 'galloping' instability mechanism discussed in Section 5.5.2. The presence of ice may change the shape of the cross

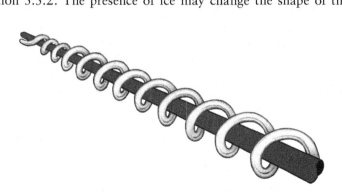

Figure 13.7 Spiral damper for mitigating aeolian (vortex-induced) vibrations of transmission line conductors.

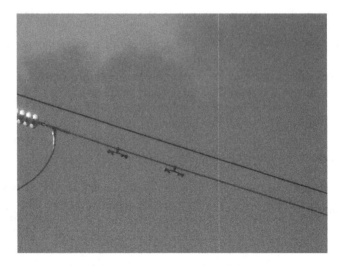

Figure 13.8 A pair of 'dogbone' dampers.

section to an asymmetric one with aerodynamic properties that make it prone to galloping instability. Galloping vibrations are in the lower modes of vibration – with a typical frequency range of 0.1–1 Hz. Large deflections amplitudes of up to 10 m can occur. This can result in line failure with a few hours from the onset of vibration.

Several devices designed to reduce or eliminate large-amplitude galloping vibrations were described by Richardson (1991). These generally act to increase the drag coefficient, C_D, and/or the lift curve slope ($dC_L/d\alpha$) and thus increasing the critical wind speed for galloping in Equation (5.46).

13.4.3 Wake-induced vibrations of bundled conductors

Wake-induced vibrations can occur when two or more circular cross sections are grouped together and have been commonly observed in 'bundled' power conductors. The motion in these vibrations involves rotation of the bundle as a group, as well as translation motion. Wake-induced vibrations typically occur in a low-frequency range (up to 10 Hz), with displacement amplitudes of up to 1 m.

A quasi-steady linear theory has been developed for a pair of circular cylinders, representing a pair of conductors, with one located in the wake for the other (e.g., Blevins, 1977). When the wind direction is such that the downwind cylinder is in the wake of the upwind one, but offset from the centerline, the downwind cylinder will develop lift as well as drag forces. For certain along-wind and cross-wind separations, this can lead to instability. No closed-form solution of the coupled equations for the deflections of the downwind cylinder, is available, and numerical solutions of a fourth-order stability polynomial are required. Given the dynamic properties of the cylinders, or conductors (i.e. mass distribution and frequencies), and the variation of lift and drag coefficients with spacing between the cylinders, stability boundaries can be determined. The system is stable – i.e. no vibrations, when either the downwind cylinder is immediately downwind of the upwind one, or it is outside the wake of the upwind cylinder (Blevins, 1977). However, if the downwind cylinder falls within certain regions in the wake, either side of the centerline, unstable oscillations can occur. The extent of these regions increases with increasing wind speed. Also, like galloping, there is a critical wind speed below which no oscillations will theoretically occur.

13.4.4 Turbulent buffeting

Like any other structure, or system, buffeting by atmospheric turbulence will lead to fluctuating response in sway of overhead lines, particularly in urban areas where the turbulence intensities are high. However, because of the high aerodynamic damping (see Section 5.5.1) on overhead lines in the along-wind direction - at high wind speeds this may exceed 20% of critical damping (Matheson and Holmes, 1981). The resonant contributions in low-frequency sway modes are generally small. However, quasi-static fluctuating stresses (i.e. 'background' response below the resonant frequencies) will still exist and over a long period of time may induce fatigue damage (see Section 5.6).

13.5 SUMMARY AND OTHER SOURCES

The available data for the specification of wind loads on transmission line structures has been critically reviewed in this chapter. Risk models that consider the risk of intersection of small intense storms such as tornadoes and downbursts with long transmission line systems are discussed. Wind-induced vibrations of transmission and other overhead lines by various mechanisms, and their mitigation are also described.

There are a number of documents published by professional industry groups to assist in the design of transmission lines and their supporting structures. The American Society of Civil Engineers has published several editions of ASCE Manual 74, *Guidelines for electrical transmission line structural loading* (ASCE, 2020). This is a comprehensive document covering all aspects of structural design including wind loading. Its limitations are that it is closely tied to the American loading Standard (ASCE 7) and may not be easily transposed to other countries.

The Electrical Power Research Institute (EPRI) has published several versions of a well-known monograph on wind-induced aeolian vibrations of conductors, and related fatigue problems (EPRI, 2017), and other industry groups like CIGRE (Conseil International des Grands Réseaux Électriques), and CEATI (Centre for Energy Advancement through Technological Innovation) also have a large number of publications related to structural design of transmission lines and their supporting structures.

REFERENCES

American Society of Civil Engineers. (2020) *Guidelines for electrical transmission line structural loading*, 4th Edition. ASCE Manual of Practice 74, ASCE, Reston, VA.

Blevins, R.D. (1977) *Flow-induced vibrations*. Van Nostrand Reinhold, New York.

CSIR. (1990) *Transmission line loading. Part I: recommendations and commentary. Part II: appendices*. Engineering Structures Programme, CSIR Building Technology, Pretoria.

Davenport, A.G. (1979) Gust response factors for transmission line loading. *Proceedings, 5th International Conference on Wind Engineering*, Fort Collins, Colorado, USA, pp. 899–909, Pergamon Press.

EPRI. (1979) *Transmission line reference book – wind-induced conductor motion*. Electric Power Research Institute, Palo Alto, CA, U.S.A.

EPRI. (2017) *Transmission line reference book – wind-induced conductor motion – revised edition*. Electric Power Research Institute, Palo Alto, CA, U.S.A.

Hawes, H. and Dempsey, D. (1993) Review of recent Australian transmission line failures due to high intensity winds. Report to Task Force on High Intensity Winds on Transmission Lines, Buenos Aires, 19–23 April 1993.

Matheson, M.J. and Holmes, J.D. (1981) Simulation of the dynamic response of transmission lines in strong winds. *Engineering Structures*, 3: 105–110.

Milford, R.V. and Goliger, A.M. (1997) Tornado risk model for transmission line design. *Journal of Wind Engineering and Industrial Aerodynamics*, 72: 469–478.

Oliver, S.E., Moriarty, W.W. and Holmes, J.D. (2000) A risk model for design of transmission line systems against thunderstorm downburst winds. *Engineering Structures*, 22: 1173–1179.

Richardson, A.S. (1991) A study of galloping conductors on a 230kV transmission line. *Electric Power Systems Research*, 21: 43–55.

de Schwarzkopf, M.L.A. and Rosso, L.C. (2001) *A method to evaluate downdraft risk*. University of Buenos Aires, Buenos Aires.

Twisdale, L.A. and Dunn, W.L. (1983) Probabilistic analysis of tornado wind risks. *Journal of Structural Engineering (ASCE)*, 109: 468–488.

Varney, T. (1926) Notes on the vibration of transmission line conductors. *AIEE Transactions*, XLV: 791–795.

Chapter 14

Other structures

14.1 INTRODUCTION

In this chapter, the wind loads on some structures not covered in Chapters 8–13, and appendages attached to buildings, will be discussed. Some of these structures may be of lesser economic importance, but are often sensitive to wind loads, fail early during a severe windstorm, and provide a source of flying debris.

In the following sections, wind loads on free-standing walls (including noise barriers along freeways or motorways), and hoardings are discussed (Section 14.2). Wind loading of free-standing roofs or canopies, including tensioned fabric roofs (Section 14.3), solar panels attached to the roofs of buildings, as well as other appendages attached to buildings such as canopies, awnings, balconies and scaffolding are also discussed (Section 14.4). Free-standing paraboloidal antennas for radio telescopes, and antennas of various geometries attached to towers or buildings (Section 14.5), and lighting frames and luminaires (Section 14.6) are considered.

Petrochemical, mining and other industrial structures are often located in high-wind regions of the world. Wind codes and standards are generally written primarily for buildings and hence industrial structures are often a secondary consideration in those documents. Section 14.7 discusses some approaches, including other sources of shape coefficients and force coefficients, for such structures, including offshore platforms.

14.2 WALLS AND HOARDINGS

14.2.1 Single walls under normal and oblique winds

In Sections 4.3.1 and 4.3.2, the mean drag coefficients on walls on the ground are discussed in the context of bluff-body aerodynamics. Discussion of wind loads on free-standing walls under normal and oblique winds will be expanded in this chapter.

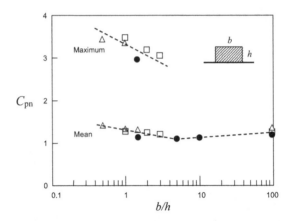

Figure 14.1 Mean and maximum pressure difference coefficients for free-standing walls (normal wind) (Letchford and Holmes, 1994).

In Figure 14.1, mean and maximum net pressure difference coefficients acting on complete walls of various breadth/height ratios are shown plotted. These values are based on boundary-layer wind-tunnel measurements (Letchford and Holmes, 1994) in simulated open country terrain (Jensen Numbers h/z_0 in the range of 50–160). The net pressure coefficient, C_{pn}, is defined in Equation (14.1) and, in this case, is equivalent to a drag coefficient.

$$C_{pn} = \frac{p_w - p_L}{\frac{1}{2}\rho_a \bar{U}_h^2} \tag{14.1}$$

where

p_w is the area-averaged pressure coefficient on the windward face of the wall,
p_L is the area-averaged pressure coefficient on the leeward face of the wall,
$\bar{U}_h$ is the mean wind speed at the top of the wall.

The maximum values are expected values for periods equivalent to 10 minutes in full scale. The mean net pressure coefficients show a small reduction in the range of b/h from 0.5 to 5, as previously shown in Figure 4.5. A larger reduction occurs for the maximum pressure coefficients – this is due to the reduction in spatial correlation for longer lengths of wall. About a 20% reduction in peak net load occurs as the wall length increases from 1 to 4 wall heights.

For a wind direction at 45° to the plane of the walls, the average net pressure coefficients are shown in Figure 14.2. In this case, the net mean pressure

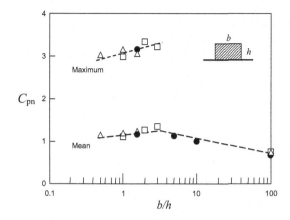

Figure 14.2 Mean and maximum pressure difference coefficients for free-standing walls (oblique wind) (Letchford and Holmes, 1994).

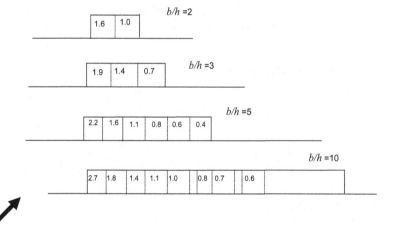

Figure 14.3 Mean pressure difference coefficients for free-standing walls (oblique wind).

coefficient reaches a maximum for a *b/h* ratio of about 3 with lower values for longer walls. For this wind direction, there is a strong separation on the leeward face of the walls of this length ratio. For longer walls, re-attachment occurs and generates lower magnitude pressures on the leeward face.

For mean wind directions normal to the wall, the net pressures do not vary much along the length of the wall. However, this is not the case for the oblique wind direction. Figure 14.3 shows how the mean net pressure coefficient varies along the wall length. The flow separation behind the windward edge generates very high pressures for the first 1–2 wall heights from

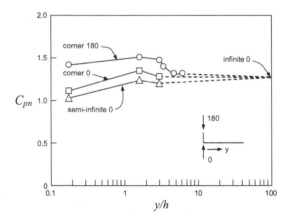

Figure 14.4 Mean pressure difference coefficients for free-standing walls with corners (normal winds) (Letchford and Holmes, 1994).

the windward edge. This also occurs for elevated hoardings (Figure 4.8 and Section 14.2.4), and is usually the critical design case for wind loads.

14.2.2 Walls with corners

Figures 14.4 and 14.5 show the effect of a right-angled corner at a free end of a wall for various wind directions on mean pressure coefficients averaged over a vertical line, at a distance y from the corner.

For a wind direction of 0°, with the corner running downwind, the effect is small; however, for 180° there is an increase in mean pressure coefficient of up to 30% (Figure 14.4). However, for the 45° wind direction (i.e. blowing from outside the corner), there is a significant reduction in mean pressure coefficients, for the region immediately adjacent to the corner (Figure 14.5).

14.2.3 Parallel walls

There is an increasing tendency to provide noise barriers along freeways and motorways when they pass through urban areas. These are generally parallel walls spaced at the width of the roadway, so that shielding effects from the opposite wall may be important for certain wind directions. The wind loads on these walls are also affected by other disturbances to the wind flow, such as topographic features and elevated bridges.

Figure 14.6 shows the variation of mean, root-mean-squared fluctuating, maximum and minimum net pressure coefficients on one wall of a pair of parallel ones, for various spacings. The pressure coefficients are based on the mean wind speed at wall height in the undisturbed flow. A negative value of

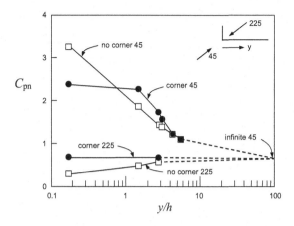

Figure 14.5 Mean pressure difference coefficients for free-standing walls with corners (oblique winds) (Letchford and Holmes, 1994).

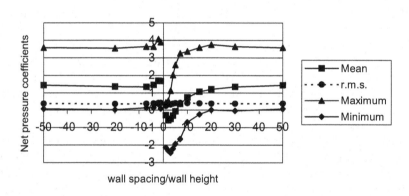

Figure 14.6 Parallel walls on flat level ground – effect of wall spacing.

wall spacing/wall height means the second wall is *downwind*. These measurements were carried out in simulated atmospheric boundary-layer flow in a wind tunnel. The values of Jensen Number, h/z_0 (Section 4.4.5) for the wall heights used in the tests were about 10–20.

The pressure tappings were arranged in vertical rows, with spacings chosen so that the pressures averaged together as a group of four (Section 7.5.2), gave a measure of the bending moment at the base of the wall. Thus, the measurements of base moment coefficients are defined as:

$$C_M = \frac{M}{\frac{1}{4}\rho_a \bar{U}^2 h^2} \tag{14.2}$$

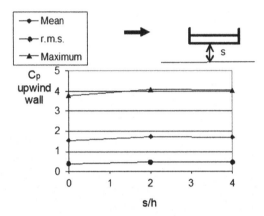

Figure 14.7 Effect of clear space, s, for parallel walls on bridges.

where M is the moment about the base, per unit length of wall. This is also an effective net pressure coefficient which, when applied uniformly over the well height, will give the correct base moment. Averaging of peak and fluctuating pressures was carried out over one wall height horizontally along the wall axis.

The mean pressure difference is negative when the upwind wall is about two wall heights away from the shielded wall; that is, it acts *upwind*. Small shielding effects are felt when the upwind wall is as much as 20 wall heights upwind.

Figure 14.7 compares the mean, maximum and root-mean-squared net effective pressure coefficients for the windward wall of the pair of parallel walls on a bridge with two different values of clear space underneath (Holmes, 2001). The thickness of the bridge deck was equal to the wall height. Values are for s/h equal to 0, 2 and 4, where s is the clear spacing under the bridge. All pressure coefficients are calculated with respect to the mean wind speed at the height of the top of the wall $(s+2h)$ in the undisturbed flow.

Figure 14.7 shows there is little difference between the net pressure coefficients for the cases of s/h equal to 2 or 4, when there is airflow beneath the bridge. However, when s/h is equal to 0 – that is when the bridge forms a flat-topped 'cliff', the mean and maximum net pressure coefficients are about 90% of the values on the elevated bridges; the root-mean-squared pressures are about 80% of those in the elevated case.

14.2.4 Elevated hoardings

The net wind pressure coefficients on elevated hoardings have generally similar characteristics to those on free-standing walls. The effect of elevation is to increase the magnitude of the net pressure coefficient for winds normal to the surface. The average mean pressure coefficient depends on the spacing to the ground beneath the hoarding. For spacing equal to the

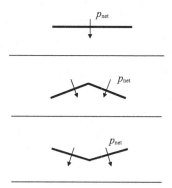

Figure 14.8 Types of free-standing roof, and sign convention for net pressures.

depth of the hoarding, a mean net pressure coefficient (with reference to the mean velocity at the top of the hoarding), of 1.5 occurs.

The oblique wind direction can produce large pressure differences near the windward end, as for free-standing walls (Figure 4.8).

Design data for elevated hoardings and signboards is given in the American (ASCE, 2016), Australian (Standards Australia, 2011) and the former British Standard (BSI, 1997).

14.2.5 Spanwise averaging

Walls, or hoardings, supported over long spans will experience lower peak wind loads than those supported over short spans. The following form was proposed for the reduction factor for peak loads on free-standing walls over spans, s (Holmes, 2001).

$$\alpha = 0.5 + 1.35 \exp\left[-\left(\frac{s}{h}\right)^{0.15}\right] \tag{14.3}$$

Equation (14.3) gives the ratio between peak net pressures for a span, s, greater than the wall height, h, and the peak net pressure on a wall with a span equal to the height. Thus, for $s=h$, α is equal to 1.0. The equation was derived for both unshielded walls and shielded parallel walls.

14.3 FREE-STANDING ROOFS AND CANOPIES

14.3.1 Pitched-free roofs

Free-standing, or 'canopy', roofs, without walls, are often used for basic shelter structures – such as those at motor vehicle service stations and railway stations, or for coverage of industrial, mineral or agricultural products. The wind loads on roofs of this type attached to buildings are discussed in Section 14.4.1. Section 14.3.3 covers the wind loads on tensioned fabric free roofs and canopies.

Free-standing roofs which are completely free of stored material underneath, allow air to flow freely underneath; this generally results in negative, or near-zero underside pressures with respect to atmospheric pressure. The addition of stored material underneath the roof in sufficient quantity will cause full or partial stagnation of the airflow, and positive pressures underneath. The nature of the upper surface pressures depends on the roof pitch and the wind direction.

Wind pressure coefficients on free-standing roofs are usually quoted in the form of net pressure coefficients, as defined in Equation (14.1). The pressures can normally be assumed to act normal to the roof surface. The usual sign convention is that positive net pressures act downwards. This sign convention and the most common three types of free-standing roof geometry are shown in Figure 14.8.

Although the pressures normal to the roof surface are the dominant ones, frictional forces acting parallel to the roof surfaces, can also be significant, and it may be necessary to consider them, when designing the bracing required to resist horizontal forces.

Free-standing roofs have been studied in both wind-tunnel tests (Gumley, 1984; Letchford and Ginger, 1992; Ginger and Letchford, 1994), and full-scale experiments (Robertson et al., 1985).

Net pressure coefficients along the centre line of a free-standing 'Dutch barn' with 15° roof pitch, measured in full scale (Robertson et al., 1985) are shown in Figure 14.9. The roof is completely empty underneath. Positive (downwards) pressure differences exist over the windward quarter of the roof for all wind directions. The largest negative loads occur near the middle of the roof for a wind direction normal to the ridge.

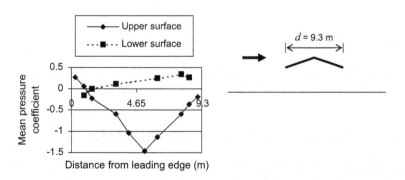

Figure 14.9 Mean pressure difference coefficients along the centre line of a free-standing roof with 15° pitch.

Letchford and Ginger (1992), and Ginger and Letchford (1994) carried out extensive wind-tunnel measurements on pitched free roofs (empty under) of approximately square plan, with a range of pitches up to 30°. Mean and fluctuating pressure measurements from single points and area-averaged (Section 7.5.2) over six panels, were also made. In addition, correlation coefficients (Section 3.3.5) were measured for the six panel pressures enabling fluctuating total forces (Section 4.6.6), and equivalent static loading distributions to be derived (Section 5.4.3).

Mean area-averaged net pressure coefficients for half the pitched roof are shown in Figure 14.10. For the 0° wind direction, the half roof is on the windward side. Figure 14.10 shows that significant positive pressures (for wind directions of 0°–30°) and negative pressures (for wind directions of 120°–180°) occur for roof pitches of 22.5° and 30°. For roof pitches of 15° or less, the net pressure difference coefficients are not large for any wind direction.

The peak (maximum and minimum) area-averaged pressure difference coefficients generally showed similar behaviour to the mean coefficients, as shown in Figure 14.10, with the 22.5° and 30° pitch roofs clearly showing larger magnitudes. When peak total uplift and horizontal forces were calculated, substantial reductions of up to 50% from values calculated from the non-simultaneous peaks on windward and leeward halves were obtained, due to the poor correlation between fluctuating wind pressures on the two surfaces (Ginger and Letchford, 1994).

Appendix F gives an example of the calculation of maximum and minimum lift and drag on a pitched-free roof, and the effective static pressures producing them, based on data from Ginger and Letchford (1994).

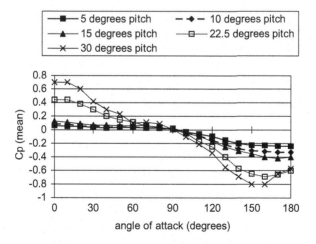

Figure 14.10 Mean pressure difference coefficients for pitched free roofs, averaged over a half roof (Letchford and Ginger, 1992).

14.3.2 Effect of porosity

The effect of porosity on mean wind loads on pitched, hipped and mono-slope free roofs was investigated, in wind-tunnel model studies, by Letchford *et al.* (2000). Two porosities of 11% and 23% were studied for comparison with the equivalent values for solid roofs (i.e. 0% porosity) for three different roof pitches. Figure 14.11 shows some of the results of Letchford *et al.* The coefficients shown are the average mean net pressure coefficients for each roof half. At the higher pitches, the main effect of porosity is on the net pressure on the leeward side. However, for the 7° pitch case, the porosity affects the net pressure on both halves.

The data in Figure 14.11 can also be interpreted in terms of mean drag and lift coefficients. For the 7° and 15° roof pitches, the mean drag force reduces with increasing porosity, whereas for the 27° pitch roof the opposite is the case. The effect of increasing porosity is to reduce the positive (upward) lift force for all roof pitches. For the 27° pitch case, the average lift force is negative (i.e. downwards) for all porosities, and this downward force increases with increasing porosity.

14.3.3 Tensioned-fabric roofs and shade sails

In hot climates, relatively cheap tensioned-fabric roofs are often used to provide shading to school playgrounds, house yards, and outdoor shopping malls and restaurants etc. These are sometimes known as 'shade sails' or 'membrane

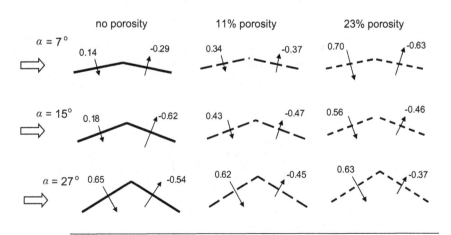

Figure 14.11 Effect of porosity on mean net pressure coefficients for pitched-free roofs – referenced to mean velocity at average roof height (adapted from data presented by Letchford *et al.*, 2000).

roofs'. Larger prestigious architect-designed tensioned-fabric roofs are sometimes used for pavilions at international exhibitions and 'EXPOS' (e.g. Sykes, 1994). Larger membrane roofs are discussed in Section 10.5.

Knitted or woven fabrics, with some porosity to air flow, are often used for shade sails in hot climates, such as northern Australia. The results of Letchford *et al.* (2000), discussed in the previous section, can be applied to determine the effects of porosity on simple pitched and mono-slope roof shapes made from such fabrics. By comparing pressure loss coefficients, Letchford *et al.* found that the porous plates with 11% and 23% solidity were equivalent to shade fabrics with about 90% ultra-violet (UV) radiation reduction.

Wind loads on tensioned fabric structures have been rarely investigated experimentally, partly due to the difficulty in installing pressure tappings in thin wind tunnel models, and partly due to the fact that fluctuating wind pressures will deform a roof and change its geometry over short time intervals.

Colliers *et al.* (2016) noted the need for data on shape factors for tension roofs of standard generic shapes, that could be codified for application in Europe and elsewhere. They collated existing (mainly European) wind-tunnel data on two basic shapes of tension fabric roofs – hyperbolic paraboloid ('hypar') and conical. For both types, roofs on enclosed buildings and in the form of free-standing roofs, or canopies were discussed. For hypar roofs on buildings with a square planform, and with a wind direction along a diagonal towards a low corner, the upwind parts of the roof are more likely to experience downwards net pressures, while the opposite is the case for wind towards a high corner. For hypar canopies, larger fluctuations in shape factors were apparent. Defining a 'shape parameter' SP, as the height differential between high and low corners, divided by the overall span across a diagonal, it was noted that for low values of SP, pressures on the upper and lower roofs surfaces tended to counteract each other, leading to low net pressure coefficients acting across the canopy. For higher values of SP (more curved hypar canopies), opposing positive and negative pressures on the upper and lower surfaces leads to larger pressure differentials over upwind and downwind zones of the roof.

For the 'conical' roofs and canopies, (similar to traditional circus tents in shape with elliptical curvature in elevation) on circular or square planforms, the highest pressure coefficients were recorded close to the upwind edges, and also near the top of the internal high point, which often may be an opening. For conical roofs with high shape factors (defined in a similar way to that for hypar roofs), the upwind zones are more prone to positive (downward acting) pressure, while the downwind zones are subject to negative net pressure.

Another class of tensioned fabric roof comprises air-supported or 'pneumatic' roofs. These are discussed in Section 10.5 in Chapter 10.

14.4 ATTACHMENTS TO BUILDINGS

14.4.1 Canopies, awnings and parapets

Several configurations of horizontal canopy attached to one wall of a low-rise building have been investigated (Jancauskas and Holmes, 1985). The width of the canopy and the height of the canopy position on the wall were the variables that were investigated. A narrow canopy mounted at the top of the wall behaves similarly to eaves on the roof.

For wind directions normal to the adjacent wall, the peak net force across the canopy is strongly dependent on the non-dimensional ratios, h_c/h, and h_c/w_c. h_c is the height of the canopy above the ground, h is the total height of the adjacent wall, and w_c is the width of the canopy.

For the peak vertical uplift force coefficient, $\hat{C}_Z$, based on the mean wind speed at the height of the canopy, the following conservative relationships were proposed, based on the wind-tunnel measurements:

$$\text{for } \frac{h}{h_c} = 1.0, \hat{C}_Z = 1.0 + 1.3\left(\frac{h_c}{w_c}\right) \text{ or } 4.0, \text{ whichever is the lesser}$$

$$\text{for } \frac{h}{h_c} = 0.75, \hat{C}_Z = 1.0 + 0.4\left(\frac{h_c}{w_c}\right) \text{ or } 4.0, \text{ whichever is the lesser}$$

$$\text{for } \frac{h}{h_c} = 0.5, \hat{C}_Z = 1.0 \tag{14.4}$$

where

$$C_Z = \frac{F_z}{\frac{1}{2}\rho_a \bar{U}_c^2 A},$$

F_z is the net vertical force on the canopy (positive upwards),
$\bar{U}_c$ is the mean wind speed at the canopy height,
A is the plan area of the canopy.

Equation (14.4) can be applied to canopies with pitch angles within 5° of the horizontal. Appropriate adjustment is required if it is applied with gust wind speeds; such adjustment has been made for the rule incorporated in the Australian wind loading standard (Standards Australia, 2011).

The relationships of Equation (14.4) are compared with the experimental data in Figure 14.12.

The higher values obtained for canopies or awnings near the top of the wall (or eaves), can be explained by the high flow velocities occurring on the upper side of the canopy producing significant negative pressures; on

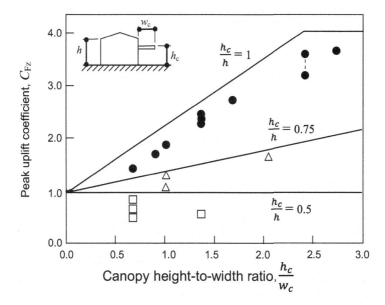

Figure 14.12 Peak uplift force coefficients for attached canopies (Jancauskas and Holmes, 1985).

the underside of the canopy, stagnation and hence positive pressures occur. When the canopy is mounted part-way up the wall, stagnation of the flow occurs on the wall, both above and below the canopy. In this situation, the mean net force coefficients are low, but turbulence produces finite peak loads in both directions.

Parapets, and their effect on roof pressures on flat roofs, have been the subject of several wind-tunnel studies. In the early work, there were some conflicting conclusions drawn by different laboratories, but the issue was largely resolved using large models and a high density of pressure tappings (Kind, 1988). With or without parapets, the worst suction peaks occur in small zones near the upwind corner of the roof, for wind directions nearly bisecting the corner. The worst suction coefficients decrease monotonically with increasing relative parapet height. The amount of the reduction depends also on the height/width ratio of the building to which the parapet is attached (see also Section 8.6).

14.4.2 Solar panels on roofs

The wind loads on solar panels attached to the roofs of a building are closely related to the flow over the roofs of the building itself, since the latter is a much larger bluff body. Figure 14.13 shows the various geometric variables that are significant in determining the wind loads on solar collector panels on a pitched roof building (Tieleman *et al.*, 1980).

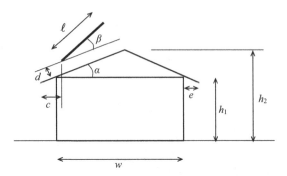

Figure 14.13 The variables affecting wind loads on solar panels (Tieleman *et al.*, 1980).

Wind-tunnel pressure tests were conducted by Wood *et al.* (2001) on a 1:100 scale model of an industrial building with solar panels mounted parallel to the flat roof. The orientation of the panels with respect to the wind flow was found to have a greater effect than panel height and lateral spacing.

A comprehensive model study for solar panels mounted on roofs of small low-rise buildings was undertaken by the James Cook Cyclone Testing Station in Townsville, Queensland, Australia (Ginger *et al.*, 2011). Area-averaged net wind pressures were measured on panels, of dimensions 1.7 m by 1.0 m in full scale, mounted in an array, parallel to pitched (gable) roofs with pitches of 7.5°, 15° and 22.5°. The gap between the panel and the underlying roof surface was either 100 or 200 mm in full scale. Measurements were also made by the JCCTS on an array of panels mounted at 15° and 30° to a flat roof. Results from the parallel-mounted panels were used as the basis of net pressure coefficients on solar panels in the Australian/New Zealand Standard on Wind Actions AS/NZS 1170.2:2011 (Standards Australia, 2011).

The following summarizes the general effects of various panel and building variables on wind loads on roof-mounted solar panels:

- *'Stand-off' spacing from the roof*: increasing stand-off appears to reduce net uplift load (normal to roof), but increases the wind force acting on the panel parallel to the roof, (Newton, 1983).
- However, the effect of the width of the gap between panel and roof, for parallel-mounted panels, on the net pressures on the panels is small (Ginger *et al.*, 2011).
- *Module shape and size*: the combined peak load on a row of panels is significantly less than that on a single panel, due to area reduction effects on the fluctuating pressures.
- *Roof pitch*: higher roof pitch produces lower uplift loads, but increases downwards wind loads (as for the loads on the roofs of low-rise buildings generally).

- *Proximity to eaves or gable ends*: the end panel (adjacent to the eaves) experiences considerably higher loads than the interior panel in a row.
- Net wind pressures on panels near the windward gable end of a pitched roof are generally similar in magnitude to the external pressures on the roof itself without the presence of panels (Ginger *et al.*, 2011).
- Net pressures on panels in the central part of the roof can be *greater than* the local external pressures on the bare roof at the same location (Ginger *et al.*, 2011). However, these net pressures are lower in magnitude than those on panels located near the edges of a roof.
- *Wind direction*: the worst uplifts occur for oblique wind directions to a row of collector panels.
- *Roof height*: the pressure *coefficients* for panels on two-storey buildings are lower than the equivalent values for single-storey buildings.
- Increasing the angle β, so that the inclination of panels is greater than that of the roof pitch, appears to increase wind loads on the panels. For arrays of inclined solar panels on flat roofs, 'guide plates', or 'guide vanes' are useful for reducing net uplift forces on the panels (e.g. Chung *et al.*, 2013a,b).

The wind loading of large arrays of solar panels forming large integrated energy supply systems, either on ground as 'solar farms' or on large flat roofs, is discussed in Sections 14.8.1 and 14.8.2.

14.4.3 Scaffolding

Scaffolding, attached to buildings under construction or renovation, often fails under wind loads as it is not designed for high wind loads, or in some jurisdictions, wind loads may not be required to be considered at all. When they are required, the annual recurrence intervals, or return periods, of design wind speeds for temporary structures are often considerably lower than those for permanent structures. Another problem for attachments to buildings, and scaffolding in particular, is that reliable shape factors (pressure and force coefficients) are hard for designers to find. There is a Eurocode – EN 12811 (British Standards Institution, 2003), and some Japanese Guidelines (SCEAJ, 1999) that contain some values, but recent wind-tunnel studies for clad scaffolding (Wang *et al.*, 2013) suggest that the values in those documents are unconservative.

Accurate wind-tunnel model studies of force coefficients on *unclad* scaffolding made from tubular members, is difficult if not impossible, due to Reynolds Number scaling issues (Section 4.2.4), and the difficulty in measuring very small forces. However, cladding of scaffolding for safety and environmental reasons is normal practice these days, and model measurements for *clad* scaffolding can be carried out using standard wind-tunnel pressure measurement techniques.

A study by Wang *et al.* (2013) is probably the most comprehensive set of measurements available. An underlying model of a generic building with rectangular planform and elevations, at a nominal geometric scale of 1:75, had four different porosity ratios from 0% to 80%, and was representative of a building under construction. It was equipped with clad scaffolding on 1, 2, 3 or 4 walls, and wind pressure coefficients were measured for five different wind directions. Positive (i.e. pressures acting towards the underlying wall) mean pressure coefficients (with respect to the mean wind speed at the top of the building) for the entire scaffolding up to +1.5, and negative values up to −1.4, were measured. These occurred when the underlying building was non-porous (0% porosity) and with cladding attached to only one or two walls of the building. Local net pressure coefficients up to 8 were measured; the worst case is near the top windward corner, with scaffolding attached to only one wall, for a wind direction oblique to that wall. Clad scaffolding covering all four walls of the building generally experienced low pressures on the interior surface, and hence low net pressures for all wind directions.

14.5 ANTENNAS

14.5.1 Radio telescopes

Wind loads on the antennas of large steerable radio telescopes – usually with dish reflectors of paraboloidal shape – are of critical importance for several design criteria (Wyatt, 1964):

* Overall strength for safety in extreme winds,
* Loads on drive system,
* Freedom from oscillations,
* Pointing accuracy,
* Distortion of the reflector.

The last four of the above conditions are serviceability criteria. Very small tolerances are required for the operation of these antennas.

The main source of wind loads is the paraboloidal dish itself. If the dish is impermeable, the pressures acting on it may be assumed to act normal to the surface, with negligible contributions from skin friction. For a paraboloid, the normal to any point on the surface passes through the generating axis, at a point $2f$ measured along the axis from that point, where f is the focal length. Therefore, it may be assumed, that the resultant aerodynamic force will act through a point on the axis, distant from the vertex by $2f$ plus half the depth of the dish, d, (Wyatt, 1964).

Figure 14.14 shows the case with the wind direction normal to the altitude axis of rotation of the dish. Resolving the aerodynamic forces in body axes (Section 4.2.2), the force coefficients are given by:

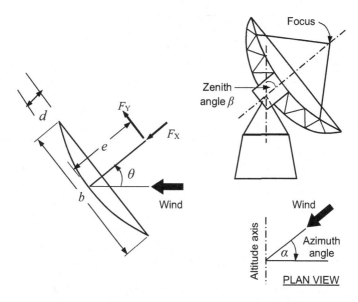

Figure 14.14 Resultant aerodynamic forces on the dish antenna of a radio telescope (Wyatt, 1964).

$$C_X = \frac{F_x}{\frac{1}{2}\rho_a \bar{U}_h^2 A} \qquad (14.5)$$

$$C_Y = \frac{F_Y}{\frac{1}{2}\rho_a \bar{U}_h^2 A} \qquad (14.6)$$

where A is the projected area normal to the dish, given by $\pi \dfrac{b^2}{4}$.

Following the arguments in the previous paragraph, the eccentricity, e, of the aerodynamic force can be closely approximated by (Wyatt, 1964):

$$e = 2f + \frac{d}{2} = 2f\left[1 + \left(\frac{b}{8f}\right)^2\right] \qquad (14.7)$$

Tests in smooth uniform flow (Wyatt, 1964) indicate maximum values of C_x of about 1.7 when the angle of attack, α, is about 45°. The transverse force coefficient C_Y is approximately constant with f/d when expressed in the form $(f/d)\, C_Y$, with a maximum value of about 0.05, for α equal to about 135°. The transverse force F_Y generates a moment about the vertex equal to $F_Y \cdot e$.

It is found that the effect of a boundary-layer mean wind profile has a relatively small effect for wind directions facing the wind. However, the effect

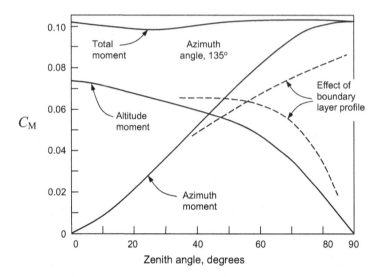

Figure 14.15 The effect of velocity profile on the aerodynamic moments on a radio telescope (Wyatt, 1964).

is greater when the wind is blowing obliquely on to the rear of the paraboloid. As shown in Figure 14.15, the effect is to increase the moment about the altitude axis and decrease it about the azimuth axis (Wyatt, 1964).

In Figure 14.15, the moment coefficients are defined as follows:

$$C_M = \frac{M}{\frac{1}{2}\rho_a \bar{U}_h^2 A d} \qquad (14.8)$$

14.5.2 Microwave dish antennas

The drag forces acting on small dish antennas used for microwave frequency transmission are of interest for the structural design of the towers supporting them. In the past, total drag forces for tower design have been obtained by simply adding the drag, measured on the antennas in isolation to that determined for the tower without antennas. This will overestimate the total drag in many cases, as usually the antennas shield part of the tower, or vice-versa; also, the drag on an antenna itself in the presence of the tower will be different to that on the antenna in isolation.

Figure 14.16 shows the drag coefficient for an impermeable unshrouded dish obtained as a function of the wind incidence angle measured from the normal to the plane of the dish, measured in both smooth (approximately 1% turbulence intensity), and turbulent flow (10% turbulence intensity) (Holmes *et al.*, 1993). The reference area is the projected area of the dish, $\pi \dfrac{b^2}{4}$.

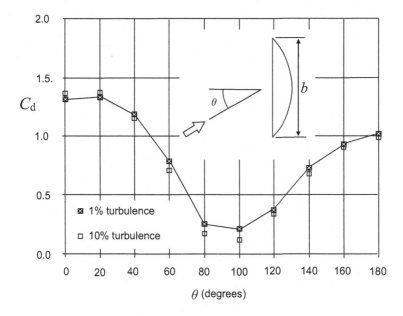

Figure 14.16 Drag coefficient as a function of angle of attack for an isolated dish antenna (Holmes *et al.*, 1993).

The drag coefficient for the isolated dish is maximum with a wind direction normal to the plane of the dish, but does not reduce much in an angular window within 30° to the normal. The maximum drag coefficient, based on the disc area is about 1.4. A large reduction occurs for wind directions from 40° to 80° to the normal. The effect of turbulence intensity is small.

The concept of *interference factor* is illustrated in Figure 14.17. The drag of an isolated antenna should be multiplied by this factor to give the measured incremental contribution to the total tower drag. The sum of the drag on the tower segment, D_t, and the incremental contribution from the antenna, $K_i \cdot D_a$, gives a total effective drag, D_e.

The interference factor for a single dish attached to a face of a lattice tower, with square cross-section and a solidity ratio of 0.3, is shown graphically, as a function of wind direction, θ, relative to the tower face in Figure 14.18 (Holmes *et al.*, 1993). The maximum interference factor of about 1.3 occurs at wind directions for which the dish accelerated the airflow over the tower, i.e. for wind directions of 90° and 270°. For wind directions of 0° and 180°, where mutual shielding occurs, interference factors as low as 0.5 can occur.

An empirical form for the interference factor, K_i, based only on the solidity and drag coefficient of the tower which fits the experimental data in Figure 14.17, and data from other cases, is Equation (14.9).

$$K_i = \exp\left[-k(C_D d)^2\right] \cdot \left[(1+t) + t\cos 2(\theta - \theta_d - 90°)\right] \tag{14.9}$$

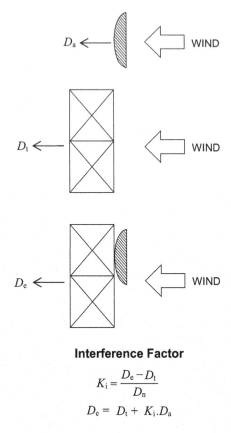

Interference Factor

$$K_i = \frac{D_e - D_t}{D_n}$$

$$D_e = D_t + K_i . D_a$$

Figure 14.17 Concept of interference factor for incremental antenna drag.

where

C_D is the drag coefficient for the tower or mast section alone, based on the projected area of members in one face, measured normal to the face,

δ is the solidity of a face of the tower,

k is a parameter equal to 1.2 for a square tower (ESDU, 1981),

t is an adjustable parameter (equal to 0.5 in Figure 14.18),

θ_d is the angle of the normal to the dish antenna relative to the tower.

As well as drag (along-wind) forces, there may be significant cross-wind forces acting for wind directions parallel, or nearly parallel, to the plane of a solid dish. These should be taken account of when designing support attachments for the dish. Basic aerodynamic force coefficients are often obtainable from the antenna manufacturers, although these would not generally include interference effects.

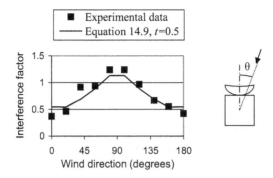

Figure 14.18 Interference factor as a function of wind direction for a single microwave dish added to a square lattice tower (Holmes *et al.*, 1993).

14.5.3 Rotating radar antennas

Aerodynamic loads on large rotating radar antennas, such as those used at large airports, pose a particular serviceability problem due to variations in the torque that arise. The operation of the antennas imposes strict limits on variations in angular velocity, and this in turn limits the variations in torque that must be overcome by the drive motor. Wind-induced variations in torque arise from two sources:

- Variations in the azimuth angle between the wind direction and the antenna
- Horizontal wind turbulence

At the rates of rotation used in practice, the first source of aerodynamic torque variation appears to be dominant.

The effect of rotation of the antenna can be treated by a quasi-steady approach. This results in the predicted variation of torque being obtained from static tests in a wind tunnel, in which the azimuth angle is varied. The effect of rotation is assumed to result in a static shift in the fluctuating torque curve, obtained from such tests (Sachs, 1978; Lombardi, 1989). However, the quasi-steady theory has been found to be only approximately correct at high rotational speeds (Lombardi, 1991).

The use of small fins on the back of the antenna has been found to be effective in reducing the aerodynamic torque. These are small lifting surfaces which produce a counter-acting torque. Figure 14.19 shows measured torque coefficients obtained from a rotating wind-tunnel model with and without fins (Lombardi, 1991).

Unfortunately, all the wind-tunnel measurements on rotating radar antenna have been carried out in smooth uniform flow. The effect of turbulent boundary-layer flow is uncertain, but the most likely effect is to

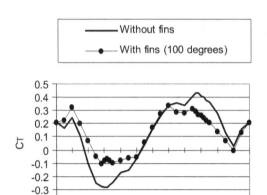

Figure 14.19 Aerodynamic torque coefficient versus yaw angle for rotating radar antennas (Lombardi, 1991).

smooth out the torque versus yaw angle graphs, such as those shown in Figure 14.19.

14.5.4 Mobile telephone antennas

Antennas for mobile telephone cells typically consist of several radiating antennas within fibreglass or plastic radomes, mounted on poles or towers, which may in turn be mounted above buildings or other structures. By their nature they are in exposed positions, and thus the interference, or shelter, effects from other structures is usually small. However, the mutual aerodynamic interference between radomes can be considerable.

Many of these antennas have been tested at full scale in large wind tunnels, for aerodynamic force coefficients but the data is usually proprietary in nature, and not freely available. The force coefficients have been found to be dependent on Reynolds Number, so that model testing at small scales will produce unreliable results. However, drag coefficients at high Reynolds Number from full-scale measurements, which illustrate the mutual interference effects are shown in Figure 14.20.

The drag coefficient for wind normal to the curved face of an antenna is around 1.1, based on the projected frontal area. This value is reduced for wind directions in which the frontal area presented to the wind is reduced, as illustrated for the value shown for a wind direction 120° from the normal.

When the antenna elements are grouped in threes, the combined drag coefficient (based on the frontal area of *one* radome element) is greatly reduced. As shown in Figure 14.20, the effect of the two downwind elements in the widely spaced (left side) cluster is neutral – i.e. the drag of the

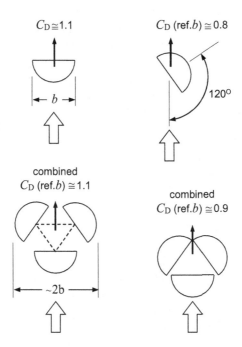

Figure 14.20 Typical drag force coefficients for mobile telephone antenna elements.

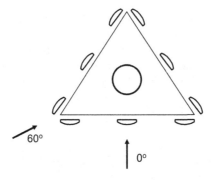

Figure 14.21 A group of mobile telephone antenna elements on a triangular frame.

upwind element in isolation is the same as the combined drag of the group of three. For the closely spaced cluster, which resembles a single bluff body with curved surfaces, the overall drag is more than 20%, but less than that of the upwind radome in isolation.

Sometimes up to nine antennas are grouped together on a triangular frame, as shown in Figure 14.21. For the case shown, the antennas on a single face are well separated to avoid large aerodynamic interference effects, but those at the corners may experience slight increases in drag due to mutual interference (Section 4.3.1, and Marchman and Werme, 1982). The six downwind antennas are shielded both by the upwind antennas, and by the supporting pole. Full-scale wind-tunnel tests on complete antenna headframes indicate an overall reduction of about 30% in the combined drag is obtained, when comparing the combined drag of the group, with that obtained by the summation of contributions from individual elements.

Isolated radome elements will also experience cross-wind forces at oblique wind directions. However, these will be largely cancelled by opposite forces from other elements, when they are part of a group.

14.5.5 UHF television antennas

Antennas for the transmission of Ultra-High-Frequency television-broadcasting signals (including digital television), consist of fibreglass or plastic radomes mounted on four or five-sided masts, and are usually quite 'solid' cross-sections. They are of the order of 1 m in breadth, and about 20 m long. They are usually mounted at the top of free-standing or guyed towers, (see Figure 7.15).

The drag coefficient for these cross-sections depends on the porosity of the cross-section – i.e. the ability of the wake to be vented to the windward side. Measurements on full-size antenna sections have yielded drag coefficients in the range of 1.1–1.6, with some dependency on wind direction.

UHF antennas have experienced significant cross-wind response due to vortex shedding (Sections 4.6.3 and 11.4). This has often occurred for antennas on guyed masts, which have a lower damping than free-standing lattice towers. Such responses occur at a critical velocity, that is quite low, and in atmospheric conditions with low turbulence intensities. The prediction of cross-wind response due to vortex shedding for *circular* cross-sections was discussed in Section 11.5. Methods also exist for *non-circular* cross-sections (e.g. ESDU, 1990). These methods require information on the Strouhal Number (rate of vortex shedding – see Section 4.6.3), and fluctuating cross-wind force coefficients (Section 4.6.5). These would not be well defined for the complex cross-sections of UHF antennas.

If vibrations occur, they can be mitigated by the use of simple damping devices, such as liquid dampers (Section 9.9.4), or *hanging chain* dampers (Koss and Melbourne, 1995).

14.6 LIGHTING FRAMES AND LUMINAIRES

Street lighting, flood lighting for railway yards, sporting grounds and industrial areas are bluff bodies of a variety of shapes and porosities. There are considerable interference effects when luminaires are arranged in groups. As for antennas, the drag of many types has been measured in wind tunnels, but such tests are usually commercially sponsored, and the results are not readily obtainable.

The largest drag coefficients of single lights for any wind direction fall in the range of 1.0–1.5, based on the largest frontal area projected vertically. The lower value applies to the more rounded types, and the higher value to sharp-edged lights.

Large rectangular headframes, with many luminaires attached, such as large floodlight systems for sporting grounds, may be treated as porous flat plates (see Section 4.3.1). A value of drag coefficient of 1.5, based on the projected 'solid' or 'wind' area is an appropriate one for solidities of 0.3–0.7.

The wind loads on, and response of, supporting poles for lighting are discussed in Chapter 11.

14.7 INDUSTRIAL COMPLEXES AND OFFSHORE PLATFORMS

Estimation of wind loads for elements of industrial complexes such as power stations, petroleum refineries, or mineral processing plants, is an extremely difficult problem. Such complexes consist of a large number of closely spaced bluff bodies, with considerable aerodynamic interference between them. It would normally be extremely conservative to estimate the total wind drag force by summing up the contributions from individual elements, as if they were isolated bluff bodies, although this is often done. The complexity and unique layouts of these plants means that is difficult or impossible to give general rules for estimating wind forces, except for some relatively common situations such as closely spaced circular cylinders. A useful source of shape factors for petrochemical, and other industrial, structures is a guide published by a committee of the American Society of Civil Engineers (ASCE, 2011).

A useful approach for densely packed industrial structures, which avoids gross overestimation of drag forces, is to treat a closely spaced complex of bodies in a 'global' way as a single 'porous' bluff body. This approach has been adopted for high-solidity open industrial structures by the ASCE guidelines (ASCE, 2011, Appendix 5B). The following formula for the maximum drag coefficient of such a structure, based on the projected frontal 'solid' area, is suggested:

$$C_D = \frac{1.4\delta\sqrt{b^2+d^2}}{2b\left(1+\dfrac{b^2}{d^2}-\dfrac{b}{d^2}\sqrt{b^2+d^2}\right)} \tag{14.10}$$

where b and d are the cross-wind breadth and along-wind depth, respectively, and δ is the solidity ratio (viewed from the front elevation, and including all equipment).

Mining equipment and equipment for loading mined ore on to ships at ports, is also in need of source data to enable wind loading calculations. Drag coefficients for open-lattice conveyor support structures can be derived from those specified for open lattice towers (see Section 11.3.2). However, when one of these is clad for environmental reasons (e.g. to prevent airborne dust), it becomes a slender bluff body, and may be subjected to cross-wind response caused by vortex shedding (see Section 11.5). Offshore platforms, used for oil exploration and production, are similar in complexity, with the topsides often exposed to severe windstorms (in many cases tropical cyclones), as well as wave action. In these cases, the overall wind forces on the above water exposed structure is of interest in the design of the underwater foundations and supporting structure.

Flare towers, on offshore platforms, or onshore in refineries or liquid natural gas (LNG) plants, are usually open lattice towers, and can be treated as other structures of this type, such as communication towers. However, the flares and risers attached to these structures often have significant frontal area, and the appropriate interference and shielding factors may not be available.

The low frequencies of 'compliant' offshore structures, such as tension leg or guyed structures in deep water locations, are of special concern because of the need to consider resonant excitation by dynamic wind forces. The frequencies of some structures of this type can be so low that they are near the peak of the spectrum of wind forces in synoptic winds (Section 3.3.4). However, it appears that hydrodynamic damping, resulting from the underwater motion of the structure (Cook *et al.*, 1986), largely mitigates resonant effects. The special problems of wind effects on compliant offshore structures are discussed in a number of specialist publications (e.g. Smith and Simiu, 1986).

14.8 LARGE ARRAYS OF PHOTOVOLTAIC COLLECTORS

14.8.1 Ground-mounted

In many countries with abundant solar energy, solar farms, with large arrays of photo-voltaic panels, have been built with many more in construction. At the end of 2016, it was reported that the total global generation capacity of utility-scale photovoltaic power stations was 96GW (source: *Wikipedia*). In Australia, in May 2019, there were 26 operating solar farms of 50MW capacity or greater, with a further 38 under construction or in planning. Large solar farms are often located in inland areas of large continental countries (such as United States, China and Australia) and may be subject to severe local thunderstorms, in which strong winds are accompanied by hail, both of which can be very damaging to solar installations (Section 1.3.3).

Since inclined solar panels are required to be tilted toward the sun, the critical wind directions, producing the greatest uplift forces, are northerly in the northern hemisphere, and southerly in the southern hemisphere. A wind-tunnel study of wind loads on four fixed arrays, each with dimensions of 19.2 m by 5 m concluded the following (Ginger *et al.*, 2019):

- Large negative (upward acting) net pressures on the panels at the leading edges of a group, for wind blowing towards the bottom surface of a group of sloping panels, ($\theta = 180°$). The largest net pressures were experienced on the leading corner panel for oblique approach winds ($\theta \sim 220°$).
- Large positive (downward acting) pressures were measured on the panels at the leading edge of a group, for wind blowing towards the top surface of the sloping panels ($\theta = 0°$). The largest net positive pressure was experienced by the bottom leading corner panel for oblique approach winds ($\theta \sim 320°$).
- Approach winds along the length of the arrays ($\theta = 90°$) generate large positive and negative net pressures on panels near the leading edges.

Deflectors, or fairings, shielding the underside of inclined panels, can be effective in minimizing the uplift wind forces for both on-ground and roof-mounted arrays.

Tracking solar panels that are centrally mounted on shafts have often been observed to experience severe vibrations due to 'stall' flutter (see Section 5.5.3 and SEAOC, 2017).

14.8.2 Roof-mounted

Large numbers of inclined solar panels have also been installed on the flat, or near-flat roofs of commercial buildings, particularly in North America. Extensive wind-tunnel studies have been carried out and have led to a practical design guide (SEAOC, 2017) and to new requirements for wind loads in the American loading Standard ASCE 7-16 (ASCE, 2016).

The number of parameters affecting wind loads on roof-mounted solar panel arrays are considerably greater than those for ground-mounted arrays, as the dimensions and shape of the building to which they are attached play a role (see also Section 14.4.2).

Banks (2013) and Kopp (2014) described extensive scale-model tests in a boundary-layer wind tunnel. The latter study noted the following conclusions:

- Wind loads on the array increase with building size; normalizing the effective wind area by the building wall size leads to enveloping curves that collapse onto a single curve for each array geometry.
- For tilt angles less than 10°, there is an approximate linear increase in the pressure coefficients as the tilt angle increases. For arrays with tilt

angles of 10° or more, the wind loads do not depend significantly on the tilt angle and are relatively constant.

• Roof zones for wind loads on solar arrays are larger than roof zones for bare roofs and depend on the array tilt angle.

Commentary in the SEAOC guide (SEAOC, 2017) states that solar panels on flat roofs should never be located closer than twice their height above the roof surface from the roof edges, unless parapets are present. This is to ensure that panels are not exposed to the separating shear layers (see Section 4.1) from the roof edge.

14.9 SUMMARY AND OTHER SOURCES

In this chapter, the wind loads on structures not covered in Chapters 8–13 have been discussed. This category includes free-standing walls and hoardings, attachments to buildings such as canopies and awnings, scaffolding and solar collectors, both roof-mounted and in arrays on ground.

Communications and broadcasting antennas of various types, particularly those impermeable enough to attract substantial wind loading, are considered in some detail. Some discussion of wind loads on elements of complex industrial structures, such as petrochemical plant, and offshore oil platforms has also been done.

Various other specialist sources for design loads have been cited in the Chapter, such as publications by the American Society of Civil Engineers (petrochemical structures), Engineering Sciences Data Unit (ESDU) (lattice structures), Structural Engineers Association of California (solar panel arrays), and the Scaffolding and Construction Equipment Association of Japan (scaffolding). Research papers on wind loading of the 'other' structures covered in this chapter can often be found in journals such as the *Journal of Wind Engineering and Industrial Aerodynamics* (Elsevier), and '*Wind and Structures*' (Techno Press).

REFERENCES

American Society of Civil Engineers (2016) *Minimum design loads and associated criteria for buildings and other structures*. ASCE/SEI 7-16. ASCE, Reston, VA.

American Society of Civil Engineers (2011) *Wind loads for petrochemical and other industrial facilities*. Prepared by the Task Committee on Wind-Induced Forces of the Petrochemical Energy Committee. ASCE, Reston, VA.

Banks, D. (2013) The role of corner vortices in dictating peak wind loads on tilted flat solar panels mounted on large flat roofs. *Journal of Wind Engineering and Industrial Aerodynamics*, 123: 192–201.

British Standards Institution (1997) *Loading for buildings. Part 2. Code of practice for wind loads.* BS 6399: Part 2: 1997.

British Standards Institution (2003) *Temporary works equipment. Part 1. Scaffolds – performance requirements and general design.* BS EN 12811: Part 1: 2003.

Chung, K.-M., Chang, K.-C., Chen, C.-K. and Chou, C.-C. (2013a). Guide plates on wind uplift on a solar collector model. *Wind and Structures*, 16: 213–224.

Chung, K.-M., Chou, C.-C., Chang, K.-C. and Chen, Y.-J. (2013b). Effect of a vertical guide plate on the wind loading of an inclined flat plate. *Wind and Structures*, 17: 537–552.

Colliers, J., Mollaert, M., Vierendeels, J. and De Laet, L. (2016). Collating wind data for doubly-curved shapes of tensioned surface structures. *Procedia Engineering*, 155: 152–162.

Cook, G.R., Kumarasena, T. and Simiu, E. (1986). Amplification of wind effects on compliant platforms. *Structures Congress '86, New Orleans, September 15–18, (Proceedings of Session: "Wind effects on compliant offshore structures"),* ASCE, New York.

ESDU (1981) *Lattice structures part 2 – Mean forces on tower-like space frames.* ESDU Data Items 81028. Engineering Sciences Data Unit (ESDU International PLC), London.

ESDU (1990) *Structures of non-circular cross section.* ESDU Data Items 90036. Engineering Sciences Data Unit (ESDU International PLC), London.

Ginger, J.D. and Letchford, C.W. (1994) Wind loads on planar canopy roofs – part 2: fluctuating pressure distributions and correlations. *Journal of Wind Engineering and Industrial Aerodynamics*, 51: 353–370.

Ginger, J.D., Payne, M., Stark, G., Sumant, B. and Leitch, C. (2011) *Investigation on wind loads applied to solar panels mounted on roofs.* Cyclone Testing Station Report, TS821. James Cook University, Townsville.

Ginger, J.D., Bodhinayake, G.G. and Ingham, S. (2019) Wind loads for designing ground-mounted solar panels. *Australian Journal of Structural Engineering*, 20: 204–218 (also Cyclone Testing Station Technical Report, TR64, James Cook University, Townsville).

Gumley, S.J. (1984) A parametric study of extreme pressures for the static design of canopy structures. *Journal of Wind Engineering and Industrial Aerodynamics*, 16: 43–56.

Holmes, J.D. (2001) Wind loading of parallel free-standing walls on bridges, cliffs, embankments and ridges. *Journal of Wind Engineering and Industrial Aerodynamics*, 89: 1397–1407.

Holmes, J.D., Banks, R.W. and Roberts, G. (1993) Drag and aerodynamic interference on microwave dish antennas and their supporting towers. *Journal of Wind Engineering and Industrial Aerodynamics*, 50: 263–269.

Jancauskas, E.D. and Holmes, J.D. (1985) Wind loads on attached canopies. *Fifth U.S. National Conference on Wind Engineering*, Lubbock, TX, November 6–8.

Kind, R.J. (1988) Worst suctions near edges of flat rooftops with parapets. *Journal of Wind Engineering and Industrial Aerodynamics*, 31: 251–264.

Kopp, G.A. (2014) Wind loads on low-profile, tilted, solar arrays placed on large, flat, low-rise building roofs. *Journal of Structural Engineering*, 140. doi: 10.1061/(ASCE)ST.1943-541X.0000825.

Koss, L.L. and Melbourne, W.H. (1995) Chain dampers for control of wind-induced vibration of tower and mast structures. *Engineering Structures*, 17: 622–625.

Letchford, C.W. and Ginger, J.D. (1992) Wind loads on planar canopy roofs – part 1: Mean pressure distributions. *Journal of Wind Engineering and Industrial Aerodynamics*, 45: 25–45.

Letchford, C.W. and Holmes, J.D. (1994) Wind loads on free-standing walls in turbulent boundary layers. *Journal of Wind Engineering and Industrial Aerodynamics*, 51: 1–27.

Letchford, C.W., Row, A., Vitale, A. and Wolbers, J. (2000) Mean wind loads on porous canopy roofs. *Journal of Wind Engineering and Industrial Aerodynamics*, 84: 197–213.

Lombardi, G. (1989) Wind-tunnel tests on a model antenna with different fin configurations. *Engineering Structures*, 11: 134–138.

Lombardi, G. (1991) Wind-tunnel tests on a model antenna rotating in a cross flow. *Engineering Structures*, 13: 345–350.

Marchman, J.F. and Werme, T.D. (1982) Mutual interference drag on signs and luminaires. *A.S.C.E. Journal of the Structural Division*, 108: 2235–2244.

Newton, J.R.H. (1983) Wind effects on buildings – recent studies at Redland wind tunnel. *Journal of Wind Engineering and Industrial Aerodynamics*, 11: 175–186.

Robertson, A.P., Hoxey, R.P. and Moran, P. (1985) A full-scale study of wind loads on agricultural canopy roof ridged structures and proposals for design. *Journal of Wind Engineering and Industrial Aerodynamics*, 21: 113–125.

Sachs, P. (1978) *Wind forces in engineering*, 2nd edition. Pergamon Press, Oxford.

Scaffolding and Construction Equipment Association of Japan (1999) *Safety technical guideline for scaffolding for wind loads* (in Japanese). SCEAJ, Tokyo.

Smith, C.E. and Simiu, E. (eds.) (1986) Wind effects on compliant offshore structures. *Proceedings of a Session at Structures Congress '86*, New Orleans, September 15–18, ASCE, New York.

Standards Australia (2011) *Structural design actions. Part 2: Wind actions*. Standards Australia, Sydney. Australian/New Zealand Standard AS/NZS1170.2: 2011.

Structural Engineers Association of California (2017) *Wind design for solar arrays*. Report SEAOC PV2-2017. SEAOC, Sacramento, CA.

Sykes, D.M. (1994) Wind loading tests on models for two structures for EXPO92, Seville. *Journal of Wind Engineering and Industrial Aerodynamics*, 52: 371–383.

Tieleman, H.W., Akins, R.E. and Sparks, P.R. (1980) *An investigation on wind loads on solar collectors*. Report VPI-E-80-1. Virginia Polytechnic Institute and State University, College of Engineering, Blacksburg.

Wang, F., Tamura, Y. and Yoshida, A. (2013) Wind loads on clad scaffolding with different geometries and building opening ratios. *Journal of Wind Engineering and Industrial Aerodynamics*, 120: 37–50.

Wood, G.S., Denoon, R.O. and Kwok, K.C.S. (2001) Wind loads on industrial solar panel arrays and supporting roof structure. *Wind and Structures*, 4: 481–494.

Wyatt, T.A. (1964) The aerodynamics of shallow paraboloid antennas. *Annals of the New York Academy of Sciences*, 116: 222–238.

Chapter 15

Wind-loading codes and standards

15.1 INTRODUCTION

Wind-loading codes and standards emerged in the second half of the twentieth century, have achieved wide acceptance, and are often the practising structural engineer's only contact with information for wind-loading calculations. Although they may be based on extensive research, they are, by necessity, simplified models of wind loading. Thus, great accuracy cannot be expected from them. Often this is consistent with the knowledge of the structure of the windstorms themselves in their country of use.

Advanced wind-loading codes and standards invariably contain the following features:

- A specification of a basic or reference wind speed for various locations, or zones, within a jurisdiction. Almost always, a reference height of 10 m in flat, open country terrain is chosen, (this is the standard location for wind measurements, specified by the World Meteorological Organization).
- Modification factors for the effects of height and terrain type, and sometimes for change of terrain, wind direction, topography, and shelter.
- Shape factors (pressure or force coefficients) for structures of various shapes.
- Some account of possible resonant dynamic effects of wind on flexible structures.

The above elements represent the first four links in the wind-loading 'chain' proposed by Davenport (1977, 1982), and shown in Figure 15.1. Each of these links contributes to the overall strength of the design process for wind loads, and weakness or inaccuracies in any one reduces the reliability of the process (see also Section 2.9).

This chapter reviews the wind-loading provisions of several prominent national, multi-national and international documents, and highlights their similarities and differences. As codes and standards are continually being revised and updated, the overview is, by necessity, time-dependent.

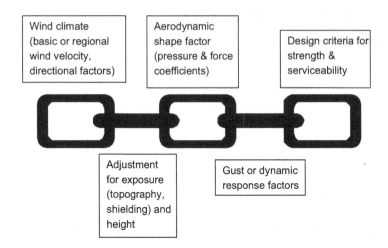

Figure 15.1   The wind loading 'chain'. (Adapted from Davenport, 1977, 1982.)

15.2 GENERAL DESCRIPTIONS

The provisions of following five codes and standards are summarized in this Chapter:

- ISO 4354:2009 – Wind actions on structures – published in 2009,
- EN 1991-1-4.6:2005, Eurocode 1: Actions on Structures–Part 1.4: General Actions - Wind Actions – published in 2005,
- ASCE Standard ASCE/SEI 7-16. Minimum Design Loads and Associated Criteria for Buildings and Other Structures – published in 2016,
- AIJ Recommendations for Loads on Buildings – published in 2004,
- Australian/New Zealand Standard. Structural design actions. Part 2: Wind actions. AS/NZS1170.2:2011 – published in 2011 with Amendments from 2012 to 2016.

The documents reviewed are those current at the time of writing. Although there are many other wind codes and standards in the world, many of these have been derived from, or are closely related to, one or the other of the above documents, so that the following comments have wide and general application.

15.2.1 ISO 4354-wind actions on structures

The current version of ISO International Standard 4354, Wind Actions on Structures, published by the International Organization for Standardization (ISO), was issued in 2009. The 2009 edition was completely re-formulated and bears little resemblance to the earlier (1997) edition. As described in the introduction to ISO 4354, the document is intended 'for use by countries without an adequate wind-loading standard and to bridge between existing standards. Most of the technical information in ISO 4354 is provided in a series of informative annexes. Two general methods of calculation of wind forces are given: one based on a peak velocity (nominally with a 3-second duration), and the other on a mean velocity (nominally with a 10-minute averaging time). However, for structures for which dynamic response effects are not important, the peak velocity method should be used.

The main part of the document is quite short, and consists largely of definitions of the terms in the expression used to calculate wind pressure:

$$p = (q_{site})(C_p)(C_{dyn}) \tag{15.1}$$

The site peak dynamic pressure, q_{site}, is derived from the peak site wind speed, which, in turn, is derived from a reference peak wind speed (in the standard exposure of 10 m height in open country terrain):

$$q_{site} = 0.5\rho(V_{site})^2 \tag{15.2}$$

$$V_{site} = V_{ref} \cdot C_{exp} \tag{15.3}$$

C_{dyn} in Equation (15.1) allows for the effects of fluctuating pressures due to upwind turbulence, in the wake of a structure, as well as forces resulting from resonant dynamic response.

Section 9 of ISO 4354 provides several methods to be used to determine pressure and force coefficients, including the use of wind-tunnel tests. Basic requirements for suitable wind-tunnel test procedures are given in *Annex H* in ISO 4354. Interestingly, *Section 9* also apparently allows the use of 'computational based data'. However, this is somewhat contradicted by *Annex I* in ISO 4354, which cautions against the use of computational fluid dynamics (CFD) techniques for this purpose, at the present stage of their development. The comments in *Annex I* on the applicability of CFD techniques are generally compatible with those in Chapter 16 of this book.

Some basic aerodynamic pressure and force coefficients for simple shapes are provided in an informative *Annex D*. This annex contains some interesting and innovative features:

    a. A clear specification of the target probability level associated with the specified pressure and force coefficients. This has been set at the 80% fractile of the extremes, and can be estimated as follows:

$$C_{p,80\%} = \bar{C}_p \pm 0.7 C_{p,rms} \tag{15.4}$$

when a reference time of 1 hour (or equivalent in a wind-tunnel test) is adopted,

$$C_{p,80\%} = \bar{C}_p \pm 2.1 C_{p,\text{rms}} \tag{15.5}$$

when a reference time of 10 minutes, or equivalent, is used.

$C_{p,\text{rms}}$ is the root-mean-square value determined from the set of the *extreme* pressure coefficients determined from multiple samples with the same reference period. (Note that this is not the same as the r.m.s. fluctuating pressure coefficient, C_p', for a complete pressure-time history, as defined in Sections 8.3.1 and 9.4.1 of this book.)

b. The use of the load-response correlation (LRC) method (see Section 5.4.3) to determine some effective pressure coefficients for overall lift and drag forces for some simple shapes of low-rise buildings. These are given in Section D10 of ISO 4354.

In *Sections D6* and *D7* of ISO 4354, external and internal pressure coefficients that originate in the Australian/New Zealand Standard, AS/NZS 1170.2, have been specified. However, these should be used with caution in conjunction with ISO 4354, as they have not been adjusted for the different gust durations in the two Standards (0.2 seconds in AS/NZS 1170.2 versus 3 seconds in ISO 4354). This could be achieved through the dynamic response factor, C_{dyn}, which should be greater than 1.0 for small structures. The suggested value of C_{dyn} of 0.85 in *Section D5* of ISO 4354 is clearly un-conservative for small low-rise buildings, when used with 3-second gust velocities and quasi-steady pressure coefficients (see Section 15.3 in this chapter, and Holmes *et al.*, 2014).

Annex E in ISO 4354 provides detailed informative advice on the dynamic response factor for dynamically sensitive structures such as tall buildings and towers. Methods are given for estimating along-wind, cross-wind and torsional response, and these are closely related to those given in the Recommendations of the Architectural Institute of Japan (AIJ).

15.2.2 EN 1991-1-4.6 Eurocode 1.
Part 1–4 wind actions

Eurocode 1 on wind loads, issued in 2005, is a European Standard (EN) which is intended for use in most European countries. The version in each member country contains a 'National Annex', applicable only to that particular country. EN 1991-1-4.6 represents several years of work by representatives from many countries of the European Union and two separate committees, and is the nearest document to a truly multi-national wind-loading standard currently in existence.

This is a lengthy document with comprehensive methods of static and dynamic design for wind loads. The code is applicable to buildings and other structures, with heights up to 200 m, and to bridges with spans less than 200 m. No basic wind speeds are provided – these are provided separately in each National Annex. However, the basic wind velocity in each country is a 10-minute mean velocity at 10 m height in open country terrain, with an annual probability of exceedance of 0.02 (50-year return period).

The mean wind velocity $v_m(z)$ at a height, z, above the terrain is given by:

$$v_m(z) = c_r(z) \cdot c_o(z) \cdot v_b \tag{15.6}$$

where v_b is the basic wind velocity at 10 m height over open country terrain; $c_r(z)$ is a roughness factor, which varies with both height and terrain type; and $c_o(z)$ is the orography (i.e. topography) factor. For the roughness factor, five different terrain types are defined with roughness lengths, z_o (see Section 3.2.1), varying from 0.003 to 1.0 m.

The peak velocity pressure, $q_p(z)$, is given by:

$$q_p(z) = c_e(z) \cdot q_b \tag{15.7}$$

where $q_b = \frac{1}{2} \rho \, v_b^2$. ρ is the density of air, given in the National Annexes, and $c_e(z)$ is an 'exposure factor' given by:

$$c_e(z) = \left[1 + 7 I_v(z)\right] \cdot \left[c_r(z) \cdot c_o(z)\right]^2 \tag{15.8}$$

$I_v(z)$ is the turbulence intensity at height, z. The term $[1 + 7 I_v(z)]$ is effectively a gust factor for the velocity pressure, and is an approximation to $[1 + g_v I_v(z)]^2$, with the peak factor g_v taking a value of 3.5.

Thus, the exposure factor, $c_e(z)$, combines gusting effects, terrain, height and topographic effects into a single height- and terrain-dependent factor, and enables the 10-minute mean wind velocity to be effectively converted to a gust wind velocity, and velocity pressure. The use of a peak factor of 3.5 with a 10-minute reference period corresponds to an equivalent gust duration of about 0.1 seconds.

The number of shape factors presented in *Section 7* of EN 1991-1-4.6 is extensive, with the number of cases covered exceeding those in most other codes and standards. External pressure coefficients on buildings, are given for loaded areas of 1 m², and 10 m², denoted by $C_{p.e1}$ and $C_{p.e10}$, respectively. For areas between 1 and 10 m², linear interpolation is applied.

A 'structural factor' that allows for reductions in effective loading due to correlation effects on large structures, and for possible increases due to resonant dynamic effects, is defined in Section 6. This is discussed in more detail in Section 15.8.

Two alternative and independent procedures for calculating parameters for along-wind dynamic response (B^2 and R^2) are given in separate Annexes (*Annex B* and *Annex C*, respectively). A separate *Annex D* provides graphical information on the structural factor as a function of height and width of the structure, with separate graphs given for concrete and steel buildings, and for chimneys, with or without liners.

The several methods given for calculation of the structural factor is potentially confusing for the user, and may have legal implications. However, the multiple alternative options may be resolved within the National Annexes for each participating country.

Annex E contains detailed information on vortex-induced response of slender structures, such as chimneys. Two different and independent approaches for calculation of the cross-wind amplitude are given in *Sections E.1.5.2* ('Approach 1') and *E.1.5.3* ('Approach 2'), respectively. Approach 1 is based on a sinusoidal excitation model (see Section 11.5.1 in this book); Approach 2 is derived from a random excitation model (see Section 11.5.2). Work for a revision of EN 1991-1-4.6 was in progress in 2019–2020.

15.2.3 ASCE/SEI standard ASCE 7-16

ASCE/SEI 7-16, 'Minimum Design Loads and Associated Criteria for Buildings and Other Structures', is a complete loading standard covering all types of loads. The wind-loading part consists of Chapters 26–31.

From 1995 onwards, ASCE 7 has incorporated a number of significant changes in the wind load provisions from the 1993 and earlier editions. This includes the use of a 3-second gust wind speed instead of the 'fastest-mile-of wind' as used in the past, a new zoning system for basic wind speeds, the use of high average recurrence intervals (300, 700 and 1,700 years) for ultimate limit states design, the incorporation of topographic factors, some new data on pressure coefficients, a simplified procedure for buildings less than 9 m in height, and a revised method for along-wind dynamic response calculation.

The wind-loading provisions of ASCE 7-16 comprise six chapters as follows:

- *Chapter 26* gives 'General Requirements', including wind hazard maps, exposure categories, topographic multipliers, gust-effect factors and internal pressure coefficients.
- *Chapter 27* provides a 'Directional Procedure', the 'main wind-force resisting system' (MWFRS) for buildings of all heights.
- *Chapter 28* gives an 'Envelope Procedure' (non-directional) for low-rise buildings (defined as having a mean roof height less than 18 m).
- *Chapter 29* describes a directional procedure for building appurtenances (such as rooftop structures and equipment) and other structures (such as freestanding walls and signage, chimneys tanks, lattice frameworks and towers).
- *Chapter 30* provides procedures for assessing wind loads on 'components and cladding'.

• *Chapter 31* describes the main requirements for the 'Wind-Tunnel Procedure'. This may be required in earlier chapters for certain wind-sensitive structures, and is available as an option for all structures or parts of structures covered by *Chapters 27–30*.

The ASCE Standard has no legal standing of its own, but its provisions are cited by many of the regional, city and county building codes. An 'International Building Code' in the United States draws on the ASCE Standard for wind load provisions.

15.2.4 AIJ recommendations for loads on buildings

The Recommendations of the Architectural Institute of Japan were revised in 2004 and 2015, and form a comprehensive loading code including the effects of dead, live, snow, seismic, temperature, earth and hydraulic pressure, as well as wind loads. Chapter 6 on wind loads comprises 54 pages, with 114 pages of Commentary. The derivation of the wind-loading section of the 2004 edition of the AIJ, and revisions from the 1993 version, were described in detail by Tamura *et al.* (2004).

Like the ASCE Standard, this is a comprehensive and advanced wind-loading document, although the Recommendations have no legally binding standing in Japan. The Building Law of Japan has a separate set of wind-loading rules – BSLJ-2000 (Ministry of Land, Infrastructure and Transport, 2000). Since the latter does not have a comprehensive set of rules for cross-wind and torsional dynamic response, the AIJ is commonly used by structural designers for buildings greater than 60 m in height.

15.2.5 Australian/New Zealand standard AS/NZS 1170.2

The joint Australian and New Zealand Standard for Wind Loads, AS/NZS 1170.2, was issued in 2002, and revised in 2011, as a combined Standard, replacing separate documents from the two countries. It is a comprehensive document of about a hundred A4 pages, and is supported by a separate Handbook (Australasian Wind Engineering Society, 2012).

AS/NZS1170.2 has an indirect legal status in Australia by being called up in the National Construction Code (Australian Building Codes Board, 2019). This document, or Part 0 of the joint Standards on Structural design actions, AS/NZS 1170.0 (Standards Australia, 2002) must be consulted to obtain the appropriate annual probability of exceedance for the importance and use of the structure, before use of AS/NZS 1170.2, Wind actions.

The nominal basic wind speed in AS/NZS1170.2 is a 0.2-second gust measured at 10 m height in open country terrain, and values are specified for a range of annual recurrence intervals from 1 to 10,000 years, for four regions. The gust duration was re-defined in 2012 from the value of 3 seconds given in earlier editions of the Standard. The justification for this was given by Holmes and Ginger (2012). For most buildings, excluding those with large numbers of occupants, and important post-disaster facilities, the annual risk of exceedance for ultimate limit states wind speeds is specified, in the Building Code of Australia, as 1/500. Tall buildings generally are assessed to have a higher 'importance level' and an annual risk of exceedance of 1/1,000 is adopted. Other structures such as temporary ones and cyclone shelters may have lower and higher values, respectively.

A draft revised edition of AS/NZS 1170.2 was issued in 2020 (Standards Australia, 2020), and is expected to be published by 2021. This will include some revision of regional boundaries for the basic regional wind speeds and directional multipliers in both Australia and New Zealand. The inland region in Australia (at least 200 km from the coastline) is dominated by non-synoptic winds generated by thunderstorms (see Sections 1.3.3 and 1.3.5), and terrain-height and topographic multipliers appropriate to these events will be specified. In the regions of Australia affected by tropical cyclones, a linear reduction in wind speed with distance from the coastline to reflect the weakening of the storms, and a 'climate change multiplier' to incorporate the apparent increase in stronger severe cyclones (see Figure 1.18) will be introduced. A number of changes to shape factors, with additional data for on-ground arrays of solar panels (solar farms), will also be included.

15.3 BASIC WIND SPEEDS OR PRESSURES

Codes and standards for wind loading are currently based on extreme wind speeds with a variety of nominal averaging times. These variations have occurred for a number of reasons, such as the type of recording systems used by meteorological services to record winds, or the type of extreme wind event that dominates designs for wind in a particular jurisdiction.

Some codes are based on wind speeds averaged over relatively long periods, such as ten minutes or one hour. However, the wind speed is often effectively converted to a gust speed within the format of the code, before calculating building pressures or forces. Gust factors, being the ratios between the expected maximum gust and the mean value, in an averaging time such as 1 hour, are therefore important for these conversions (see Section 3.3.3). The use of a gust wind speed as a basis is particularly recommended for jurisdictions with extreme winds caused by transient, non-synoptic wind events, for which hourly-averages are not appropriate.

Most national wind codes and standards are based on a maximum gust wind speed, with a defined gust duration; it is most common to find it stated as a '3-second' gust. The reason for this value is partly historical, with a perception of this being a typical averaging time of anemometers used to record historical data on gust wind speeds. However, as discussed by Holmes *et al.* (2014), the effective frontal area associated with a gust of this duration, at typical design wind speeds in synoptic wind events, is equivalent to that of a tall building. For smaller structures, a code or standard based on a 3-second gust should therefore incorporate a 'gust effect factor', or amplification factor, somewhat greater than 1.0, to allow for the lack of reduction due to correlation effects over small frontal areas.

Table 15.1 summarizes the basic wind speed characteristics used, or recommended, in the five documents surveyed in this chapter. In all cases, the standard meteorological reference position of 10 m height in flat, open country is used.

The ISO Standard, as previously discussed, does not give basic wind speeds or dynamic pressures. *Annex B* in ISO 4354 provides peak factors and gust factors for conversion of wind data with various other averaging times, to the 3-second gust and 10-minute velocities used as a basis for calculation of wind loads in ISO 4354. For synoptic winds, more comprehensive terrain- and wind speed-dependent conversions are given in ESDU 83054 (ESDU International, 1983), and by Holmes *et al.* (2014).

The Eurocode EN 1991-1-4.6, also does not give basic wind speeds, although a previous 1994 draft (CEN, 1994) gave 'reference wind velocities', for 18 countries in Europe in an informative Annex. National Annexes now provide basic wind speed information for individual countries in Europe.

The US Standard (ASCE 7-16) contains maps with wind-speed contours, with closely spaced ones for Alaska and the coastal regions adjacent to the Gulf of Mexico and the Atlantic Ocean. In the latter case, the effects of hurricanes are of particular concern. Maps of basic wind speeds for the islands of Hawaii incorporate topographic effects. The values of basic wind speed given on these maps, are peak gust wind speeds, with mean recurrence intervals (MRI) of 300, 700, 1,700 and 3,000 years. The MRI required to be used depends on the importance of the building, or other structure. The basic wind speed values, for the non-hurricane regions of the continental United States, have changed significantly in the 2016 edition of ASCE-7

Table 15.1 Definitions of basic wind speeds

Code	Averaging time	Return periods/annual recurrence intervals[a]
ISO 4354:2009	3 seconds, (10 minutes)	Not specified
EN 1991-1-4.6	10 minutes	50 years
ASCE 7-16	3 seconds	350–700–1,700–3,000 years
AIJ (2004)	10 minutes	100 years
AS/NZS1170.2:2011	0.2 seconds	500–1,000 years[b]

[a] For ultimate limit states design.
[b] A wide range of annual recurrence intervals are provided in AS/NZS 1170.2 for various limits states, and importance levels.

compared with those in the 2010 version, following extensive re-analyses of anemometer data.

The Recommendations of the Architectural Institute of Japan (AIJ) gives a detailed map showing contours of the basic wind speed (10-minute mean with 100-year return period). Single values are given for the outlying territories such as Okinawa. A map of 500-year return period values is also given to enable users to interpolate for intermediate return periods.

In the Australian/New Zealand Standard, the 500–1,000 years return periods shown in Table 15.1 apply to the majority of buildings in Australia (Importance Level 2 or 3 in the National Construction Code of Australia) for assessment of ultimate limit states design criteria. Basic wind speeds are given in the form of maps with five regions, denoted by A, B, C, D and W. Two of these regions (C and D) comprise a coastal strip exposed to the effects of tropical cyclones (Section 1.3.2). Regional wind speeds are specified for each Region as a function of annual probability of exceedance. The analysis of extreme wind speeds for Region A, covering most of Australia, in the 2002 Australian Standard was described by Holmes (2002). Some analysis of wind speeds for Regions C and D was discussed by Dorman (1984).

15.4 MODIFICATION FACTORS

All the documents include modifiers for the effect of terrain/height and topography, although in the case of ASCE 7, these act on the dynamic *pressure*, rather than wind *speed*.

In the Eurocode, the mean wind speed is modified for terrain and height (roughness factor c_r), and for topography (described as 'orography'), c_o, then converted into a *gust* dynamic pressure at the height of interest, by a factor involving turbulence intensity (i.e. a gust factor acting on the dynamic pressure). The exposure coefficient $c_e(z)$ includes terrain/height and topographic (orographic) effects within Equation (15.8).

EN 1991-1-4.6 and AS/NZS1170.2 use a logarithmic law (or a modification for gust speeds) to define the terrain/height variation, ASCE 7 and AIJ use a power law variation, and ISO 4354 gives parameters for both. AS/NZS1170.2 allows for averaging of terrain roughness upwind of the site, with an interpolation of terrain/height multipliers.

All five documents provide factors or multipliers for topographic, or orographic, speed-up effects on wind speeds. None of the documents allow for any shielding effects of topography. However, there are significant differences in the magnitudes of the speed ups predicted by the various documents for the same topographic geometry, as discussed by Holmes *et al.* (2005a).

Table 15.2 Calculation formats for velocity, dynamic pressures and building pressure

Code	Velocity	Dynamic Pressure	Building Pressure/Force
ISO4354	$V_{site}=V_{ref} C_{exp}$	$q_{site}=(1/2)\,\rho\,(V_{site})^2$	$p=q_{site}\,C_p\,C_{dyn}$
EN 1991-1-4.6	$v_b=c_{dir}\,c_{season}v_{b,0}$	$q_p(z)=c_e(z)\,(1/2)\,\rho\,v_b^2$	$w_e=q_p(z)\,c_{pe}$
ASCE 7	V	$q_z=(1/2)\,\rho\,K_z K_{zt} K_d V^2 I$	$p=q\,(GC_p)$
AIJ	$U_H=U_0 K_D E_H k_{rW}$	$q_H=(1/2)\,\rho\,U_H^2$	$W_f=q_H\,C_f\,G_f\,A$ [a]
AS/NZS1170.2	$V_{sit,\beta}=VM_d M_{(z,cat)}\,M_s\,M_t$	$q_z=(1/2)\,\rho_{air}V_{des,\theta}^2$	$p=q_z\,C_{fig}\,C_{dyn}$

[a] The subscript f denotes D (for walls) or R for roofs in the AIJ.

The Australian/New Zealand Standard, AS/NZS1170.2, is unique in having a 'Shielding Multiplier', which allows for reductions in velocity when there are buildings upwind of greater or similar height.

Table 15.2 summarizes the formats for calculation of design wind velocities and dynamic pressures in the various documents.

15.5 BUILDING EXTERNAL PRESSURES

Table 15.2 also shows the general format for calculation of external pressures on wall or roof surfaces of enclosed buildings.

The formulae (in the right-hand column) appear to be quite different from each other, but they all contain quasi-steady or mean pressure coefficients (C_p, c_{pe}, C_f, C_{fig}) and factors to adjust the resulting pressures to approximate peak values. In the case of ISO 4354 and AIJ, there are gust factors on pressure (C_{dyn} and G_f); in the case of EN 1991-1-4.6, the gust factor is incorporated into the exposure coefficient, $c_e(z)$, as discussed previously.

In ASCE 7, the quantities G and C_p are usually combined together as (GC_p) in tables. In AS/NZS1170.2, the 'aerodynamic shape factor', C_{fig}, consists of pressure coefficients, multiplied by factors for area reduction, combination of roof and wall surfaces, local pressure effects and porous cladding. The local pressure factor K_l is always greater than 1, and the area reduction factor K_a, which allows for correlation effects over large areas in separated flow regions, is less than one. AS/NZS1170.2 is alone in having a factor (K_p) for porous cladding.

The tables of shape factors and pressure coefficients of exterior surfaces of buildings given in the various documents are also sources of significant differences. In all cases, the nominal wind directions are normal to the walls of buildings of rectangular plan.

ASCE 7 and AS/NZS1170.2 require alternative positive roof pressure coefficients to be considered. These are important values for the design of frames, especially for those in colder climates where dead loads are often high, as pointed out by Kasperski (1993).

EN 1991-1-4.6 gives tables of external pressure coefficients c_{pe} which are comparable to those in ASCE-7 and AS/NZS1170.2, since they are effectively applied to gust dynamic pressure through the use of the exposure coefficient $c_e(z)$. The tables give two values: $c_{pe,1}$, intended for tributary areas less than $1\,m^2$, i.e. local cladding design, and $c_{pe,10}$ intended for major structural members. It appears that the numerical values for flat and gable ('duopitch') roofs in EN 1991-1-4.6 are comparable to those in ASCE-7 and AS/NZS1170.2, and alternative (positive or lower negative) values are given for most roof pitches. However, no variation with height/width ratio is given.

The factors incorporated into the shape factor in the Australia/New Zealand Standard AS/NZS1170.2 for flat and gable-roofed buildings have already been discussed. However, it should also be mentioned that the effect of tributary area and correlation effects are dealt with by the use of the three factors: K_a (area reduction factor), K_c (action combination factor) and K_l (local pressure factor). The action combination factor, K_c, in the Australian and New Zealand Standard, allows for a reduction when wind pressures from more than one building surface, e.g. walls and roof, contribute significantly to a load effect.

The AIJ Recommendations also separate the specification of loads on the structural frames and on the 'components and cladding' of buildings. The specification of pressure coefficients is separated from the specification of the gust factor. Unlike any of the other documents, the gust factor, G_R, for the loads on the roofs of low-rise buildings has a dependency on natural frequency. Buildings are classified as those with heights less than, or greater than 45 metres, a somewhat greater height than used in the other documents.

15.6 BUILDING INTERNAL PRESSURES

The treatment of internal pressures varies considerably from one document to another.

For buildings with uniformly distributed openings, EN 1991-1-4.6 gives a graph of c_{pi}, varying from +0.35 to −0.5, as a function of an opening ratio, μ. For a dominant opening, the internal pressure coefficient is expressed as a fraction of the external pressure coefficient on the face with the opening. This document also gives fairly detailed guidance on pressures on walls and roofs, with more than one skin.

ASCE 7-16 (in *Clause 26:13*), specifies four different situations: open, partially open, partially enclosed, and enclosed buildings, and specifies values of GC_{pi} between +0.55 and −0.55. A feature, not found in the other standards, is a reduction factor, R_i, for large building volumes.

AS/NZS1170.2 gives two tables with various positive and negative values of internal pressure coefficients, $C_{p,\,i}$. For one of these tables, the values depend on the ratio of dominant openings on the windward wall to the total open area on other walls and roof. ISO 4354 gives similar internal pressure coefficients to those in AS/NZS 1170.2.

The AIJ Recommendations do not specify a positive internal pressure, i.e. the possibility of dominant openings is not considered. For buildings without dominant openings, values of C_{pi} of 0 or -0.5 are specified.

15.7 OTHER SHAPES AND SECTIONAL FORCE COEFFICIENTS

Apart from the AIJ Recommendations, which is intended exclusively for buildings, all the surveyed documents contain shape, or force, coefficients for a variety of structure shapes and cross sections. Table 15.3 summarizes the data given.

The data in all these documents appear to be based on modern wind-tunnel measurements for the most part. EN 1991-1-4.6 clearly contains the most comprehensive set of data. ISO 4354 only provides a limited amount of data on shape factors in Annex D, but suggests that other sources may be used in conjunction with standard, but with appropriate adjustment for gust averaging time and exposure. These sources include other codes or standards.

15.8 DYNAMIC RESPONSE CALCULATIONS

The five standards contain procedures for the calculation of dynamic response for wind-sensitive structures, such as slender, flexible, lightly damped tall buildings. ISO 4354, in *Clause E.2.2*, has a relatively complex set of numerical criteria to determine whether a structure is 'dynamically sensitive' in the along-wind, cross-wind and torsional modes. EN 1991-1-4:2005, in *Clause 6.2*, also has a set of empirical criteria to determine whether a structure need not be treated as dynamic; however, generally buildings with height to along-wind depth (i.e. h/d) greater than four, require calculation of the structural factor. ASCE 7 and AS/NZS1170.2 classify wind-sensitive structures as those with a first-mode natural frequency less than 1 Hz; ASCE 7 also requires a height-to-breadth (or -depth) ratio greater than four.

In ISO 4354, information for calculation of along-wind response is given in two alternative formats – one for use with a (10-minute) mean dynamic pressure (*Clause E3*), and a second for use with the (3-second) peak dynamic pressure (*Clause E4*). These are respectively in the *gust response factor* discussed in Section 5.3.2 of this book, and the *dynamic response factor* format of Section 5.3.4. However, the user should be aware that,

Table 15.3 Shape factors contained in five documents (excluding rectangular enclosed buildings)

Type	ISO 4354	EN 1991	ASCE 7	AIJ	AS/NZS1170.2
Stepped roofs	No	No	Yes	No	No
Free-standing walls, hoardings	Yes (walls only)	Yes	Yes	No	Yes
Parapets	No	Yes[b]	Yes	No	No
Free-standing roofs (canopies)	Yes	Yes	Yes	Yes	Yes
Attached canopies	No	No	No	No	Yes
Multispan roofs (enclosed)	No	Yes	Yes	Yes[a]	Yes
Multispan canopies	No	Yes	No	No	No
Arched roofs	No	Yes	Yes	Yes[a]	Yes
Domes	No	Yes	Yes	Yes[a]	Yes[c]
Bins, silos, tanks	Yes	Yes	Yes	No	Yes
Circular sections	Yes	Yes	Yes	Yes	Yes
Polygonal sections	No	Yes	Yes	No	Yes
Structural angle sections	Yes	Yes	No	No	Yes
Bridge decks	No	Yes	No	No	No
Lattice sections	Yes	Yes	Yes	No	Yes
Flags	No	Yes	No	No	Yes
Sphere	No	Yes	No	No	Yes
Roof-mounted solar panels	No	No	Yes	No	Yes

[a] Given in commentary section.
[b] Treated as free-standing walls.
[c] Treated as arched roofs.

in the latter formulation, the specified peak factor, g_v, of 3.4 in *Equation (E.4.3)* of ISO 4354 is too high when used with gust pressures derived from a 3-second gust, and is incompatible with Table B.1 in *Annex B* of ISO 4354 (see also Table 3.4 in this book) – potentially leading to un-conservative values of C_{dyn}.

In *Section E6.1* of ISO 4354, some recommended values of structural damping are given as a function of construction material and building height. These values are apparently for ultimate limit states design criteria, as 75% of the specified values are recommended for 'habitability to horizontal vibrations'.

ISO 4354 recognizes the importance of vortex shedding in causing dynamic cross-wind effects in slender prismatic and cylindrical structures. For rectangular cross-sections, the critical wind speed at which large amplitude motions may result, can be calculated, as a function of side ratio and Scruton number (see Sections 5.5.2 and 11.5.1).

A 'structural factor', in EN 1991-1-4.6, denoted by $c_s c_d$, incorporates the combined effects of size (c_s) and resonant dynamic response (c_d) for overall structural loads, or loads on major structural elements. The structural factor is equivalent to the 'gust effect factor' for dynamic structures in ASCE 7, and the 'dynamic response factor' discussed in Section 5.3.4.

This factor is given in the Eurocode by:

$$c_s c_d = \frac{1 + 2k_p I_v\left(z_s\right)\sqrt{B^2 + R^2}}{1 + 7 \cdot I_v\left(z_s\right)}$$ (15.9)

z_s is a reference height. For buildings and towers, this is normally taken as 60% of the roof height.
k_p is a peak factor for the response.
$I_v(z_s)$ is the longitudinal turbulence intensity at the reference height.
B^2 is a background factor.
R^2 is a resonance response factor.

As discussed in *Clause 6.3.1* in EN 1991-1-4.6, the structural factor is actually specified as a combination of a size factor, c_s with a dynamic factor, c_d, to form the combined structural factor, $c_s c_d$. The size factor separately takes account of correlation effects.

Alternative methods of calculation of the factors, k_p, B^2 and R^2, are given in *Annexes B* and *C*, with a simplified graphical method for estimation of $c_s c_d$, given in *Annex D*. Alternative expressions for the standard deviation of along-wind acceleration, for serviceability limit states, are also given in *Annexes B* and *C*.

Comprehensive information, including working equations, regarding vortex excitation and other aeroelastic effects such as vortex-induced large amplitude lock-in type vibrations, galloping (Section 5.5.2 in this book), various types of interference excitations, and flutter (Section 5.5.3) are included in Annex E of EN 1991-1-4.6. Recommended calculation procedures for dynamic structural properties, including natural frequencies, mode shapes, equivalent masses and logarithmic decrement, are given in *Annex F* of EN 1991-1-4.6.

In ASCE 7-16, an analytical procedure for determination of a 'gust effect factor', G_f, for the along-wind vibrations of flexible buildings and other structures is presented in the commentary section of that Standard. The development of this factor was described by Solari and Kareem (1998). The gust effect factor is, in fact, a *dynamic response factor* (Section 5.3.4), defined in the same way as the structural factor, $c_s c_d$, in EN 1991-1-4.6, i.e. it is based on a 3-second gust wind speed, as adopted by ASCE-7. The calculation procedure is nearly identical to that in EN 1991-1-4.6, making use of the closed-form equations of Solari (1983). Expressions for maximum along-wind displacement and standard deviation and maximum along-wind acceleration are also given. However, no analytical procedure for cross-wind response is given.

In the AIJ Recommendations, a detailed procedure is applied to estimate the dynamic response of wind-sensitive structures. For along-wind response, a standard gust response factor approach along the lines of Equation (15.3) is used to determine a gust effect factor G_D. A mode shape correction for prediction of peak base bending moments for buildings with nonlinear mode shapes is provided (Tamura *et al.*, 2004).

Vortex-induced cross-wind vibration and wind loads can also be determined from the AIJ, based on r.m.s. cross-wind base moment data obtained from wind-tunnel tests. Expressions for effective cross-wind load distributions, displacement and acceleration are given. However, the cross-wind response calculations are restricted to prismatic cross-sections with a height-to-breadth ratio no greater than six, and to wind directions normal to a face of the building. Expressions for torsional angular acceleration and torsional wind load distribution are also given. Guidelines for assessing potential aeroelastic instabilities including lock-in type vortex resonance and galloping instabilities are presented.

The dynamic along-wind response of tall buildings and towers is dealt with through a dynamic response factor, C_{dyn}, in *Section 6* of AS/NZS1170.2; this corresponds with the definition of dynamic response factor given in Section 5.3.4 of this book. The methodology is a greatly modified version of the description given by Vickery (1971).

Cross-wind base overturning moment and acceleration can be determined from cross-wind force spectrum coefficients, derived from wind-tunnel test data for a series of square and rectangular section buildings, with the incident wind normal to a face. Suggested values of damping for a range of steel and concrete structures under different stress levels are given. The importance of aeroelastic instabilities, such as lock-in, galloping, flutter and interference are discussed separately in the Handbook for AS/NZS1170.2. However, a 'diagnostic' method for the crosswind response of chimneys, masts and poles of circular cross-section is provided in the Standard itself.

15.9 INTER-CODE COMPARISONS

Several numerical comparisons have been made of wind load calculations by various international codes and standards. The results from some of those comparisons are summarized here.

General comparisons between major wind-loading codes and standards have been made by Cook (1990), Mehta (1998), and Zhou *et al.* (2002) for dynamic effects. A special issue of the journal *Wind and Structures* comprised five papers in which all aspects of codification for wind loads were reviewed (Holmes *et al.*, 2005a, b; Tamura *et al.*, 2005; Letchford *et al.*, 2005; Kasperski and Geurts, 2005).

Holmes *et al.* (2009) described an extensive comparison of calculations of wind loads on three buildings from fifteen codes and standards in the Asia-Pacific region. The three buildings, comprising a generic low-rise, a medium-rise and a high-rise building, are shown in Figure 15.2. The low-rise building was a typical steel-framed portal frame structure located in a rural area; the medium-rise building was a 48 m high office building located in an urban area, and the high-rise building was 183 m tall and also located in an urban area. In all three cases, design wind speeds at the top of each building were pre-specified; wind speeds with averaging times of 3 seconds, 10 minutes and 1 hour were all specified, and the participants selected an appropriate time according to the stated averaging time in their own code or standard at the time of the comparison.

The comparisons for the low-rise building showed coefficients of variation of 20%–31% in the *net* pressures across the building surfaces. These relatively high values were partly caused by differences in the treatment of *internal* pressures, since a large opening in the windward wall of the building was specified. Somewhat smaller coefficients of variation of 13%–26% were obtained for the wind loads on a large door and a small window in the building.

For the medium-rise building, the coefficients of variation for the predictions of base shear and bending moments, and cladding pressures were consistent at 22%–23%. The coefficients of variation for along-wind and cross-wind base shears and moments on the 183 m high rise building were relatively small at 14%–17%. This was attributed to the common origin of many of the code provisions for dynamic response to wind.

Holmes (2014) described a comparison of the responses of a generic tall building as calculated by three codes and standards, with the consensus along-wind response determined by several wind-tunnel groups using the high-frequency base balance technique (see Section 7.6.2). The comparisons of Holmes (2014) are interesting because of the direct comparison of code values with wind-tunnel data. For wind acting normal to the wide face of the tall building in question (similar in dimensions to the high-rise building shown in Figure 15.2), AS/NZS 1170.2:2011 produced along-wind base bending moments that are within −4% to +8% of the averages of the wind-tunnel data (for three different wind speeds). ASCE 7-10 produced values 8%–17% below the average of the wind-tunnel data, and the Hong Kong Code of Practice at the time (HK Buildings Department, 2004) produced values 27%–33% below the average of the wind-tunnel data. The paper gave explanations for the discrepancies; in the case of the Hong Kong Code of Practice, the differences were directly associated with low drag coefficients specified in the code. That discrepancy has been addressed in the

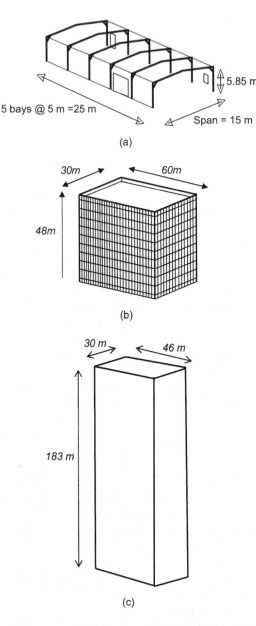

Figure 15.2 Three buildings used for an inter-code comparison in the Asia-Pacific region (Holmes et al., 2009): (a) low-rise industrial building, (b) medium-rise office building and (c) high-rise building.

2019 revision of the Hong Kong Code (Buildings Department, Hong Kong, 2019).

15.10 OTHER CODES, STANDARDS AND DESIGN GUIDES

The codes and standards discussed in earlier sections of this chapter are primarily intended for wind-loading design of habitable buildings, which, in most developed countries, is controlled by legislation. The extent to which other structures are covered in the major codes and standards varies greatly. For example, shape factors for the wind loading of bridges are included in Eurocode 1, but not in the other documents. Another difference is the extent to which wind-tunnel testing has been standardized. In ASCE 7, a separate chapter (*Chapter 31*) covers the 'Wind-Tunnel Procedure'. This mainly restricts the reduction in design loads that can be obtained from wind-tunnel testing compared to those from the main procedures of the standard.

In many cases, other standards and design guides have been issued for structures apart from buildings. Some of these are discussed in the following.

As discussed in Chapter 12, the aerodynamics of long-span bridges is a specialist topic and normally involves extensive wind-tunnel testing, which only a relatively few wind tunnels are capable of undertaking. As mentioned in Section 12.6, a basic standard published in the United Kingdom, BD 49/01 (Highways Agency, 2001), sets out some design rules for the aerodynamic design of bridges with useful limiting parameters for dynamic instabilities such as flutter and galloping of bridge decks. Requirements for wind-tunnel testing, including boundary-layer modelling, topographic effects and section, and full aeroelastic modelling are given in this document.

Eurocode 3 for 'Design of Steel Structures' (CEN, 2006) includes wind-loading information for steel communication towers and masts, such as sectional drag coefficients for lattice tower sections of various solidities. A similar document, AS 3990 (Standards Australia, 1994), was issued in Australia, although the wind-loading information in that document has now been updated and included in the main wind-loading standard, AS/NZS 1170.2.

Transmission line structures – both high-voltage and local distribution towers and poles – have particular vulnerability to small local non-synoptic storms. This is reflected in Australian/New Zealand Standard AS/NZS 7000:2016, (Standards Australia, 2016) a standard for design of overhead lines, which provides special rules for extreme winds from convective downdrafts, including increased span reduction factors (see Section 13.2.3). The ASCE Guidelines (ASCE, 2020) perform a similar function, although technically the document is not an enforceable standard.

The proliferation in a short time of large-scale energy harvesting by arrays of solar panels on inclined frames has left wind codes and standards struggling to provide up-to-date wind-load information for the supporting structures, the design of which is governed by wind loads. For panels on flat or near-flat roofs, the guidelines of the Structural Engineers Association of California have provided very useful design information (SEAOC, 2017). This document also includes loads for arrays of panels on ground – i.e. 'solar farms' (see also Section 14.7).

Wind-tunnel testing, although widely used throughout the world as a method of determining wind loads (see Chapter 7), has been lightly regulated. Two documents that provide minimum recommended criteria for modelling and processing of data are an American Standard, ASCE/SEI 49-12 (ASCE, 2012), and a Quality Assurance Manual of the Australasian Wind Engineering Society, AWES-QAM-1–2019 (AWES, 2019). These documents are widely used and quoted, but are, up to now, not mandatory.

15.11 GENERAL COMMENTS AND FUTURE DEVELOPMENTS

This chapter has reviewed the provisions of five major and current (at the time of writing) standards for wind loading. Considerable differences exist in both format and the type of information presented in these documents.

At present, the international standard on wind loading, ISO 4354:2009, is not generally used for structural design. It is difficult to use it as an operating standard for design, without an accompanying set of wind climate data – i.e. the first link in the wind-loading chain of Figure 15.1. ISO 4354 has been adopted by Ethiopia, but without the necessary basic design wind speeds.

Eurocode 1 (EN 1991-1-4.6), however, has been adopted in nearly all of the European countries. It has also been adopted in South Africa and in parts of Asia. ASCE 7 is also widely used outside of the United States, such as in the Middle East, and earlier versions have been adapted for certain Asian jurisdictions (such as the Philippines and Taiwan). AS/NZS 1170.2 is widely used in the islands of the South Pacific, and an earlier version has been adapted for Malaysia.

It seems that a true international standard, although desirable for wind loads, remains an unachieved objective so far. The first requirement is a common format and notation. As indicated by Table 15.2, there are significant differences in the notation used across the codes, for what are essentially the same parameters.

For wide international acceptance in tropical and sub-tropical, as well as temperate, climates, the special requirements of regions affected by typhoons (tropical cyclones or hurricanes) and thunderstorms will need to be incorporated. Some greater consistency in the treatment of effects such as directionality, topography and internal pressures could be achieved, as there is general agreement amongst researchers on the underlying physical and statistical principles behind these effects.

One problem appears to be that in some jurisdictions, loading codes are the responsibility of government agencies, and regulators without the necessary wind engineering expertise, or practical experience that comes from using the documents for design. In other cases, the users of the document (i.e. practicing structural engineers) are the dominant 'players' in the development of a document. In further cases, academic wind engineers may be dominant, possibly leading to a document that is generally technically 'correct', but may contain ambiguities and is generally user-unfriendly. Ideally all three groups should play a role in the development of a wind code or standard to achieve a satisfactory user-friendly, non-ambiguous document, that is also technically acceptable.

REFERENCES

American Society of Civil Engineers. (2020) *Guidelines for electrical transmission line structural loading*, 4th Edition. ASCE Manual of Practice 74, ASCE, Reston, VA.

American Society of Civil Engineers. (2012) *Wind tunnel testing for buildings and other structures*. ASCE Standard ASCE/SEI 49-12. ASCE, Reston, VA.

American Society of Civil Engineers. (2016) *Minimum design loads and associated criteria for buildings and other structures*. ASCE/SEI 7-16. ASCE, Reston, VA.

Architectural Institute of Japan. (2004) *AIJ recommendations for loads on buildings*. AIJ, Tokyo.

Australian Building Codes Board. (2019) *National Construction Code of Australia*. ABCB, Canberra.

Australasian Wind Engineering Society. (2012) Wind loading handbook for Australia and New Zealand – Background to AS/NZS 1170.2 Wind actions. AWES-HB-001-2012.

Australasian Wind Engineering Society. (2019) Wind engineering studies of buildings. AWES-QAM-1-2019.

Buildings Department, Hong Kong. (2019) *Code of practice on wind effects in Hong Kong*, Hong Kong Special Administrative Region, China, September 2019.

CEN (European Committee for Standardization). (1994) Eurocode 1: basis of design and actions on structures. Part 2–4: Wind actions (draft). ENV-1991-2-4, CEN, Brussels, Belgium.

CEN (European Committee for Standardization). (2005) Eurocode 1: Actions on structures - Part 1–4: General actions - Wind actions. EN 1991-1-4, CEN, Brussels, Belgium.

CEN (European Committee for Standardization). (2006) Eurocode 3: Design of steel structures - Part 3-1: Towers, masts and chimneys. EN 1993-3-1:2006, CEN, Brussels, Belgium.

Cook, N.J. (1990) *The designer's guide to wind loading of building structures. Part 2 Static structures*. Building Research Establishment and Butterworths, London.

Davenport, A.G. (1977) The prediction of risk under wind loading. *Proceedings, 2nd International Conference on Structural Safety and Reliability*, Munich, Germany, September 19–21, pp. 511–596.

Davenport, A.G. (1982) Chapter 12: The interaction of wind and structures. In *Engineering Meteorology*. ed. E. J. Plate, pp. 527–572. Elsevier, Amsterdam.

Dorman, C.M.L. (1984) Tropical cyclone wind speeds in Australia. *Civil Engineering Transactions, Institution of Engineers, Australia*, CE26: 132–139.

ESDU International. (1983) Strong winds in the atmospheric boundary layer. Part 2: discrete gust speeds, ESDU Data Item 83045, ESDU International, London, UK, (revised 2002).

Highways Agency (UK). (2001) Design rules for aerodynamic effects on bridges, BD 49/01.

Holmes, J.D. (2002) A re-analysis of recorded wind speeds in Region A. *Australian Journal of Structural Engineering*, 4: 29–40.

Holmes, J.D. (2014) Along-wind and cross-wind response of a generic tall building – comparison wind-tunnel data with codes and standards. *Journal of Wind Engineering and Industrial Aerodynamics*, 132: 136–141.

Holmes, J.D. and Ginger, J.D. (2012) The gust wind speed duration in AS/NZS 1170.2. *Australian Journal of Structural Engineering*, 13: 207–217.

Holmes, J.D., Baker, C.J., English, E.C., and Choi, E.C.C. (2005a) Wind structure and codification. *Wind and Structures*, 8: 235–250.

Holmes, J.D., Kasperski, M., Miller, C.A., Zuranski, J.A. and Choi, E.C.C. (2005b) Extreme wind structure and zoning. *Wind and Structures*, 8: 269–281.

Holmes, J.D., Tamura, Y. and Krishna, P. (2009) Comparison of wind loads calculated by fifteen different codes and standards, for low, medium and high-rise buildings. *11th Americas Conference on Wind Engineering*, San Juan, Puerto Rico, 22–26 June.

Holmes, J.D., Allsop, A. and Ginger, J.D. (2014) Gust durations, gust factors and gust response factors in wind codes and standards. *Wind and Structures*, 19: 339–352.

Hong Kong Buildings Department. (2004) *Code of practice on wind effects in Hong Kong*. Government of the Hong Kong Special Administrative Region, Buildings Department, Mongkok.

International Standards Organization. (2009) Wind actions on structures. ISO International Standard. ISO 4354.

Kasperski, M. (1993) Aerodynamics of low-rise buildings and codification. *Journal of Wind Engineering and Industrial Aerodynamics*, 50: 253–263.

Kasperski, M. and Geurts, C. (2005) Reliability and code level. *Wind and Structures*, 8: 295–307.

Letchford, C.W., Holmes, J.D., Hoxey, R.P. and Robertson, A.P. (2005) Wind pressure coefficients on low-rise structures and codification. *Wind and Structures*, 8: 283–294.

Mehta, K.C. (1998) Wind load standards. Proceedings, Jubileum Conference on Wind Effects on Buildings and Structures, Porto Alegre, Brazil, 25–29 May.

Ministry of Land, Infrastructure and Transport, Japan. (2000) Building Standard Law of Japan, Enforcement Orders Regulations and Notifications.

Solari, G. (1983) Gust buffeting II: dynamic along-wind response. *Journal of Structural Engineering (ASCE)*, 119: 383–398.

Solari, G. and Kareem, A. (1998) On the formulation of ASCE-7–95 gust effect factor. *Journal of Wind Engineering and Industrial. Aerodynamics*, 77–78: 673–684.

Standards Australia. (1994) *Design of steel lattice towers and masts.* Australian Standard AS 3995-1994. Standards Australia, Sydney.

Standards Australia and Standards New Zealand. (2002) *Structural design actions. Part 0: general principles.* Standards Australia, Sydney, and Standards New Zealand, Wellington. Australian/New Zealand Standard AS/NZS1170.0:2002 (amended 2003 to 2011).

Standards Australia. (2011) Structural design actions. Part 2: wind actions. Standards Australia, Sydney. Australian/New Zealand Standard AS/NZS1170.2:2011

Standards Australia (2016) Overhead line design. Standards Australia, Sydney. Australian/New Zealand Standard AS/NZS 7000:2016

Standards Australia. (2020) Structural design actions. Part 2: wind actions. Standards Australia, Sydney. Draft Australian/New Zealand Standard DR AS/NZS1170.2:2020

Structural Engineers Association of California. (2017) Wind design for solar arrays. Report SEAOC PV2–2017, SEAOC, Sacramento, California, USA.

Tamura, Y., Kawai, H., Uematsu, Y., Okada, H. and Ohkuma, T. (2004) Documents for wind resistant design of buildings in Japan. *Workshop on Regional Harmonization of Wind Loading and Wind Environmental Specifications in Asia-Pacific Economies*, Atsugi, Japan, 19–20 November 2004.

Tamura, Y., Kareem, A., Solari, G., Kwok, K.C.S., Holmes, J.D. and Melbourne, W.H. (2005) Aspects of the dynamic wind-induced response of structures and codification. *Wind and Structures*, 8: 251–268.

Vickery, B.J. (1971) On the reliability of gust loading factors. *Civil Engineering Transactions, Institution of Engineers, Australia*, CE13: 1–9.

Zhou, Y., Kijewski, T. and Kareem, A. (2002) Along-wind load effects on tall buildings: a comparative study of major international codes and standards. *Journal of Structural Engineering (ASCE)*, 128: 788–796.

Chapter 16

Application of computational fluid dynamics to wind loading

16.1 INTRODUCTION

The advance and development of computer technology has led to more research on and usage of numerical methods. Analytical solutions for physical phenomena are limited to very few cases. This difficulty opens the door to numerical approximation by translating continuous phenomena to discrete approximations. The use of discrete approximations has been facilitated by the increased use of computers and advancement of numerical methods. The use of numerical methods for fluid structure problems has been a research subject for many years, but recently the emphasis has moved towards application to day-to-day engineering problems. The subject of Computational Fluid Dynamics (CFD) is now a standard course in many higher education institutes. The application of CFD for external fluid flow problem is more challenging due to high Reynolds number, turbulence and unbounded domain. The branch of Computational Fluid Dynamics applied to wind engineering has been named *Computational Wind Engineering* (CWE).

CWE has been progressively introduced to the modelling of wind loads, pollutant dispersion, pedestrian-level wind predictions, wind energy, ventilation and in several areas of comfort, as discussed by a number of researchers such as, Clannachan (2009), Blocken and Stathopoulos (2013), and Mohotti *et al.* (2019). One of the attractive features of CWE simulation is the ability to simulate problems without scaling in full-scale atmospheric boundary layer (ABL) wind flows. The application of CWE can be considered as an alternative, or complimentary, option to a wind-tunnel test procedure.

CWE is widely used in pedestrian-level wind studies for development approval and improving landscape design and wind mitigation strategies. These applications have been supported and strengthened by the development of quality control guidelines (Architectural Institute of Japan, 2006; Tominaga *et al.*, 2008; Franke *et al.*, 2007).

When performing a CWE simulation, one has to take several aspects into account: the first one is the physical model of the flow, which defines the set of equations to be solved. Second, the fluid volume in which the flow is to be computed must be defined. This fluid volume is called the computational fluid domain. The computational domain has to be discretised by the computational grid, which defines the spatial resolution of the numerical solution. For the discretisation of the equations on this grid, appropriate numerical approximations have to be used. For the solution of the discretised equations, residual criteria for limiting the calculations have to be set in the iterative scheme. Then, the resulting solution must be analysed, and if necessary, some of the preceding steps should be repeated with adaption to the solution (Franke *et al.*, 2004, 2007; Clannachan, 2009).

16.2 APPLICATION OF CWE TO STRUCTURAL WIND LOADING

The application of CWE to a structural wind-loading problem involves a combination of bluff body aerodynamics, inflow (atmospheric) turbulence, wake turbulence, grid generation and high Reynolds number (up to the order of 10^7–10^8). All these aspects require special attention in the application of numerical methods to structural wind loading.

A structural wind-load study may require the calculation of both mean and fluctuating loads. In addition, local pressure distributions may be required, and/or the global effect of wind loads on a whole structure.

A structural load study usually requires determination of peak values of wind loads. These values are time dependent, and thus transient simulations are required. General simulations used in practice are based on the solution of Reynolds-Averaged Navier-Stokes (RANS) equations. Using RANS turbulence modelling in transient simulations does not provide a true time variation of flow characteristics, due to the inherent averaging technique used. Thus, dependable structural load predictions for the design of structures cannot rely on CWE alone. Due to the errors incurred and the risks involved in using this method alone, it is still in the development phase. However, many designers use RANS and URANS simulations to assist with optimum designs for many structures. The development of the large-eddy simulation (LES), which is based on eddy sizes, and not on averaging technique like RANS, makes it more attractive for structural load applications.

LES turbulence modelling is a transient formulation that can produce the time history of fluid-structure interaction. The method is based on resolving the large eddies, while modelling the small eddies. However, the method has its own challenges numerically, in terms of small-scale eddy modelling and transient inflow generation. Another drawback of LES is the high computational resource requirement. The method requires a very high spatial resolution for which a high memory capacity is needed, along with small-time steps, which require high-capacity processors. With the present computing capacity, it is possible to implement this method for studies of local pressure, as demonstrated by many practitioners, but is only of limited availability for structural load studies.

The high demand on computational resources described above has led to a hybrid method of LES and RANS. LES is used to resolve the near-wall region of the inner layer with a very fine mesh, leading to a very big mesh size for LES simulation. The inner-wall region is then modelled using a wall function of the RANS method to reduce the high mesh requirement. A form of this hybrid method used in practice is called 'detached-eddy simulation' (DES).

The various methods, discussed above, for computing turbulent flow, representative of that in the natural wind around a structure, are described in more detail in Section 16.4.4.

Further development of the LES numerical method, in respect of small eddy modelling and inflow generation, should expand the adoption of CWE and make it a dependable tool for structural load studies. Recently, some vortex-flow structural studies using CWE, with fluid-structure interaction (FSI), have complimented wind-tunnel simulations in a number of cases.

16.3 GOVERNING EQUATIONS

The fundamental governing equations of fluid dynamics contain the continuity, momentum and energy equations. The mathematical descriptions of the three physical phenomena are the bases for fluid dynamics. CFD is based on these equations. CWE is a part of CFD that is specific to wind engineering applications. A firm understanding of these equations is important in the use of CWE in solving fluid-structure interaction problems.

The physical principles of the governing equations are based on the following relations: the continuity equation based on mass conservation, the momentum equation is based on Newton's Second Law and energy equations of conservation of energy. Because CFD is based mainly on these equations, they are essential to understand, so as to have a good grip on the numerical methods and connection between the continuous and discrete representations of the physical process.

The governing equations are presented in two forms, based on the type of flow considered: compressible and incompressible flows. In compressible flow, changes in density are included, while for incompressible flows, the density is constant with a known value. The governing equations presented below are for an incompressible flow, which is appropriate for external wind flow around structures.

Conservation of mass – the continuity equation: The continuity equation for incompressible flow is time independent. The velocity field should be divergence free. The equation in the Cartesian coordinate system is given by Equation (16.1).

$$\frac{\partial u}{\partial x} + \frac{\partial v}{\partial y} + \frac{\partial w}{\partial z} = 0 \tag{16.1}$$

Momentum equations: The momentum equations for incompressible flow and constant viscosity are given by Equations (16.2)–(16.4).

$$\rho_a \left[\frac{\partial u}{\partial t} + \frac{\partial u}{\partial x} u + \frac{\partial u}{\partial y} v + \frac{\partial u}{\partial z} w \right] = -\frac{\partial p}{\partial x} + \mu \left(\frac{\partial^2 u}{\partial x^2} + \frac{\partial^2 u}{\partial y^2} + \frac{\partial^2 u}{\partial z^2} \right) + \rho_a g_x \tag{16.2}$$

$$\rho_a \left[\frac{\partial v}{\partial t} + \frac{\partial v}{\partial x} u + \frac{\partial v}{\partial y} v + \frac{\partial v}{\partial z} w \right] = -\frac{\partial p}{\partial y} + \mu \left(\frac{\partial^2 v}{\partial x^2} + \frac{\partial^2 v}{\partial y^2} + \frac{\partial^2 v}{\partial z^2} \right) + \rho_a g_y \tag{16.3}$$

$$\rho_a \left[\frac{\partial w}{\partial t} + \frac{\partial w}{\partial x} u + \frac{\partial w}{\partial y} v + \frac{\partial w}{\partial z} w \right] = -\frac{\partial p}{\partial z} + \mu \left(\frac{\partial^2 w}{\partial x^2} + \frac{\partial^2 w}{\partial y^2} + \frac{\partial^2 w}{\partial z^2} \right) + \rho_a g_z \tag{16.4}$$

Energy equation: Equation (16.5) is the energy equation for constant thermal conductivity and kinematic viscosity.

$$\rho_a c_p \left(\frac{\partial T}{\partial t} + u \frac{\partial T}{\partial x} + v \frac{\partial T}{\partial y} + w \frac{\partial T}{\partial z} \right) = k \left(\frac{\partial^2 T}{\partial x^2} + \frac{\partial^2 T}{\partial y^2} + \frac{\partial^2 T}{\partial z^2} \right) + \varnothing \tag{16.5}$$

where c_p is the specific heat of air at constant pressure, and k is the thermal conductivity.

In Equation (16.5), the last term, $\varnothing$, is a dissipation function which is important only when viscous heating is significant. The dissipation function is given by Equation (16.6).

$$\varnothing = \mu \left\{ \begin{array}{l} 2\left[\left(\frac{\partial u}{\partial x}\right)^2 + \left(\frac{\partial v}{\partial y}\right)^2 + \left(\frac{\partial w}{\partial z}\right)^2 \right] + \left(\frac{\partial v}{\partial x} + \frac{\partial u}{\partial y}\right)^2 + \left(\frac{\partial w}{\partial y} + \frac{\partial v}{\partial z}\right)^2 \\ + \left(\frac{\partial u}{\partial z} + \frac{\partial w}{\partial x}\right)^2 \end{array} \right\} \tag{16.6}$$

16.4 NUMERICAL METHODS

Numerical methods are an approximate method for solving mathematical problems. The fluid flow governing equations and Navier-Stokes equations are solvable analytically for only a limited number of flows under certain assumptions. Thus, the alternative is to solve the governing partial differential equations (PDEs) using numerical approximation.

The numerical approximation starts by changing the PDEs to equivalent algebraic equations. There are a number of numerical methods used for solving the PDEs. The available numerical methods are finite difference, finite volume, finite element, boundary element, spectral methods and vorticity-based methods.

Finite difference method: This is the oldest numerical method. It was first published by Richardson (1910), and later developed upon by Courant *et al.* (1928), Thom (1933) and Harlow and Fromm (1965). The method is very simple to use, and is restricted largely to a simple mesh. The set of algebraic equations for the governing equations are obtained by the Taylor series expansions and truncated to different levels of accuracy. The error produced by truncating during the change of the differential equation to algebraic equation is called truncation error or discretization error. The system of algebraic equations is solved iteratively or through a coupled system.

Finite volume method: This is another numerical method used to change the governing differential equations to a system of algebraic equations. It was introduced into numerical methods by McDonald (1971) and MacCormack and Paullay (1972). The method uses a volume-integral formulation of the problem with a finite partitioning set of volumes to discretise the equations. It is widely used for modelling fluid dynamics problems. Similar to the finite difference method, the system of algebraic equations in this method can be solved by an iterative or coupled system.

Finite element method: This method forms a set of algebraic equations by partitioning bounded domains into a number of small, non-overlapping subdomains and finite elements, over which functions are approximated, generally to polynomials. This method is useful for very complex geometry.

Spectral method: The use of spectral methods for solving meteorological problems was first proposed by Blinova (1944). The method is characterized by the expansion of the solution in terms of global and orthogonal polynomials. At present, there are a number of versions of the method which are used for different applications.

Boundary element method: To use this method, the fluid flow governing equations of PDEs must be transformed into boundary integral forms. One of the early adopters of this method was Brebbia (1978). Discretization is only required on the boundary for this method. As a result, it avoids full-body discretization and also reduces the dimensions from three dimensions to two dimensions, and from two dimensions to one dimension. This reduction simplifies the solution process.

16.4.1 Discretization

There are two type of discretization: equation discretization and solution domain discretization. The domain discretization will be presented under 'meshing' in Section 16.4.3.

Equation discretization is the process of transferring continuous forms of equations into discrete forms. The equation discretization method varies according to the type of numerical method. The discretization methods for the three types of numerical methods are as follows:

Finite difference method: The partial differential equations are changed to a set of algebraic equations using the Taylor series expansion or polynomial fitting. The method can be illustrated by considering a simple one-dimensional differential equation, as shown in Equation 16.7.

$$\frac{dv}{dx} + v(x) = 0; \quad v'(x) + v(x) = 0 \qquad 0 \le x \le 1; \quad v(0) = 1 \qquad (16.7)$$

To solve the above equations, the differential term must be changed to a discrete point algebraic term using the Taylor series expansion as shown in Equation (16.8). The discrete point representation in this one-dimensional expansion can be represented as given in Figure 16.1.

$$v(x + \Delta x) = v(x) + \Delta x v'(x) + \frac{\Delta x^2}{2!} v''(x) + \frac{\Delta x^3}{3!} v'''(x) + \cdots \qquad (16.8)$$

$$v(x - \Delta x) = v(x) - \Delta x v'(x) + \frac{\Delta x^2}{2!} v''(x) - \frac{\Delta x^3}{3!} v'''(x) + \cdots \qquad (16.9)$$

Rearranging Equation (16.8) will produce Equation (16.10). This is a first-order approximation, since the terms of order Δx, $(O(\Delta x))$ are truncated from the series. This truncated term is the approximation error introduced in the process of changing the differential equation to an algebraic equation. This approximation is also called first-order approximation of forward differencing for the term $v'(x)$. Rearranging Equation (16.8) leads to the first-order backward difference approximation of the term $v'(x)$.

$$\frac{v(x + \Delta x) - v(x)}{\Delta x} = v'(x) + \frac{\Delta x}{2!} v'' + \frac{\Delta x^2}{3!} v''' + \cdots = v'(x) + O(\Delta x) \qquad (16.10)$$

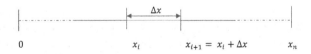

Figure 16.1 One-dimensional discrete point locations.

Combining Equations (16.8) and (16.9), a second-order approximation of the differential equation can be constructed, as shown in Equation (16.11). This approximation is called 'central differencing'.

$$\frac{v\left(x+\Delta x\right)-v\left(x-\Delta x\right)}{2\Delta x} = v'\left(x\right)+\frac{\Delta x^2}{3!}v''' +\cdots = v'\left(x\right)+O\left(\Delta x^2\right) \quad (16.11)$$

Using a similar method of combining the Taylor series expansion, higher order approximation can be derived. The finite difference method uses the above method to change the governing fluid flow differential equation to a system of algebraic equations.

Finite volume method: The finite volume method is based on the integral conservation law. The solution domain is subdivided into a finite number of attached control volumes or cells. The conservation equations are applied to each cell (control volume). Each control volume is defined by its boundary and a cell centre at which the variable values are approximated. The variable values at the cell centre are expressed using interpolation at the control volume surface. Algebraic equations for each control volume are constructed using an approximation of the surface and volume integral.

The integral conservation laws for a control volume can be written as shown in Equation (16.12).

$$\frac{\partial}{\partial t}\int_{\Omega} Ud\Omega+\oint_{s} \vec{F}\cdot d\vec{S} = \int_{\Omega} Qd\Omega \quad (16.12)$$

In the above equation, Ω is a control volume, S is the boundary surface of the control volume and Q is a source term. Equation 16.12 for control volume Ω_A with a source term of $Q=0$, and control volume average assumed by Equation (16.13), will produce Equation (16.14).

$$U_A = \frac{1}{A_s}+\int_{\Omega_A} Ud\Omega \quad (16.13)$$

$$A_s\frac{dU_A}{dt}+\oint_{s} \vec{F}\cdot\vec{n}ds = 0 \quad (16.14)$$

If we consider a 2D control volume shown in Figure 16.2 and rewriting Equation (16.14) for control volume Ω_A.

$$A_s\frac{dU_A}{dt}+\int_{1}^{2} H\left(U,\vec{n}_{AB}\right)ds+ \int_{2}^{3} H\left(U,\vec{n}_{AC}\right)ds+ \int_{3}^{4} H\left(U,\vec{n}_{AD}\right)ds$$

$$+\int_{4}^{1} H\left(U,\vec{n}_{AE}\right)ds = 0 \quad (16.15)$$

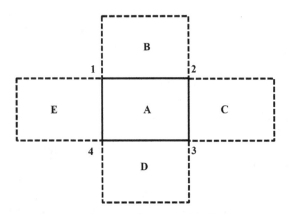

Figure 16.2 Two-dimensional control volume.

$H(U,\vec{n}) = \vec{F} \cdot \vec{n}$, $\vec{n}_{AB}$ is a unit normal from control volume A to control volume B. The discrete form of Equation (16.15) is constructed based on the following assumptions: (i) The flux interface is determined by the upwind method, as shown for face AB in Equation (16.16). (ii) Secondly, by adopting a second-order discretization for the time derivative, we obtain the algebraic form, Equation (16.17), for each control volume.

$$H(U,\vec{n}_{AB}) = H_D(U_A,U_B,\vec{n}_{AB}) = \frac{1}{2}\vec{U}_{AB} \cdot \vec{n}_{AB}(U_A + U_B)$$

$$-\frac{1}{2}\left|\vec{U}_{AB} - \vec{n}_{AB}\right|(U_B - U_A)$$

(16.16)

$$A_s \frac{U_A^{i+1} - U_A^{i-1}}{2\Delta t} + H_D\left(U_A^i,U_B^i,\vec{n}_{AB}\right)\Delta s_{AB} + H_D\left(U_A^i,U_C^i,\vec{n}_{AC}\right)\Delta s_{AC}$$

$$+ H_D\left(U_A^i,U_D^i,\vec{n}_{AD}\right)\Delta s_{AD} + H_D\left(U_A^i,U_E^i,\vec{n}_{AE}\right)\Delta s_{AE} = 0$$

(16.17)

Finite element method: The finite element method is similar to the finite volume method. The domain is broken into a set of discrete volumes or finite elements that are generally unstructured; in two dimensions (2D), they are usually triangles or quadrilaterals, while in three dimensions (3D), tetrahedra or hexahedra are most often used. In finite element methods, the equations are multiplied by a weighting function before they are integrated over the entire domain. In the simplest finite element methods, the solution is approximated by a linear shape function within each element in a way that guarantees continuity of the solution across element boundaries. Such a function can be constructed from its values at the corners of the elements. The weighting function is usually of the same form.

One of the advantages of finite element methods is the ability to handle arbitrary geometries. Finite element methods are relatively easy to analyse mathematically, and can be shown to have optimality properties for certain types of equations. The meshes are easily refined; each element is simply subdivided. Literature on the construction of meshes for finite element methods has been published by many researchers. The principal drawback of this method is that the matrices of the linearized equations are not as well structured as those for regular meshes, making it more difficult to find efficient solution methods.

16.4.2 Computational domain

A computational domain is one of the geometrical setups required to solve CWE problems. A computational domain is bounded by faces and edges in 3D and 2D geometries, respectively. The domain can be bounded and constrained by physical boundaries in internal flow or can be unbounded domain as in external flow. The domain for an external flow with an infinite domain is converted to a finite domain by using an artificial boundary. The boundary used to construct finite domains should not significantly affect the properties found in the infinite domain.

The size of a computational domain should be selected with consideration for various parameters of the CWE modelling. Domain boundaries should be located far enough from the areas of interest to minimize the impact of the boundary conditions on the results. A domain size sensitivity check should be conducted to determine the extent of the artificial

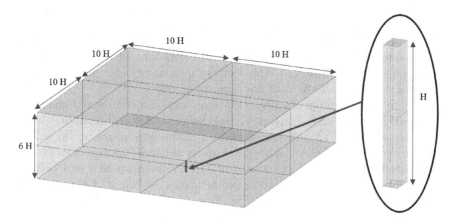

Figure 16.3 Computational domain for a single building.

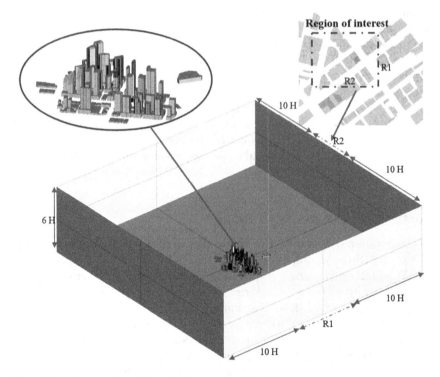

Figure 16.4 Computational domain for a group of buildings.

boundary condition. Examples of computational domain for external flow are shown in Figures 16.3 and 16.4.

There are recommendations on the size of computational domain from various organizations and researchers, with some variation in their advice. However, there is consistency across the literature in the importance of minimizing the influence of the artificial boundary condition. To reduce the computational costs, geometric details are often omitted and the computational domain is minimized. This leads to uncertainty in a simulation and increases the error in the fluid flow approximation.

The following comprises advice on sizing the computational domain for external flow in terms of lateral, vertical height and flow direction:

Vertical height: There are a number of recommendations on the extent of the vertical height. A height of $5H$ above the building height of H was suggested by Hall (1997), Cowan *et al.* (1997) and Scaperdas and Gilham (2004), while a blockage-based recommendation of height of $4H$ for small blockage (3% blockage) and $10H$ for large blockage (10% blockage), was suggested by VDI (2005).

Lateral dimension: The lateral dimension is a compromise between the vertical height and the selected blockage ratio. For a 3% blockage ratio, the extent of lateral dimension can be 2.3*H*. However, Hall (1997), Cowan *et al.* (1997), Scaperdas and Gilham (2004) and Bartzis *et al.* (2004) recommended a blockage of 1.5% and a lateral distance of 5*H*.

Domain in flow direction: Similar to the other directions, there are various recommendations for the dimension in the flow direction. For a single building of a height *H*, the inlet boundary is recommended to be 2*H*, with a distance of 15*H* from the building to the outflow boundary. Furthermore, it is also recommended that the distance of the building to the inlet boundary be kept short to minimize the inflow boundary condition distortion, before approaching the test building.

Taking H_{max} as the maximum building height in the region of interest (Franke *et al.*, 2010) recommended a domain size of $6H_{max}$ on the vertical direction and $10H_{max}$ from the building edge or region of interest in any direction if the domain is for a multi-direction flow. This is illustrated in Figures 16.3 and 16.4 for a single building and region of interest, respectively. In the case where the flow is in one direction, the inflow domain can be reduced to $5H_{max}$.

The above recommendations are based on general consideration and a generic simulation. However, the final choice of a size for a simulation must take into consideration other simulation requirements and objectives.

16.4.3 Meshing

Discretization of the governing equation of fluid flow was presented in Section 16.4.1. This section will discuss discretization of the domain and meshing. The computational domain must be filled with discrete points and divided into discrete control volumes to relate to the discrete, algebraic equations. This is an important part of the computational wind engineering process in solving a problem. The effort and knowledge required in meshing to a large extent defines the final solution that can be achieved by a CWE simulation. Discretization of the computational domain is a critical stage, because it affects not only the accuracy of the results but also the calculation time and numerical stability of the model (Gnatowska, 2019; Sosnowski *et al.*, 2019).

The type of mesh used in CWE can be structured, unstructured or hybrid. Generally, these meshes can have three forms, hexahedral (HEX), tetrahedral (TETRA) or polyhedral (POLY) meshes, as illustrated in Figure 16.5. Each type of shape has its own characteristics in ease of use and numerical diffusion which lend it to a given simulation. The following is a brief description on each type of mesh:

Hexahedral (HEX) meshes: The computational domain is divided into hexahedral (HEX) shapes. This mesh is known for its low numerical diffusion. The low numerical diffusion occurs only when the flow is aligned with the mesh, and increases with the deviation from alignment. This mesh

Figure 16.5 Hexahedral, tetrahedral and polyhedral meshes.

is very difficult to apply to complex geometry. The general shape of hexahedral mesh is shown in Figure 16.5.

Tetrahedral (TETRA) meshes: A tetrahedral mesh can be applied to complex geometry and can be automated by defining a few parameters. This method of generating mesh is adapted by most commercial software. One of the disadvantages of this method is its high numerical diffusion compared to the hexahedral flow. Difficulties may be encountered when computing gradients with tetrahedral meshes, causing a convergence difficulty (Spiegel *et al.*, 2011). In addition, this mesh will have more cells than hexahedral meshes, and therefore a longer computing time. The higher numerical diffusion of tetrahedral meshes and difficulty of computing gradients can lead to reduced accuracy (Sosnowski *et al.*, 2019). An illustration of tetrahedral meshes is shown in Figure 16.5.

Polyhedral (POLY) meshes: Polyhedral mesh can be easily applied to complex geometry and adopted to automatic mesh generation. The method is also known for its lower numerical diffusion. Polyhedral mesh has many faces, resulting in better gradient approximation than that with tetrahedral mesh, and reduced numerical diffusion than the application of hexahedral meshes to the same cases (Sosnowski *et al.*, 2019). The polyhedral meshes for a computational domain in Figure 16.4 are shown in Figure 16.6. The same domain meshed with tetrahedral mesh is shown in Figure 16.7.

Comparisons between the above three types of meshes for CWE have been made by Iqbal and Chan, (2015), with their study showing that polyhedral meshes perform much better than hexahedral and tetrahedral meshes, in terms of computing time, accuracy and low-mesh density.

The mesh density inside the domain varies according to the flow characteristics, simulation objective and local geometry inside the domain. The meshes are usually stretched away from the wall boundary condition, increasing the mesh closer to the wall. The mesh density is also increased where a high gradient of flow parameters is expected. Therefore, a good understanding of the flow and prediction of the high-gradient area will lead to construction of proper mesh, and thus a more accurate solution, less computational time and stable simulation.

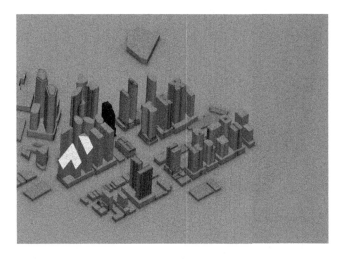

Figure 16.6 Polyhedral mesh for flow around buildings.

Figure 16.7 Tetrahedral mesh for flow around buildings.

Meshes that are stretched away from the wall and high-density mesh around the high-gradient area illustration are shown in Figure 16.8. High mesh stretching can increase the area and lead to an unrealistic solution.

Advice on stretching techniques and recommended values given by researchers suggests that mesh stretching/compression must be small in regions of high gradients, to keep the truncation error small. The expansion ratio between two consecutive cells should be below 1.3 in these regions. Scaperdas and Gilham (2004) and Bartzis *et al.* (2004) recommended a

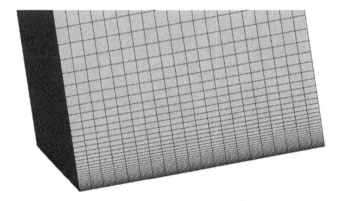

Figure 16.8 Mesh stretching.

maximum of 1.2 for the expansion ratio. If mesh refinement is not possible due to resource limitation, local refinement in regions of high expected flow gradient, areas of interest and around sharp geometrical shapes is recommended. For finite volume methods, keeping the angle between the normal vector of a cell surface and the line connecting the midpoint of neighbouring cells to a parallel provides a better-quality mesh (Ferziger and Perić, 2002).

16.4.4 Turbulence models

Selection of the turbulence model mainly depends on the objective of a physical phenomenon that needs to be resolved. Reynolds–averaged Navier Stokes (RANS) and Large Eddy Simulation (LES) are two widely-used methods for calculating the effect of wind loads on buildings. A choice of a given model has a significant contribution on both accuracy and the computational cost. Therefore, the selection of the appropriate turbulence model is a very important decision in CWE simulations.

Reynolds-averaged Navier-Stokes equations (RANS): This is a widely-used model for most simulations due to its reduced computational resource requirement, and also provides a reasonable approximation for many simulations. RANS involves averaging of the turbulence terms, with various levels of approximations. The following are some of the variations in RANS turbulence models:

Spalart-Allmaras (one-equation model): This is the simplest RANS model that solves a modelled transport equation for the kinematic eddy (turbulent) viscosity (Spalart and Shur, 1997). The method is found to be effective at low Reynolds numbers, and allows the application of the model, independent of near-wall y^+ resolution (ANSYS Inc., 2020).

K-epsilon, $K - \varepsilon$ (two-equation model): The standard K-epsilon model involves solution of two transport equations based on turbulent kinetic energy and dissipation rate (Launder and Spalding, 1972). The two transport equations for K and ε are given in Equations (16.18) and (16.19), respectively. This model is used widely in many engineering applications due to its simplicity, computational time, and reasonable solutions for most applications. However, the model has difficulty in resolving some flow regions.

$$\frac{\partial}{\partial t}(\rho_a K) + \frac{\partial}{\partial X_i}(\rho_a K u_i) = \frac{\partial}{\partial X_j}\left[\left(\mu + \frac{\mu_t}{\sigma_K}\right)\frac{\partial K}{\partial X_j}\right] + G_K + G_b - \rho\varepsilon$$

$$-Y_M + S_K \qquad (16.18)$$

$$\frac{\partial}{\partial t}(\rho_a \varepsilon) + \frac{\partial}{\partial X_i}(\rho_a \varepsilon u_i) = \frac{\partial}{\partial X_j}\left[\left(\mu + \frac{\mu_t}{\sigma_\epsilon}\right)\frac{\partial \varepsilon}{\partial X_j}\right] + C_{1\varepsilon}\frac{\varepsilon}{K}(G_K + C_{3\varepsilon}G_b)$$

$$-C_{2\varepsilon}\rho_a\frac{\varepsilon^2}{K} - S_\varepsilon \qquad (16.19)$$

In the above equations, $C_{1\varepsilon}$, $C_{2\varepsilon}$, and $C_{3\varepsilon}$ are constants, σ_K is the turbulent Prandtl number for kinetic energy, and σ_ϵ is the turbulent Prandtl number for dissipation rate. G_K is the generation of turbulent energy due to mean velocity gradient, G_b is the generation of turbulent kinetic energy due to buoyancy. S_K and S_ε are source terms. The turbulent viscosity is parameterized by $\mu_t = \rho C_\mu \frac{K^2}{\varepsilon}$.

There have been a number of proposed improvements to the standard $K - \varepsilon$ model. The most used variations of the standard model are: renormalisation group (RNG) and realizable $K - \varepsilon$ models.

- *K-omega, $K - \omega$ (two-equation model)*: The standard K-omega model requires solution of two transport equations based on turbulence kinetic energy and specific dissipation rate (Wilcox, 1998). The model includes low Reynolds number effect, compressibility and shear flow spreading. It is sensitive for values outside the shear layer (ANSYS Inc., 2020). There are variations of the original model which improves its performance.
- *Transition $K - K_l - \omega$ (three-equation model)*: This model uses three transport equations to predict boundary layer development. The transport equation uses turbulent kinetic energy (K), laminar kinetic energy (K_l) and inverse turbulent time scale (ω) to establish the transition of the boundary layer from a laminar to a turbulent regime.

- *Transition SST (four-equation model)*: Shear-stress transport (SST) is a four-equation model which is an extension of the Menter (1992, 1994) two-equation model (Langtry and Menter, 2009). The model combines the SST K-ω equations with gamma (γ) and Re theta (Re_θ). The two additional transport equations are for intermittency and transition. The model gained a wide application and improvement. However, there are some limitations, as listed below (ANSYS Inc., 2020):
  - The transition SST model is only applicable to wall-bounded flows. The model is not applicable to transition in free-shear flows and predicts free -shear flows as fully turbulent.
  - The model should not be applied to a surface that moves relative to the coordinate system for which the velocity field is computed.
  - The transition SST model is for flows with non-zero free-stream velocity like boundary layer flows.
  - The model is not suitable for wall jet flows.
- *Reynolds stress model (RSM) (seven-equation model)*: This is a higher-level turbulence closures model also called 'second-order closure'. The RSM does not use the eddy viscosity approach used in the RANS models described previously, but computes the Reynolds stress tensor directly (Chou, 1945; Rotta, 1951). The model accounts for the effect of streamline curvature, rapid change in strain rate, swirl and rotation to provide better prediction in complex flow. However, the results are limited to the assumptions made to form the transport equation for the Reynolds stress. The additional equations in RSM increase the required amount of computational resources. In addition, due to the assumption on closure model, it may not lead to accurate result in all cases. Thus, justification of its use may be required for simulation efficiency. However, for some flows, such as cyclone flows, highly swirling flows in combustors, rotating flow passages and the stress-induced secondary flows in ducts, it may provide a better result (ANSYS Inc., 2020).
- *Large eddy simulation (LES)*: LES is a method based on flow scaling, where large scales are resolved, while small scales are modelled. This method is different from the RANS model which is based on time averages or ensemble averaging. In direct numerical simulation (DNS), all the scales are resolved, and no modelling is required. However, this is computationally very expensive and not practical for high Reynolds number flows. In LES, large eddies are resolved directly, while small eddies are modelled. As turbulent modelling and range of resolved scales, LES can be considered to fall between DNS and RANS. Since LES models the large eddies, the requirement of mesh sizes is much lower than DNS. However, the requirement is impractical for most industrial applications.

- *Detached eddy simulation (DES)*: The DES model combines the RANS model with LES. This combination allows LES to be used for high Reynolds number by employing RANS for boundary layer resolution. The model still requires a considerable computational resource but much less extent than modelling with LES. The RANS model can be any one of the models discussed – one-, two-, three- or four-equation models.

16.4.5 Boundary conditions

Unbounded computational domain changed to finite computational area by introducing artificial boundaries. The characteristics of these artificial boundaries need to be specified in order to solve the discretised algebraic equations. How the boundary conditions specified have a direct relation with the accuracy of the solution, computational time and numerical stability, is an area for study. Some common boundary conditions are presented below:

Wall boundary condition: This is the boundary condition used to represent the solid part of the computational domain. In external flow, the ground and the surfaces of buildings, or any solid structure, are represented by a wall boundary condition. In general, the wall boundary condition can be stationary or moving. The parameters specified by this boundary condition are no slip or slip condition, surface roughness, shear stress for velocity, stability, and unbounded domain as in external flow.

The wall boundary is one of the most difficult boundaries to handle, and the result of the difference between turbulent models. In turbulent flow near the wall, viscous damping reduces the tangential velocity fluctuations, while kinematic blocking reduces the normal fluctuations. Due to large gradients in mean velocity, the turbulence is rapidly increased by the production of turbulence kinetic energy toward the outer part of the near-wall region. Large changes of the variables gradient and momentum occur in the near-wall region. The near-wall region is divided into three layers: viscous sublayer, buffer layer and full turbulent region (log law region).

Viscous sublayer ($y^+ < 5$): This layer is the innermost layer where the flow is almost laminar, and viscosity plays a dominant role in momentum and heat or mass transfer. It can be assumed that the Reynolds shear stress is negligible. The "linear velocity law" is given by the following relation.

$$u^+ = y^+ \tag{16.20}$$

where $y^+ = yu^* / v$ (u^* is the friction velocity and v is the kinematic viscosity), and $u^+ = u / u^*$, a non-dimensional velocity near the surface.

Buffer layer ($5 < y^+ < 30$): This layer is the transition region between the viscosity-dominated region and turbulence-dominated part of the flow.

Viscous and turbulent stresses are of similar magnitude, and since it is complex, the velocity profile is not well defined and the original wall functions avoid the first cell centre located in this region.

Log-law region ($y^+ > 30$): The turbulent stress dominates the flow in this region, and the velocity profile varies very slowly with a logarithmic function along the coordinate, y. The relation for this region, with a value of von Karman's constant, k, of 0.41, is as follows:

$$u^+ = \frac{1}{k}\ln\left(y^+\right) + 5.2 \qquad (16.21)$$

The near-wall region can be modelled by two different methods. In the first method, the viscous affected region is resolved by a fine mesh all the way to the wall. The second method involves semi-empirical formula used to represent the viscous sublayer and buffer layer. This method is called the 'wall-function' method. These two methods are illustrated in Figure 16.9.

A wall function can be used to compute the wall shear stress instead of resolving the near-wall region. This approach reduces computational cost. However, the solution accuracy is reduced, or it can lead to instability if the model is used with a very fine mesh. Thus, the recommended value of y^+ for first mesh point is between 30 and 500 (Franke *et al.*, 2007).

Inlet boundary condition: Information on velocity and turbulent quantities is required to be specified at the inlet boundary. For steady-state simulation of synoptic-scale atmospheric boundary layers, the standard velocity profile described using power law, logarithmic law, friction velocity and roughness of the upwind terrain can be used. The turbulence characteristics can be specified by the turbulent kinetic energy, dissipation rate, intensity, specific dissipation rate, and length scale. The choice of turbulence parameters for the boundary conditions can be directly related to the turbulence

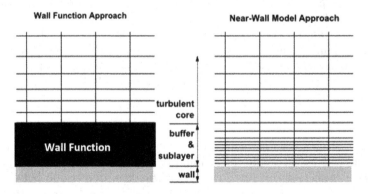

Figure 16.9 Near-wall region modelling.

model used in the flow modelling. For example, the following relationships for velocity and turbulence parameters can be used:

$$U(z) = \frac{u^*}{k} \ln\left(\frac{z + z_0}{z_0}\right), \text{ velocity profile} \tag{16.22}$$

$$K(z) = \frac{u^{*2}}{\sqrt{C_\mu}}, \text{ turbulent kinetic energy} \tag{16.23}$$

$$\hat{I}(z) = \frac{u^{*3}}{k(z + z_0)}, \text{ turbulent dissipation rate} \tag{16.24}$$

u^* is the friction velocity in the atmospheric boundary layer (ABL); k is von Karman's constant (Section 3.2.1), C_μ is the model constant for the $K - \varepsilon$ model, z is the height, and z_0 is the roughness length (see Section 3.2.1).

A problem in steady-state simulation is that the inlet boundary conditions specified for mean velocity and turbulence will dissipate, and be modified, due to numerical dissipation and the wall function. Blocken *et al.* (2007) have discussed the boundary layer and wall function problem and provided some suggestions.

For unsteady simulations, an instantaneous inlet profile should be specified to reproduce the unsteady behaviour of the flow. There are a number of suggestions by researchers to impose unsteady inlet boundary conditions. Commercial software also provides alternate ways of generating unsteady inlet boundary condition (ANSYS Inc., 2020).

Outflow boundary condition: The boundary at downstream, leaving the computational domain, can be specified as an outflow or pressure outlet boundary condition.

With an outflow boundary condition, the derivatives of all flow variables are forced to vanish. Since no flow variable is specified in this condition, convergence can be a problem for the same flows. The pressure outlet boundary condition uses a constant static pressure at the outflow boundary. This boundary should be set far enough away from the region of interest during the computational domain construction, so that backflow does not affect the solution. For LES convective outflow boundary conditions, see Ferziger and Peric (2002).

Top boundary condition: The top boundary is specified as symmetry or inlet boundary condition. In the case of a symmetrical boundary, the assumption is that there is no convective flux across the boundary. The normal velocity component at the boundary is zero. There is no diffusion flux across the boundary, and the normal gradient of all flow variables is zero. There is a problem of sustaining equilibrium boundary layer profiles, as reported by Blocken *et al.* (2007). Thus, use of the inflow velocity and turbulent profile at the top boundary is recommended.

Lateral boundary condition: The sides of the computational domain are specified by a symmetrical boundary condition when these boundaries are parallel to the flow direction. They can be specified as outflow or pressure outlet boundary condition if they are located at the downstream of the flow. These boundaries should be away from the region of interest, with a blockage of 3% or less in order not to have any influence on solution variables.

16.4.6 Simulations

CWE studies are usually done using commercial software such as ANSYS FLUENT, CFX, Star-CD, and others, or publicly-available codes, such as OpenFOAM, and rarely with codes developed 'in-house'. There are three parts in every CWE study. The first part is called pre-processing. In the pre-processing stage, the definition of the problem, the method of simulation, the computational domain and the mesh generation will take place. The second step is simulation, computing. At this stage, some of the following steps are required to be implemented:

- *Computational resources*: It is necessary to know the computing resources available. Without a proper consideration of the resources available, it is not possible to have a simulation. The size of the mesh directly goes to the memory capacity, and the processing time goes to the processor capacity, parallel computing capability and available processors. Adjusting the simulation requirement and what can be achieved with the available resources is the first step of a simulation. This decision is part of the pre-processing.
- *Time dependence*: The study can be steady or unsteady simulation. Unsteady simulation is advanced in time using explicit or implicit time stepping. The order of accuracy can be of first order at the initial stage of the simulation for stability consideration and then be changed to a second-order approximation.
- *Level of accuracy*: Numerical approximation introduces various errors at different levels of modelling and solution process. It is necessary to determine acceptable error for simulation and to tune the model accordingly. The numerical error can be the round-off error depending on whether single or double precision is used; the iteration error is needed to determine an acceptable convergence level; the mesh dependent error requires the solution to be independent of mesh density.
- *Initialization*: The computational domain needs to be initialized to start an iterative process towards a solution. This initial value is very important in reducing computing time, stability and arriving at an accurate solution any approximate value can be used for the initial condition extending the boundary condition. First stage modelling

uses first-order discretization, and one- or two-equation turbulence models can be used as initial condition for higher level modelling or unsteady simulation to achieve more stable solution and reduced time.

- *Solution monitoring*: During the iterative process in steady or unsteady state simulation, the changes in solution variables are recorded to determine the convergence level. However, reducing the iteration error by order of magnitude may not always give accurate solution. Thus, monitoring solution variables at a few locations in the domain will add the confidence level and protect from unrealistic solution.

- *Data recording*: During a solution process, solution variables can be recorded. For steady state simulation, recording during the solution process will help to use the data to continue the simulation, in case of unforeseen interruption or any other situation. For unsteady state simulations, recording of selected data, in addition to some full data, will help to analyse the time dependence of the solution variables.

16.4.7 Data analysis and results (post processing)

The advantage of CWE over experiments is the amount of data available. Processing data at the end of a simulation, or during the solution analyses, is called post-processing. Post-processing provides an insight into results of a CWE simulation. Post-processing can be the presentation of data by vector and contour plots, such as, 2D plots, animation, graphs and charts. The following are the main post-processing activities, according to the desired flow characteristics that need to be extracted from a solution:

- Calculation of integral parameters,
- Calculation of derived quantities,
- Data analysis by statistical tools,
- 1D data plots of a variable trend,
- 2D contour and vector plots,
- 3D plots of iso surfaces and volumes,
- Particle tracing and animation.

16.4.8 Quality assurance

Quality assurance is another important step in any CWE study, and the uncertainty in solutions is reduced by the use of a quality assurance procedure. There are a number of sources of advice on each step of a CFD study such as those by AIAA G-077-1998 (2002), AIJ (2006) and Franke *et al.* (2004, 2007). In addition, the procedures in Sections 16.3 and 16.4 can be adopted to ensure the quality of a CWE results. The following are the main considerations for a quality assurance process:

- *Computational domain*: The computational domain needs to be in the recommended range.
- *Mesh*: The density of mesh in a high-gradient area is critical, with very high stretching of a mesh often necessary. The solution should be mesh independent; thus, solutions of different mesh sizes need to be compared.
- *Level of uncertainty and error*: The simulation incurs different levels of error, round-off, iterative, modelling and discretization. All errors have to be listed.
- *Assumptions*: The modelling and solution process contain various assumptions. These assumptions needs to be listed.
- *Turbulence model*: The choice of a turbulence model for a given problem can be very important. A solution should not be highly influenced by the type of turbulent model used.
- *Boundary conditions*: Boundary conditions have a big influence in a solution. Thus, the solution needs to be scrutinized for the extent it is influenced by a boundary condition and if it is within the acceptable level.
- *Time step*: Unsteady simulation marching in time and the total computational time has to be reported. Is the total time of simulation sufficient enough to characterize a flow? Is the time step taken reasonable? This question raised at the initial stage of modelling needs to be considered again.
- *Verification*: The above points for quality assurance can be established through numerical verification. In addition to the numerical verification, comparing them with experimental data should be considered in all cases when opportunity is available.
- *Independent evaluation (peer review)*: The use of an independent reviewer, not involved in a modelling and solution process for the full modelling, solutions and post-processing process is important.

16.5 CASE STUDIES

In this section, some studies are presented to illustrate the capabilities of CWE techniques to be applied to the wind loading of structures, as well as the difficulties encountered.

16.5.1 Computation of forces on a two-dimensional flat plate

Two-dimensional (2-D) CWE modelling requires less computational resources and time than three-dimensional (3-D) modelling. It also provides valuable qualitative and quantitative information in many areas. With an understanding of its limitations, 2-D modelling can help refine a more detailed 3-D study. The computation of a blocked flow normal to a 2-D flat plate by Lasher (2001), in a channel flow of width B, and a plate width, c, illustrates both the problems with, as well as the quality of, predictions that can be obtained using 2-D simulations. Lasher's study was conducted for various blockage ratios, with blockage defined as c/B. The study conducted quasi-steady and transient modelling with six variations of the $K - \varepsilon$ model, for several different blockage ratios. The 2-D predictions of drag force coefficients from quasi-steady and time-averaged drag forces coefficients of transient simulations for all turbulent models were compared with experimental values.

The quasi-steady predictions show an under-prediction of the experiment, with constant deviation for different blockage ratios. However, the the transient simulation over-predicted and the results did not deviate from the experimental values by a constant amount, as seen in quasi-steady predictions. Lasher noted that the transient simulations are more sensitive to the choice of turbulence model, and the boundary conditions, than quasi-steady simulations.

16.5.2 Simulation of wind flow over topography

Wind flow over topography is an area where CFD can serve as an important and useful tool. Wind energy, pollutant dispersion and transmission-line wind loads are possible applications. Application of CFD to flow over topography and complex terrain was pioneered by Taylor (1977) and Ayotte *et al.* (1994), amongst others.

A more recent CFD study of wind flow over natural complex terrain was conducted by Blocken *et al.* (2015). An area of 25×20.5 km and modelling of the geometry with GIS data of 10 m horizontal resolution was studied. The simulation used 3D steady RANS and the realisable $K - \varepsilon$ turbulent model.

The predicted values of wind speed, ratio and direction were used to compare with field measurements. A wind-tunnel simulation of an area as large as 25 km is quite difficult. However, the CFD simulation of the flow over a complex terrain predicted wind velocities which deviated by 10%– 20% from the field measurements, indicating that CFD is a useful tool for studying wind flow over topography, if that accuracy is acceptable.

16.5.3 Mean pressure distribution on a medium-height building

Murakami and Mochida (1987, 1988) were amongst the first to calculate pressures on a cubic shape, approximating that of a building, using both a RANS solution with a $K - \varepsilon$ turbulence model, and a LES calculation.

Pressure distributions for the simple geometry of a building of rectangular plan and elevation (120 m height, and a floor plan of 60 m by 60 m) were calculated by Baskaran and Stathopoulos (1989), using both the standard and a modified $K - \varepsilon$ model. The standard $K - \varepsilon$ model was modified by introducing a streamline curvature correction and computing C_μ (Section 16.4.4) for each mesh point. The modified turbulence model showed positive improvement in the front, back and side mean pressure coefficient distributions. This study showed the impact of the turbulence model assumptions and, the effect of the constants used for the predictions for different parts of the fluid structure interaction.

16.5.4 Pressures on a low-rise building

One of the difficulties in using wind tunnels for experimental investigation of pressure distributions on buildings is the need for scaling. A comparison of pressure distributions of the Texas Tech (TTU) Building (Section 8.2.2) between full-scale values and wind-tunnel-model scale measurements presented a challenge for many researchers (e.g. Tieleman et al., 2001). This challenge was extended to numerical modelling – e.g. Selvam (1992), Paterson and Holmes (1992), He and Song (1997), Bekele and Hangan (2002) and Bekele (2004).

Comparisons of mean pressure distributions were made by Bekele and Hangan (2002), using an unsteady Reynolds stress turbulence model, for both normal and oblique wind directions, with good agreement with the full-scale data. However, comparison of the peak values showed a deviation of these full-scale values from the CWE predictions for both wind directions.

The comparison of predicted values of peak pressures showed the difficulty in reproducing high negative pressure peak values by RANS turbulent models. However, predictions using a LES turbulence model provided better comparisons (Bekele, 2004).

16.5.5 Pressures on complex roof shapes

CWE can play a complementary role, in structural design of large roofs of complex geometry. Holmes and Paterson (1993) described the calculation of mean pressure distributions on arched roofs of varying rise/span ratios; turbulence was modelled using the $K - \varepsilon$ model (Section 16.4.4). These results are discussed in Section 10.3.1.

A study of pressure distributions on a long-span complex roof structure for the Shenzhen railway station, China, was conducted using LES by Lu et al. (2012). The station building is 450 m long and 408 m wide with long cantilever roofs of an undulating shape.

The LES simulation of the station, with inflow turbulent generation, and a proposed one-equation sub-grid-scale model, predicted the pressure distributions in more detail than the wind-tunnel test of the station. The mean pressure coefficient comparison showed a reasonable agreement between the wind-tunnel and the LES results. However, there were significant differences reported by the authors at the bottom surface of the canopy and the station, where the local highest suction pressure coefficient predicted by LES was –1.1, while the wind-tunnel test predicted a coefficient of only –0.22.

A comparison of the root-mean-square fluctuating pressure coefficients showed the biggest difference at the edges and middle of the roof, with approximately 30% difference, while the other locations were in better agreement, with only 15% difference.

16.5.6 Wind loads on tall buildings

The study of wind loads on tall buildings using scale model wind-tunnel testing is a well-established tool in current practice (Section 9.12). However, developments in CWE now make it a viable tool for use in wind-loading assessment for new tall building developments. It should be viewed as a complementary tool at present, with promising progress for the future.

The computational evaluation of wind loads on a standard tall building using LES by Dagnew and Bitsuamlak (2014) is a significant contribution to this (see also: Dagnew, 2012). The LES simulation on the CAARC benchmark building (Section 9.12) was conducted using a number of inflow turbulence generation techniques. The predictions of the LES simulation were compared with wind-tunnel data for mean pressure coefficients, mean drag, lift and torsional moments, and fluctuating drag, lift and torsion.

The mean pressure coefficients showed good agreement with the wind-tunnel test data in most locations, for the two inflow generation techniques used, while a deviation was observed for one of the inflow generations. These comparisons showed the importance of inflow generation on the predicted pressure distribution. The mean force coefficients also generally showed a good agreement between the LES simulations and the wind-tunnel data, despite one of the inflow generation techniques showing similar deviations to those observed in the mean pressure coefficient comparison.

16.6 SUMMARY

CWE is rapidly progressing and gaining more acceptance and application in wind engineering. However, the tool still has to expand to its full extent into all areas of wind engineering. With the current progress in hardware development and numerical methods, it is highly likely that a reliable CWE study of a structure in a transient fluid flow will be possible in a few years.

To safeguard the correct application of the method in structural design, quality control in CWE and well-trained practitioners is necessary. The availability of commercial software has created more flexibility and is time-saving for both researchers and practitioners. This availability may have led to some abuse and unqualified use of the method. However, hopefully this will disappear as development progresses further.

REFERENCES

American Institute of Aeronautics and Astronautics (1998) *Guide for the Verification and validation of computational fluid dynamics simulations.* AIAA-G077-1998. AIAA, Reston, VA.

ANSYS Inc. (2020) *FLUENT 2020R1 theory guide.* ANSYS Inc., Canonsburg, PA.

Architectural Institute of Japan (2006) *AIJ guidelines for practical applications of CFD to pedestrian wind environment around buildings.* AIJ, Tokyo.

Ayotte, K.W., Xu, D. and Taylor, P.A. (1994) The impact of turbulence closure schemes on predictions of the mixed spectral finite difference model for flow over topography. *Boundary-Layer Meteorology*, 68: 1–33.

Bartzis, J.G., Vlachogiannis, D. and Sfetsos, A. (2004) Thematic area 5: best practice advice for environmental flows. *The QNET-CFD Network Newsletter*, 2: 34–39.

Bekele, S.A. (2004) *Numerical and physical investigation of roof corner vortex dynamics.* PhD thesis, University of Western Ontario, Canada.

Bekele, S.A. and Hangan, H. (2002) A comparative investigation of the TTU pressure envelope. Numerical versus laboratory and full-scale results. *Wind and Structures*, 5: 337–346.

Blinova, E. N. (1944) *Hydrodynamic theory of pressure and temperature waves and center of action of the atmosphere.* Patterson Field, Dayton, OH.

Baskaran, A. and Stathopoulos, T. (1989). Computational analysis of wind effects on buildings. *Building and Environment*, 24: 325–333.

Blocken, B. and Stathopoulos, T. (2013). CFD simulation of pedestrian-level wind conditions around buildings: past achievements and prospects. *Journal of Wind Engineering and Industrial Aerodynamics*, 121: 138–145.

Blocken, B., Stathopoulos, T. and Carmeliet, J. (2007) CFD simulation of the atmospheric boundary layer: wall function problems, *Atmospheric Environment* 41: 238–252.

Blocken, B., van der Hout, A., Dekker, J. and Weiler O. (2015) CFD simulation of wind flow over natural complex terrain: case study with validation by field measurements for Ria de Ferrol, Galicia, Spain. *Journal of Wind Engineering and Industrial Aerodynamics*, 147: 43–57.

Brebbia, C.A. (1978) *The boundary element method for engineers.* Pentech Press, London.

Chou, P. (1945) On velocity correlations and the solutions of the equations of turbulent fluctuation. *Quarterly of Applied Mathematics*, 3: 38–54.

Clannachan, G. (2009) Practical application of CFD for wind loading on tall buildings. (https://www.researchgate.net/publication/228544098_Practical_Application_of_CFD_for_Wind_Loading_on_Tall_Buildings).

Courant, R., Friedrichs, K. and Lewy, H. (1928) On partial difference equations of mathematical physics. *IBM Journal*, 11: 215–234 (1967). (English translation of the original work, "Uber die Partiellen Differenzengleichungen der Mathematischen Physik," Math. Ann. 100: 32–74, 1928.)

Cowan, I.R., Castro, I.P. and Robins, A.G. (1997) Numerical considerations for simulations of flow and dispersion around buildings. *Journal of Wind Engineering and Industrial Aerodynamics*, 67 & 68: 535–545.

Dagnew, A. (2012) *Computational evaluation of wind loads on low and high-rise buildings*. PhD Thesis, Florida International University, Miami, FL.

Dagnew, A.K. and Bitsuamlak, G.T. (2014) Computational evaluation of wind loads on a standard tall building using LES. *Wind & Structures*, 18: 567–598.

Ferziger, J.H. and Perić, M. (2002) *Computational methods for fluid dynamics*, 3rd edition. Springer Verlag, Berlin.

Franke, J., Hellsten, A. and Schlunzen, H. (2007) Best practice guideline for the CFD simulation of flows in the urban environment. COST Action 732, Belgium.

Franke, J., Hirsch, C., Jensen, G., Krüs, H. W., Miles, S. D., Schatzmann, M., Westbury, P.S., J.A. Wisse, J.A. and Wright, N. (2004). Recommendations on the use of CFD in wind engineering. *Proceedings of the International Conference on Urban Wind Engineering and Building Aerodynamics*, Rhode-Saint-Genèse, Belgium, 5–7 May 2004, pp. C.1.1–C1.11. (https://bwk.kuleuven.be/bwf/projects/annex41/protected/data/Recommendations%20for%20CFD%20in%20wind%20engineering.pdf).

Gnatowska R. (2019) Wind-induced pressure loads on buildings in tandem arrangement in urban environment. *Environmental Fluid Mechanics*, 19: 699–718. doi: 10.1007/s10652-018-9646-0.

Hall, R.C. (ed.) (1997) *Evaluation of modelling uncertainty. CFD modelling of near-field atmospheric dispersion*. Project EMU final report, European Commission Directorate–General XII Science, Research and Development Contract EV5V-CT94-0531. WS Atkins Consultants Ltd., Surrey.

Harlow, F.H. and Fromm, J.E. (1965) Computer experiments in fluid dynamics. *Scientific American*, 213: 104–110.

He, J. and Song, C.C.S. (1997) A numerical study of wind flow around the TTU building and the roof corner vortex, *Journal of Wind Engineering and Industrial Aerodynamics*, 67 & 68: 547–558.

Holmes, J.D. and Paterson, D.A. (1993) Mean wind pressures on arched-roof buildings by computation. *Journal of Wind Engineering and Industrial Aerodynamics*, 50: 235–242.

Iqbal, M. Z., and Chan, A. (2015) A study of the effect of element types on flow and turbulence characteristics around an isolated high-rise building. *Eleventh International Conference on CFD in the Minerals and Process Industries*, Melbourne, Australia, 7–9 December 2015.

Langtry, R.B. and Menter, F.R. (2009) Correlation-based transition modeling for unstructured parallelized computational fluid dynamics codes. *AIAA Journal*, 47: 2894–2906.

Lasher, W.C. (2001) Computation of two-dimensional blocked flow normal to a flat plate, *Journal of Wind Engineering and Industrial Aerodynamics*, 89: 493–513.

Launder, B.E. and Spalding, D.B. (1972) *Lectures in mathematical models of turbulence*. Academic Press, London.

Lu, C.L., Li, Q.S., Huang, S.H., Chen, F.B. and Fu, X.Y. (2012) Large eddy simulation of wind effects on a long-span complex roof structure. *Journal of Wind Engineering and Industrial Aerodynamics*, 100: 1–18.

MacCormack R.W. and Paullay, A.J. (1972) Computational efficiency achieved by time splitting of finite difference operators. *10th Aerospace Sciences Meeting*, San Diego, CA, January 17–19, AIAA Paper 72-154.

McDonald, P.W. (1971) *The computation of transonic flow through two-dimensional gas turbine cascades*, ASME Paper 71-GT-89.

Menter, F.R. (1992) *Improved two-equation k-ω turbulence model for aerodynamic flows*. NASA Technical Memorandum TM-103975. NASA Ames Research Center, Mountain View, CA.

Menter, F.R. (1994) Two-equation eddy-viscosity turbulence models for engineering applications. *AIAA Journal*, 32: 1598–1605.

Mohotti, D., Wijesooriya, K. and Dias-da-Costa, D. (2019). Comparison of Reynolds Averaging Navier-Stokes (RANS) turbulent models in predicting wind pressure on tall buildings. *Journal of Building Engineering*, 21: 1–17.

Murakami, S. and Mochida, A. (1987) Three-dimensional numerical simulation of air flow around a cubic model by means of large eddy simulation. *Journal of Wind Engineering and Industrial Aerodynamics*, 25: 291–305.

Murakami, S. and Mochida, A. (1988) 3-D numerical simulation of airflow around a cubic model by means of the k-e model. *Journal of Wind Engineering and Industrial Aerodynamics*, 31: 283–303.

Paterson, D.A. and Holmes, J.D. (1992). Computation of wind pressures on low-rise structures, *Journal of Wind Engineering and Industrial Aerodynamics*, 41–44: 1629–1640.

Richardson, L.F. (1910) The approximate arithmetical solution by finite differences of physical problems, *Transactions of the Royal Society A (London)*, 210: 307–357.

Rotta, J. (1951) Statistische theorie nichthomogener turbulenz. *Zeitschrift für Physik A*, 129: 547–572.

Scaperdas, A. and Gilham, S. (2004) Thematic Area 4: best practice advice for civil construction and HVAC. *The QNET-CFD Network Newsletter*, 2: 28–33.

Selvam, P.R. (1992) Computation of pressure on the Texas Tech Building. *Journal of Wind Engineering and Industrial Aerodynamics*, 41–44: 1619–1627.

Sosnowski, M., Gnatowska, R., Grabowska, K., Krzywanski, J. and Jamrozik, A. (2019) Numerical analysis of flow in building arrangement: computational domain discretization. *Applied Sciences*, 9: 941–958.

Spalart, P.R. and Shur, M.L. (1997) On the sensitization of turbulence models to rotation and curvature. *Aerospace Science and Technology* 1: 297–302.

Spiegel, M., Redel, T., Zhang, Y.J., Struffert, T., Hornegger, J., Grossman, R.G., Doerfler, A. and Karmonik, C. (2011) Tetrahedral vs. polyhedral mesh size evaluation on flow velocity and wall shear stress for cerebral hemodynamic simulation. *Computer Methods in Biomechanics and Biomedical Engineering*, 14: 9–22.

Taylor, P.A. (1977) Some numerical solutions of surface boundary-layer flow over gentle topography. *Boundary-Layer Meteorology*, 11: 439–465.

Thom, A. (1933). The flow past circular cylinder at low speeds. *Proceedings of the Royal Society of London. Series A*, 141: 651–669.

Tieleman, H.W., Reinhold, T.A. and Hajj, M.R. (2001) Detailed simulation of pressure in separated/reattached flows. *Journal of Wind Engineering and Industrial Aerodynamics*, 89: 1657–1670.

Tominaga, Y., Mochida, A., Yoshie, R., Kataoka, H., Nozu, T., Yoshikawa, M. and Shirasawa, T. (2008). AIJ guidelines for practical applications of CFD to pedestrian wind environment around buildings. *Journal of Wind Engineering and Industrial Aerodynamics*, 96: 1749–1761.

Verein Deutscher Ingenieure (VDI) (2005) *Environmental meteorology – prognostic microscale windfield models – evaluation for flow around buildings and obstacles*. VDI 3783 Part 9:2005, Beuth Verlag, Berlin.

Wilcox, D.C. (1998) *Turbulence modeling for CFD*. DCW Industries Inc., La Canada, CA.

Appendix A: Terminology

Aerodynamic admittance: Transfer function relating the gust spectral density to the spectral density of an aerodynamic force (Sections 5.3, 5.3.1 and 12.3.3).

Aerodynamic damping: Aerodynamic forces proportional to the velocity of a structure, and additional to the structural damping (Section 5.5.1).

Average recurrence interval: The average time interval between exceedances of an extreme windspeed or structural response (Sections 2.2.2 and C.6). It is also the reciprocal of the up-crossing rate.

Background response: That part of dynamic response to wind excluding the effects of resonant amplifications.

Bernoulli's equation: Equation describing irrotational and inviscid fluid flow (Section 4.2.1).

Blockage effect: Distortion effect of wind-tunnel walls on measurements, particularly force and pressure measurements (Section 7.7).

Bluff body: Body with a large frontal dimension, from which the airflow separates.

Body axes: Axes defined by the body or structure (Section 4.2.2).

Boundary layer: Region of reduced air velocities near the ground or the surface of a body (Sections 3.1, 4.1).

Cauchy number: Ratio of internal forces in a structure to inertial forces in the air (Chapter 7).

Coriolis force: Apparent force on moving air due to the rotation of the earth.

Correlation: Statistical relationship between two fluctuating random variables (Section 3.3.5).

Downburst: Severe downdraft of air occurring in thunderstorms (Sections 1.3.5, 3.2.7 and 3.3.7).

Drag: Along-wind force.

Dynamic response factor: Ratio of expected maximum structural response including resonant and correlation effects, to that ignoring both effects (Section 5.3.4).

Ekman spiral: Turning effect of the wind vector with height in the atmospheric boundary layer (Section 3.1).

Flutter: One, or two, degree-of-freedom aeroelastic instability, involving rotational motion (Section 5.5.3).

Friction velocity: Non-dimensional measure of surface shear stress (Section 3.2.1).

Froude number: Ratio of inertial forces in the air to gravity forces (Chapter 7).

Galloping: Single-degree-of-freedom translational aeroelastic instability (Section 5.5.2).

Geostrophic drag coefficient: Ratio of friction velocity to geostrophic wind speed (Section 3.2.5).

Gradient wind: Upper-level wind that can be calculated from the gradient wind equation (Section 1.2.4).

Gust factor: Ratio of expected maximum to mean value of wind speed (Section 3.3.3).

Gust response factor: Ratio of expected maximum to mean structural response (Section 5.3.2).

Helmholtz resonance: Resonance in internal pressure fluctuations associated with the compressibility of the air within a building, and the mass of air moving in and out of a large opening (Section 6.2.3).

Influence coefficient: Value of a structural response, such as a reaction, bending moment, shearing force or a deflection, resulting from a unit pressure at a point or acting on a panel (Section 5.3.5).

Inviscid: Fluid flow in which the effects of viscosity are non-existent or negligible (Section 4.2.1).

Isotach: Contour of constant basic design wind speed.

Jensen number: Ratio of building dimension (usually height) to roughness length in atmospheric boundary-layer flow (Sections 4.2.3 and 4.4.5).

Lift: Cross-wind force, usually but not necessarily, vertical (Section 4.2.2).

Limit states design: A method of structural design, which separately considers structural failure through collapse or overturning, from the functional (serviceability) aspects (Section 2.9.1).

Lock-in: The enhancement of fluctuating forces produced by vortex shedding due to the motion of the vibrating body (Sections 4.6.3 and 5.5.4).

Logarithmic law: A mathematical representation of the profile of mean velocity with height in the lower part of the atmospheric boundary layer (Section 3.2.1).

Manifold: A device for averaging pressure measurements from several measurement positions (Section 7.5.2).

Mechanical admittance: Transfer function relating the spectral density of aerodynamic forces to the spectral density of structural response (Section 5.3.1).

Peak factor: Ratio of maximum (or minimum) minus mean value, to standard deviation, for wind velocity, pressure, force or response (Sections 3.3.3 and 5.3.3).

Peak gust: Maximum value of wind speed in a defined time period (Section 3.3.3).

Pressure coefficient: Surface pressure made non-dimensional by the dynamic pressure in the wind flow (Section 4.2.1).

Quasi-steady: A model of wind loading that assumes that wind pressures on buildings fluctuate directly with the fluctuations in wind speed immediately upstream (Section 4.6.2).

Return period: Inverse of the probability of exceedance of an extreme value, such as an annual extreme wind speed (Sections 2.2.2 and C.6).

Reynolds number: Ratio of inertial forces to viscous forces in fluid flow (Section 4.2.4).

Roughness length: A measure of the aerodynamic roughness of a surface, which affects the boundary-layer flow over it (Section 3.2.1).

Safety index: A measure of probability of failure of a structure. 'Reliability Index' is also used (Section 2.9.2).

Scruton number: A non-dimensional parameter incorporating the ratio of structural mass to fluid mass, and structural damping, which is a measure of the propensity of a structure to resonant dynamic response (Sections 5.5.2, 11.5.1).

Shear stress (fluid flow): The force per unit area exerted by a layer of moving fluid on the adjacent layer, or ground surface (Section 3.2.1).

Spectral density: A measure of the contribution to a fluctuating quantity (e.g. wind velocity, wind pressure, deflection) within a defined frequency bandwidth (Sections 3.3.4 and 5.3.1)

Stagnation point: Point on a body where the approaching flow is brought to rest (Section 4.2.1).

Stationary: Description of a random process whose statistical properties do not change with time.

Strouhal number: Non-dimensional vortex-shedding frequency (Sections 4.6.3 and 11.5.1).

Synoptic winds: Winds created by large-scale meteorological systems, especially gales produced by extra-tropical depressions (Section 1.3.1).

Thunderstorm: Thermally driven local storm, capable of producing strong downdraft winds (Section 1.3.3).

Tornado: Local intense storm formed from thunderclouds, with intense winds rotating around a vortex structure (Sections 1.3.4 and 3.2.8).

Tropical cyclone: An intense tropical storm which can occur over warm tropical oceans. A generic name including 'hurricane' (used for Caribbean and north-west Atlantic storms) and 'typhoon' (used in the north-west Pacific) (Section 1.3.2).

Turbulence: Fluctuations in fluid flow. In meteorology, the term 'gustiness' is also used (Section 3.3).

von Karman's constant: Dimensionless constant in the logarithmic law for the profile of mean velocity in a turbulent boundary layer (Section 3.2.1).

Vortex-shedding: The periodic shedding of eddies formed from the rolling-up of the boundary layer shed from a bluff body (Section 4.6.3 and 11.5).

Vulnerability curve: A graphical relationship between the damage produced by a severe wind event and a representative wind speed – usually a peak gust in the event (Section 1.6.2).

Wake: The region of low velocity and turbulent flow in the region downstream of a body (Section 4.1).

Wind axes: Axes parallel and normal to the mean wind direction (Section 4.2.2).

Appendix B: List of symbols

Note: symbols that are specific to particular wind loading codes and standards described in Chapter 15 are not listed in this appendix.

a i. Coriolis acceleration (Chapter 1)
        ii. Scale factor in probability distribution (Chapter 2)
        iii. Empirical constant (Equation 3.10)
        iv. Modal (generalized) coordinate
        v. Parameter in wide-band fatigue (Equation 5.59)
        vi. Parameter in line-length multiplier (Chapter 13)
a_s Speed of sound
b i. Dimensional parameter (Equation 1.26)
        ii. Cross-wind breadth of bluff body
        iii. Parameter in wide-band fatigue (Equation 5.59)
        iv. Ratio of bulk modulus of air to bulk modulus of building (K_A/K_B) (Chapter 6)
        v. Parameter in line-length multiplier (Chapter 13)
        vi. Diameter of antenna dish (Chapter 14)
b' Breadth of a wake behind a bluff body (Chapter 4)
c i. Scale factor in damage index function (Equation 1.24)
        ii. Translation speed of a hurricane (Chapter 3)
        iii. Ground clearance of elevated hoarding (Figure 4.8)
        iv. Chord of a flat plate (Section 4.6.3)
        v. Damping constant
        vi. Scale factor in the Weibull distribution
        vii. Distance of solar panel from roof edge (Chapter 14)
        viii. Plate width (Section 16.5.1)
c_p Specific heat at constant pressure (Chapter 16)

d	i. Effective diameter of rod-type objects (Chapter 1)
	ii. Along-wind dimension of building or bluff body, chord of bridge deck
	iii. Stand-off of solar panel from roof surface (Chapter 14)
	iv. Depth of antenna dish
	v. Diameter of pole (Appendix E)
$d()$	Drag force per unit length
e	i. Eccentricity (Chapter 14 and Appendix E)
	ii. Eaves overhang
	iii. Wall thickness of pole (Appendix E)
f	i. Coriolis parameter (= $2\Omega \sin \lambda$)
	ii. Force per unit length
	iii. Focal length of parabolic antenna dish (Chapter 14)
$f()$	Probability density function
$f(\phi)$	Function of mode shape (Equation 11.17)
g	Peak factor
g_0	Gravitational constant
h	Gradient height (Equation 3.7)
$h(t)$	Unit impulse response function (Chapter 5)
h_c	Height of canopy (Chapter 14)
h_e	Height to eaves
h_i	Depth of inflow layer in tornado (Chapter 7)
h_p	Height of parapet (Chapter 8)
i, j	Indices
$i()$	Influence coefficient/line
k	i. $\left(\rho_a C_F\right)\big/\left(2\rho_m \ell\right)$ (Chapter 1)
	ii. shape factor in generalized extreme value distribution (Chapter 2, Appendix C)
	iii. von Karman's constant (Chapters 3 and 16)
	iv. Decay constant for coherence (Equation 3.35)
	v. Spring stiffness
	vi. Average surface roughness height (Chapter 4)
	vii. Orifice constant (Chapter 6)
	viii. Mode shape parameter (Equation 11.15)
	ix. Parameter (Equation 14.9)
	x. Thermal conductivity (Chapter 16)
	xi. Shape factor in generalized Pareto distribution (Appendices C and G)
k'	Loss coefficient for flow through leakage (Chapter 6)
k_t	Constant for type of topographic feature
k_x	Exponent (Equation 9.22)
k_y	Exponent (Equation 9.23)
ℓ	i. Characteristic dimension for compact objects

ii. Correlation length

iii. Length of tornado path (Chapter 13)

iv. Length of solar panel (Chapter 14)

ℓ_e Effective length of air 'slug' (Chapter 6)

ℓ_u Turbulence length scale (Chapters 2 & 6)

$\ell()$ Lift (cross-wind) force per unit length (Chapter 11)

m i. Mass or mass per unit length

ii. Exponent in fatigue s-N relationship

iii. Mean value (Chapter 9)

iv. Exponent (Equation 9.18)

n i. Exponent (Equation 3.12)

ii. Frequency

iii. Stress cycle

iv. Number of events, e.g. number of tornado occurrences in a region (Chapter 13)

v. Exponent (Equations 5.64 and 9.17)

vi. Shape factor for generalized extreme value distribution (Chapter 6)

n_c Characteristic frequency for internal pressure fluctuations

n_s Vortex shedding frequency

p i. Pressure

ii. Probability (Chapter 2)

iii. Exponent (Equation 5.66)

p_o i. Central pressure of a tropical cyclone (Chapter 1)

ii. Ambient (static) atmospheric pressure

p_f Probability of failure (Chapter 2)

p_n i. Atmospheric pressure at the edge of a storm (Chapter 1)

ii. Net pressure (Chapter 14)

p_L Leeward face pressure

p_w Windward face pressure

q Dynamic pressure

r i. Radius of curvature – of isobars (Chapter 1), or square section (Chapter 4)

ii. Risk (Chapter 2)

iii. Radius in downburst (Equation 3.17)

iv. General structural response, or load effect

v. Ratio A_L/A_W (Chapter 6)

vi. Roughness factor $(= 2I_u)$

vii. Radius of gyration (Chapter 12)

viii. Rate of intersection of tornadoes with a transmission line

r_c Radius of core in tornado (Chapter 7)

r_u Radius of updraft region in tornado (Chapter 7)

s i. Position factor (Equation 3.38)

ii. Stress

iii. Height for calculation of load effects
iv. Span length of a transmission line or free-standing wall (Chapters 13 and 14)
v. Clear space under bridge (Chapter 14)
vi. Skewness (Appendix C)

t i. Time
ii. Thickness of sheet objects (Chapter 1)
iii. Parameter for interference factor (Equation 14.9)

u, v, w Orthogonal velocity components

u Location parameter of extreme value distribution and damage index function

u_o Wind speed level (Chapter 2)

u_* Friction velocity (Chapters 3 and 16)

u^+ Non-dimensional velocity near a solid surface (Chapter 16)

v Fractional change of internal volume (Chapter 6)

v_m Velocity of flying debris

w i. Shape factor in the Weibull distribution (Chapter 2)
ii. Shortest distance from the smoothed coastline (Chapter 3)
iii. Width of tornado or downburst path (Chapter 13)
iv. Width of building (Figure 14.13)

w_a Average width of tower (Appendix E)

w_b Base width of tower (Appendix E)

w_c Width of canopy (Chapter 14)

w_o Assumed wind load per unit height (Equation 7.11)

x Horizontal distance travelled by a windborne missile (Chapter 1)
Distance travelled inland by a hurricane (Chapter 3)

$x(t)$ Random process, structural response

x_i Distance to inner boundary layer

x, y, z Cartesian coordinate system (z is vertical)

y^+ Non-dimensional distance from a solid surface (Chapter 16)

z i. Variable of integration, or transformed random variable (Appendix C)
ii. Dummy variable $= -\left(R_I^{-k} \right)$ (Appendix G)

z_h Zero-plane displacement

z_o Roughness length

z^* Characteristic height (Equation 3.17)

A i. Scaling parameter (Equation 1.10)
ii. Constant for hill type (Equation 3.39)
iii. Reference or frontal area
iv. Parameter in cross-wind response (Equation 11.19)
v. Area of a region (Chapter 13)

A_i	Flutter derivative for rotational motion (Chapters 5 and 12)
A_i^*	Normalised flutter derivative for rotational motion (Chapter 12)
A_L	Area of openings on leeward wall
A_W	Area of openings on windward wall
B	i. Exponent (Equation 1.10)
	ii. Background factor (also B_s)
	iii. Bandwidth parameter (Equation 11.18)
	iv. Channel width (Section 16.5.1)
C	i. Term in generalized extreme value distribution (Chapter 2, Chapter 6, Appendix G)
	ii. Decay constant (Equation 3.34)
	iii. Modal damping
C_d	Coefficient of drag force per unit length
C_D	Drag coefficient
C_f	Coefficient of aerodynamic force per unit length
C_F	Aerodynamic force coefficient
C_I	Inertial coefficient $= \ell_e/\sqrt{A}$ (Chapter 6)
C_k	Equivalent glass design coefficient
C_M	Moment coefficient
C_N	Normal force coefficient
C_p	Pressure coefficient
C_p^*	Effective peak pressure coefficient (Equation 9.7)
C_{pn}	Net pressure coefficient (Chapter 14)
C_{ps}	Equivalent pressure coefficient for glass loading (Equation 9.20)
C_{pe}	External pressure coefficient (Chapter 6)
C_{pi}	Internal pressure coefficient (Chapter 6)
C_{pio}	Initial value of internal pressure coefficient (Section 6.2.2)
$C_{pi,eff}$	Risk-consistent internal pressure coefficient (Chapter 6)
C_{pW}	Windward wall pressure coefficient (Chapter 6)
C_{pL}	Leeward wall pressure coefficient (Chapter 6)
C_T	Torque coefficient
C_X	Coefficient of X force
C_Y	Coefficient of Y force
C_Z	Coefficient of Z force
$C_{1\epsilon}$	Model constant for turbulent energy dissipation (Chapter 16)
$C_{2\epsilon}$	Model constant for turbulent energy dissipation (Chapter 16)
$C_{3\epsilon}$	Model constant for turbulent energy dissipation (Chapter 16)
C_μ	Model constant for turbulent kinetic energy (Chapter 16)
$Co(\ )$	Co-spectral density

D	i. Damage index (Chapter 1)
	ii. Term in generalized extreme value distribution (Chapter 2, Chapter 6, Appendix G)
	iii. Nominal dead load (Chapter 2)
	iv. Drag force
	v. Accumulated damage (Chapters 5 and 9)
D_a	Antenna drag (Chapter 14)
D_e	Effective tower drag with antenna attached (Chapter 14)
D_t	Tower drag (Chapter 14)
E	i. Young's modulus
	ii. Non-dimensional spectral density (Chapter 11)
F	Force
$F(\ )$	Cumulative probability distribution function
F_i	Parameter in along-wind response (Chapter 11)
G	i. Generalized mass
	ii. Gust factor, gust response factor
	iii. Shear modulus
G_b	Generation of turbulent energy due to buoyancy (Chapter 16)
G_K	Generation of turbulent energy due to a mean velocity gradient (Chapter 16)
$G(\ )$	Complementary cumulative probability distribution (Appendix C)
H	Height of hill (Chapter 3); height of building (Chapter 16)
H_i	Flutter derivative for vertical motion (Chapters 5 and 12)
H_i^*	Normalized flutter derivative for vertical motion (Chapter 12)
$H(\ )$	Dynamic amplification factor; square root of mechanical admittance
I	i. Fixing strength integrity parameter (Chapter 1)
	ii. Mass moment of inertia (Chapters 7 and 12)
	iii. Second moment of area
I_u, I_v, I_w	Turbulence intensities
Je	Jensen number
K	i. $\dfrac{1}{2}\dfrac{\rho_a}{\rho_m}\dfrac{U^2}{gl}\dfrac{\ell}{t}$ Tachikawa number (Chapter 1)
	ii. Constant (Equation 3.12)
	iii. Modal stiffness
	iv. Constant in fatigue s-N relationship
	v. Bulk modulus (Chapter 6)
	vi. Constant (Equation 7.1)
	vii. Constant (Equation 9.18)
	viii. Mode shape factor (Equation 11.16)
	ix. Turbulent kinetic energy (Chapter 16)
K_{ao}	Parameter for negative aerodynamic damping (Equation 11.19)

K_i Interference factor

K_p Porosity factor (Chapter 4)

K_A Bulk modulus of air (Chapter 6)

K_B Bulk modulus of building (Chapter 6)

K_w Correlation length factor (Equation 11.16)

K_θ Wind incidence factor (Equation 11.5)

L i. Lifetime of a structure (Chapter 2)
 ii. Lift (cross-wind) force
 iii. Characteristic length (Chapter 7)
 iv. Length of building (Figure 10.4)
 v. Length of a transmission line (Chapter 13)
 vi. Likelihood function (Appendix G)

L_N Parameter to calculate frequency of lattice tower (Appendix E)

L_S Span of bridge (Appendix E)

M Moment

M_b Base bending moment

M_t Topographic multiplier

M_L Line length multiplier (Chapter 13)

N i. Number of wind direction sectors (Chapters 2 and 13)
 ii. Cycles to failure by fatigue
 iii. Non-dimensional parameter for internal pressure (Chapter 6)
 iv. Number of patch loads for a guyed mast (Chapter 11)
 v. Number of samples of a random variable (Appendix C)
 vi. Number of thresholds (Appendix G)

Q i. Generalized force
 ii. Volume flow rate (Chapter 6)
 iii. Source term (Equation 16.12)

R i. Structural resistance (Chapter 2)
 ii. Characteristic radius (Equation 3.17)
 iii. Radius of maximum wind in a tornado (Figure 3.7)
 iv. Reduction factor for hurricane wind speeds (Section 3.6)
 v. Resonant response factor
 vi. Radius of liquid damper (Chapter 9)
 vii. Rise of arch (Figure 10.4)

R_c Combined return period for winds from more than one storm type

R_j Structural response (load effect) due to unit modal coordinate, in mode, j

R_I Average recurrence interval (Chapter 2, Appendix G)

R_p Return period (Chapter 2)

R_L Average recurrence interval for impact of gust windspeed on a transmission line (Chapter 13)

Re	Reynolds number
Re_b	Reynolds number based on diameter
Re_k	Reynolds number based on roughness height
Ro	Rossby number
S	i. Structural load effect (Chapter 2)
	ii. Size factor
	iii. Span of arch (Chapter 10)
	iv. Boundary surface of control volume (Equation 16.12)
S^*	Non-dimensional parameter for internal pressure (Chapter 6)
S_K	Source term for turbulent kinetic energy (Chapter 16)
S_ε	Source term for turbulent energy dissipation (Chapter 16)
$S(\ )$	Spectral density
Sc	Scruton number
St	Strouhal number
St^*	Universal wake Strouhal Number
T	i. Time of flight of missile
	ii. Sample time (Chapter 3)
	iii. Time period
U	Wind speed
U_b	'Background' wind speed (Section 3.6)
U_0	i. Maximum wind speed before hurricane landfall (Section 3.6)
	ii. Free-stream velocity (Chapter 4)
	iii. Threshold wind speed (Appendix G)
U_f	Wind speed for threshold of flight of debris
U_H	Mean wind speed on hill top (Equations 3.42 and 3.43)
U_R	Wind gust speed corresponding to average recurrence interval, R_I
$\bar{U}_s$	Velocity of flow outside a shear layer (Chapter 4)
U_V	Mean wind speed at valley bottom (Equations 3.42 and 3.43)
V	Wind speed (in some code notations – Chapter 15)
V_o	Internal volume
W	i. Nominal wind load (Chapter 2)
	ii. Weighting factor
X	General random variable (Appendix C)

Superscripts

- Mean (time averaged) value
′ Fluctuating value
˙ Differentiation with respect to time
^ Maximum value, optimum value (Appendix G)
˅ Minimum value
* i. Friction velocity (Chapters 3 and 16)
  ii. Characteristic height for downburst wind profile (Section 3.2.7)

Subscripts

$_0$ Initial value, free-stream value

$_1$ Natural frequency (Chapter 5)

$_{10}$ 10 m height

$_a$ Air

$_{av}$ Average

$_b$ i. Base of building
  ii. Tower or pole
  iii. Breadth or diameter
  iv. Background

$_{back}$ Background response (Chapter 10)

$_c$ i. Combined return period (Chapter 2)
  ii. Tornado core
  iii. Canopy
  iv. Cable
  v. Conductor
  vi. Continuous averaging (Chapter 7)

$_d$ i. Drag force per unit length
  ii. Antenna (dish)
  iii. Discrete averaging (Chapter 7)

$_e$ i. External
  ii. Effective (Chapter 6)
  iii. Eaves height (Chapter 10)

$_{eff}$ Effective

$_{env}$ Envelope

$_f$ i. Flight speed (Chapter 1)
  ii. Failure (Chapter 2)

$_g$ Geostrophic (Chapter 3)

h	Zero-plane displacement (Chapter 3)
i	i. Index of position or wind direction
	ii. Internal
	iii. Inflow layer of tornado (Chapter 7)
j	i Wall-opening scenario (Chapter 6)
	ii Index of mode of vibration (Chapter 12)
k	Roughness height
ℓ	Lift (cross-wind) force per unit length
lat	Lateral
m	i. Missile (Chapter 1)
	ii. Area on members (Chapter 4)
	iii. Model (Chapter 7)
	iv. Bending moment (Chapter 11)
max	Maximum
n	Net (pressure)
ni	Normal to face i (Chapter 11)
p	i. Peak
	ii. Porosity (Chapter 4)
	iii. Prototype (Chapter 7)
	iv. Parapet (Chapter 8)
q	Shearing force
r	i. Radial wind velocity component in a tornado (Chapter 3)
	ii. Ratio of model to prototype value (Chapter 7)
ref	Reference value
s	i. Local gust speed (Equation 1.26)
	ii. Sound (speed of)
	iii. Structure
	iv. Shedding
	v. Sloshing
t	Topographic, total area (Chapter 4), top of building, tower or pole
u	Updraft region in tornado (Chapter 7)
v	Vertical wind velocity component in a tornado (Chapter 3)
x	x-direction response
y	y-direction response
B	i. Background response
	ii. Barrier (Figure 7.5)
D	Drag force
F	Flutter (Chapter 12)
H	Helmholtz resonance

I Average recurrence interval (Chapter 2)

L Leeward, lift (cross-wind) force

N i Net force (Chapter 4)
      ii Number of wall-opening scenarios (Chapter 6)

P Return period (Chapter 2)

Pl Patch load (Equation 11.27)

R resonant response (Chapter 5)

T i. Top of wind tunnel (Figure 7.4)
      ii. Torsional (frequency) (Chapter 12)

TL Turbulence length scale (Equation 11.28)

V Vertical bending (frequency) (Chapter 12)

W Windward

1 First mode of vibration

θ Tangential wind velocity component in a tornado (Chapter 3)

Greek symbols

α i. Exponent in power law (Equation 3.8)
       ii. Decay parameter for hurricanes (Equation 3.47)
       iii. Angle of attack
       iv. Span reduction factor for transmission lines or walls (Chapters 13 and 14)
       v. Roof pitch angle

α_j Coefficient for influence of resonant mode, j, on a structural load effect

β i. Safety index (Chapter 2)
       ii. Decay constant for hurricanes (Equation 3.49)
       iii. Mode shape exponent (Chapter 7)
       iv. Angle of intersection of tornado path width with transmission line (Chapter 13)
       v. Angle of solar panel to roof surface (Figure 14.13)

γ Ratio of specific heats of air (Chapter 6)

γ_D Dead load factor

γ_W Wind load factor

δ Solidity of porous body

ϵ i. Characteristic height (Equation 3.17)
       ii. Spectral bandwidth parameter (Equation 5.62)
       iii. Load effect (Chapter 9)
       iv. Rate of dissipation of turbulent kinetic energy (Chapter 16)

ϕ Upwind slope of topographic feature

$\phi(\ )$ Mode shape

η Ratio of damping to critical

φ i. Resistance factor (Chapter 2)

       ii. Wall porosity (Chapter 6)

       iii. Phase angle (Chapter 11)

       iv. Transmission line direction (Chapter 13)

κ Surface drag coefficient

λ i. Angle of latitude (Chapter 1)

       ii. Crossing rate (Chapter 2, Appendix G)

       iii. Scaling factor (Equation 3.17)

       iv. Parameter in wide-band fatigue (Equation 5.59)

       v. Factor for guyed mast response (Equation 11.28)

       vi. Parameter in Poisson distribution (Appendix C)

       vii. Parameter for pole frequency calculation (Appendix E)

μ i. Dynamic viscosity

       ii. Moment of spectral density (Chapter 5)

ν i. Kinematic viscosity of air

       ii. Up-crossing rate

       iii. Cycling rate or rate of occurrence (Appendix C)

ξ Empirical factor in approximate formula for natural frequency of steel chimneys (Appendix E)

π i. Ratio of circumference to diameter of a circle

       ii. (with subscript) non-dimensional group

θ i. Angular rotation

       ii. Angle of incidence

       iii. (As a subscript) tangential velocity component in a tornado (Chapter 3)

       iv. Angle of downburst path to transmission line (Chapter 13)

$\rho, \rho(y)$ Correlation coefficient

$\rho(n)$ Normalized co-spectral density

ρ_a Air density

ρ_b Average building density

σ i. Scale factor (Chapter 2)

       ii. Standard deviation

       iii. Parameter in lognormal distribution (Appendix C)

       iv. Scale factor in generalized Pareto distribution (Appendices C and G)

σ_ϵ Turbulent Prandtl number for the turbulence dissipation rate (Chapter 16)

σ_K Turbulent Prandtl number for the turbulent kinetic energy (Chapter 16)

$\hat{\sigma}$ Optimum value of scale factor (Appendix G)

τ i. Equilibrium, or response, time for internal pressure (Chapter 6)

       ii. Dummy time variable for integration (Chapter 5)

       iii. Averaging time or gust duration (Chapters 3, 7)

τ_o	Surface shear stress
v	Rate of occurrence of tornadoes per unit area (Chapters 2 and 13)
ω	Circular frequency
Δ	Mean deflection (Figure 9.1)
Φ	Non-dimensional parameter (Chapter 6)
$\Phi(\;)$	Cumulative distribution function of a normal (Gaussian) random variable (zero mean and unit standard deviation)
$\Gamma(\;)$	Gamma function
Γ	Imposed circulation on tornado (Chapter 7)
Ω	i. Angular velocity of rotation of the earth (Chapter 1)
	ii. Control volume (Chapter 16)
Π	Repeated multiplication
Σ	Repeated summation
$X^2(n)$	Aerodynamic admittance
$\varnothing$	Dissipation function (Chapter 16)

Appendix C: Probability distributions relevant to wind engineering

C.I INTRODUCTION

Probability distributions are an essential part of wind engineering as they enable the random variables involved such as wind speeds, wind directions, surface pressures and structural response (e.g. deflections and stresses), to be modelled mathematically. Some of these variables are random *processes*, i.e. they have time-varying characteristics, as shown in Figure C.1. The probability density describes the distribution of the magnitude or amplitude of the process, without any regard to the time axis.

The appendix first covers some basic statistical definitions. Second a selection of probability distributions for the complete population of a random variable – the normal (Gaussian), lognormal, Weibull, Poisson, are considered. Third, the three types of Extreme Value distributions and the closely related Generalized Pareto Distributions are discussed. The distributions considered in Sections C.3 and C.4 include all those introduced in the main text.

The final two sections of this appendix discuss up-crossing and out-crossing rates of random processes, and the relationship between 'return period' and 'average recurrence interval'.

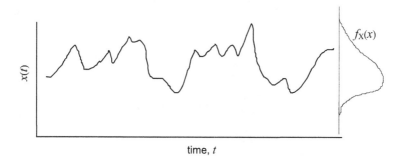

Figure C.1 A random process and amplitude probability density.

525

C.2 BASIC DEFINITIONS

C.2.I Probability density function (p.d.f.)

The probability density function, $f_X(x)$ is the limiting probability that the value of a random variable, X, lies between x and $(x+\delta x)$. Thus, the probability that X lies between a and b is:

$$\Pr\{a < X < b\} = \int_a^b f_X(x)dx \tag{C.1}$$

Since any value of X must lie between $-\infty$ and $+\infty$:

$$\int_{-\infty}^{\infty} f_X(x)dx = \Pr\{-\infty < X < \infty\} = 1$$

Thus, the area under the graph of $f_X(x)$ versus x must equal 1.0.

C.2.2 Cumulative distribution function (c.d.f.)

The cumulative distribution function $F_X(x)$ is the integral between $-\infty$ and x of $f_X(x)$.

$$\text{i.e. } F_X(x) = \int_{-\infty}^{x} f_X(x)dx = \Pr\{-\infty < X < x\} = \Pr\{X < x\} \tag{C.2}$$

The complementary cumulative distribution function, usually denoted by $G_x(x)$ is:

$$G_X(x) = 1 - F_X(x) = \Pr\{X > x\} \tag{C.3}$$

$F_X(a)$ and $G_X(b)$ are equal to the areas indicated on Figure C.2.

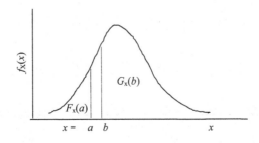

Figure C.2 Probability density function and cumulative distribution functions.

Note that:

$$f_X(x) = \frac{dF_X(x)}{dx} = -\frac{dG_X(x)}{dx}. \tag{C.4}$$

C2.3 Moments of the p.d.f.

The following basic statistical properties of a random variable are defined, and their relationship to the underlying probability distribution is given.

Mean

$$\bar{X} = (1/N)\sum_i x_i$$

$$= \int_{-\infty}^{\infty} x \, f_X(x) \, dx \tag{C.5}$$

Thus, the mean value is the first moment of the probability density function (i.e. the x coordinate of the centroid of the area under the graph of the p.d.f.)

Variance and standard deviation

$$\text{Variance: } \sigma_X^2 = (1/N)\sum_i \left[x_i - \bar{X}\right]^2 \tag{C.6}$$

σ_X (the square root of the variance) is called the *standard deviation*

$$\sigma_X^2 = \int_{-\infty}^{\infty} (x - \bar{X})^2 f_x(x) \, dx \tag{C.7}$$

Thus, the variance is the second moment of the p.d.f. about the mean value. It is analogous to the second moment of area of a cross-section about a centroid.

Skewness

$$s_X = \left[1/(\sigma_X^3)\right]\sum_i \left[x - \bar{X}\right]^3$$

$$= (1/\sigma_X^3)\int_{-\infty}^{\infty} (x - \bar{X})^3 \, f_X(x) \, dx \tag{C.8}$$

The skewness is the normalised third moment of the probability density function. Positive and negative skewness are illustrated in Figure C.3. A distribution that is symmetrical about the mean value has a zero skewness.

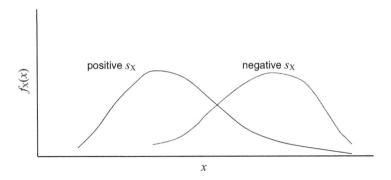

Figure C.3 Positive and negative skewness.

C.3 PARENT DISTRIBUTIONS

C.3.1 Normal, or Gaussian, distribution

For $-\infty < X \, \infty$

$$f_x(x) = \frac{1}{\sqrt{2\pi}\sigma_x} \exp\left[\frac{-(x-\overline{X})^2}{2\sigma_x^2}\right] \tag{C.9}$$

where $\overline{X}$ and σ_X are the mean and standard deviation of X, respectively.

This is the most commonly used distribution. It is a symmetrical distribution (zero skewness) with the familiar bell-shaped curve (Figure C.4).

$$F_X(x) = \Phi\left(\frac{x-\overline{X}}{\sigma_x}\right) \tag{C.10}$$

where $\Phi(\ )$ is the cumulative distribution function of a normally distributed variable with a mean of zero and a unit standard deviation,

$$\text{i.e. } \Phi(u) = \left(\frac{1}{\sqrt{2\pi}}\right)\int_{-\infty}^{u} \exp\left(\frac{-z^2}{2}\right)dz \tag{C.11}$$

Tables of $\Phi(u)$ are readily available in statistics textbooks.

If $Y = X_1 + X_2 + X_3 + \ldots X_N$

where $X_1, X_2, X_3 \ldots X_N$ are random variables with any distribution, then the distribution of Y tends to become normal as N becomes large. If X_1, X_2 … themselves have normal distributions, then Y has a normal distribution for any value of N.

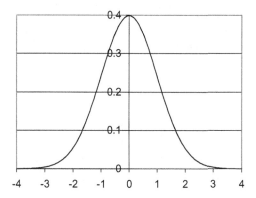

Figure C.4 Normal distribution.

In wind engineering, the normal distribution is applicable to turbulent velocity components, and for response variables (e.g. deflection) of a structure undergoing random vibration. It should be used for variables that can take both negative and positive values, so it would not be suitable for scalar wind speeds that can only be positive.

C.3.2 Lognormal distribution

$$f_X(x) = \frac{1}{\sqrt{2\pi}\sigma\, x} \exp\left[\frac{-\left\{ \log_e\left(\frac{x}{m}\right)\right\}^2}{2\sigma^2} \right] \tag{C.12}$$

where

the mean value, $\bar{X}$, is equal to $m \exp(\sigma^2/2)$,
and the variance σ_X^2 is equal to $m^2 \exp(\sigma^2)\,[\exp(\sigma^2) - 1]$,
$\log_e X$ has a normal distribution with a mean value of $\log_e m$ and a variance of σ^2.

If a random variable $Y = X_1 . X_2 . X_3 \ldots X_N$, where $X_1, X_2, X_3 \ldots X_N$ are random variables with any distribution, the distribution of Y tends to become lognormal as N becomes large. Thus, the lognormal distribution is often used for the distribution of a variable that is itself the product of a number of uncertain variables – for example wind speed factored by multipliers for terrain, height, shielding, topography etc.

The lognormal distribution has a positive skewness equal to $[\exp(\sigma^2) + 2][\exp(\sigma^2) - 1]^{1/2}$.

C.3.3 'Square-root-normal' distribution

Consider the distribution of $X = Z^2$, where $Z = \sqrt{X}$, has a normal distribution. Then, X has a distribution, which, by analogy with the 'lognormal' distribution, we can call a 'square-root-normal' distribution.

$$f_X(x) = \frac{1}{2\left(\frac{\sigma_Z}{\overline{Z}}\right)\sqrt{2\pi x}} \left\{ \exp\left[-\left(\frac{1}{2}\right)\left(\frac{\sqrt{x}-1}{\left(\frac{\sigma_Z}{\overline{Z}}\right)}\right)^2 \right] + \exp\left[-\left(\frac{1}{2}\right)\left(\frac{\sqrt{x}+1}{\left(\frac{\sigma_Z}{\overline{Z}}\right)}\right)^2 \right] \right\}$$

(C.13)

and the c.d.f. is:

$$F_X(x) = \Phi\left(\frac{\sqrt{x}-1}{\left(\frac{\sigma_Z}{\overline{Z}}\right)}\right) + \Phi\left(\frac{\sqrt{x}+1}{\left(\frac{\sigma_Z}{\overline{Z}}\right)}\right) - 1$$

(C.14)

This distribution is useful for modelling the pressure fluctuations on a building which, by the quasi-steady assumption, are closely related to the square of the upwind velocity fluctuations, which can be assumed to have a normal distribution (e.g. Holmes, 1981).

C3.4 Folded-normal distribution

If X has a normal distribution, then the absolute value of X, $|X|$, has a 'folded' normal distribution. The usual case of interest (see Section C5.3) is when the mean of X is zero. In that case, the p.d.f. of the distribution of $|X|$ is:

$$f_{|X|}(x) = \sqrt{\frac{2}{\pi}} \frac{1}{\sigma_X} \exp\left[\frac{-x^2}{2\sigma_X^2}\right]$$

(C.15)

where σ_X is the standard deviation of X.
Then the mean value of $|X|$ is:

$$\overline{|X|} = \sigma_X \sqrt{\frac{2}{\pi}}$$

(C.16)

The standard deviation of $|X|$ is:

$$\sigma_{|X|} = \sigma_X \sqrt{1 - \left(\frac{2}{\pi}\right)} \qquad (C.17)$$

C.3.5 Weibull distribution

$$f_X(x) = \left(\frac{wx^{w-1}}{c^w}\right)\exp\left[-\left(\frac{x}{c}\right)^w\right] \qquad (C.18)$$

$$F_X(x) = 1 - \exp\left[-\left(\frac{x}{c}\right)^w\right] \qquad (C.19)$$

where
   c (> 0) is known as the scale parameter, with the same units as x,
   w (> 0) is the shape parameter (dimensionless).

The mean value is:

$$c\sqrt{\left[\Gamma(1 + 1/w)\right]}$$

where $\Gamma()$ is the gamma function.
   The standard deviation is given by:

$$c\sqrt{\Gamma(1 + 2/w) - \left[\Gamma(1 + 1/w)\right]^2}$$

The shape of the p.d.f. for the Weibull distribution is quite sensitive to the value of the shape factor, w, as shown in the Figure C.5. The Weibull distribution can only be used for random variables that are always positive. It is often used as the parent distribution for wind speeds, with w in the range of about 1.5–2.5. The Weibull Distribution with $w=2$ is a special case known as the Rayleigh distribution. When $w=1$, it is known as the 'exponential' distribution.

C.3.6 Poisson distribution

The previous distributions are applicable to *continuous* random variables, i.e. x can take any value over the defined range. The Poisson distribution is applicable only to positive *integer* variables, e.g. number of cars arriving at an intersection in a given time, number of exceedances of a defined pressure level at a point on a building during a windstorm.

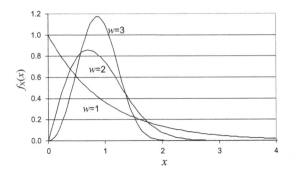

Figure C.5 Probability density functions for Weibull distributions (c = 1).

In this case, there is no probability density function but instead a probability mass function:

$$p_X(x) = \lambda^x \frac{\exp(-\lambda)}{x!} \tag{C.20}$$

where λ is the mean value of X. The standard deviation is $\lambda^{1/2}$.

The Poisson distribution is used quite widely in wind engineering to model exceedances or up-crossings of a random process such as wind speed, pressure or structural response, or events such as number of storms occurring at a given location. It can also be written in the form:

$$p_X(x) = (vT)^x \frac{\exp(-vT)}{x!} \tag{C.21}$$

where v is now the mean rate of occurrence per unit time, and T is the time period of interest.

C.4 EXTREME VALUE DISTRIBUTIONS

In wind engineering, as in other branches of engineering, we are often concerned with the largest value of a random variable (e.g. wind speed) rather than the bulk of the population.

If a variable Y is the maximum of n random variables, $X_1, X_2 \ldots X_n$ and the X_i are all independent,

$$F_Y(y) = F_{X1}(y) \cdot F_{X2}(y) \ldots F_{Xn}(y),$$

since $P[Y < y] = P[\text{all } n \text{ of the } X_i < y] = P[X_1 < y] \cdot P[X_2 < y] \ldots P[X_n < y].$

In the special case that all the X_i are identically distributed with c.d.f. $F_X(x)$,

$$F_Y(y) = \left[F_X(x) \right]^n \tag{C.22}$$

If the assumptions of common distribution and independence of the X_i hold, the shape of the distribution of Y is insensitive to the exact shape of the distribution of the X_i. In this case, three limiting forms of the distributions of the largest value, Y, as n becomes large, may be identified. (Fisher and Tippett, 1928; Gumbel, 1958). However, they are all special cases of the Generalized extreme value distribution.

C.4.1 Generalized extreme value distribution

The c.d.f. may be written,

$$F_Y(y) = \exp\left\{ -\left[1 - \frac{k(y-u)}{a} \right]^{1/k} \right\} \tag{C.23}$$

In this distribution, k is a shape factor, a is a scale factor, and u is a location parameter. There are thus three parameters in this generalized form.

The three special cases are:

Type I ($k=0$). This is also known as the Gumbel distribution.
Type II ($k<0$). This is also known as the Frechet distribution.
Type III ($k>0$). This is a form of the Weibull distribution.

The Type I can also be written in the form:

$$F_Y(y) = \exp\left\{ -\exp\left[-(y-u)/a \right] \right\} \tag{C.24}$$

The GEV is plotted in Figure 2.1 in Chapter 2, with k equal to -0.2, 0 and 0.2 such that the Type I appears as a straight line, with a reduced variate, z, given by:

$$z = -\log_e\left\{ -\log_e\left[F_Y(y) \right] \right\}$$

As can be seen, the Type III ($k=+0.2$) curve approaches a limiting value at high values of the reduced variate (low probabilities of exceedance). Thus, the Type III distribution is appropriate for phenomena that are limited in magnitude for geophysical reasons, including many applications in wind engineering. The Type I ($k=0$) distribution can be assumed to be a conservative limiting case of Type III, and it has only two parameters (a and u), since k is predetermined to be 0. For that reason, the Type I (Gumbel distribution) is easy to fit to actual data, and is very commonly used as a model of extremes for wind speeds, wind pressures and structural response. The Type III GEV is also easy to fit, if the shape factor, k, is fixed at a small positive value, and several methods of fitting based on the closely-related Generalized Pareto distribution (see Section C.4.2), are discussed in Appendix G.

C.4.2 Generalized Pareto distribution

The complementary cumulative distribution function is:

$$G_X(x) = \left[1 - \left(\frac{kx}{\sigma} \right) \right]^{\frac{1}{k}}$$

(C.25)

The p.d.f. is:

$$f_X(x) = \left(\frac{1}{\sigma} \right) \left[1 - \left(\frac{kx}{\sigma} \right) \right]^{\left(\frac{1}{k} \right) - 1}$$

(C.26)

k is the shape parameter and σ is a scale parameter. The range of X is $0 < X < \infty$ when $k < 0$ or $k = 0$. When $k > 0$, $0 < X < (\sigma/k)$. Thus, positive values of k only apply when there is a physical upper limit to the variate, X. The mean value of X is as follows:

$$\bar{X} = \frac{\sigma}{k+1}$$

(C.27)

The probability density functions for three values of k are shown in Figure C.6.

The generalized Pareto distribution has a close relationship with the Generalized Extreme Value distribution (Hosking and Wallis, 1987), so that the three types of the GEV are the distributions for the largest of a group of n variables, that have a Generalized Pareto parent distribution with the same shape factor, k. It also transpires that the Generalized Pareto distribution is the appropriate one for the excesses of independent observations above a defined threshold (Davison and Smith, 1990). This distribution is used for the excesses of maximum wind speeds in individual storms over defined thresholds (Holmes and Moriarty, 1999, and Section 2.4). From the mean rate of occurrence of these storms, which are assumed to occur with a Poisson distribution, predictions can be made of wind speeds with various annual exceedance probabilities.

C.5 LEVEL CROSSING RATES AND EXPECTED MAXIMUM VALUES

C.5.1 Introduction

As previously discussed in Section C.1, the probability density, or distribution, gives no information on the rate of change of a random process, the

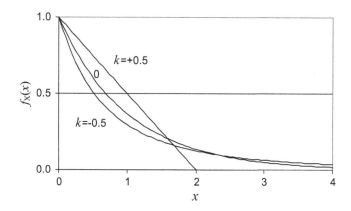

Figure C.6 Probability density functions for Generalized Pareto distributions (σ = 1).

rate of crossing of a given level, or of the maximum value in a time *T*. For example, imagine that the sample of a random process previously shown in Figure C.1 represents a year's variation of hourly mean wind speeds at a measurement site, as in Figure C.7.

Now imagine that the process is 'sped-up' so that the year's variation is now compressed into 6 months (Figure C.8). The value that was the maximum value in 1 year, is now the maximum value in 0.5 years. However, the probability density and distribution for the 'sped-up' process are exactly the same as that for the original process. Thus, it is clear that the probability density/distribution alone is not sufficient information to determine values like 'expected annual maximum', or the 'value that is expected to be exceeded once per year'. Additional information is required on the time rate of change (derivative) of the process.

C.5.2 Level crossing rates – Rice's formula

The expected rate of crossing of defined levels by a random process is itself useful, and it is also a necessary intermediate step in the determination of the expected maximum value in a defined period *T*. Figure C.9 shows five 'up-crossings' of the value *U*=*a* by the random process *U*(*t*) in a time period of *T* equal to 1 year.

A formula for the rate of crossing of a defined level, *a*, by a stationary random process was published by Stephen O. Rice in 1944 (Rice, 1944–1945). Although originally developed for application in telecommunications, it has also found many applications in wind engineering.

Rice's formula can be derived by considering the joint probability of *U*(*t*) and $\dot{U}(t)$ under the conditions that a crossing of the level *U*=*a* occurs with a positive slope within a short time interval.

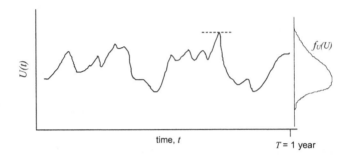

Figure C.7 Long-term wind variation at a site.

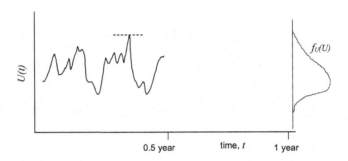

Figure C.8 'Compressed' random variation of wind speed.

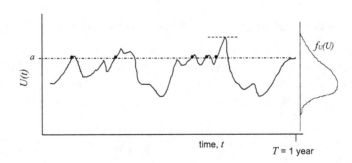

Figure C.9 Up-crossings of the level $U=a$ by the random process $U(t)$.

Making further assumptions that the processes $U(t)$ and $\dot{U}(t)$ are uncorrelated with each other, and that $\dot{U}(t)$ has a normal (Gaussian) probability distribution, Rice's formula, for the rate of crossing of the level $U=a$, can be written, in its simplest form, as:

$$v_a^+ = \sqrt{2\pi} v_U \sigma_U \; f_U(a) \tag{C.28}$$

where σ_U is the standard deviation of the process $U(t)$, and $v_U = \dfrac{\sigma_{\dot{U}}}{2\pi\sigma_U}$ is known as the 'cycling rate' of the process, where $\sigma_{\dot{U}}$ is the standard deviation of the process, $\dot{U}(t)$ – i.e. the rate of change of $U(t)$.

$f_U(a)$ is the probability density function of U evaluated for the level $U=a$.

The full derivation of Rice's formula and the application to the special case of a Gaussian random process were given by Newland (1993). Application to a process with a Weibull distribution, used to represent long-term wind speeds, was made by Gomes and Vickery (1977).

For the Weibull distribution, substituting for $f_U(a)$ from Equation (C.18) in Equation (C.28):

$$v_a^+ = \sqrt{2\pi} v_U \sigma_U \left(\frac{wU^{w-1}}{c^w} \right) \exp\left[-\left(\frac{U}{c} \right)^w \right] \tag{C.29}$$

where c is the Weibull scale parameter, and w is the shape parameter.

For a Gaussian random process, $x(t)$, with a zero mean, the up-crossing rate is:

$$v_a^+ = \frac{1}{2\pi} \frac{\sigma_{\dot{x}}}{\sigma_x} \exp\left[-\frac{a^2}{2\sigma_x^2} \right] \tag{C.30}$$

Note that for this special case, $v_0^+ = v_x$. That is the crossing rate of the mean value (i.e. $a=0$), is equal to the cycling rate for the process. Note that this relationship does not apply to processes with other probability distributions.

The expected largest value of a random process can be obtained by determining the level, a, which is expected to have *zero* crossings in a time T. This was derived for Gaussian random processes by Davenport (1964).

Example C5.2 – application of Rice's formula: An analysis of 10-minute average wind speeds for the Melbourne area recorded over a 40-year period gave the following values for all-direction wind speeds at standard conditions (flat, open country terrain, 10 m height):

$\sigma_U = 3.02$ m/s (standard deviation of 10-minute means)
$w = 1.91$ (Weibull shape factor)
$c = 6.25$ m/s (Weibull scale factor)

The value of the cycling rate, v_U, was found to be about 600 cycles per annum.

The wind speed crossed on average once per year, obtained from Equation (C29), can be obtained by trial and error, by finding the value of a that makes v_a^+ equal to 1. This is 18.9 m/s.

To check this, substitute the above values and $a = 18.9$ m/s into the right-hand side of Equation (C.29):

$$v_a^+ = \sqrt{2\pi}\,(600)(3.02)\left(\frac{1.91 \times 18.9^{0.91}}{6.25^{1.91}}\right)\exp\left[-\left(\frac{18.9}{6.25}\right)^{1.91}\right] = 0.97 \cong 1.0$$

Similarly, the value of a exceeded, on average, once every 50 years can be obtained by finding the value of a that gives a v_a^+ value of 0.02; this value is 23.4 m/s. This agrees quite well with the value obtained from a direct analysis of daily gust extremes (see Chapter 2), when thunderstorm gusts are excluded from the analysis.

C.5.3 Two-dimensional outcrossing rates

Some methods of prediction of the extreme wind speeds and responses are based on an 'out-crossing' formula, which gives the rate of crossing by a two-dimensional process of a defined boundary (see Section 9.11).

With certain assumptions, Equation (C.31) for this rate was derived (Davenport, 1977) as:

$$N_R = \sqrt{2\pi}v_U\sigma_U\int_0^{2\pi}\sqrt{1+\left(\frac{1}{r_R}\frac{dr_R}{d\theta}\right)^2}\,f_U\left(r_R,\theta\right)d\theta \tag{C.31}$$

where

> r, θ are the polar coordinates which define the two-dimensional process, and the response boundary, R.
> $f_U(U,\theta)$ is the joint probability density of wind speed and direction.

For the special case of a circular response boundary (i.e. r_R is constant with θ), the term $\dfrac{dr_R}{d\theta}$ becomes zero, the integrand reduces to $f_U\left(r_R\right)$ and Equation (C.31) reduces to Equation (C.28).

An alternative expression for the outcrossing rate is Equation (C32) (Lepage and Irwin, 1985):

$$N_R = \left(\frac{1}{2}\right)\int_0^{2\pi}\overline{|\dot{U}|}\sqrt{1+\left(\frac{\overline{|\dot{\theta}|}}{\overline{|\dot{U}|}}\frac{dr_R}{d\theta}\right)^2}\,f_U\left(r_R,\theta\right)d\theta \tag{C.32}$$

where $\left|\dot{U}\right|$ and $\left|\dot{\theta}\right|$ are the mean modulus rates of change of wind speed and direction, respectively (note that the modulus operation is applied before the averaging).

If $\dot{U}$ has a normal, or Gaussian, probability distribution, then $\left|\dot{U}\right|$ has a 'folded' normal distribution (Section C.3.4), for which the mean value, $\overline{\left|\dot{U}\right|}$, is $\sigma_{\dot{U}}\sqrt{\dfrac{2}{\pi}}$. The term outside the integration in Equation (C.31) can then be written as:

$$\sqrt{2\pi}v_U\sigma_U = \sqrt{2\pi}\,\frac{\sigma_{\dot{U}}}{2\pi\sigma_U}\,\sigma_U = \frac{\sigma_{\dot{U}}}{\sqrt{2\pi}} = \frac{\overline{\left|\dot{U}\right|}}{2} \tag{C.33}$$

Thus, the leading term in Equation (C32) is the same as that in Equation (C31), except that $\left|\dot{U}\right|$ is included in the integration in Equation (C32), but not in Equation (C31). Lepage and Irwin found that $\overline{\left|\dot{U}\right|}$, varies with wind speed, U, in which case it should be within the integrand.

C.6 RETURN PERIOD AND AVERAGE RECURRENCE INTERVAL

In Section 2.2.2, it was noted that there is a difference between the reciprocal of the *probability* of exceedance of an extreme variate (for example, an annual maximum wind speed, U), which is denoted as 'return period', R_P (Equation 2.3; Gumbel, 1958), and the 'average (or mean) recurrence interval', R_I. The latter is the reciprocal of the *rate* of exceedance, or its equivalent, the average interval between exceedance of high values, and can be determined by calculation of level crossing rates (see Section C.5.2). In the following, the relationship between R_P and R_I (Equation 2.4) is derived.

Assume a Poisson distribution (Section C.3.6) for the number of exceedances, n, of wind speed U in a given time period T:

$$P(n,\mu) = e^{-\mu} \cdot \frac{\mu^n}{n!}$$

where μ is the mean number of exceedances.

In this case: $\mu = v.T = \dfrac{T}{R_I}$

where R_I is the average recurrence interval, by definition.

Probability of U being the maximum value in time T = probability of *zero* exceedances in time T

$$= P(0,\mu) = e^{-\mu} \cdot \frac{\mu^0}{0!} = e^{-\mu} = \exp\left(-\frac{T}{R_I}\right)$$

By definition of return period, R_P,

Probability of non-exceedance of U in 1 year $= 1 - \dfrac{1}{R_p}$

Hence, $1 - \dfrac{1}{R_p} = \exp\left(-\dfrac{1}{R_I}\right)$, or

$$\frac{1}{R_p} = 1 - \exp\left(-\frac{1}{R_I}\right) \tag{C.34}$$

The two sides of Equation (C.34), the same equation as Equation (2.4) in Chapter 2, are plotted in Figure C.10. Although the two plots appear to converge completely at high values of R_P and R_I, actually there remains a small difference between the lines.

The limit,

$$\lim_{R_I}\{R_P - R_I\} = \lim_{R_I}\left\{\left[1 - \exp\left(-\frac{1}{R_I}\right)\right]^{-1} - R_I\right\} = 0.5 \tag{C.35}$$

This limit can be proved mathematically by a series expansion of $\left[1 - \exp\left(-\dfrac{1}{R_I}\right)\right]^{-1}$, but it can also be easily demonstrated by calculation (for example, using EXCEL).

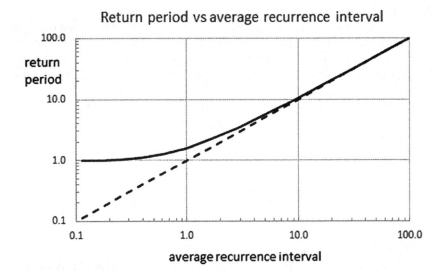

Figure C.10 Return period versus average recurrence interval.

C.7 OTHER PROBABILITY DISTRIBUTIONS

There are many other probability distributions apart from the ones discussed in this appendix. The properties of the most common ones are listed by Hastings and Peacock (1974).

The general application of probability and statistics in civil and structural engineering was discussed in specialized texts by Benjamin and Cornell (1970) and Ang and Tang (1975).

The theory and application of random processes are covered in detail by Bendat and Piersol (2010) and Newland (1993).

REFERENCES

Ang A.H-S. and Tang, W.H. (1975) *Probability concepts in engineering planning and design – volume I – basic principles.* John Wiley, New York.

Bendat, J.S. and Piersol, A.G. (2010) *Random data: analysis and measurement procedures*, 4th Edition. Wiley, New York.

Benjamin J.R. and Cornell, C.A. (1970) *Probability, statistics and decision for civil engineers.* McGraw-Hill, New York.

Davenport A.G. (1964) Note on the distribution of the largest value of a random function with application to gust loading. *Proceedings, Institution of Civil Engineers*, 28: 187–196.

Davenport A.G. (1977) The prediction of risk under wind loading. *Proceedings, 2nd International Conference on Structural Safety and Reliability*, Munich, Germany, September 19–21, pp. 511–596.

Davison, A.C. and Smith, R.I. (1990) Models for exceedances over high thresholds. *Journal of the Royal Statistical Society, Series B*, 52: 393–442.

Fisher, R.A. and Tippett, L.H.C. (1928) Limiting forms of the frequency distribution of the frequency distribution of the largest or smallest member of a sample. *Proceedings of the Cambridge Philosophical Society*, 24(pt.2): 180–190.

Gumbel, E.J. (1958) *Statistics of extremes.* Columbia University Press, New York.

Gomes, L. and Vickery, B.J. (1977) On the prediction of extreme winds from the parent distribution. *Journal of Industrial Aerodynamics*, 2: 21–36.

Hastings, N.A.J. and Peacock, J.B. (1974) *Statistical distributions.* Wiley, New York.

Holmes, J.D. (1981) Non-gaussian characteristics of wind pressure fluctuations. *Journal of Wind Engineering and Industrial Aerodynamics*, 7: 103–108.

Holmes J.D. and Moriarty, W.W. (1999) Application of the generalized Pareto distribution to extreme value analysis in wind engineering. *Journal of Wind Engineering and Industrial Aerodynamics*, 83: 1–10.

Hosking, J.R.M. and Wallis, J.R. (1987) Parameter and quantile estimation for the generalized Pareto distribution, *Technometrics*, 29: 339–349.

Lepage, M.F. and Irwin, P.A. (1985). A technique for combining historical wind data with wind tunnel tests to predict extreme wind loads. *Proceedings 2B-71-78. 5th U.S. National Conference on Wind Engineering*, Lubbock, TX, November 6–8.

Newland, D.E. (1993) *An introduction to random vibrations, spectral and wavelet analysis*, 3rd Edition. Longman Group Limited, London and New York.

Rice, S.O. (1944–1945) Mathematical analysis of random noise. *Bell System Technical Journal*, 23: 282–332 (1944) and 24: 46–156. Reprinted in N. Wax, *Selected papers on noise and stochastic processes*, Dover, New York, 1954.

Appendix D: Extreme wind climates – a world survey

In this appendix, an attempt has been made to describe the general type of extreme wind climate, and to catalogue reliable design wind speed information available, from many countries in the world. Classification is done on a national basis, although of course extreme wind climates do not follow national boundaries. For small countries without wind loading standards, or building codes with wind loading information, it would be appropriate to use information from neighbouring countries.

It should be noted that wind loading codes and standards are constantly under revision, and the values of design wind speed, zoning systems, etc. given in this appendix may change periodically.

D.1 SEVERE WIND STRENGTH CLASSIFICATION SYSTEM

There have been cases where major errors have been made in the general level of design wind speeds used for a particular country or region by engineers from other parts of the world. This is most likely to happen in the tropical and equatorial regions, where the interface between very severe winds, produced by tropical cyclones (typhoons, hurricanes), and the low extreme winds near the Equator where tropical cyclones do not occur, may not be clearly defined. It is very useful to have a general idea of the level of design wind speeds in a country or parts of a large country. This information may be sufficient for the design of small buildings, and less important structures such as signs or poles.

Table D.1 presents a simple classification system, which can be used to 'grade' any country or region in terms of its general level of extreme wind speed. Nothing is stated in Table D.1 with regard to the type of windstorm that is dominant in a country. A dominant storm type in one country can produce similar extreme value statistics to another storm type in a different country. Note that for some storm types, such as convective downdrafts

Table D.1 A classification system for design wind speeds

Level	1-second gust[a] (m/s)	10-minute mean (m/s)
I	<35	<22
II	35–45	22–30
III	45–55	30–35
IV	55–65	35–40
V	>65	>40

Wind speeds with 50-year average recurrence interval (ARI) at 10 m height.

[a] This is a nominal value, based on a moving-average definition of gust duration (see Section 3.3.3). Peak gusts with durations ranging from 0.2 to 3 seconds are used, or are in consideration, for national wind codes and standards.

generated by thunderstorms, a peak gust may be a more relevant indicator than a 10-minute mean wind speed. Although a 50-year ARI is specified in this table, several countries (e.g. USA, Australia, Japan) use higher values of ARI, or return period, for ultimate limit states design, in their national wind codes or standards.

D.2 COUNTRY BY COUNTRY SURVEY

There is often a great deal of information publicly available, for many countries, on average wind speeds, or the 'parent' population probability distribution of winds. This information should be treated with caution by structural designers. While this is generally useful information for wind *energy* applications, it is usually not useful for structural designers, who require *extreme* wind speeds generated by infrequent storms such as gales, thunderstorms and tropical cyclones, typhoons or hurricanes. The information that is available on the latter has been used for the classifications in the following sections.

However, for many jurisdictions covered in the following sections, the information on wind extremes is 'sketchy' and the proposed classifications should be treated as a general guidance only. It should also be noted that the assessments in this appendix are based on historical data only, and the possible effects of future climate change on extreme wind speeds (Section 1.8) have not been considered.

Unless stated otherwise, all design wind data in the following sections are referred to a 10 m high, flat, and an open country terrain.

For major towns and cities in 33 European countries, an interactive map for basic wind speeds (10-minute mean, 50-year return period), for use with the Eurocode (C.E.N., 2010), is available at: https://www.dlubal.com/en/load-zones-for-snow-wind-earthquake/wind-onorm-b-1991-1-4.html

D.2.1 Afghanistan

Much of Afghanistan is over 1500 m above sea level. Extreme wind speeds are expected to be similar to those in Iran. The Department of Defense of the United States (2010) specified a design wind speed of 35 m/s (3-second gust, 50-year ARI) for the capital, Kabul.

Extreme wind classifications: II, III

D.2.2 Algeria

Although there are detailed regulations for earthquake-resistant design for Algeria in north Africa, and comprehensive information on *average* wind speeds for wind energy potential exists, a code, or other information, for extreme wind speeds for structural designs appears to be lacking. However, extreme wind speeds should be similar to neighbouring Morocco with higher values in mountainous regions.

Extreme wind classifications: II (including the capital, Algiers), III

D.2.3 Angola

This country in south-west Africa has a coastline exposed to the South Atlantic Ocean, but fortunately tropical cyclones almost never form in that basin. Only a weaker tropical storm affected Angola in 1991. The International Federation of Red Cross and Red Crescent Societies gives a 10-minute wind speed, based on a 50-year return period of 24 m/s for the whole of Angola.

Extreme wind classification: II

D.2.4 Antarctica

All of the continent of Antarctica experiences persistent, very strong, easterly winds throughout the year. The highest wind gust recorded at Mawson Station (Australia) in east Antarctica in the period from 1954 to 2001 was 69 m/s.

Extreme wind classification: V

D.2.5 Antigua (see Leeward Islands, D.2.74)

D.2.6 Argentina

Argentina is a large country and is affected by a range of different types of windstorms, although tropical cyclones do not occur (Vallis *et al.*, 2018). Large extra-tropical depressions are the dominant winds in the south (Patagonia and Tierra del Fuego). In the north east (Cordoba region), the dominant winds are caused by severe thunderstorms; tornadoes and down-bursts ('tormentas') have caused failures of several high voltage transmission lines. Downslope 'zonda' winds with severe gustiness occur in the Andes.

Early extreme value analyses (Riera and Reimundin, 1970; Viollaz *et al.*, 1975) used the Frechet (Type II) Extreme Value Distribution to fit data from 63 stations in Argentina. This distribution (Section 2.2.1) is known to give excessively conservative predictions at high return periods. More recent extreme value analysis, based on the Gumbel distribution, for six stations in the north-east of the country, gave 50-year return period gusts of 44–47 m/s. Thunderstorm winds were dominant in these records (de Schwarzkopf, 1995).

Extreme wind classifications: II, III

D.2.7 Ascension (U.K.)

In 2010, the Department of Defense of the United States specified a design wind speed of 30 m/s (3-second gust, 50-year ARI) for Ascension Island, a small island in the Atlantic Ocean.

Extreme wind classification: I

D.2.8 Australia

This large continental country has a variety of severe wind types, with large extra-tropical gales along the south coast and Tasmania, moving from the west, and 'East Coast lows' in the Tasman Sea affecting the eastern coastline. Thunderstorm-generated downbursts originating from local convection are the dominant windstorms in the interior. The strongest recorded winds, at 10 m height, in the four major capitals of Sydney, Melbourne, Adelaide and Brisbane are also caused by local downbursts from thunderstorms. Severe tropical cyclones can affect the coastline within about 100 km from the sea between 25° and 10° S latitudes. The most common and most severe occur on the west coast between 18° and 23° S.

Analyses of extreme wind speeds for Australia have been carried out by Whittingham (1964), Gomes and Vickery (1976a,b), Dorman (1983), Holmes (2002), Cechet and Sanabria (2011) and Wang *et al.* (2013), with the latter two studies carried out in the context of climate change effects. Wind speeds for structural design are given in a draft Australian Standard AS/NZS 1170.2 (Standards Australia/Standards New Zealand, 2020). Several regions are defined; these are labelled from A to D with increasing basic design wind speeds.

Regions C to D are considered dominated by severe to very severe tropical cyclones. Wind direction multipliers vary for Regions A0 to A5. In Region B2, incorporating a tropical coastal strip between 50 and 100 km inland, weakening tropical cyclones can produce high wind speeds (Table D.2).

D.2.9 Austria

The basic wind velocities (10-minute mean, 50-year annual recurrence interval) for Austria in the National Annex to the Eurocode, EN-1991-1-4,

Table D.2 Australia

Regions (2020)	Description	Classification
A0 to A5	Thunderstorm downbursts and synoptic winds (gales)	II
B1, B2	Severe thunderstorms or weakening tropical cyclones	III
C	Moderately severe to severe tropical cyclones	IV
D	Severe tropical cyclones	V

range from 17.6 to 28.3 m/s (Kray and Paul, 2015). For Vienna, the velocity of 25.1 m/s is specified.

Extreme wind classifications: I, II

D.2.10 Azerbaijan

The capital of the oil-rich republic of Azerbaijan, Baku on the Caspian Sea, is known for its strong winds, although several wind-sensitive tall buildings have been constructed there recently. However, seismic loading is dominant for most structures in Azerbaijan.

Extreme wind classification: III

D.2.11 Azores (Portugal)

In 2010, the Department of Defense of the United States specified a design wind speed of 53 m/s (3-second gust, 50 year ARI) for the Azores, a small island group in the mid-North Atlantic Ocean.

Extreme wind classification: III

D.2.12 Bahamas

This island group is subjected to frequent Atlantic hurricanes. The U.S. Department of Defense (2010) gave 50-year return period gust speeds of 66–72 m/s. However, a report by Applied Research Associates (Vickery and Wadhera, 2008) showed a significantly lower value of about 54 m/s.

Extreme wind classification: IV

D.2.13 Bahrain

Analysis of combined gust data from Bahrain and Doha (Qatar) by the author (unpublished) gave a 50-year return period peak gust of 35 m/s.

Extreme wind classification: II

D.2.14 Bangladesh

The International Federation of Red Cross and Red Crescent Societies gives a 10-minute wind speed, based on a 50-year return period, of 24 m/s for the

whole of Bangladesh. However, there is a risk of high winds from tropical cyclones in the Bay of Bengal.
Extreme wind classification: III

D.2.15 Barbados (see Windward Islands, D.2.141)

D.2.16 Belarus

The basic wind velocities (10-minute mean, 50-year annual recurrence interval) for Belarus in eastern Europe, in the Eurocode, EN-1991-1-4, are either 21 or 23 m/s. For the capital Minsk, the specified value is 23 m/s.
Extreme wind classification: II

D.2.17 Belgium

The basic wind velocities (10-minute mean, 50-year annual recurrence interval) for Belgium, in the National Annex to the Eurocode, EN-1991-1-4, are 23 to 26 m/s (Kray and Paul, 2015). For Brussels the specified value is 25 m/s.
Extreme wind classification: II

D.2.18 Belize

Belize in Central America experiences severe winds from hurricanes. Analysis of extreme wind speeds for the Commonwealth Caribbean was carried out by Shellard (1972). These results were used by the Caribbean Uniform Building Code (Caribbean Community Secretariat, 1986). This code specified a 50-year return period 10-minute mean wind speed of 36 m/s for the north of the country, and 30.5 m/s for the south.
Extreme wind classifications: IV (north); III (south)

D.2.19 Benin

The extreme wind zones for Benin in west Africa may be assumed to be similar to that of Ghana to the west.
Extreme wind classifications: I – south of 7° N latitude;
II – north of 7° N latitude

D.2.20 Bolivia

In the wind loading code of Bolivia (APNB 1225003-1) basic 3-second gust speeds, with 50-year return periods, varying between 24 and 44 m/s are given for nine different cities. This large variability is probably due to sampling errors resulting from processing of short records.
Extreme wind classification: II

D.2.21 Botswana

This country in southern Africa can be assumed to have similar design wind speeds as the northern part of the neighbouring South Africa – for which a 10-minute mean wind speed with 50-year return period of 28 m/s is specified (Goliger et al., 2017).

Extreme wind classification: II

D.2.22 Brazil

In Brazil, extreme winds are produced primarily by a mixture of large extratropical depression systems ('synoptic') and local thunderstorm downdrafts ('non-synoptic') (Riera and Nanni, 1989; Loredo-Souza, 2012; Vallis et al., 2018; Vallis, 2019). The coastline of the South Atlantic normally does not experience tropical cyclones due to the low water temperature; however, a cyclonic system (named 'Catarina') with the characteristics of a weak hurricane formed off the coastline of southern Brazil in January 2004 (Loredo-Souza and Paluch, 2005).

Salgado Vieira Filho (1975) carried out extreme value analyses of wind speeds for 49 Brazilian stations, but used the conservative Frechet Distribution for predictions, and apparently the data were not separated by storm type. Later analyses by Riera and Nanni (1989) indicate that thunderstorm winds are dominant in most locations. The Brazilian wind loading code of 1987 (NBR-6123, 1988) gave isotachs of 3-second gust speeds with 50-year return period varying from 30 m/s (north half of country) to 50 m/s (extreme south).

Vallis (2019) processed data from nearly 700 weather stations across Brazil, separated the non-synoptic gusts from the synoptic ones, and proposed seven zones with distributions of 3-second non-synoptic gust speeds versus average recurrence interval. For synoptic wind gusts, six zones with different boundaries have been proposed. However, for nearly all of Brazil, non-synoptic events dominate, with values of 3-second gusts, for a 50-year average recurrence interval, ranging from 30 m/s in the far north (equatorial) zone, to 48 m/s for a small zone in the south near the border with Paraguay and Argentina.

Extreme wind classifications: I (north of 10° S); II (central); III (far south)

D.2.23 Bulgaria

The National Annex to the Eurocode, EN-1991-1-4, for Bulgaria in eastern Europe specifies basic wind velocities (10-minute mean, 50-year annual recurrence interval) of 24–35.8 m/s (Kray and Paul, 2015). For the capital Sofia, the specified value is 27.7 m/s.

Extreme wind classification: II, III

D.2.24 Cambodia

The handbook HB 212, 'Design Wind Speeds for the Asia-Pacific Region', (Holmes and Weller, 2002), suggests that extreme winds in the east part of Cambodia are similar to those in southern Vietnam. In the west of Cambodia, they are similar to Thailand.

Extreme wind classifications: I – east of 104° E longitude;
II – west of 104° E longitude

D.2.25 Canada

Extreme winds in Canada are primarily generated by large-scale synoptic systems, and surface extreme winds can be quite well predicted from gradient wind observations (Davenport and Baynes, 1972). Appendix C to the National Building Code of Canada (NRC, 2015) gives climatic information, including values of dynamic pressures for structural design. The equivalent hourly mean wind speeds with a 30-year return period, range from 24 to 28 m/s in the main populated area around the Great Lakes (including Toronto, Montreal and Ottawa), to 30 to 35 m/s in Newfoundland, and in the Hudson's Bay area.

Extreme wind classifications: III (Newfoundland and north);
II (rest of Canada)

D.2.26 Caroline Islands (U.S.)

Typhoon 'Ophelia' devastated much of the small atoll of Ulithi in the Caroline group in 1960. The Department of Defense of the United States (2010) specified a design wind speed of 42 m/s (3-second gust, 50-year ARI) for Koror and Paulau.

Extreme wind classification: III

D.2.27 Chad

For Chad in central Africa, the International Federation of Red Cross and Red Crescent Societies gives a 3-second gust speed, based on a 50-year return period of 50 m/s.

Extreme wind classification: III

D.2.28 Chile

The Peru Current, flowing north, prevents the formation of tropical cyclones in the south-eastern Pacific Ocean. However, southern Chile is in the 'roaring forties' latitudes and subjected to strong westerly winds.

The Chilean wind loading code, NCh432.Of71(INN, Chile, 1971), specifies basic wind velocities (at 10 m height) of 33 and 42 m/s for cities and open terrain, respectively (Vallis *et al.*, 2018). These are apparently applicable to the whole country. However, the averaging time for these wind speeds is not specified.

Extreme wind classifications: II (north of 40° S); III (south of 40° S)

D.2.29 China

China is a large country with a range of extreme wind types ranging from severe gales arising from synoptic systems in Siberia in the north-west, to typhoons along the southern coastline. There is a region with downslope winds.

A combined loading code published in 1994 by the Department of Standards and Norms, Ministry of Construction, includes a wind loading section. This was also available with English translation (GBJ-9 – Department of Standards and Norms, 1994). A map was included with this standard which gives contours of dynamic pressure in kN/m² (kPa). The standard stated that the 'wind reference pressure' is calculated from the 10-minute mean wind speed at 10 metres height by the formula, $w_o = v_o^2/1600$. These values had a 50 year return period. Values of dynamic pressure on these contours ranged from 0.30 to 0.90 kPa. For most of the country, the values were in the range of 0.30–0.50 kPa (Table D.3).

D.2.30 Colombia

Most of Colombia in South America is within 10° latitude from the Equator; hence design wind speeds are generally low, and structural loading is dominated by seismic loads in most locations. However, the loading code (NSR-10) contains detailed rules for wind loading, based on ASCE 7, with a contour map showing five levels of design wind speed.

Extreme wind classifications: II – Zones 4 and 5 in NSR-10 (including Medellin, and Caribbean coastline); I – remainder of the country (including Bogota)

D.2.31 Croatia

The extreme wind climate in Croatia was described by Bajic and Peros (2005). Most of the country was assigned a value of 25 m/s for the 10-minute mean

Table D.3 China

Region	Description	Classification
Central Mainland	Pressure contours from 0.30 to 0.50 kPa	II
North west and inner southern coast	Pressure contours from 0.60 to 0.70 kPa	III
Outer southern coast and islands, Hainan	Pressure contours from 0.80 to 0.90 kPa	IV
Taiwan	Severe typhoons	V

wind speed with a return period of 50 years. The exceptions are some valleys and mountain passes with accelerated *bora* (downslope) winds (see Section 1.3.6). In these regions, it is stated that values of 38–55 m/s were obtained. The National Annex to the Eurocode, EN-1991-1-4, for Croatia specifies basic wind velocities (10-minute mean, 50-year annual recurrence interval) of 20–48 m/s (Kray and Paul, 2015).

For the capital, Zagreb, the specified value is 25 m/s.

Extreme wind classification: II – most of country; **IV–V** – downslope wind areas

D.2.32 Cuba

Cuba is affected by Caribbean hurricanes. A report by Applied Research Associates (Vickery and Wadhera, 2008) showed effective 50-year return period peak gust speeds of 36 m/s in the east of the country increasing to 54 m/s in the west.

Extreme wind classifications: II (east side); **III** (west side)

D.2.33 Cyprus

For the island of Cyprus in the Mediterranean, the interactive map for the Eurocode, EN-1991-1-4, gives three zones with values of basic wind speed (10-minute mean, 50-year average recurrence interval) of 24 m/s (yellow), 30 m/s (green) and 40 m/s (red).

Limassol and Nicosia are both in the 24 m/s (yellow) zone.

Extreme wind classifications: II – yellow zone; **III** – green zone;
IV – red zone

D.2.34 Czechia (formerly Czech Republic)

Kral (2007) carried out extreme wind analyses for Czechia in eastern Europe. The National Annex to the Eurocode, EN-1991-1-4, gives several zones with values of basic wind speed (10-minute mean, 50-year average recurrence interval) between 22.5 and 36 m/s. The value is 22.5 m/s for the capital Prague. These values include the effect of altitude above sea level.

Extreme wind classifications: II, III

D.2.35 Denmark

Wind speed observations have been made in Denmark since the 1870s. The dominant source of extreme winds in Denmark is severe extra-tropical depressions moving in from the North Atlantic Ocean. Extreme value analyses of extreme wind speeds have been made by Jensen and Franck (1970) and Kristensen *et al.* (2000).

The National Annex to the Eurocode, EN-1991-1-4, for Denmark specifies basic wind velocities (10-minute mean, 50-year annual recurrence interval) of 24–27 m/s (Kray and Paul, 2015).

Extreme wind classification: II

D.2.36 Diego Garcia

Tropical cyclones often form to the south of Diego Garcia, a small atoll in the Indian Ocean. In 2010, the Department of Defense of the United States specified a design wind speed of 47 m/s (3-second gust, 50-year ARI) for the military base on Diego Garcia.

Extreme wind classification: III

D.2.37 Djibouti

In 2010, the Department of Defense of the United States specified a design wind speed of 40 m/s (3-second gust, 50 year ARI) for Djibouti, a small state on the east coast of Africa (Department of Defense, 2010).

Extreme wind classification: II

D.2.38 Dominican Republic

The Dominican Republic, occupying the eastern part of the island of Hispaniola in the Caribbean, is affected by hurricanes. A report by Applied Research Associates (Vickery and Wadhera, 2008) showed effective 50-year return period peak gust speeds of 36 m/s in the north west of the country increasing to 54 m/s in the south.

Extreme wind classifications: II and III

D.2.39 Ecuador

In Ecuador, on the west coast of South America, seismic loading is more important than wind loading for structural design. With an equatorial climate, Ecuador experiences low extreme wind speeds from thunderstorms. The national loading code (CPE INEN-NEC-SE-CG 26-1) specifies a single basic wind speed, with an unspecified averaging time, of 21 m/s.

Extreme wind classification: I

D.2.40 Egypt

Basic extreme wind speeds (hourly mean and 3-second maximum gusts) for Egypt were reported by Bakhoum *et al*. Locations on the coast have similar extreme winds to other Mediterranean locations – e.g. the 50-year gust for Alexandria is given as 38 m/s by the U.S. Department of Defense.

Inland values appear to be significantly lower. Cairo was calculated to have a 50-year return period gust speed of less than 30 m/s by Bakhoum *et al.* Extreme wind classifications: I (inland); II (coastal)

D.2.41 Eritrea

Eritrea in East Africa has a mountain part and a coastal plain with different climates. The capital, Asmara, is in elevated terrain where extreme winds are produced by intermittent thunderstorms; for this location the lowest classification (I) applies. At the port of Massawa on the Red Sea coastline, extreme winds are higher (II).
Extreme wind classifications: I, II

D.2.42 Estonia

As for Latvia and Lithuania, the coastline of Estonia is exposed to gales from the Baltic Sea. The National Annex to the Eurocode, EN-1991-1-4, gives a single value of basic wind speed (10-minute mean, 50-year average recurrence interval) of 21 m/s for the whole country (Kray and Paul, 2015).
Extreme wind classification: II

D.2.43 Ethiopia

Ethiopia is a land-locked country in the tropics with elevated terrain between 1,800 and 2,400 m above sea level in the central part. Extreme winds in the capital, Addis Ababa are generated by thunderstorms and are low.
Extreme wind classification: I

D.2.44 Falkland Islands (U.K.)

The highest recorded gust speed in Falkland Islands, in the South Atlantic Ocean, is 37 m/s. The extreme wind climate is similar to southern Argentina.
Extreme wind classification: II

D.2.45 Fiji

The Fijian islands are subject to periodic visits from tropical cyclones (e.g. 'Kina' (1993), 'Ami' (2003), 'Daman' (2007), 'Evan' (2012), 'Winston' (2016, a Category 5 storm)), and consequently very high design wind speeds.
Extreme wind classification: IV

D.2.46 Finland

The National Annex to the Eurocode, EN-1991-1-4, gives a range of values of basic wind speed (10-minute mean, 50-year average recurrence interval) of 21–26 m/s for Finland on the Baltic Sea (Kray and Paul, 2015).
Extreme wind classification: II

D.2.47 France

Like other Western European countries, the extreme wind climate of France is dominated by synoptic gales from large depression systems moving in from the Atlantic Ocean. Sacré (2002) calculated iso-vents for France based on 50-years of recorded extremes from 1949, and following the severe storms of 1999 (see Table 1.3). Values of 10-minute mean speeds, with 50-year return periods, varied from 24 m/s for inland locations to 30 m/s near the coastlines, and the east of Corsica.

The original draft Eurocode (C.E.N., 1994) specified four values of 10-minute mean wind speed with a 50-year return period, for four zones in metropolitan France, ranging from 24 to 30.5 m/s. The highest values were given for Zone 4, which included parts of Brittany and Normandy, the Mediterranean coastline, and eastern Corsica.

However, the French National Annex to the current Eurocode, EN-1991-1-4, now gives three zones for mainland France with a lower range of values of basic wind speed (10-minute mean, 50-year average recurrence interval) of 22–26 m/s (Kray and Paul, 2015). The specified value for Paris is 24 m/s. For the east part of Corsica, 28 m/s is now specified.

Extreme wind classification: II

D.2.48 French Caribbean (see also Leeward and Windward islands)

The original draft Eurocode (C.E.N., 1994) specified a value of 10-minute mean wind speed with a 50-year return period, of 34 m/s, for the French territories of Guadeloupe and Martinique in the southern Caribbean. These territories experience visits from hurricanes, like other islands of the Caribbean.

Extreme wind classification: III (Martinique); IV (Guadeloupe)

D.2.49 Germany

The original draft Eurocode ENV-1991-2-4 (C.E.N., 1994) gave a map with a system of four zones. The highest wind speed zone, three, was on the North Sea coast. The main source of strong winds is gales accompanying large-scale depressions moving into Germany from the west. The zone system was different from an earlier zoning system for the Federal Republic (West Germany) by Caspar (1970).

Analyses by Schueller and Panggabean (1976) for stations in the former West Germany gave distributions for gust speeds, which give 50-year return period values between 35 and 50 m/s. An exception was Feldberg with 60 m/s; this is a mountain station, with topographic influences.

Kasperski (2002) carried out an analysis of hourly mean wind speeds for 183 stations in Germany. This analysis included separation of winds from different storm types – namely frontal depressions, thunderstorms and gust fronts. He proposed five wind zones with 50-year return period with hourly mean speeds between 22.5 and 32.5 m/s. His Zones I–III correspond to Classification II in this Appendix; his Zones IV and V correspond to Classification III in Table D.1.

The German National Annex to the Eurocode, EN-1991-1-4, gives four zones with a range of values of basic wind speed (10-minute mean, 50-year average recurrence interval) of 22–30 m/s (Kray and Paul, 2015). The red zone with the highest value of 30 m/s applies to the north-east coastline, and a small area on the Baltic Sea. The specified value for Berlin is 25 m/s (Table D.4).

D.2.50 Ghana

The south of Ghana, on the coast of west Africa, is subject to severe thunderstorms which may generate strong winds for short periods. Ghana has recently issued a Building Code with a contour map of gust wind speeds, with 50-year average recurrence interval, and listed values for the major centres in Part 5 (Ghana Standards Authority, 2018). The values are stated to be 'maximum 3-second gusts', but may have a shorter duration if they are based on earlier Dines pressure-tube anemometer records. The capital, Accra has a stated value of 29 m/s.

Extreme wind classifications: I – south of 7° N latitude;
II – north of 7° N latitude

D.2.51 Greece

The National Annex for Greece to the Eurocode, EN-1991-1-4, gives a range of values of basic wind speed (10-minute mean, 50-year average recurrence interval) of 27–33 m/s (Kray and Paul, 2015).

Extreme wind classifications: II, III

D.2.52 Greenland (Denmark)

The U.S. Department of Defense recommended 50-year gust wind speeds between 54 and 74 m/s for U.S. Air Force bases in Greenland.

Extreme wind classification: IV

Table D.4 Germany

Zone (Kasperski, 2002)	Description	Classification
I	Central	II
II	Northern Germany & southern alpine region	II
III	North west & Saxony	II
IV	North Sea & Baltic coasts	III
V	North Sea offshore	III

D.2.53 Guam (U.S.)

This Pacific Island has experienced some of the strongest recorded tropical cyclones. ASCE 7-16 specifies a 3-second gust speed, with a 300-year mean recurrence interval, of 80 m/s.

Extreme wind classification: V

D.2.54 Guyana

This country has an equatorial climate with low wind speeds. An analysis of extreme wind speeds for the Commonwealth Caribbean was carried out by Shellard (1972). These results have been used by the Caribbean Uniform Building Code (Caribbean Community Secretariat, 1986) and by the Code of Practice of the Barbados Association of Professional Engineers (Gibbs, 1972; BAPE, 1981). The former specified a 50-year return period 10-minute mean speed of 18 m/s, and the latter gave a 50-year return period 3-second gust speed of 22 m/s.

Extreme wind classification: I

D.2.55 Haiti

Haiti, occupying the western part of the island of Hispaniola, is occasionally affected by Caribbean hurricanes. A report by Applied Research Associates (Vickery and Wadhera, 2008), showed effective 50-year return period peak gust speeds (3 seconds duration) of 36 m/s in the north of Haiti increasing to 49 m/s in the south east.

Extreme wind classifications: II and III

D.2.56 Hong Kong and Macau (China)

As for the rest of the south China coastline, Hong Kong and Macau are subject to frequent visits from moderate to severe typhoons. Hong Kong has good quality recorded wind speed data extending more than a hundred years from 1884 to 1957, from the Royal Hong Kong Observatory, and since 1957, from Waglan Island. Analyses of extreme winds from typhoons has been carried out by a number of authors including Faber and Bell (1967), Chen (1975), Davenport et al. (1984), and Melbourne (1984), Jeary (1997) and Holmes et al. (2009). Most of these studies have normalised the wind speeds to a height of 50 m, rather than 10 m. Design wind speeds in Hong Kong and Macau are set by the respective building departments. The wind pressure in the Hong Kong Code of Practice (HK Buildings Department, 2019) implies a 50-year return period gust wind speed at 10 m height of about 57 m/s.

Extreme wind classification: IV

D.2.57 Hungary

For Hungary in eastern Europe, National Annex to the Eurocode, EN-1991-1-4, gives a single value of basic wind speed (10-minute mean, 50-year average recurrence interval) of 23.6 m/s for the whole country (Kray and Paul, 2015).

Extreme wind classification: II

D.2.58 Iceland

Iceland is subject to strong Atlantic gales. The original draft Eurocode ENV-1991-2-4 specified a 50-year return period 10-minute mean at 10 m height, of 39 m/s for coastal areas within 10 km of the coastline. For inland areas, the value was 36 m/s. However, the National Annex to the Eurocode, EN-1991-1-4, now gives a single value of basic wind speed (10-minute mean, 50-year average recurrence interval) of 36 m/s for the whole country (Kray and Paul, 2015).

Extreme wind classification: IV

D.2.59 India

India, a large sub-continental tropical country, has a range of extreme wind zones, with extreme tropical cyclones being dominant on the east (Bay of Bengal) coast, and less frequent ones on the west coast. In inland areas, thunderstorms and monsoon winds are prevalent.

India has a good network of meteorological stations, and there have been a number of extreme value analyses of wind speeds for the country. The Indian Standard for Wind Loads IS875 Part 3 (Bureau of Indian Standards, 2015) divides the country into six zones, giving 50-year return period gust wind speeds ranging from 33 to 55 m/s (Table D.5).

Table D.5 India

Zone	Description	Classification
1	Tripura, Mizoram, Ladakh	IV
2	Coastal strips of Tamil Nadu (including Madras), Andhra Pradesh, Orissa, Gujarat, West Bengal (including Calcutta), Assam	III
3	Northern India including Delhi, central Tamil Nadu	III
4	Coastal strip on Arabian Sea, including Bombay, inland Madhya Pradesh, Orissa	II
5	Most of southern India	II
6	Inland Karnataka, including Bangalore	I

Refer to map in IS 875: Part 3, for details of zones.

D.2.60 Indonesia

Like Malaysia and Singapore, Indonesia is entirely in the Equatorial zone, and so does not experience typhoons, and design wind speeds from weak thunderstorms and monsoonal winds are low. Structural design is generally governed by earthquakes.

Extreme wind classification: I

D.2.61 Iran

Iran is a large country, with most of the landmass and major cities at elevations of 1,000 m, or more, above sea level. Gusts of over 60 m/s have occasionally been recorded in some locations. Rajabi and Modarres (2008) describe an analysis of limited data from Isfahan province near the centre of the country. They predicted a 15-minute average with an average recurrence interval of 50 years of about 36 m/s.

Extreme wind classification: III

D.2.62 Iraq

The Department of Defense of the United States (2013) specified a design wind speed of 45 m/s (3-second gust, 300-year ARI) for Baghdad and Basra.

Extreme wind classification: II

D.2.63 Ireland

Ireland is exposed to severe Atlantic gales on its west coast. Ex-Atlantic hurricane 'Ophelia' did significant damage in the country in 2017, and produced the highest recorded gust of 53 m/s off the coast of County Cork.

The National Annex for Ireland to the Eurocode, EN-1991-1-4, gives a range of values of basic wind speed (10-minute mean, 50-year average recurrence interval) of 25–28 m/s, (Kray and Paul, 2015). For the capital, Dublin, the specified value is 25.2 m/s.

Extreme wind classifications: II, III (west coast)

D.2.64 Israel

Israel experiences several different wind storm types, including the 'sharqiya' easterly downslope winds, similar to the 'bora' of Croatia (Saaroni et al., 1998). The highest wind gust recorded at Jerusalem was 44 m/s on 20/1/1974.

A contour map of 10-minute average wind speeds with 50-year return period gives values between 24 and 36 m/s (Goldreich, 2003).

Extreme wind classifications: II, III

D.2.65 Italy

Extreme value analyses of wind speeds for Italy have been carried out by Ballio *et al.* (1999) and several others.

In the National Annex to the Eurocode, EN-1991-1-4, four basic wind speeds are specified. These are 10-minute mean speeds with a 50-year return period, ranging from 25 to 30 m/s. For Rome, the specified value is 27 m/s (Table D.6).

D.2.66 Ivory Coast

In the absence of definitive extreme wind data for the Ivory Coast (Cote d'Ivoire) in west Africa, the extreme wind zones may be assumed to be similar to that of Ghana to the east.

Extreme wind classifications: I – south of 7° N latitude; II – north of 7° N latitude

D.2.67 Jamaica

Jamaica is in a region of hurricane tracks in the Caribbean, and experiences severe winds from these events. An analysis of extreme wind speeds for the Commonwealth Caribbean was carried out by Shellard (1972). These results were used by the Caribbean Uniform Building Code (Caribbean Community Secretariat, 1986) and by the Code of Practice of the Barbados Association of Professional Engineers (1981). The former specified a 50-year return period 10-minute mean wind speed of 36.5 m/s, and the latter a 56 m/s peak gust. A more recent report by Applied Research Associates (Vickery and Wadhera, 2008) showed an effective 50-year return period peak 3-second gust speed of about 45 m/s.

Extreme wind classification: III

D.2.68 Japan

Japan is subject to typhoons from the Pacific in Kyushu and Okinawa, and temperate synoptic systems in the north of the country. The Architectural Institute of Japan has a contour map of design wind speeds (10-minute mean, 100-year return period) in its wind load recommendations. Values

Table D.6 Italy

Description	Classification
Northern Italy (25 m/s)	II
Central and southern Italy (27 m/s)	II
Sardinia and Sicily and Liguria (28 m/s)	II
Trieste and islands (30 m/s)	III

range from 26 to 44 m/s on the main islands, to 50 m/s on Okinawa which is subject to frequent severe typhoons.

Extreme wind classifications: II, III, IV, V

D.2.69 Kenya

The whole of Kenya is within 5° of the Equator, and should have an equatorial extreme wind climate dominated by thunderstorms; this would be modified by the elevated land of most of the country.

Ong'ayo *et al.* (2014) applied an extreme value analysis to hourly mean wind data from eight stations in Nairobi County; however, the corrections for terrain and altitude were large, and the use of hourly averaged winds may have 'filtered' out the thunderstorms.

Extreme wind classification: I

D.2.70 Korea

The coastline of South Korea has some influence from typhoons on the south and east coasts and the island of Cheju, but these are relatively infrequent.

The Architectural Institute of Korea has a map of 10-minute mean 100-year return period wind speeds varying from 25 m/s in the inland centre to 40 m/s at some points on the eastern and southern coastline. Seoul is specified as 30 m/s.

Extreme wind classifications: II, III, IV

D.2.71 Laos

The handbook HB 212, 'Design Wind Speeds for the Asia-Pacific Region', (Holmes and Weller, 2002), suggests that extreme winds in the east part of Laos are similar to those in south Vietnam. In the west of the country, they are assumed to be similar to Thailand.

Extreme wind classifications: I – (east of 104° E longitude); II – (west of 104° E longitude)

D.2.72 Latvia

The Latvian Building Code LNBN-003-01, (Republic of Latvia 2001, 2003) shows recorded maximum gusts – experienced at major centres during a 45-year period. For the capital, Riga, and Liepaja on the Baltic coast, these values were 31 and 48 m/s, respectively. The west coast is exposed to winter westerlies from the Baltic Sea.

The basic wind velocities (10-minute mean, 50-year annual recurrence interval) for Latvia in the National Annex to the Eurocode, EN-1991-1-4, range from 21 to 27 m/s (Kray and Paul, 2015).

Extreme wind classifications: I – east of 24° E longitude (including Riga); III – coastal strip (west of 22° E longitude); II – remainder of country

D.2.73 Lebanon

The extreme wind climate of Lebanon at the eastern end of the Mediterranean Sea should be similar to Israel.

Extreme wind classification: II

D.2.74 Leeward Islands

This group of islands is affected by hurricanes in the Caribbean. An analysis of extreme wind speeds for Commonwealth countries in the Caribbean was carried out by Shellard (1972). These results were used by the Caribbean Uniform Building Code (Caribbean Community Secretariat, 1986) and by the Code of Practice of the Barbados Association of Professional Engineers (1981). The latter specified a 50-year return period, 3-second gust speed of 64 m/s, based on studies for Antigua. This value is also applicable to St. Kitts-Nevis, Montserrat, and the Virgin Islands. However, a more recent report by Applied Research Associates (Vickery and Wadhera, 2008), showed effective 50-year return period peak gust speeds of 53–58 m/s for the latter islands.

Extreme wind classification: IV

D.2.75 Lesotho

This small country is completely surrounded by the Republic of South Africa, and can be assumed to have similar design wind speeds – for which a 10-minute mean wind speed with 50-year return period of 28 m/s is specified in the northern part (Goliger *et al.*, 2017).

Extreme wind classification: II

D.2.76 Liberia

No definitive information is available, but Liberia on the west African coast can be assumed to have a similar extreme wind climate as southern Ghana and Ivory Coast, with thunderstorm winds dominant.

Extreme wind classification: I

D.2.77 Libya

Abohedma and Alshebani (2010), after analysing historical wind data from 22 stations in Libya in north Africa, proposed dividing the country into four zones of extreme winds for structural design. The proposed design wind speeds (10-second gusts with 50-year return period) varied from 29 to 34 m/s. The equivalent range of 1-second gusts is 33–39 m/s.

Extreme wind classification: I (A & A, Zones 1 and 2); II – (Zones 3 and 4)

D.2.78 Lithuania

Data on extreme winds in Lithuania were reviewed in detail by Vaidogas and Juocevicius (2011). They noted that there were 35 storms with 10-minute mean wind speeds greater or equal to 35 m/s (10 m height, open country terrain) in the period 1962 to 2005. The Lithuanian Loading Code, STR 2.05.04:2003, showed three zones with basic wind speeds (10-minute mean, 50-year annual recurrence interval) of 24, 28 and 32 m/s. The latter applied only to a narrow strip about 20 km wide adjacent to the Baltic coast. These values are also in the Lithuanian National Annex to the Eurocode, EN-1991-1-4 (Kray and Paul, 2015).

Extreme wind classifications: III – coastal strip (within 20 km of the west coast); II – remainder of country

D.2.79 Luxembourg

The basic wind velocity (10-minute mean, 50-year annual recurrence interval) for Luxembourg in the National Annex to the Eurocode, EN-1991-1-4, is 24 m/s (Kray and Paul, 2015).

Extreme wind classification: II

D.2.80 Madagascar

The island of Madagascar, particularly the northern part, has exposure to regular landfalling, and damaging, tropical cyclones from the southern Indian Ocean. Two of the severest of these were 'Kamisy' (1984) and 'Gafilo' (2004).

Extreme wind classification: IV – northern part;
III – remainder of Madagascar

D.2.81 Malawi

The strongest winds in Malawi in southern Africa are from severe thunderstorms. However, the southern part can also be subjected to winds from weakening, land-falling tropical cyclones.

Extreme wind classifications: I – north of 15° S latitude;
II – south of 15° S latitude

D.2.82 Malaysia

Malaysia is entirely in the Equatorial zone, and so does not experience typhoons, and it has very low extreme winds from weak thunderstorms and monsoonal winds. Monthly maximum wind data is available from more than 30 stations in the country, including Miri and Kuching in East Malaysia (Sarawak).

Analysis of this data for 50-year return period gust values for 20 stations by the Malaysian Meteorological Service gave values between 24 and 32 m/s. There is evidence of higher wind speeds in the highland stations away from the coastal plains.

Extreme wind classification: I

D.2.83 Mali

Mali in west Africa suffers from severe convective downdrafts with gust fronts that often produce severe dust storms ('haboobs'). For the capital Bamako, the International Federation of Red Cross and Red Crescent Societies gives a 3-second gust speed, based on a 50-year return period, of 61 m/s.

Extreme wind classification: III

D.2.84 Malta

Malta in the central Mediterranean had not produced a national Annex to the Eurocode EN-1991-1-4 at the time of writing. However, a wind gust of 37 m/s was reported to have occurred near the capital, Valletta in 2019.

Extreme wind classification: II

D.2.85 Marshall Islands (U.S.)

In 2010, the Department of Defense of the United States specified a design wind speed of 47 m/s (3-second gust, 50 year ARI) for Marshall Islands, a small island group in the Pacific Ocean, that are subject to occasional tropical storms and weak tropical cyclones.

Extreme wind classification: III

D.2.86 Mauritania

An investigation of design wind speeds for offshore oil and gas platforms, indicated strong winds caused by severe thunderstorm events with a 50-year return period gust wind speed of around 55 m/s, although the International Federation of Red Cross and Red Crescent Societies gives a lower 3-second gust speed over land of 43 m/s.

Extreme wind classification: III

D.2.87 Mauritius

Like neighbouring Reunion, Mauritius in the Indian Ocean experiences land fall from a tropical cyclone about once every five years (Sites and Peterson, 1995).

Extreme wind classification: III

D.2.88 Mexico

Mexico experiences extreme winds from hurricanes on both its Pacific and Caribbean coasts. For inland areas, thunderstorms are dominant. Extreme value analyses were carried out by Vilar *et al.* (1991), and Lopez and Vilar, (1995) for the Mexican Electrical Utility (CFE) using the Generalized Extreme Value distribution, for data from 57 stations. An isotach map resulting from this study shows 50-year return period 3-second gusts ranging widely from 28 m/s in the Mexico City area to 61 m/s on the Pacific coast. The table below is not official, but describes zones based on the isotach map (Table D.7).

D.2.89 Morocco

The 50-year gust for Casablanca on the Atlantic coast was given as 40 m/s by the U.S. Department of Defense. Inland mountainous regions may have higher values.

Extreme wind classifications: II, III

D.2.90 Mozambique

The coastline of Mozambique is subject to impact from tropical cyclones from the South Indian Ocean. Cyclone 'Idai' (2019) was the worst recorded of these (see Table 1.3 in Chapter 1). Fearon (2014) using synthetic track modelling, with a wind-field model, generated 50-year return period, 1-minute wind speeds of up to 35 m/s at locations along the Mozambique east coast. Inland extreme winds of similar magnitude to those in northern South Africa may be expected.

Extreme wind classifications: III – within 50 km from the east coast;
II – elsewhere in Mozambique

D.2.91 Namibia

This country in south-west Africa can be assumed to have similar design wind speeds as the northern part of the neighbouring South Africa – for which a 10-minute mean wind speed with 50-year return period of 28 m/s is currently specified (Goliger *et al.*, 2017).

Extreme wind classification: II

Table D.7 Mexico

Description	Classification
South of 24° S excluding coastline	I
North of 24° S excluding coastline	II
Within 50 km of Caribbean coast, 50–100 km from Pacific coast	III
Within 50 km of Pacific coast	IV

D.2.92 Netherlands

The basic wind velocities (10-minute mean, 50-year annual recurrence interval) for the Netherlands in the National Annex to the Eurocode, EN-1991-1-4, vary between 24.5 and 29.5 m/s (Kray and Paul, 2015). Amsterdam and Rotterdam both have a value of 27.5 m/s
Extreme wind classification: II

D.2.93 New Caledonia

New Caledonia in the South Pacific Ocean is affected by tropical cyclones, such as 'Beti' in 1996, and 'Erica' in 2003, both of which produced significant damage (Holmes and Weller, 2002).
Extreme wind classification: IV

D.2.94 New Zealand

Turner *et al.* (2012) described some damaging extreme wind events in New Zealand; these include localized tornados and downslope winds. However, the main extreme winds affecting most of New Zealand are temperate synoptic systems, although the north of the country can experience the effects of decaying tropical cyclones. The map of basic wind speeds in the 2020 draft of the Australian/New Zealand Standard (Standards Australia/ Standards New Zealand, 2020) shows four wind regions. Regions N1 and N2 have the same all-directional basic wind speeds (500 year return period 0.2-second gusts) of 45 m/s, but differ in their directional wind speeds. Regions N3 and N4 have 500-year values of 53 and 50 m/s, respectively. There are a number of mountain areas, especially in the South Island, where downslope winds occur – for these the wind speed may be increased by a 'Lee Multiplier' of up to 1.35.
Extreme wind classifications: II, III (some mountain areas)

D.2.95 Niger

For Niger in west Africa, the International Federation of Red Cross and Red Crescent Societies gives a 3-second gust speed, based on a 50-year return period, of 54 m/s.
Extreme wind classification: III

D.2.96 Nigeria

The extreme winds in the south of Nigeria are produced by thunderstorms. A gust speed with a 50-year return period of about 63 knots (32.5 m/s) was found by Okulaja (1968) for the capital of Nigeria, Lagos, using the standard Gumbel method of fitting.

Soboyejo (1971) produced isopleth maps of Nigeria showing gust wind speeds with 50-, 75- and 100-year return periods. The 50-year values varied from 35 m/s in the south west to 55 m/s in the north east of the country. These are probably gusts of about 0.2-second duration, as Dines pressure-tube-float anemometers were used in Nigeria (Okulaja, 1968).

Extreme wind classifications: I – south of 10° N latitude;
II – north of 10° N latitude

D.2.97 Norway

The basic wind velocities (10-minute mean, 50-year annual recurrence interval) in the Norwegian National Annex to the Eurocode, EN-1991-1-4, vary between 22 and 31 m/s (Kray and Paul, 2015).

Extreme wind classifications: II; III

D.2.98 Oman

Oman has a coastline facing the Arabian Sea and Gulf of Oman. In winter it suffers from north-westerly 'shamal' winds. A study by Almaawali *et al.* (2008) determined 50-year return period gust ranging from 36.5 m/s (Masirah – offshore island) to 52.6 m/s for Nizwa (inland city). The region around the capital, Muscat, has a value of about 40 m/s. The Department of Defense of the United States (2013) specified design gust wind speeds of 55–60 m/s but these are 300-year ARI values.

Extreme wind classification: II

D.2.99 Pakistan

Structural design in Pakistan in south Asia is governed by the need to design for earthquakes. The Pakistan building code specifies basic design wind speeds of 33 m/s for inland structures and 35 m/s for structures 'along the coast', (unspecified averaging time or recurrence interval). However, high winds can occur from thunderstorms, with gusts of 45 m/s reported at Rawalpindi in 2017.

Extreme wind classification: II

D.2.100 Panama

In 2010, the Department of Defense of the United States specified a design wind speed of 42 m/s (3-second gust, 50-year ARI) for the Panama Canal Zone.

Extreme wind classification: II

D.2.101 Papua-New Guinea

The majority of Papua-New Guinea (including Port Moresby) is in the equatorial zone, and the design winds, originating from thunderstorms produced by local convective activity, are quite low. An extreme value (Gumbel) analysis for Port Moresby by Whittingham (1964) using only 11 years of data, gives a 50-year return period gust of 31 m/s. The addition of some extra years gives even lower values. The P-NG loading code gives a contour map with 50-year return period gust wind speeds ranging from 24 to 32 m/s. For the south-east tip 40 m/s is specified. Values for major centres are: Port Moresby 28 m/s, Lae 23 m/s, and Rabaul 26 m/s.

The islands on the south east are occasionally exposed to developing Coral Sea cyclones, and should have higher design wind speeds.

Extreme wind classifications: I (most of country), II (south-east tip)

D.2.102 Paraguay

As is the case for neighbouring southern Brazil and north-eastern Argentina, non-synoptic wind events (thunderstorm convective downdrafts) are dominant in Paraguay (Vallis *et al.*, 2018). The national wind loading code (La Norma Paraguaya di Viento NP 196) specifies basic wind speeds of up to 55 m/s.

Extreme wind classification: III

D.2.103 Peru

Peru, on the Pacific coast of South America, is subject to severe earthquakes, which produce the dominant environmental loading for most structures. The Peru Current carries cool water northwards and prevents the formation of tropical cyclones in the eastern South Pacific Ocean. The national loading code (Norma Técnica de Edificación E.020 Cargas) specifies a basic wind speed, with unspecified duration, of 'not less than' 75 km/h (21 m/s).

Extreme wind classifications: I (north of 10° S); II (south of 10° S)

D.2.104 Philippines

The Philippines experiences typhoons from the south-west Pacific Ocean, which often cross the northern Philippines (Luzon) and re-form in the south China sea. On the other hand, the southern island of Mindanao has little or no influence from typhoons, and effectively has an equatorial extreme wind climate. An extreme value analysis of one-minute average extreme wind speeds in the Philippines was carried out in the early nineteen-seventies by Kintenar (1971). This gave widely ranging 50-year return period values, and probably suffering from sampling errors due to short records.

The National Structural Code of the Philippines has previously specified three extreme wind zones (I, II, and III) with one-minute sustained wind speeds of 200km/h (55.5 m/s), 175 km/h (48.6 m/s) and 150 km/h (41.7 m/s), respectively.

Simiu and Marshall (1977), from the National Bureau of Standards of the United States, analysed annual maximum wind data (1-minute averages) from 16 stations in the Philippines. They concluded that the wind speed requirements of the National Structural Code of the Philippines were generally adequate for Zones I and II, but were conservative for Zone III. They also concluded that Zone II would be better divided into two, with higher wind speeds in the northern section than in the southern part.

The 7th edition of the National Structural Code (2015) contains three new maps with numerous smaller wind zones. The zones provide 3-second gust wind speeds for three different classes of structure, each with wind speeds based on different average recurrence intervals for ultimate limit states design (i.e. a similar format to ASCE 7 and AS/NZS 1170.2).

For this appendix, a simplified table is given below; this is based on the zoning in the earlier editions of the National Structural Code. For more accuracy, users should refer to the current National Structural Code of the Philippines (Table D.8).

D.2.105 Poland

The former Polish wind loading standard PN-77/B-02011 gave 'characteristic' wind speeds of 20, 24, 27 and 30 m/s for four zones. In the largest zone, 20 m/s was specified. These values are 10-minute mean speeds, with a return period of 50 years. In the National Annex to the Eurocode, EN-1991-1-4, values range from 22 to 26 m/s (Kray and Paul, 2015).

Extreme wind classification: II

D.2.106 Portugal

The basic wind velocities (10-minute mean, 50-year annual recurrence interval) for Portugal in the National Annex to the Eurocode, EN-1991-1-4, range from 27 to 30 m/s (Kray and Paul, 2015). The higher values apply to the Azores, Madeira, and the 5 km coastal strip of the mainland.

Extreme wind classification: II

Table D.8 Philippines

Zone in structural code	Description	Classification
I	Eastern Luzon	V
II	Remainder of Philippines	IV
II	Eastern Mindanao	III
III	Western Mindanao, Palawan	II

D.2.107 Puerto Rico (U.S.)

As for other Caribbean islands, Puerto Rico is subjected to hurricane winds. Hurricane 'Maria' in 2017 produced damage estimated as in excess of US$90 billion. A report by Applied Research Associates (Vickery and Wadhera, 2008), showed effective 50-year return period with peak 3-second gust speeds of 54–58 m/s.

Extreme wind classification: IV

D.2.108 Qatar

Analysis of combined gust data from Doha (Qatar) and Bahrain by the author (unpublished) gave a 50-year return period peak gust of 35 m/s.

Extreme wind classification: II

D.2.109 Reunion I

This small French island in the southern Indian Ocean has a design wind speed (10-minute mean, 50-year return period) of 34.0 m/s as specified in the draft Eurocode (C.E.N., 1994). According to Sites and Peterson (1995), Reunion experiences landfall of a tropical cyclone about once every five years.

Extreme wind classification: III

D.2.110 Romania

The Romanian Standard STAS 10101/20-78 on Actions on Structures specified five zones for design wind pressures. These pressures correspond to peak gust wind speeds (10-year return period) ranging from 27 to 37 m/s. The basic wind velocities (10-minute mean, 50-year annual recurrence interval) for Romania in the National Annex to the Eurocode, EN-1991-1-4, range from 27 to 35 m/s (Kray and Paul, 2015).

Extreme wind classifications: II, III

D.2.111 Russia

Russia has a vast land area, with a range of extreme wind climates. The Russian loading code SniP 2.01.07.85 specified 8 zones for design wind pressures. The specified values were 5-year return period pressures with a 10-minute averaging time, ranging from 240 Pa for the central part of the country to 1200 Pa on the coastal part of the Far East, and the islands of the Barents Sea (Popov, 2001).

Extreme wind classifications: II, III, IV, V

D.2.112 Samoa

Samoa in the South Pacific (latitude 13°–14° S) is affected by tropical cyclones (Holmes and Weller, 2002). Cyclones 'Ofa' (1990) and 'Val' (1991) did

significant damage. The U.S. Department of Defense (2010) recommended a 50-year gust wind speed of 67 m/s for the capital, Apia. ASCE 7–16 specifies a 3-second gust of 300-year MRI of 67 m/s for American Samoa.
Extreme wind classification: IV

D.2.113 Saudi Arabia

The national oil company has an Engineering Standard (Saudi Aramco, 2005) containing meteorological and seismic design data intended for use with the American Loading Standard ASCE-7; this contains 50-year return period gust data for many locations in the country. These range from 41 to 51 m/s, with a value of 46 m/s for the capital, Riyadh.
Extreme wind classifications: II, III

D.2.114 Senegal

For Dakar, the capital of Senegal on the coast of west Africa, the International Federation of Red Cross and Red Crescent Societies gives a 3-second gust speed, based on a 50-year return period, of 42 m/s. However, a higher value of 50 m/s is specified for St. Louis.
Extreme wind classifications: II – Dakar; III – St. Louis

D.2.115 Singapore

Like Malaysia, Singapore in the Equatorial zone, does not experience typhoons, and has very low extreme winds from weak thunderstorms and monsoonal winds (Choi, 1999). Good quality corrected monthly maximum extreme gust data is available from Tengah and Changi airfields. A Gumbel extreme value analysis for data up to 1997 from these data (unpublished) gives 50-year return period gusts of 33 and 25 m/s, respectively. However, a gust of 40.1 m/s was recorded at Tengah in 1984.
Extreme wind classification: I

D.2.116 Slovakia

The range of basic wind velocities (10-minute mean, 50-year annual recurrence interval) in the National Annex to the Eurocode, EN-1991-1-4, for Slovakia, in eastern Europe, is 24–26 m/s (Kray and Paul, 2015).
Extreme wind classification: II

D.2.117 Slovenia

Slovenia is affected by large-scale systems approaching from north-west Europe; 'bora' winds can occur at the Adriatic coast (near Trieste). The basic wind velocities (10-minute mean, 50-year annual recurrence interval) in the Slovenian National Annex to the Eurocode, EN-1991-1-4, vary between 20 and 30 m/s (Kray and Paul, 2015).

Extreme wind classification: II

D.2.118 Solomon Islands

The Solomon Islands in the South Pacific Ocean are occasionally exposed to developing tropical cyclones, which then move to the south. The Papua -New Guinea loading code specifies a 50-year gust speed of 34 m/s for the capital Honiara.

Extreme wind classification: II

D.2.119 South Africa

Kruger *et al.* (2010, 2012) described the various mechanisms producing strong wind gusts in the Republic of South Africa. Goliger *et al.* (2017) used this data to propose new zonings for design winds in the country. The country is subject to severe thunderstorms on the inland high plains, and synoptic winds in the south, with the sources of the latter being subdivided into troughs, ridges and cold fronts. The 1989 Code of Practice for Loading of the South African Bureau of Standards (SABS 0160-1989) had a map showing design wind speeds for the country (50-year return period, 3-second gust). This map was based on the analysis of annual maximum wind speeds by Milford (1987). The value given for the majority of the country was 40 m/s. This value is specified for the main cities of Johannesburg, Pretoria, Cape Town and Durban. A small zone around Beaufort West had a value of 50 m/s.

The 2010 design standard (SANS 10160-3:2010) is based on 10-minute mean wind speeds to enable compatibility with the Eurocode. The zone map in this document has three zones with mean wind speeds of 28, 32 and 36 m/s, with the high values centred around Beaufort West in the Western Cape Province.

Extreme wind classifications: III – southern part of Northern Cape province; IV – northern part of Western Cape province; II – remainder of RSA

D.2.120 Spain

The basic wind velocities (10-minute mean, 50-year annual recurrence interval) in the Spanish National Annex to the Eurocode, EN-1991-1-4, range from 26 to 29 m/s (Kray and Paul, 2015). There are some downslope wind areas in the Pyrenees.

Extreme wind classification: II

D.2.121 Sri Lanka

The east coast of Sri Lanka is exposed to relatively weak tropical cyclones. For other parts of Sri Lanka, the extreme wind speeds are benign. A building code was prepared by an Australian consulting group in the nineteen-seventies and three design wind zones were specified. These were later used by the Ministry of Local Government, Housing and Construction with 50-year 3-second gust speeds of 49, 42 and 33 m/s for Zones 1, 2 and 3, respectively. Maduranga and Lewangamage (2018) processed recent recorded data from 24 automatic weather stations, and proposed a revised wind map. However, the length of data processed from every station was quite short (Table D.9).

D.2.122 Sweden

The basic wind velocities (10-minute mean, 50-year annual recurrence interval) in the Swedish National Annex to the Eurocode, EN-1991-1-4, vary from 21 to 26 m/s (Kray and Paul, 2015). For Stockholm, the value is 24 m/s.
Extreme wind classification: II

D.2.123 Switzerland

The basic wind velocities (10-minute mean, 50-year annual recurrence interval) in the Swiss National Annex to the Eurocode, EN-1991-1-4, range between 20 and 24 m/s (Kray and Paul, 2015). There are a number of mountain areas where downslope winds occur, and for which higher values of 30 m/s or greater may be applicable.
Extreme wind classifications: II (most of the country);
III (some mountain areas)

D.2.124 Taiwan

On average nearly three typhoons make landfall per year on Taiwan, usually between May and October. The building code gives contours of 10-minute mean speeds between 25 m/s and 45 m/s, with higher values for some outlying islands.
Extreme wind classifications: III, IV, V (west coast and outlying islands)

D.2.125 Tanzania

Lwambuka (1992) processed extreme gust data from several stations in Tanzania in east Africa. Using the Type I (Gumbel) extreme value

Table D.9 Sri Lanka

Zone in building code	Description	Classification
I	50 km from east coast	III
2	Inland strip	II
3	South and west (including Colombo)	I

distributions he found 50-year ARI values of 27, 29, 30 and 38 m/s for Dar es Salaam, Zanzibar, Tanga and Mtwara, respectively.

Extreme wind classifications: I, II

D.2.126 Thailand

Thailand has a particularly mixed wind climate. Most of the country appears to be dominated by extreme winds from thunderstorms and monsoons. However, occasionally typhoons have impacted on southern Thailand, as did Typhoon 'Gay' in 1989, inflicting considerable damage. Post-landfall, typhoons can also affect north-west Thailand.

An analysis of historical gust data for 60 meteorological stations is described by Davenport *et al.* (1995), using Type I (Gumbel) Extreme Value Distributions. There were apparently siting problems for many of the anemometers, and although extreme winds caused by typhoons were separated, those from thunderstorms apparently were not.

The analysis by Davenport *et al.* (1995) proposed two design wind speeds based on 50-year return period 10-minute means of 26.5 and 30 m/s. The latter value applies to small zones on the east and north-west of Thailand. In proposals for a new Thailand loading code (Lukkunaprasit, 1997), these values were converted to nominal mean hourly speeds of 24.9 and 28.2 m/s, respectively.

Extreme wind classification: II

D.2.127 Togo

The extreme wind zones for Togo (Togolese Republic) in west Africa may be assumed to be similar to that of Ghana to the west.

Extreme wind classifications: I – south of 7° N latitude;
II – north of 7° N latitude

D.2.128 Tonga

Tonga, an island group in the South Pacific, suffers regular visits by tropical cyclones (Holmes and Weller, 2002), including 'Isaac' (1982), 'Hina' (1997), 'Cora' (1998), 'Waka' (2001) and 'Gita' (2018) all of which did significant damage. 'Gita', the strongest on record for Tonga with estimated maximum gusts of 77 m/s, destroyed or damaged more than 1,000 houses and the parliament building.

Extreme wind classification: IV

D.2.129 Trinidad and Tobago

An analysis of extreme wind speeds for the Commonwealth Caribbean was carried out by Shellard (1972). These results were used by the Caribbean Uniform Building Code (Caribbean Community Secretariat, 1986) and by the Code of Practice of the Barbados Association of Professional Engineers (1981). The former specified 50-year return period 10-minute mean speeds of 20–28 m/s, and the latter gave a 50-year return period 3-second gust speed of 45 m/s for Trinidad, and 50 m/s for Tobago. The latter values were based on a Frechet (Type II) Extreme value distribution (Section 2.2.1), and may be conservative.

Extreme wind classification: II

D.2.130 Turkey

The extreme wind climate of coastal Turkey can be assumed to be similar to that of Greece. A national wind code exists which is similar to the former German DIN code. The U.S. Department of Defense (2010) recommended a gust of 44 m/s with 50-year ARI for Ankara.

Extreme wind classifications: II – west of 32° E longitude;
III – remainder of Turkey

D.2.131 Uganda

Uganda is bisected by the Equator, and should have an equatorial extreme wind climate dominated by thunderstorms (similar to Kenya); this would be modified by the elevated land of most of the country.

Extreme wind classification: I

D.2.132 Ukraine

Horokhov and Nazim (2001) gave a zoning map of wind velocities for Ukraine. The values have an averaging time of 2-minutes, with a 10-year recurrence interval. Values range between 20 and 45 m/s. The higher values are primarily in mountainous regions or adjacent to the Black or Azof Seas. The basic wind velocities (10-minute mean, 50-year annual recurrence interval) in the Ukrainian National Annex to the Eurocode, EN-1991-1-4, range between 24 and 31 m/s (Kray and Paul, 2015).

Extreme wind classifications: II, III

D.2.133 United Arab Emirates

U.A.E. along the south-eastern tip of the Arabian Peninsula is a federation of seven emirates: Abu Dhabi, Dubai, Sharjah, Ajman, Umm al-Qaiwain, Ras al-Khaimah and Fujairah. Analyses by various wind-tunnel laboratories for building projects in Dubai gives predictions of 50-year gust speed between 35 and 39 m/s. The dominant wind directions for the extreme winds are in the north to west quadrant.

Extreme wind classification: II

D.2.134 United Kingdom

The U.K. has a close network of meteorological stations, and high-quality data. The main strong wind source is severe gales moving in from the Atlantic on the west. Design winds are generally stronger on the west, reducing further east.

Analyses of extreme winds for the U.K. were carried out by Shellard (1958, 1962) and Cook and Prior (1987). The latter work was used for the design wind speed data in the British Standard BS6399:2. A further analysis of hourly extremes for the U.K. was undertaken by Miller *et al.* (2001); however, this was only based on data from 1970 to 1980.

BS6399:2 contained a map of 1-hour mean wind speeds (50-year return period) ranging from 20 to 30 m/s. The latter values occur only for the Shetland Islands in the north. The map also covers the whole of Ireland. The basic wind velocities (10-minute mean, 50-year annual recurrence interval) in the U.K. National Annex to the Eurocode, EN-1991-1-4, range between 22 and 32 m/s with iso-vents at 0.5 m/s intervals (Kray and Paul, 2015).

Extreme wind classifications: II, III

D.2.135 United States

The U.S.A. has a vast array of meteorological stations operated by the U.S. Weather Bureau, and other agencies, such as those involved in defence. Until 1995, the standard extreme wind was the 'fastest mile of wind', calculated from the time taken by a cup anemometer to rotate through one mile. The introduction of automatic weather stations has seen this measure replaced by a peak gust wind speed, which is nominally of 3-seconds (moving-average) duration. There have been many extreme value analyses for the United States, including those by Thom (1960, 1968), Simiu *et al.* (1979), Peterka and Shahid (1998) and Lombardo (2012). The latter study separated thunderstorm wind speeds from non-thunderstorm extremes for the first time.

Outside the coastal regions affected by hurricanes in the continental United States, the 2016 ASCE Loading Standard (ASCE, 2016) contains contours of 3-second wind gust speeds, with mean recurrence intervals of 300 years, varying between 40 and 47 m/s. The Atlantic Ocean and Gulf of Mexico coastlines have iso-vents ranging from 49 to 76 m/s. Alaska has contours from 47 to 67 m/s. Hawaii has 300-year contours between 45 and 89 m/s; however, the higher values are influenced by the large-scale topography of the Hawaiian Islands (Table D.10).

Table D.10 United States

Description	Classification
Non-hurricane mainland	II
Atlantic and Texas coasts, most of Alaska, Hawaii	III
Southern Florida and Louisiana, Alaska coasts, elevated parts of Hawaii	IV

D.2.136 Uruguay

Uruguay in South America lies between the latitudes of 30° and 35° S. The extreme wind climate of Uruguay has been discussed in detail by Duranona (2011, 2013). Extreme winds are produced by extra-tropical cyclones formed over the South Atlantic Ocean, and severe convective downdrafts from thunderstorms, with the latter affecting mainly the inland areas. The national wind code UNIT 50-84 (Instituto Uruguayo de Normas Tecnicas, 1984) gives a 3-second gust speed of 158 km/h (43.9 m/s) for a return period of 10 years; however, this may be an overestimate since it was based on measurements above a 43-m high building in Montevideo (Duranona, 2011, 2013).

Extreme wind classification: II

D.2.137 Vanuatu

Vanuatu is an island group in the South Pacific, often affected by tropical cyclones (Holmes and Weller, 2002), including 'Uma' (1987), 'Dani' (1999), 'Ivy' (2004) and 'Harold' (2020) all of which did significant damage.

Extreme wind classification: IV

D.2.138 Venezuela

Like Colombia, extreme winds in Venezuela are low, and seismic loading is more important for structures. The wind load code (Norma COVENIN 2003-89), which is based on an earlier version of the American Standard ASCE-7, provides a contour map of design wind speeds of 50-year return period; these appear to be 'fastest mile' wind speeds. Values vary between 70 and 100 km/h.

Extreme wind classification: I

D.2.139 Vietnam

Vietnam is influenced by typhoons over most of its coastline, although the influence is weaker on the southern provinces. For design wind speeds, Vietnam is divided into five zones with 20-year return period, and gust speeds ranging from 33 to 55 m/s, in the national loading code TCVN-2737 (values of dynamic pressure are given in the code). The zones of higher wind speeds occur close to the coast and reflect different degrees of influence from typhoons (Table D.11).

D.2.140 Virgin Islands

The US and UK Virgin Islands, in the north east of the Caribbean Sea, are, like Puerto Rico (D.2.107), in the 'front-line' of hurricanes from the

Table D.11 Vietnam

Zone in loading code	Description	Classification
I	Inland north and south	II
II	Inland north and southern coast	II
III	Central and northern coastline	III
IV,V	Offshore islands in north	IV

Atlantic Ocean. In recent years, they have been affected by 'Marilyn in 1995, and 'Irma' and 'Maria' in 2017. The latter two, both occurring in September 2017, caused damage in excess of US$10 billion, in the Virgin Islands alone.

ASCE 7-16 gives a value of 3-second gust with a 300-year MRI of 67 m/s for the US Virgin Islands.

Extreme wind classification: IV

D.2.141 Windward Islands

These islands in the Caribbean are visited by developing hurricanes, and weaker tropical storms. An analysis of extreme wind speeds for the former British colonies in the Caribbean was carried out by Shellard (1972). These results were used by the Caribbean Uniform Building Code (Caribbean Community Secretariat, 1986) and by the Code of Practice of the Barbados Association of Professional Engineers (1981). The latter specified a 50-year return period 3-second gust speed of 58 m/s, based on studies for Barbados. This value was also applicable to St. Vincent, St. Lucia, Grenada and Dominica (Gibbs, 1972). However, a more recent report by Applied Research Associates (Vickery and Wadhera, 2008), showed effective 50-year return period peak 3-second gust speeds of 41 m/s for Barbados and St. Vincent, and 45–47 m/s for St. Lucia and Dominica. The same report gives 38 m/s for Grenada, but this seems too low, given that island was struck by the very damaging Hurricane 'Ivan' in 2004.

Extreme wind classification: III

D.2.142 Zambia

Strong winds in Zambia in central Africa are caused by severe thunderstorms. The specifications for a United Nations-funded warehouse in Lusaka specified a basic (gust) wind speed of 42 m/s.

Extreme wind classification: II

D.2.143 Zimbabwe

Zimbabwe is an elevated land-locked country with most land at 1000 metres above sea level or greater. The country falls between 15° and 22° South in latitude. The expected dominant windstorm in this environment would be thunderstorm winds created by local convection. A code of practice for wind loads (Central African Standards Institution, 1977) specified a basic design wind speed (50 year return period gust) of 35 m/s, for the whole country. An analysis by Lewis (1983) for five different locations found higher and lower values than this.

Extreme wind classifications: I and II

REFERENCES

Abohedma, M.B. and Alshebani, M.M. (2010) Wind load characteristics in Libya. *International Journal of Civil and Environmental Engineering*, 4: 88–91.

Almaawali, S.S.S., Majid, T.A. and Yahya, A.S. (2008) Determination of basic wind speed for building structures in Oman. *Proceedings, International Conference on Construction and Building Technology*, Kuala Lumpur, Malaysia, June 16–20, pp. 235–244.

American Society of Civil Engineers (2016) *Minimum design loads and associated criteria for buildings and other structures*. ASCE/SEI 7-16. ASCE, New York.

Associação Brasileira de Normas Tecnicas (2008) *Forças devidas ao vento em edificações*. NBR 6123. Associação Brasileira de Normas Tecnicas, Rio de Janeiro.

Bajic, A. and Peros B. (2005) Meteorological basis for wind loads calculation in Croatia. *Wind and Structures*, 8: 389–405.

Bakhoum, M.M., Samaan, S. and Shafiek, H.S. (1998) A model for calculating wind pressures for design of structures at different locations in Egypt based on actual records of wind data. *8th International Colloquium on Structural and Geotechnical Engineering*, Cairo, Egypt, 15–17 December 1998.

Ballio, G., Lagomarsino, S., Piccardo, G. and Solari, G. (1999) Probabilistic analysis of Italian extreme winds: reference velocity and return criterion. *Wind and Structures*, 2: 51–68.

Barbados Association of Professional Engineers (1981) *Code of practice for wind loads for structural design*. Consulting Engineers Partnership Ltd. and Caribbean Meteorological Institute, Roseau.

British Standards Institution (1997) *Loading for buildings. Part 2. Code of practice for wind loads*. BS 6399: Part 2: 1997.

Buildings Department (2019) *Code of practice on wind effects in Hong Kong*. Buildings Department, Hong Kong Special Administrative Region.

Bureau of Indian Standards (2015) *Design loads (other than earthquake) for buildings and structures. Part 3. Wind loads*. Indian Standard IS: 875 (Part 3). Bureau of Indian Standards, New Delhi.

Caribbean Community Secretariat (1986) *Caribbean uniform building code. Part 2. Section 2. Wind load*. Caribbean Community Secretariat, Georgetown.

Caspar, W. (1970) Maximale windgeschwindigkeiten in der BRD. *Bautechnik*, 47: 335–340.

Cechet, R.P. and Sanabria, L.A. (2011) Australian extreme wind baseline climate investigation project: intercomparison of time series and coincident Dines and cup anemometer observations, Geoscience Australia, GA Record 2011/23, Australian Government Canberra, Australia.

C.E.N. (European Committee for Standardization) (1994) *Eurocode 1: basis of design and actions on structures. Part 2–4: wind actions (draft).* ENV-1991-2-4. C.E.N., Brussels. *Part 1–4: general actions – wind actions.* EN 1991-1-4. C.E.N., Brussels.

C.E.N. (European Committee for Standardization) (2010) *Eurocode 1: actions on structures. Part 1–4: general actions – wind actions.* EN 1991-1-4:2005+A1. C.E.N., Brussels.

Central African Standards Institution (1977) *Wind loads on buildings.* CASI 160, Part 2. Central African Standards Institution, Harare.

Chen, T.Y. (1975) Comparison of surface winds in Hong Kong. Hong Kong Royal Observatory, Technical Note 41.

Choi, E.C.C. (1999) Extreme wind characteristics over Singapore – an area in the equatorial belt. *Journal of Wind Engineering and Industrial Aerodynamics*, 83: 61–69.

Cook, N.J. and Prior, M.J. (1987) Extreme wind climate of the United Kingdom. *Journal of Wind Engineering and Industrial Aerodynamics*, 26: 371–389.

Davenport, A.G. and Baynes, C.J. (1972) An approach to the mapping of the statistical properties of gradient winds over Canada. *Atmosphere*, 10: 80–92.

Davenport, A.G., Georgiou, P.N., Mikitiuk, M., Surry, D. and Lythe, G. (1984) The wind climate of Hong Kong. *Proceedings, Third International Conference on Tall Buildings*, Hong Kong and Guangzhou, 10–15 December 1984.

Davenport, A.G., Lukkunaprasit, P., Ho, T.C.E., Mikitiuk, M. and Surry, D. (1995) The design of transmission line towers in Thailand. *Proceedings, Ninth International Conference on Wind Engineering*, New Delhi, 9–13 January, pp. 57–68.

Department of Defense (U.S.A.) (2010 and 2013) *Unified facilities criteria.* UFC 3-301-01. Department of Defense, Washington, DC.

Department of Standards and Norms (China) (1994) *Load code for the design of building structures.* (English TranslationGBJ-9-87. New World Press, Beijing.

Dorman C.M.L. (1983) Extreme wind gust speeds in Australia, excluding tropical cyclones. *Civil Engineering Transactions, Institution of Engineers, Australia*, 25: 96–106.

Duranona, V. (2011) Wind impact on Uruguay: vulnerability to extreme winds and estimation of their risk. *Proceedings of the 13th International Conference on Wind Engineering*, Amsterdam, Netherlands, July 10–15.

Duranona, V. (2013) Highest wind gusts in Uruguay: characteristics and associated meteorological events. *Proceedings of the 12th Americas Conference on Wind Engineering*, Seattle, WA, June 16–20.

E.C.C.S. (European Committee for Structural Steelwork) (1978) *Recommendations for the calculation of wind effects on buildings and structures.* Technical Committee T12, E.C.C.S., Brussels.

Faber, S.E. and Bell, G.J. (1967) Typhoons in Hong Kong and building design. Hong Kong Royal Observatory, Reprint No. 37.

Fearon, G. (2014) *Extreme wind speeds for the south-west Indian Ocean using synthetic tropical cyclone tracks.* M.Sc. Thesis, Stellenbosch University, South Africa.

Ghana Standards Authority (2018) *Ghana building code, part 5, structural loads and design.* GSA, Accra.

Gibbs, A. (1972) An introduction to the BAPE Code 'Wind loads for structural design'. *Journal of the Barbados Association of Professional Engineers.*

Goldreich, Y. (2003) *The climate of Israel: observation, research and application.* Kluwer Academic/Plenum Publishers, New York.

Goliger, A.M., Retief, J.V. and Kruger, A.C. (2017) Review of climatic input data for wind load design in accordance with SANS 10160-3. *Journal of the South African Institution of Civil Engineering,* 59: 2–11.

Gomes, L. and Vickery, B.J. (1976a) On thunderstorm wind gusts in Australia. *Civil Engineering Transactions, Institution of Engineers, Australia,* 18: 33–39.

Gomes, L. and Vickery, B.J. (1976b) Tropical cyclone gust speeds along the northern Australian coast. *Civil Engineering Transactions, Institution of Engineers, Australia,* 18: 40–48.

Holmes, J.D. (2002) A re-analysis of recorded wind speeds in Region A. *Australian Journal of Structural Engineering,* 4: 29–40.

Holmes, J.D., Kwok, K.C.S. and Hitchcock, P. (2009) Extreme wind speeds and wind load factors for Hong Kong. *7th Asia-Pacific Conference on Wind Engineering,* Taipei, Taiwan, November 8–12.

Holmes, J.D. and Weller, R. (2002) *Design wind speeds for the Asia-Pacific Region.* Handbook, HB212. Standards Australia, Sydney.

Horokhov Y.V. and Nazim, Y.V. (2001) Wind and sleet loads on the aerial power lines in Ukraine. *Journal of Wind Engineering and Industrial Aerodynamics,* 89: 1409–1419.

Instituto Uruguayo de Normas Tecnicas (1984) Code wind actions on structures. (in Spanish). UNIT 50-84.

Instituto Nacional de Normalización, Chile (1971) Cálculo de la acción del viento sobre las construcciones. NCh432.Of71.

Jeary, A.P. (1997) The wind climate of Hong Kong. *Journal of Wind Engineering and Industrial Aerodynamics,* 72: 433–444.

Jensen, M. and Franck, N. (1970) *The climate of strong winds in Denmark.* Danish Technical Press, Copenhagen.

Kasperski, M. (2002) A new wind zone map of Germany. *Journal of Wind Engineering and Industrial Aerodynamics,* 90: 1271–1287.

Kintenar, R.L. (1971) *An analysis of annual maximum wind speeds in the Philippines.* UNESCO, Manila.

Kral, J. (2007) Ten-minute wind speeds and gusts in the Czech Republic. Journal of Wind Engineering and Industrial Aerodynamics, 95: 1216–1228.

Kray, T. and Paul, J. (2015) Comparative study of effects on peak velocity pressure calculated by thirty-four European wind loading standards. *14th International Conference on Wind Engineering,* Porto Alegre, Brazil, June 21–26.

Kristensen, L., Rathmann, O., and Hansen, S.O. (2000) Extreme winds in Denmark. *Journal of Wind Engineering and Industrial Aerodynamics*, 83: 61–69.

Kruger, A.C., Goliger, A.M., Retief, J.B. and Sekele, S. (2010) Strong wind climatic zones in South Africa. *Wind and Structures*, 87: 147–166.

Kruger, A.C., Goliger, A.M., Retief, J.B. and Sekele, S. (2012) Clustering of extreme winds in the mixed wind climate of South Africa. *Wind and Structures*, 15: 87–109.

Lewis, G. (1983) Probabilistic estimation of extreme climatological parameters over Zimbabwe. *Proceedings, Institution of Engineers, (U.K.)*, 75(Pt. 2): 551–555.

Lombardo, F.T. (2012) Improved windspeed estimation for wind engineering applications. *Journal of Wind Engineering and Industrial Aerodynamics*, 104–106: 278–284.

Loredo-Souza, A.M. (2012) Meteorological events causing extreme winds in Brazil. *Wind and Structures*, 15: 177–188.

Loredo-Souza, A.M. and Paluch, M.J. (2005) Brazil storm Catarina: hurricane or extratropical cyclone? *Proceedings, Tenth Americas Conference on Wind Engineering*, Baton Rouge, LO, May 31–June 4.

Lopez, A. and Vilar, J.I. (1995) Basis of the Mexican wind handbook for the evaluation of the dynamic response of slender structures. *Proceedings of the Ninth International Conference on Wind Engineering*, New Delhi, January 9–13, pp. 1890–1900.

Lukkunaprasit, P. (1997) Seismic and wind loading codes in Thailand. *International Workshop on Harmonization in Performance-Based Building Structural Design*, Tsukuba, Japan, December 1–3.

Lwambuka, L. (1992) Evaluation of wind data records to predict extreme wind speeds in Tanzania, *Uhandsi Journal*, 16: 95–108.

Maduranga, W.L.S and C. S. Lewangamage, (2018) Development of wind loading maps for Sri Lanka for use with different wind loading codes. *Journal of the Institution of Engineers, Sri Lanka*, 51: 47–55.

Melbourne, W.H. (1984) Design wind data for Hong Kong and surrounding coastline. *Proceedings of the Third International Conference on Tall Buildings*, Hong Kong and Guangzhou.

Milford, R.V. (1987) Annual maximum wind speeds for South Africa. *The Civil Engineer in South Africa*, January, 15–19.

Miller, C.A., Cook, N.J. and Barnard, R.H. (2001) Towards a revised base wind speed map for the United Kingdom. *Wind and Structures*, 4: 197–212.

Ministerio del Desarrollo Urbano (Venezuela) (2003) Acciones del viento sobre las construcciones. Norma Venezolana, COVENIN 2003-89.

National Research Council (Canada) (2015) Climatic information for building design in Canada. Appendix C to the National Building Code of Canada. N.R.C. Ottawa.

Okulaja, F.O. (1968) The frequency distribution of Lagos/Ikeja wind gusts. *Journal of Applied Meteorology*, 7: 379–383.

Ong'ayo, E.O., Mwea, S.K. and Abuodha, S.O. (2014) Determination of basic mean hourly wind speeds for structural design in Nairobi County. *International Journal of Engineering Sciences & Emerging Technologies*, 7:631–640.

Peterka, J.A. and Shahid, S. (1998) Design gust wind speeds in the United States. *Journal of Structural Engineering (A.S.C.E.)*, 124: 207–214.

Popov, N.A. (2001) Wind load codification in Russia and some estimates of a gust load accuracy provided by various codes. *Journal of Wind Engineering and Industrial Aerodynamics*, 88: 171–181.

Rajabi, M.R. and Modarres, R. (2008) Extreme value frequency analysis of wind data from Isfahan, Iran. *Journal of Wind Engineering and Industrial Aerodynamics*, 96: 78–82.

Republic of Latvia (2001) *Latvian building code – construction climatology.* LBN-003-01. Ministry of Economics, Riga.

Republic of Lithuania (2003) Poveikiai ir apkrovos (Actions and loads), STR 2.05.04:2003, Vilnius, Lithuania.

Riera, J.D. and Nanni, L.F. (1989) Pilot study of extreme wind velocities in a mixed climate considering wind orientation. *Journal of Wind Engineering and Industrial Aerodynamics*, 32: 11–20.

Riera, J.D. and Reimundin, J.C. (1970) *Sobre la distribucion de velocidades maximas de viento en la Republica Argentina, Simposio sobre Acciones en Estructuras.* National University of Tucuman, Tucuman.

Sacré, C. (2002) Extreme wind speed in France: the '99 storms and their consequences. *Journal of Wind Engineering and Industrial Aerodynamics*, 90: 1163–1171.

Salgado Vieira Filho, J.M. (1975) *Velocidades maximas do vento no Brasil.* Master's Thesis, Federal University of Rio Grande do Sul, Brazil.

Saaroni, H., Ziv, B., Bitan, A., and Alpert, P. (1998) Easterly wind storms over Israel. *Theoretical and Applied Climatology*, 59: 61–77.

Saudi Aramco (2005) *Meteorological and seismic data.* Engineering Standard SAES-A-112. Environmental Protection Department, Saudi Aramco, Dhahran.

Schueller, G.I. and Panggabean, H. (1976) Probabilistic determination of design wind velocity in Germany. *Proceedings, Institution of Engineers, (U.K.)*, 61(Pt. 2): 673–683.

de Schwarzkopf, M.L.A. (1995) Meteorological weather patterns and wind storm types. Course notes. 'Design Loadings on Transmission Lines', Brisbane, July 5–7.

Shellard, H.C. (1958) Extreme wind speeds over Great Britain and Northern Ireland. *Meteorological Magazine*, 87: 257–265.

Shellard, H.C. (1962) Extreme wind speeds over the United Kingdom for periods ending 1959. *Meteorological Magazine*, 91: 39–47.

Shellard, H.C. (1972) Extreme wind speeds in the Commonwealth Caribbean. *Journal of the Barbados Association of Professional Engineers*, December 1–8.

Simiu, E. and Marshall, R.D. (1977) *Building to resist the effect of wind. Volume 2. Estimation of extreme wind speeds and guide to the determination of wind forces.* Building Science Series 100. National Bureau of Standards, U.S. Department of Commerce, Washington, DC.

Sites, J.S. and Peterson, R.E. (1995) Climatology of Southwest Indian Ocean tropical cyclones – (1962–1987). *Proceedings of the Ninth International Conference on Wind Engineering*, New Delhi, 9–13 January.

Simiu, E., Changery, M.J. and Filliben, J.J. (1979) *Extreme wind speeds at 129 stations in the contiguous United States.* NBS Building Science Series 118. National Bureau of Standards, Washington, DC.

Soboyejo, A.B.O. (1971) Distribution of extreme winds in Nigeria. *The Nigerian Engineer*, 7: 21–34.

South African Bureau of Standards (1989) *The general procedures and loadings to be adopted in the design of buildings.* South African Standard SABS 0160-1989.

Standards Australia/Standards New Zealand (2020) *Structural design actions. Part 2: wind actions.* Standards Australia, Sydney, draft Australian/New Zealand Standard DR-AS/NZS1170.2:2011.

Thom, H. (1960) Distributions of extreme winds in the United States. *ASCE Journal of the Structural Division*, 85: 11–24.

Thom, H. (1968) New distributions of extreme winds in the United States. *ASCE Journal of the Structural Division*, 94: 787–1801.

Turner, R., Revell, M., Reese, S., Moore, S. and Reid, S. (2012) Some recent extreme wind events in New Zealand. *Wind and Structures*, 15: 163–176.

Vaidogas, E.R. and Juocevicius, V. (2011) A critical estimation of data on extreme winds in Lithuania. *Journal of Environmental Engineering and Landscape Management*, 19: 178–188.

Vallis, M.B., Loredo-Souza, A.M., Watrin, L.C. and Dolz-Bênia, M.C. (2018) Extreme winds east of the Andes. *38th South American Conference on Structural Engineering*, Lima, Peru, 24–26 October 2018.

Vallis, M.B. (2019) Modelo climatico para ventos extremos do Brasil. Doctorate thesis, Universidade Federal do Rio Grande do Sul, Porto Alegre.

Vickery, P.J. and Wadhera, D. (2008) *Wind speed maps for the Caribbean for application with the wind load provisions of ASCE 7.* ARA Report 18108-1. Applied Research Associates, Raleigh, NC, prepared for the Pan American Health Organization.

Vilar, J.I. et al. (1991) *Analisis estadistico de datos de vientos maximos.* Reporte Interno 42/2929/I/02/P. Departamento de Ingenieria Civil. Instituto de Investigaciones Electricas, Mexico.

Viollaz, A., Riera, J.D. and Reimundin, J.C. (1975) *Estudio de la distribucion de velocidades maximas de viento en la Republica Argentina.* Informe I-75-1, Structures Laboratory. National University of Tucuman, Tucuman.

Wang, C-H., Wang, X. and Khoo, Y.B. (2013) Extreme wind gust hazard in Australia and its sensitivity to climate change. *Natural Hazards*, 67: 549–567.

Whittingham, H.E. (1964) *Extreme wind gusts in Australia.* Bulletin No. 46. Bureau of Meteorology, Melbourne.

Appendix E: Some approximate formulas for structural natural frequencies

A necessary pre-requisite for dynamic response estimation is knowledge of the natural frequencies in the lowest sway modes of the structure. It is also useful to know these values to determine whether or not dynamic response calculations to wind are, in fact, necessary.

Most modern frame-analysis or finite element computer programs will of course give this information. However, if the structure is still in the early design stage, application of simple empirical formulae may be useful. Some of these are given here.

- for multistorey buildings that are nearly uniform in cross section (Jeary and Ellis, 1983):

$$n_1 \approx 46 / h \tag{E.1}$$

where h is the height of the building in metres. There are many similar formulae to (E.1) with 'constants' different to 46, and there are equivalent formulae based on number of floors instead of building height.

An alternative approximate formula for tall buildings is by Hirsch and Ruscheweyh, (1971):

$$n_1 \approx 0.4 \left(100 / h\right)^{1.58} \tag{E.2}$$

- for cantilevered masts or poles of uniform cross-section (in which bending action dominates):

$$n_1 = \frac{0.56}{h^2} \sqrt{\frac{EI}{m}} \tag{E.3}$$

where
EI is the bending stiffness of the section,
m is the mass/unit height.

This is an exact formula for uniform masts or towers; it can be used for those with a slight taper, with average values of EI and m.

- an approximate formula for cantilevered, *tapered*, circular poles (European Convention for Structural Steelwork, 1978):

$$n_1 \approx \frac{\lambda}{2\pi h^2}\sqrt{\frac{EI}{m}} \qquad (E.4)$$

where h is the height, and E, I and m are calculated for the cross-section *at the base.*

λ depends on the wall thicknesses at the tip and base, e_t and e_b, and external diameter at the tip and base, d_t and d_b, according to the following formula:

$$\lambda = \left[1.9\exp\left(\frac{-4d_t}{d_b}\right)\right] + \left[\frac{6.65}{0.9+\left(\dfrac{e_t}{e_b}\right)^{0.666}}\right] \qquad (E.5)$$

Note that for $(d_t / d_b) = (e_t / e_b) = 1.0$, i.e. a uniform cylindrical tube, $\lambda = 3.52$, and Equation (E.3) results.

- For steel, tubular chimneys, Hirsch and Bachmann (1997a) give the following approximate formula:

$$n_1 = \xi \cdot 1010 \cdot \left(\frac{b}{h}\right)^2 \qquad (E.6)$$

where b is the external diameter in metres, h is the height in metres, and ξ is an empirical factor, approximately equal to 1.15 times (mass of steel part/total mass of the chimney, including insulation, etc.)$^{1/2}$.

- for free-standing lattice towers (without added ancillaries such as antennas, lighting frames, etc.) (Standards Australia, 1994), the following formula may be used:

$$n_1 \approx 1500\, w_a/h^2 \qquad (E.7)$$

where
w_a is the average width of the structure in metres,
h is tower height.

An alternative formula for lattice towers (with added ancillaries) is (Wyatt, 1984):

$$n_1 \approx \left(\frac{L_N}{h}\right)^{2/3}\left(\frac{w_b}{h}\right)^{1/2} \qquad (E.8)$$

where
w_b=tower base width
L_N=270 m for square base towers,
or 230 m for triangular base towers.

- A formula which seems to fit data on bridges with spans between 20 and 1,000 m (Jeary, 1997) is:

$$n_1 \approx 40\left(L_s\right)^{-3/4} \qquad \text{(E.9)}$$

where L_s is the span in metres (main span in the case of a multi-span structure).

An alternative formula for cable-stayed bridges is (Hirsch and Bachmann, 1997b):

$$n_1 \cong 110 \,/\, L_s \qquad \text{(E.10)}$$

where L_s is the length of the main span in metres.

REFERENCES

ECCS (European Convention for Structural Steelwork) (1978) *Recommendations for the calculation of wind effects on buildings and structures*. Technical Committee T12. ECCS, Brussels.

Hirsch, H. and Bachmann, H. (1997a) *Wind-induced vibrations: chimneys and masts, in vibration problems in structures*, 2nd Edition. Birkhäuser, Verlag, Basel.

Hirsch, H. and Bachmann, H. (1997b) *Wind-induced vibrations: suspension and cable-stayed bridges, in vibration problems in structures*, 2nd Edition. Birkhäuser, Verlag, Basel.

Hirsch, G. and Ruscheweyh, H. (1971) Newer investigations of non-steady wind loadings and the dynamic response of tall buildings and other constructions. *Proceedings of the 3rd International Conference on Wind Effects on Buildings and Structures*, Saikon Co. Ltd, Tokyo, Japan, 6–9 September, pp. 811–823.

Jeary, A.P. (1997) *Designer's guide to the dynamic response of structures*. E.F. and N. Spon, London.

Jeary, A.P. and Ellis, B.R. (1983) On predicting the response of tall buildings to wind excitation. *Journal of Wind Engineering and Industrial Aerodynamics*, 13: 173–182.

Standards Australia (1994) *Design of steel lattice towers and masts*. Australian Standard AS3995-1994. Standards Australia, North Sydney.

Wyatt, T.A. (1984) Sensitivity of lattice towers to fatigue induced by wind gusts. *Engineering Structures*, 6: 262–267.

Appendix F: Example of application of the LRC method for the effective static wind load on a simple structure

F.I INTRODUCTION

In this appendix, the LRC formula of Kasperski (1992) is applied to a simple structure – a pitched free roof – to illustrate the method of determining the effective static wind pressures. Data was obtained from wind tunnel tests carried out by Ginger and Letchford (1991). The calculations make use of Equations (5.34), (5.36) and (5.37) in Chapter 5, but because there are only two panels, the calculations are easily carried out by hand.

F.2 WIND PRESSURE DATA

A model of a pitched free roof (i.e. no walls), with a roof pitch of 22.5°, at a geometric scaling ratio of 1/100, was tested in a boundary-layer wind tunnel by Ginger and Letchford (1991). Net area-averaged pressures across the windward and leeward roof slopes were measured. Three panels per roof-half were used, but the data used here applies to the central panels, that is the central third of the roof.

Figure F.1b shows the mean and standard deviation pressure coefficients for a wind direction normal to the ridge as shown; the latter values are in brackets. Maximum and minimum panel pressure coefficients were also recorded, and are shown in Figure F.1(c). The directions for positive net panel pressures are shown on the figure.

F.3 EFFECTIVE STATIC LOADS FOR TOTAL LIFT AND DRAG

At first, one might assume that the maximum total lift force should be obtained from the two recorded minimum pressures on the two roof panels. Similarly, the maximum drag could be obtained from the maximum

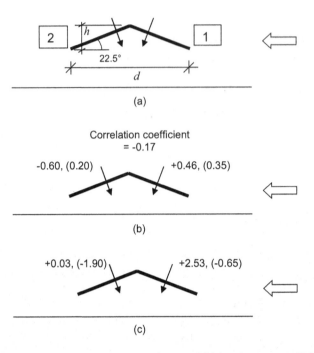

Figure F.1 Pressure coefficients for a pitched free roof. (a) Roof geometry. (b) Mean (standard deviation) pressure coefficients and correlation coefficient. (c) Maximum and (minimum) pressure coefficients.

on panel 1 and the minimum on panel 2. However, this would be incorrect, and conservative, as these values do not occur simultaneously. The *expected* pressure coefficients coinciding with the maximum and minimum lift and drag are derived in the following section.

F.3.1 Mean lift and drag

The mean lift force (positive upwards) is obtained as follows:

$$\bar{L} = (-1)\,(0.46)q_h(d/2)+(-1)(-0.60)q_h(d/2)=0.14q_h(d/2)$$

where

q_h is the reference mean dynamic pressure at roof height $\left(=\dfrac{1}{2}\rho_a\bar{U}_h^2\right)$,
d is the along-wind length of the roof.

In this case, the *influence coefficients* for the *lift force* are both equal to $-(d/2)$.

The mean drag force is given by:

$$\bar{D} = (+1)(0.46)q_h(h) + (-1)(-0.60)q_h(h) = 1.06q_h(h) = 0.44q_h(d/2)$$

since, $h/(d/2) = \tan 22.5° = 0.414$

The *influence coefficients* for the *drag force* are equal to $+h = (d/2)\tan 22.5°$ for panel 1, and $-h = -(d/2)\tan 22.5°$, for panel 2.

F.3.2 Standard deviations of lift and drag

The r.m.s. fluctuating, or standard deviation, lift and drag forces can be obtained by covariance integration (Holmes and Best, 1981; Ginger and Letchford, 1991, 1994).

The standard deviation of the lift force, σ_L, is obtained as follows:

$$\underline{\sigma}_L = q_h(d/2)\left[(0.35)^2 + (0.20)^2 + 2(-0.17)(0.35)(0.20)\right]^{1/2}$$

$$= \underline{0.372\ q_h(d/2)}$$

The standard deviation of the drag force, σ_D, is:

$$\sigma_D = q_h(d/2)\tan 22.5°\left[(0.35)^2 + (0.20)^2 - 2(-0.17)(0.35)(0.20)\right]^{1/2}$$

$$= 0.432q_h(d/2)\tan 22.5° = \underline{0.179q_h(d/2)}$$

F.3.3 Effective pressures for peak lift force

The expected pressure on panel 1 when the *lift* is a maximum is given by (Kasperski, 1992):

$$(p_1)_{\hat{L}} = q_h[\bar{C}_{p1} + g\rho_{p1,L}\sigma_{Cp1}]$$

where g is a peak factor for the lift (it will be taken as 4), and $\rho_{p1,L}$ is the correlation coefficient between the pressure $p_1(t)$ and the lift $L(t)$. The covariance between the pressure $p_1(t)$ and the lift $L(t)$ is given by:

$$-q_h^2(d/2)\left[\sigma_{Cp1}^2 + \overline{p_1'p_2'}\right] = -q_h^2(d/2)\left[(0.35)^2 + (-0.17)(0.35)(0.20)\right]$$

$$= -(0.111)q_h^2(d/2)$$

Then,

$$\rho_{p1,L} = \frac{-0.111}{(0.35)(0.372)} = -0.853$$

Hence,

$$(p_1)_{\hat{L}} = q_h \left[\bar{C}_{p1} + g\rho_{p1,L}\sigma_{Cp1} \right] = q_h \left[(0.46) + 4(-0.853)(0.35) \right] = -0.73q_h$$

Similarly, the covariance between the pressure $p_2(t)$ and the lift $L(t)$ is given by:

$$-q_h^2(d/2)\left[\sigma_{Cp2}^2 + \overline{p_1'p_2'} \right] = -q_h^2(d/2)\left[(0.20)^2 + (-0.17)(0.35)(0.20) \right]$$

$$= -(0.028)q_h^2(d/2)$$

Then,

$$\rho_{p2,L} = \frac{-0.028}{(0.20)(0.372)} = -0.376$$

Hence,

$$(p_2)_{\hat{L}} = q_h \left[\bar{C}_{p2} + g\rho_{p2,L}\sigma_{Cp2} \right] = q_h \left[(-0.60) + 4(-0.376)(0.20) \right]$$

$$= -0.90q_h$$

Thus, the expected pressure coefficients corresponding to the maximum lift (acting upwards) are:

$$(C_{p1})_{\hat{L}} = -0.73 \quad (C_{p2})_{\hat{L}} = -0.90$$

The pressures corresponding to the *minimum* lift force (downwards) are also of interest.

In this case,

$$(p_1)_{\hat{L}} = q_h \left[\bar{C}_{p1} - g\rho_{p1,L}\sigma_{Cp1} \right] = q_h \left[(0.46) - 4(-0.853)(0.35) \right] = +1.65q_h$$

and,

$$(p_2)_{\hat{L}} = q_h \left[\bar{C}_{p2} - g\rho_{p2,L}\sigma_{Cp2} \right] = q_h \left[(-0.60) - 4(-0.376)(0.20) \right] = -0.30q_h$$

Hence,

$$(C_{p1})_{\hat{L}} = +1.65 \quad (C_{p2})_{\hat{L}} = -0.30$$

These pressure coefficients are shown in Figure F.2a and b.

F.3.3 Effective pressures for maximum drag force

The expected pressures for the maximum *drag* force can be determined in a similar way as the lift force, as follows.

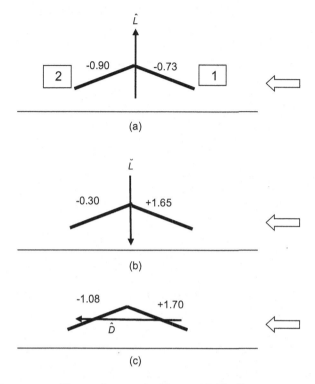

Figure F.2 Pressure coefficients for a pitched free roof. (a) Effective pressures for maximum lift force. (b) Effective pressures for minimum lift force. (c) Effective pressures for maximum drag force.

The covariance between the pressure $p_1(t)$ and the drag $D(t)$ is given by:

$$q_h^2\left(d/2\right)\tan 22.5°\left[\sigma_{Cp1}^2 - \overline{p_1'p_2'}\right] = q_h^2\left(d/2\right)\tan 22.5°$$
$$\times\left[(0.35)^2 - (-0.17)(0.35)(0.20)\right]$$
$$= (0.134)q_h^2\left(d/2\right)\tan 22.5°$$

Then,

$$\rho_{p1,D} = \frac{0.134}{(0.35)(0.432)} = 0.886$$

Hence,

$$\left(p_1\right)_{\hat{D}} = q_h\left[\overline{C}_{p1} + g\rho_{p1,D}\sigma_{Cp1}\right] = q_h\left[(0.46) + 4(0.886)(0.35)\right] = \underline{1.70q_h}$$

(again taking a peak factor of 4)

Similarly, the covariance between the pressure $p_2(t)$ and the drag $D(t)$ is given by:

$$-q_h^2(d/2)\tan 22.5°\left[\sigma_{Cp2}^2 - \overline{p_1'p_2'}\right] = -q_h^2(d/2)\tan 22.5°$$

$$\times\left[(0.20)^2 - (-0.17)(0.35)(0.20)\right]$$

$$= -(0.052)q_h^2(d/2)$$

Then,

$$\rho_{p2,D} = \frac{-0.052}{(0.20)(0.432)} = -0.602$$

Hence,

$$(p_2)_{\hat{D}} = q_h\left[\overline{C}_{p2} + g\rho_{p2,D}\sigma_{Cp2}\right] = q_h\left[(-0.60) + 4(-0.602)(0.20)\right] = \underline{-1.08q_h}$$

Thus, the expected pressure coefficients corresponding to the maximum drag are:

$$(C_{p1})_{\hat{D}} = +1.70 \quad (C_{p2})_{\hat{D}} = -1.08$$

These pressure coefficients are shown in Figure F.2c.

F.4 DISCUSSION

The effective pressure coefficients for maximum and minimum lift, and maximum drag, as summarized in Figure F2, are clearly quite different to each other, and indicate the difficulty in specifying a single set of pressure coefficients in a code or standard, for a structure such as this.

It can be checked that the values obtained in the previous section will, in fact, give the correct values of the peak load effects. For example, the maximum lift can be obtained in two ways as follows.

From the effective static pressure coefficients:

$$\hat{L} = (-1)(-0.73)q_h(d/2) + (-1)(-0.90)q_h(d/2) = 1.63q_h(d/2)$$

Directly from the mean and standard deviation:

$$\hat{L} = \overline{L} + 4\sigma_L = 0.14q_h(d/2) + 4\times0.372q_h(d/2) = 1.63q_h(d/2)$$

The effective static pressure coefficients for each panel should lie between the limits set by the maximum and minimum pressure coefficients for each panel. This is the case here (see Figures F.1 and F.2), except that the value on panel 1 for $\hat{L}$, -0.73, is slightly more negative that the measured minimum value of -0.65. This could result from a sampling error in the measured peak, or the choice of a slightly conservative peak factor of 4 for the lift force.

F.5 CONCLUSIONS

This example has explained, using a simple 2-panel case, the LRC methodology for determining the expected pressure distributions corresponding to peak load effects, based on correlations. More complex cases, such as large roofs, require a large number of panels, and a matrix of correlation coefficients, but the principles of the calculation remain the same.

REFERENCES

Ginger, J.D. and Letchford, C.W. (1991) *Wind loads on canopy roofs*. University of Queensland, Department of Civil Engineering, Research Report, CE132, June.

Ginger, J.D. and Letchford, C.W. (1994) Wind loads on planar canopy roofs – part 2: fluctuating pressure distributions and correlations. *Journal of Wind Engineering and Industrial Aerodynamics*, 51: 353–370.

Holmes, J.D., and Best, R.J. (1981) An approach to the determination of wind load effects on low-rise buildings. *Journal of Wind Engineering and Industrial Aerodynamics*, 7: 273–287.

Kasperski, M. (1992) Extreme wind load distributions for linear and nonlinear design. *Engineering Structures*, 14: 27–34.

Appendix G: Fitting of the generalized Pareto distribution to peaks-over-threshold data

G.1 INTRODUCTION

In this appendix, several methods of fitting peaks-over-threshold wind-speed data to the Generalized Pareto Distribution with a fixed shape factor are discussed with examples of their application.

G.2 SIMPLE METHOD BASED ON A SINGLE THRESHOLD

An equation which gives the extreme wind speed for a given average recurrence interval, R_I, can be derived as follows.

The average crossing rate of the windspeed U_R can be estimated by: average crossing rate of $U_0 \times$ the probability of the excess above U_0 exceeding $(U_R - U_0)$

i.e. $\left(\dfrac{1}{R_I}\right) = \lambda \times \left[1 - \dfrac{k(U_R - U_0)}{\sigma}\right]^{\frac{1}{k}}$ (G.1)

where the GPD, with positive shape factor, k, (>0) and scale factor (standard deviation), σ, has been assumed for the probability distribution of excesses of wind speed above V_0.

Re-arranging Equation (G.1), $\dfrac{k(U_R - U_0)}{\sigma} = 1 - \left(\dfrac{1}{\lambda R_I}\right)^k$

$U_R = U_0 + \dfrac{\sigma}{k}\left[1 - (\lambda R_I)^{-k}\right]$ (G.2)

For the GPD, the mean value is equal to $\left(\dfrac{\sigma}{1+k}\right)$ (Appendix C – Equation C.27).

The estimate of the mean value is E, the calculated average of all positive excesses above the threshold.

Then substituting for σ in Equation (G.2) gives Equation (G.3) as follows:

$$U_R = U_0 + \left(\frac{1+k}{k} \right) E \left[1 - (\lambda R_I)^{-k} \right] \tag{G.3}$$

Equations (G.2) and (G.3) can be written in the general form:

$$U_R = C - D \left(R_I \right)^{-k} \tag{G.4}$$

$$\text{where } C = U_0 + \frac{\sigma}{k} = U_0 + \left(\frac{1+k}{k} \right) E \tag{G.5}$$

$$\text{and } D = \frac{\sigma}{k} (\lambda)^{-k} = \left(\frac{1+k}{k} \right) E(\lambda)^{-k} \tag{G.6}$$

Equation (G.4) is the simplified form of the Generalized Extreme Value Distribution (GEV) discussed in Chapter 2, Section 2.2.2. – i.e. Equation (2.6). This illustrates that the PoT/GPD approach is, in fact, a method of fitting the GEV to extremes.

G.3 MEAN EXCEEDANCE PLOTS

The method described above is based on fitting the GPD to excesses over a *single* high threshold, U_0. An 'optimum' fit, based on several thresholds, can be obtained by making use of the observation, (e.g. Davison and Smith, 1990; Coles, 2001), that the GPD, with the same parameters (k and σ), applies to excesses above any threshold, if it is high enough, (see also Holmes and Moriarty, 1999).

Then, the mean excess, E_i, above any high threshold, U_i greater than U_0, is given by:

$$E_i = \frac{\sigma - k \left(U_i - U_0 \right)}{1 + k} \tag{G.7}$$

The plot of E_i against $(U_i - U_0)$ is known as a 'mean exceedance plot' and should be a straight line, if Equation (G.7) holds.

When the shape factor, k, is fixed, it can be shown that an optimum (least-squares) fit of the straight line with known slope, $-k/(1+k)$, has an intercept on the $(U_i - U_0)$ axis of $\hat{\sigma}/(1+k)$, where $\hat{\sigma}$ is given by:

$$\hat{\sigma} = \overline{E_i} \left(1 + k \right) + k \cdot \overline{\left(U_i - U_0 \right)} \tag{G.8}$$

where

$$\overline{E_i} = \frac{1}{N} \sum_{i=1}^{N} E_i,$$

the average of the mean excesses over all N thresholds,

$$\overline{(U_i - U_0)} = \frac{1}{N} \sum_{i=1}^{N} (U_i - U_0),$$

i.e. the average of the increments of the N thresholds, $U_{i,}$ above the lowest, U_0.

Then Equation (G.2), with σ replaced by $\hat{\sigma}$, can be used for predictions of U_R. Note that λ remains the rate of crossing of the lowest threshold, U_0, in this approach.

G.4 LEAST SQUARES FITTING TO RAW P.O.T. DATA

Another fitting approach is to fit Equation (G.4) to the 'raw' data – i.e. to all wind speeds above the threshold, U_0.

Equation (G.4) can be re-written in the form: $y = C + Dz$, where $y = U_R$ and $z = -\left(R_I^{-k}\right)$.

Then standard linear regression techniques can be used to determine the intercept, C and slope, D.

Unbiased estimates of the average recurrence intervals for the recorded data can be estimated by ordering, from smallest to largest, $(j=1, N)$, the recorded values of gust speed, U_j, above the threshold, U_0, determining an empirical probability of exceedance, $1 - \left[j/(N+1) \right]$, and then dividing the reciprocal by the average rate, λ, of excesses above the threshold, U_0.

$$R_{U,j} \cong \frac{1}{\lambda \left(1 - \dfrac{j}{N+1} \right)} \tag{G.9}$$

G.5 MAXIMUM LIKELIHOOD APPROACH

The maximum likelihood method, regarded as the most rigorous method of fitting probability distributions to a set of measured data, is a preferred method for statisticians. The method involves maximizing a likelihood function, or its natural logarithm. The likelihood function is a measure of the joint probability of the individual data points being estimated by the chosen distribution; maximizing it enables optimum values of the parameter(s) of the distribution to be determined.

In the case of the GPD, the log-likelihood function is (Hosking and Wallis, 1987):

$$\log L\left(U; \sigma, k\right) = -N \log + \left(\frac{1}{k}-1\right)\sum_{j=1}^{N}\log\left[1-\frac{k\left(U_j-U_0\right)}{\sigma}\right] \qquad (G.10)$$

The summation in Equation (G.10) is over the N data points representing the individual values of peaks-over-threshold (U_j-U_0). The maximum value of $\log L$ corresponding to an optimum value of shape factor, $\hat{\sigma}$, can be determined numerically.

Alternatively, the derivative of $\log L$ with respect to σ can be set to zero:

$$\text{i.e.} \quad \frac{d\left(\log L\right)}{d\sigma} = -\frac{N}{\sigma} + \left(1-k\right)\sum_{j=1}^{N}\frac{\left(U_j-U_0\right)}{\sigma^2\left[1-\dfrac{k\left(U_j-U_0\right)}{\sigma}\right]} = 0 \qquad (G.11)$$

$$\text{for maximum } \log L, \quad N = \left(1-k\right)\sum_{j=1}^{N}\frac{\left(U_j-U_0\right)}{\left[\hat{\sigma}-k\left(U_j-U_0\right)\right]} \qquad (G.12)$$

An iterative solution is required to solve Equation (G.12) for the optimum value, $\hat{\sigma}$.

G.6 EXAMPLE

The following example illustrates the application and results of the four methods described in the previous sections, applied to extreme wind gust data.

Automatic weather station gust data from non-synoptic storms (thunderstorms) is available from Woomera, South Australia, for the years from 1991 to 2017. After correction for terrain and gust duration, 23 non-synoptic gusts above 25 m/s (U_0) in the 26 years were identified, giving an annual rate (λ) of 23/26=0.885 per annum. The average excess above 25 m/s is 4.496 m/s. The rate of crossings and average excesses for other thresholds are tabulated in Table 2.5.

First method (Section G.2)

Applying the first method (Section G.2), based on the average crossing rate and excess above U_0 of 25 m/s, and assuming a shape factor, k, of 0.1:

$$\text{Equation (G.3) gives } U_R = 25 + \left(\frac{1+0.1}{0.1}\right)4.496\left[1-\left(0.885R_I\right)^{-0.1}\right].$$

This gives $U_R = 74.45 - 50.06\left(R_I\right)^{-0.1}$ from which U_1=24.4 m/s and U_{500}=47.6 m/s

A plot of U_R versus average recurrence interval R_I, using the method of Section 2.4.1, is given in Figure G.1.

Alternative predictions can be obtained by varying the threshold from 25 to 26, 27, 28, or 29 m/s. Then use of values from Table G.1 gives the values of C, D, U_1 and U_{500} shown in Table G.2.

The values of C and D have been averaged in the right-hand column, giving 'average' predictions of U_1 and U_{500}.

Second method (Section G.3)
Applying the second approach,

$$\overline{E_i} = \tfrac{1}{5}\left(4.496 + 3.871 + 3.676 + 3.608 + 3.182\right) = 3.767 \text{ m/s}$$

$$\overline{(U_i - U_0)} = \tfrac{1}{5}\left(0 + 1 + 2 + 3 + 4\right) = 2.0$$

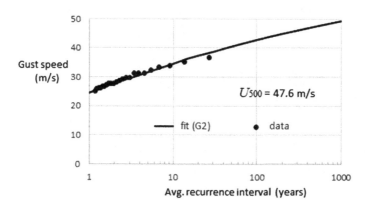

Figure G.I Fitted line using the method of Section G.2 applied to peaks-over-threshold data (U_0 =25 m/s).

Table G.I Peaks-over threshold statistics for Woomera, South Australia

Threshold (U_0) m/s	25	26	27	28	29
Number of crossings	23	21	17	13	11
Rate (per annum)	0.885	0.808	0.654	0.500	0.423
Average excess (m/s)	4.496	3.871	3.676	3.608	3.182

Table G.2 Predictions for various thresholds

Threshold (U_0) m/s	25	26	27	28	29	Average
C (m/s)	74.5	68.6	67.4	67.7	64.0	68.4
D (m/s)	50.1	43.5	42.2	42.5	38.1	43.3
U_1 (m/s)	24.4	25.1	25.2	25.2	25.9	25.1
U_{500} (m/s)	47.6	45.2	44.8	44.8	43.5	45.1

Hence, from Equation (G.8), $\hat{\sigma} = 3.767\,(1.1)+0.1(2.0) = 4.343$.

Then, $C = 25 + \dfrac{4.343}{0.1} = 68.43$ m/s, and $D = \dfrac{4.343}{0.1}(0.885)^{-0.1} = 43.97$ m/s.

Hence, $U_R = 68.43 - 43.97\,(R_I)^{-0.1}$ from which $U_1 = 24.5$ m/s and $U_{500} = 44.8$ m/s.

The mean exceedance plot, with the line fitted by the least-squares method, with a (forced) slope of $-k/(1+k) = -(1/11)$ is shown in Figure G.2.

The fit to the original data and predictions for higher values of R_I are shown in Figure G.3.

Third method (Section G4)

Table G.3 shows the tabulated original data for the application of Method 3. Note that the values Column 4 (i.e. the estimates of R_I) are obtained by taking

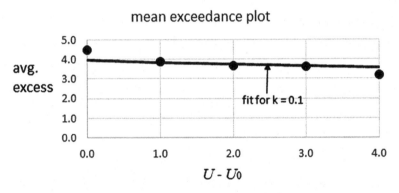

mean exceedance plot

Figure G.2 Mean exceedance plot with fitted line using the method of Section G.3.

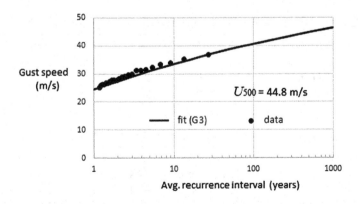

Figure G.3 Fitted line using the method of Section G3 (least squares fit to mean exceedance plot) applied to peaks-over-threshold data.

Table G.3 Tabulated data for method 3

U_R (m/s)	rank, j	$1 - j/(N+1)$	R_I (years)	z
25.2	1	0.958	1.18	−0.9836
25.9	2	0.917	1.23	−0.9793
26.2	3	0.875	1.29	−0.9747
26.2	4	0.833	1.36	−0.9700
26.7	5	0.792	1.43	−0.9650
26.7	6	0.750	1.51	−0.9598
27.3	7	0.708	1.60	−0.9543
27.7	8	0.667	1.70	−0.9486
27.8	9	0.625	1.81	−0.9425
27.8	10	0.583	1.94	−0.9360
28.2	11	0.542	2.09	−0.9291
28.7	12	0.500	2.26	−0.9217
29.2	13	0.458	2.47	−0.9137
29.7	14	0.417	2.71	−0.9050
29.8	15	0.375	3.01	−0.8955
31.3	16	0.333	3.39	−0.8850
31.3	17	0.292	3.88	−0.8733
31.4	18	0.250	4.52	−0.8599
32.3	19	0.208	5.43	−0.8444
33.3	20	0.167	6.78	−0.8258
33.8	21	0.125	9.04	−0.8024
35.1	22	0.083	13.57	−0.7705
36.8	23	0.042	27.13	−0.7189

the reciprocals of Column 3, then dividing by λ (i.e. 0.885). As previously noted, z in Column 5 is $- \left(R_I \right)^{-0.1}$.

A least-squares fit, based on $U_R = C + Dz$, gives an intercept, C, of 69.34 m/s and a slope, D, of 44.15 m/s. The fit is shown in Figures G.4 and G.5.

The resulting gust speed versus average recurrence interval relationship is:

$U_R = 69.34 - 44.15 \left(R_I \right)^{-0.1}$, from which $U_1 = 25.3$ m/s and $U_{500} = 45.6$ m/s

The fit to the original data and predictions for higher values of R_I are shown in Figure G.6.

Fourth method (Section G5)

Finally, the method of maximum likelihood can be applied to determine an optimum value of scale factor, $\hat{\sigma}$, which maximizes the log-likelihood function of Equation (G.10).

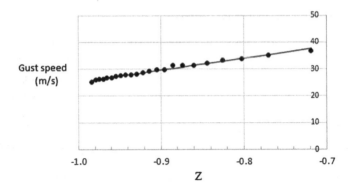

Figure G.4 U_R versus z, with least squares fit.

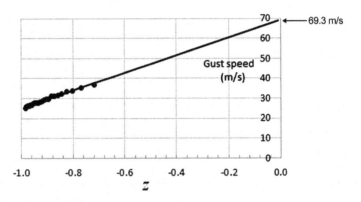

Figure G.5 U_R versus z, showing intercept on the z=0 axis.

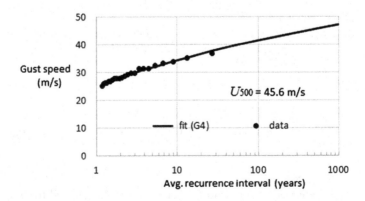

Figure G.6 Fitted line obtained using the method of Section G.4 (least squares fit to the 'raw' peaks-over-threshold data).

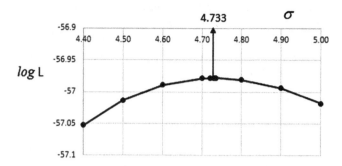

Figure G.7 Log likelihood function.

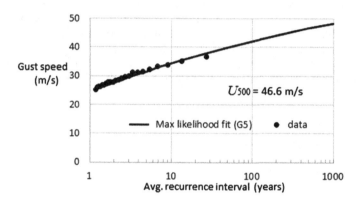

Figure G.8 Fitted line using the maximum likelihood method.

Figure G.7 shows the plot of $log\ L$, for various values of σ, calculated by the right-hand side of Equation (G.10), with $N=23$; $U_0=25$ m/s; $k=0.1$. The optimum value, $\hat{\sigma}$, is 4.733.

Then, $C = U_0 + \dfrac{\sigma}{k} = 25 + \dfrac{4.733}{0.1} = 72.33$ m/s

$D = \dfrac{\sigma}{k}(\lambda)^{-k} = \dfrac{4.733}{0.1}(0.885)^{-0.1} = 47.91$ m/s

Figure G.8 shows the fit of the distribution, based on the above values of C and D.

G7 DISCUSSION

Comparing Figures G.1, G.3, G.6 and G.8 in the above example, it may be noted that the predictions from the fitted lines for average recurrence intervals up to 10 years are very close – that is within about 1 m/s. However, not surprisingly, the extrapolations to higher values of ARI results in a range of predictions for U_{500} of nearly 3 m/s. However, the fits from methods of Sections G.4 and G.5 will give more emphasis to 'outliers' (i.e. values that exceed the trend shown by the majority of the data) if they are present – than the prediction lines from methods in Sections G.2 or G.3.

REFERENCES

Coles, S. (2001) *An introduction to statistical modeling of extreme values*. Springer-Verlag, London, Berlin and Heidelberg.

Davison, A.C. and Smith, R.L. (1990) Models for exceedances over high thresholds. *Journal of the Royal Statistical Society, Series B*, 52: 339–442.

Holmes, J.D. and Moriarty, W.W. (1999) Application of the generalized Pareto distribution to extreme value analysis in wind engineering. *Journal of Wind Engineering and Industrial Aerodynamics*, 83: 1–10.

Hosking, J.R.M. and Wallis, J.R. ((1987) Parameter and quantile estimation for the generalised Pareto distribution. *Technometrics*, 29: 339–49.

Index

Printed in the United States
By Bookmasters